SHEEP PRODUCTION AND MANAGEMENT

C. V. Ross

PRENTICE HALL, Englewood Cliffs, New Jersey 07632

Library of Congress Cataloging-in-Publication Data

Ross, C. V.
Sheep production and management / C. V. Ross.
p. cm.
Includes index.
ISBN 0-13-808510-2
1. Sheep. I. Title.
SF375.R74 1989 88-6396
636.3—dc19 CIP

Editorial/production supervision: Raeia Maes
Cover design: Joel Mitnick Design, Inc.
Manufacturing buyer: Robert Anderson

A Division of Simon & Schuster
Englewood Cliffs, New Jersey 07632

Printed in the United States of America

10 9 8 7 6 5 4 3 2 1

ISBN 0-13-808510-2

Prentice-Hall International (UK) Limited, *London*
Prentice-Hall of Australia Pty. Limited, *Sydney*
Prentice-Hall Canada Inc., *Toronto*
Prentice-Hall Hispanoamericana, S.A., *Mexico*
Prentice-Hall of India Private Limited, *New Delhi*
Prentice-Hall of Japan, Inc., *Tokyo*
Simon & Schuster Asia Pte. Ltd., *Singapore*
Editora Prentice-Hall do Brasil, Ltda., *Rio de Janeiro*

Contents

Preface

When I graduated from high school and became a freshman at the University of Missouri, it seemed there was little for me to learn about farm livestock, but then I came under the influence of A. J. Dyer, my first instructor in animal husbandry, and Luke McGlothan, the Scotch shepherd in charge of the college flock. They soon exposed my ignorance, but they also infected me with an incurable interest in sheep that has lasted to this day. Three years as an instructor of vocational agriculture in a high school convinced me that teaching students to produce livestock, especially sheep, was the most satisfying thing I had ever done. When I returned from World War II after five years in the U.S. Navy, Professor L. A. Weaver convinced me that there was still much to learn about livestock. He hired me to teach animal husbandry courses and to do research with sheep at the University of Missouri. Later, after I received my Master's degree, I did the same thing at the University of Illinois to complete a Ph.D. degree.

After five years of practical experience with Swift and Co. as supervisor of their lamb and cattle feeding operations, I returned to the University of Missouri and was in charge of sheep teaching and research for the next 27 years.

In addition to teaching and research, it was always a pleasure and privilege to devote time to working with commercial sheep producers and purebred breeders. I felt deeply indebted to a whole host of sheep producers, businessmen, colleagues in our department and at other universities, extension agents, and college undergraduate and graduate students who helped with my program.

Teaching the Sheep Production and Management course was always a special challenge because so few students had worked with the species. Hundreds of inexperienced students in that course developed skills and learned the fundamentals of production and management. During the years of teaching the course, I accumulated a great deal of material from my research, reading, travel, experience in the field, and yes from my students. Much of the current information was not covered in textbooks and thus was available to students and others only by handouts and lectures. My collection of information was equally useful in teaching an extension course for adults in the interior of Missouri and in preparing presentations for extension meetings throughout Missouri and adjoining states.

One of the principle objectives of this book was to share my information in order to at least partially repay the sheep industry and the wonderful people whose help was so needed and unselfishly given throughout my career.

This book is written for those interested in practical sheep production and management. It is meant to be used as a text by college undergraduates, but is stated simply enough for 4-H club members, high school students, and producers to understand. I hope that the information contained herein will be so useful that practical producers will want to keep a copy near at hand at all times.

The author wishes to acknowledge the assistance given by Linda and Ken Coffey, my student assistant and graduate assistant, for reading the original version and suggesting changes. Extension workers Cynde McDonald, James E. Ross, and James M. Thompson read the second version and made excellent suggestions. Reviews by Dr. Don Ely, Dr. Robert Jordan, Dr. Lyle McNeil, and Dr. Jim Outhouse were timely and useful. Jack Rhoades, University of Missouri Shepherd, and Roland Rench, owner and operator of a lamb feedlot, were helpful in obtaining pictures and information. Larry Mead, editor and publisher of *Sheep Breeder and Sheepman Magazine* provided inspiration and furnished most of the pictures of breeds of sheep and other material. The Border Leicester, Finnsheep, and Polypay associations generously provided pictures of their breeds. I am indebted to Professor Ken McClure of the Ohio Agricultural Research and Development Center; Dr. Kreg Leymaster and Dr. Larry Young of the U.S. Meat Animal Research Center, Clay Center, Nebraska; Dr. Jeffrey Green, Dr. Charles F. Parker, and Dr. John N. Stellflug, of the USDA Sheep Experiment Station at DuBois, Idaho; and Dr. A. M. Fahmy of the Lennoxville Quebec Research Station for photographs of facilities and sheep breeds. Photos and other information were provided by Dr. J. M. Asplund, Dr. Harold Hedrick, Dr. Duane Keisler, and Dr. Dale Vogt of the University of Missouri, T. G. Nash of the University of Illinois, Kenneth Johnson of the National Livestock and Meat Board, Guy Connolly of the Animal and Plant Inspection Service, USDA at Twin Falls, Idaho, and Mike Beaucage, USDA ARS at the University of California, Davis.

Some of the drawings and illustrations were made by Stephanie Smith of Columbia, Missouri, and Dr. E. M. Constantinesque of the University of Missouri, College of Veterinary Science. Illustrations and information were provided by Richard Haslem of ASPC and SID, Denver, Colorado, and by the Missouri Conservation Commission.

Special thanks is given to the National Academy of Sciences for permission to use data from *Nutrient Requirements of Sheep,* 6th ed., 1985. The staff at Ellis Library, University of Missouri, and Professor Glenn Grimes of the University of Missouri and Don Bay, Missouri State Statistician, were especially helpful in finding statistical information used in the book.

Typing of the manuscript was done by Joan Stauffer, Cynde Glascock, and Mary Smith.

Finally, I wish to express my gratitude to my son, Bob, who helped me with photographs, my two daughters, Laure and Betsy, and my wife, Helen, who have shown faith, loyalty, encouragement, and, above all, patience with my efforts to complete this work.

C. V. Ross

1 History and Development of the Sheep Industry

1.1 ORIGIN AND DOMESTICATION

It is not clear just when or where the origin of civilization occurred. Undoubtedly, civilization was related to the domestication of animals, and perhaps the dog was the first. Surely not far behind came sheep, goats, and cattle. Apparently, these species were serving people as long as 10,000 years ago when the Neolithic people gained ascendency over their CroMagnon counterparts.

The relationship between humans and sheep has always been beneficial for both. Sheep are unsurpassed at converting grass and other herbage into the highest quality food products, meat and milk, and unmatched textiles and hides. Our contribution consists of care and protection for the sheep. Still today sheep faithfully produce their precious products. If people keep their part of the bargain, there is no more productive species.

Where the domestication of sheep took place has not been firmly established, but the greatest weight of evidence indicates it was probably in central Asia. The keeping of sheep on the plains of Mesopotamia by Hebrew shepherds far antedates written history. There are 179 references to sheep, 137 to a ram or rams and 167 to a lamb or lambs, in the Old Testament. The first domestic animal named in the Bible (Genesis 11:2) is the sheep. Abel, the son of Adam and Eve, was a shepherd. Jacob served Laban seven years as a shepherd to win the hand of Rachel. Laban, however, reneged on his agreement and required seven more years of service. Jacob acquired a wife but no wages. The enterprise had prospered under his guidance and he agreed to continue if he would be given ownership of all sheep that were speckled, spotted, or ring straked rather than the normal brown or dingy black. By selection and using mutations for breeding, Jacob produced the first sheep of a different color. Jacob was the owner of numerous flocks when he left the service of Laban. His strain of sheep was continued by his descendants, who by selecting the sheep with the most white in their fleeces finally achieved a pure white color. David and Solomon both noted the existence of white sheep in their writing. Jacob can be remembered as probably the first sheep breeder. The descendants of his sheep are the Jacob sheep breed, which have more than one pair of horns.

The world may never know how sheep were domesticated or what was the parent stock. There is reason to believe that they may have sprung from something similar to the Moufflon, which originated in Europe and Asia, or a similar sheep, the Asiatic Urial. Both are true sheep and will cross readily with their domestic counterparts, but the wild sheep have horns, short tails, and are covered with hair, rather than wool, thus indicating the descent was not recent.

1.2 SHEEP IN EUROPE

The Greeks at the height of their power (3000–1000 B.C.) brought sheep and wool technology from the Middle East to Europe. They developed superior husbandry practices, and they taught many people to spin and weave cloth. Later, when Rome ruled the world, the Romans further developed sheep husbandry and distributed the beasts throughout their empire. Calumella and Virgil both described sheep culture in their writings.

The Romans developed a strain of sheep called the Tarenton that had a superior fleece, but they lacked hardiness and thus required special care. In Spain, during the reign of Claudius (1437–1440), the Tarenton sheep were crossed with the Laodician sheep of Asia Minor and were the parent stock of the Merino breed. Merinos or breeds derived from them lead in world numbers. They have contributed more to wool production in the world than any other breed or combination of breeds.

In addition to distributing breeding stock and knowledge of sheep husbandry, the Romans also taught the indigenous populations in their empire how to spin and weave wool into textiles. They established a woolen mill in Winchester, England, in 50 A.D. But the sheep industry developed by the Romans almost disappeared during the Dark Ages. It was probably the monks of the Catholic church who continued to keep sheep while the rest of the world was in chaos.

The wool industry was revived when the Saracens conquered Spain during the early part of the eighth century. Spain became the leader in wool trade under the Saracens until they were overthrown. The sheep industry was an important source of income for the royal family and the Catholic church in Spain for hundreds of years. The voyages by the conquistadors and Columbus were partially financed by money derived from wool.

Spain was careful not to export Merinos. In fact, it was a capital offense to ship ewes out of the country until, in 1768, Louis XVI imported 385 head for his research farm at Rambouillet in France. These sheep were improved by breeding and became the Rambouillet breed.

England was a close competitor with Spain in the wool trade. They imported Flemish weavers and concentrated on their textiles. Historians claim wool made up two-thirds of the foreign trade of England in the sixteenth century. During the reigns of Henry VIII and Elizabeth I, flocks belonging to monasteries were seized and distributed to favorites of the court, who fenced their lands and left many shepherds unemployed and destitute. These were some of the immigrants who came to the new colonies in North America and helped build a sheep industry.

1.3 SHEEP IN THE AMERICAS

Sheep were brought to the West Indies by Columbus on his second voyage in 1493. They were first introduced to Santa Domingo and Cuba and then to other islands in the Carribean and to Florida.

In 1519, Cortez defeated the Aztec Indians and Mexico was open to exploitation. Once Mexico was conquered, missions were established with flocks of sheep and herds of cattle, a vital part of the organization to colonize and exploit new worlds. The Indians were converted to Christianity and taught to care for sheep and to spin and weave woolen products.

Spanish colonists moved into New Mexico, Arizona, Texas, and California. By the late 1790s, missions had been established as far north as San Francisco. At the end of the eighteenth century, it was estimated that there were millions of sheep in the southwestern part of what ultimately became the United States.

Sheep were also brought to the East Coast of North America by the English and Dutch colonists. They were kept primarily for their wool. The raising of sheep was encouraged in the colonies, and there was a woolen mill in operation in Watertown, Massachusetts, as early as 1662. But England was jealous of her wool trade and placed severe restriction on trading in sheep and wool either as export or between colonies. These oppressive laws contributed materially to the American Revolution and a great buildup of sheep numbers. Even before the Revolutionary War, Americans were convinced they must establish a wool industry. George Washington and Thomas Jefferson were inaugurated in suits of American wool. In 1808, Spain sold some of her finest Merinos to the United States. By 1811, thousands of head had been imported and the strain firmly established.

1.4 DEVELOPMENT OF THE SHEEP INDUSTRY IN THE UNITED STATES

During the War of 1812, there was a Merino craze in the United States. Washington, Jefferson, and Madison all had sheep on their farms. Madison encouraged importation of Merinos. Prices of wool and sheep advanced to unheard of levels. Unfortunately for the speculators, the end of hostilities in 1815 brought a great supply of woolen goods from Europe, at much reduced prices, which burst the bubble and demoralized the industry. It was not to recover until protective tariffs were established in 1828.

As the population increased in the East, grazing land became more difficult to obtain and there was a constant movement to the West. In 1810, the center of the sheep population of the nation was New England and New York. In 1840, the population of sheep was estimated at 19 million head and the center had moved westward. In 1850, the center of population was in Ohio, and there were many sheep in states as far west as Missouri. The census of 1850 indicated only 377,271 sheep in New Mexico and 17,574 in California, but these figures appear to be grossly in error since many more sheep than that were driven out of those states after the Civil War when the territorial ranges were opened in Wyoming and Montana.

After the Civil War, sheep spread into Idaho, Wyoming, and Montana from New Mexico, California, and Oregon. The sheep were trailed in by well-marked routes from 1880 to 1900. Wentworth's book, *America's Sheep Trails,* will be informative for those interested in sheep history in the Western United States.

1.5 CHANGES IN SHEEP INDUSTRY IN THE UNITED STATES

By 1870, the center of sheep population was located in the Far West. By 1884, the expansion of numbers had reached a high point of 51.1 million head. Most of the western sheep were at least partly of Merino (Rambouillet) breeding because wool was relatively more important than meat.

As sheep production spread throughout the nation, the structure of the industry changed. Since wool could be produced at an advantage in the western states and the increasing population in the northeastern part of the country demanded more lamb, the emphasis changed in the midwestern and eastern states to production of more meat and less wool. American consumers showed a preference for younger, more highly finished lambs. Thus the size, age, and condition of lambs were affected by market demands. Lamb finishing in feedlots became an important factor in the overall plan to provide year-round supplies to the market. Many feedlots sprang up in the Midwest and East and lambs were shipped great distances from the western ranges to the source of feed and on to slaughter.

The invention of the refrigerator car made it possible for lambs to be slaughtered in Midwest packing centers and the chilled carcasses shipped by rail to the heavy consumption centers on the East Coast. National packing companies such as Armour, Swift, and Wilson had many packing plants and distribution systems throughout the United States. There were also large plants in the East that specialized in kosher slaughter. Central markets developed along the Mississippi and Missouri rivers at rail centers such as Kansas City and St. Joseph, Missouri; East St. Louis, Illinois; Omaha, Nebraska; St. Paul, Minnesota; and, Sioux City, Iowa. Each had an important lamb division. Markets nearer the source of western lamb production were Denver, Ogden, and San Angelo, which handled the bulk of lambs from the range states. There were also large packing plants that slaughtered lambs in Portland, Oregon; Los Angeles and San Francisco, California; and Ogden, Utah, for the West Coast states. The bulk of lambs were handled by four or five packers competing throughout the United States. Production patterns were well understood by the packers. Climatic conditions were an important factor affecting lambing seasons, availability and duration of the pasture season, and readiness for market. Lamb feeding in open lots developed in Colorado, Kansas, Nebraska, and Iowa to fatten lambs that came off summer ranges either too small or thin for slaughter. Many feedlots had the capacity for thousands of lambs and were mechanized.

Lamb feeding allowed the packers to provide a constant supply of lambs to the trade. The events leading up to World War II caused a great increase in sheep numbers, and an all-time peak in stock sheep on farms occurred in 1942 with 56 million head recorded.

1.6 DECLINE IN THE SHEEP INDUSTRY

Sheep numbers declined steadily after World War II, except for a slight increase during the Korean War, until they reached an all time low in 1985. In 1986, it appears as if numbers have begun to increase.

There have been dramatic changes in livestock marketing and slaughter and merchandizing of the meat in the last 40 years. The large central markets have mostly disappeared, rail transportation is seldom used, and the leading packers of yesteryear are mostly out of the business.

As sheep numbers have decreased, so has the per capita consumption of lamb. It was 4.8 lb (2.2 kg) in 1960 and 1.4 lb (0.64 kg) in 1985. Imports of frozen lamb amounted to approximately 9% of the total in 1985. Wool consumption in 1986 was approximately 1 lb (0.45 kg) per capita, with imports accounting for approximately 75% of the total.

1.7 CAUSES FOR THE DECLINE IN THE SHEEP INDUSTRY

Causes for the decline in sheep numbers have never been positively indentified. It seems reasonable that it may be a combination of many factors. It is interesting that most of them have always haunted the industry, but the good sheep herder has found ways to overcome adversity and achieve success.

Some of the constraints to the sheep industry are:

1. *Predation*. Terrill (1981) made a study of the reasons farmers left the sheep business and concluded that predation was the most important one. According to his data, death losses, mostly from coyotes, rose 250% from 1957 to 1975 and amounted to an annual loss of 1.1 million head with a value of $12.5 million. Coyotes are a problem over most of the land area of the continental United States. Predation has always been a major factor in sheep production. It was solved in ancient times by primitive means, and it can be accomplished much more easily today if the producer is willing to make the effort.

2. *Internal parasites*. This too has always been a problem and will remain one as long as sheep graze intensively, but the means exist to control parasites effectively.

3. *Lack of prestige or glamour*. In the past, cowboy boots, hat, and levis were the trappings of cattlemen, some of whom felt they were superior to others in the livestock business. In recent years, western wear has been adopted by people in all walks of life. That "cowboy" you see may very well be the driver of an 18-wheel truck or work in a bank. The younger generation seems more inclined to pay attention to their financial situation than personal appearance. They still want to work outdoors and with nature, but they want it to pay. Sheep make money for the good operator.

4. *Diseases*. Foot rot, pneumonia, blue tongue, infectious abortion, and other diseases can cause losses, but they can be prevented or controlled by following approved health practices.

5. *Poisonous plants cause losses*. This is no new problem. Losses can be controlled by proper management to prevent contact with the noxious plants or by eradication of them.

6. *Low production efficiency*. Sheep operations, because of the nature of the species, have made less progress in increased efficiency of production than swine or poultry. This is because those two species have a much higher rate of reproduction and consume high-energy diets, which can be handled mechanically with much less labor. They also lend themselves to confinement with restricted space requirements. These comparisons are not always valid since much of the diet of sheep is composed of herbage that only an active ruminant can harvest. But new systems of sheep production are being developed to increase reproductive efficiency by accelerating lambing and using confinement rearing of lambs to increase labor, feed, and facility efficiency. Genetic improvement is gaining momentum.

7. *Sheep require skilled labor*. Seldom has there been a time when farm or ranch laborers could demand as much pay and fringe benefits as unionized factory or industrial workers. But there are many people who love the outdoors and find that working with responsive animals is much more satisfying than the monotony and confinement associated with a job in a mill or factory. It is true that skill is required, but any reasonably intelligent, energetic, and ambitious person can soon acquire that by attending school or working with a successful producer.

8. *There is severe competition for land resources*. It appears inevitable that more and more political pressure will be brought to bear to divert range land to recreational use. Ranchers must improve their methods and be organized to justify their use of the resources. If they cannot, then sheep production may shift to other regions of the country.

9. *Lack of markets*. Marketing of sheep and lambs is undergoing major changes. Traditional methods of moving livestock into market channels are no longer available. It is more difficult to obtain fair value for sheep, lambs, and wool, but the progressive producers are doing it and actually receiving a premium for their products. Markets do exist, but they are different than they were and they will likely change even more in the future.

10. *High expense of slaughtering, processing, and transportation*. The traditional way of slaughtering lambs and shipping carcasses to the retail outlet is too wasteful of resources. New methods of slaughter, processing, and transporting meat are being used. When these innovations are put into effect, lamb should be competitive with other meats.

11. *Income from wool has declined dramatically*. It is true wool prices have been relatively low, but a good producer can still realize a considerable income from the sale of the fleece, plus an incentive payment. Top-flight producers of wool realize much more for their fleeces than the ones who have an inferior product to sell.

12. *Per capita consumption indicates Americans do not like lamb*. It is true the per capita consumption of lamb is only 1.4 lb, an all-time low, but that information may be misleading. Actually, Americans consume all the lamb that is produced domestically, plus frozen meat from other countries. The scarcity has prevented many people from exposure to one of our most palatable meats. A whole generation has grown up with little or no knowledge of lamb. They have been bombarded with pro-

paganda from nutritional "experts" who continually warn people against eating red meat because of the buildup of blood cholesterol from saturated fatty acids in animal fat. They often fail to mention the close relationship between consumption of unsaturated fatty acids and certain forms of cancer. The fact is that lean lamb is a safe, delicious, wonderfully nutritious food. It is rich in essential amino acids, minerals, and vitamins, and it is low in calories relative to other protein-rich foods. Our young, affluent generation represents a great market potential for the sheep industry. If they are properly introduced to delicious, satisfying lamb, they, like people throughout the ages, will be enthusiastic consumers. All the constraints can be overcome by ambitious, intelligent, industrious producers. Not only can the problems be solved, but favorable factors exist that are grounds for optimism for the future. They should be reviewed before becoming discouraged by the problems the industry faces.

1.8 FACTORS FAVORABLE TO THE SHEEP INDUSTRY

It should be said that there are favorable factors that may be strong enough to swing the balance toward a buildup of sheep numbers. Some factors that may have a strong influence in the future are:

1. Energy from nonrenewable sources will undoubtedly continue to be a problem. Sheep rank high in energy efficiency among ruminants, and they also have the ability to produce high-quality protein food and excellent textile fibers with a minimum use of energy from fossil sources. An additional advantage is the insulating value of wool clothing. If dwellings are to be maintained at lower temperatures in winter, wool is the preferred apparel by most people.

2. Crops grown on much of the land area of the United States, because of the rough terrain and climatic conditions, can only be harvested by grazing animals. Much of the vegetative growth of plants on mountains, deserts, and untillable land can be utilized best by grazing animals. Domesticated and feral (wild) ruminants have the ability to find and consume great quantities of grasses, legumes, weeds, and browse. Thus, when a lamb is shipped to market, it represents hundreds of pounds of unsalable forage that has been condensed by the ewe and lamb into highly nutritious human food that sells at high prices. Sheep are especially adapted to grazing steep hillsides on grasses, legumes, weeds, and browse. Few if any species is equal to them in utilizing the forages on rough terrain.

3. During the 1960s and 1970s, the world markets for cereal grains and soybeans were so favorable that there was a great emphasis on these cash crops. Large machinery and cheap fuel made it possible for a single farmer to expand operations manyfold. Some farmers essentially became grain specialists. They required a large acreage to justify the equipment. As a consequence, many of the soil conservation practices that had been so painfully learned in the 1930s were ignored. In many cases, farmers plowed up terraces, discontinued strip cropping, and destroyed soil-saving dams. In the early 1980s, the public suddenly became conscious that there was an alarming increase in soil erosion. Citizens are becoming aware that our precious soil resources are being squandered and are advocating legislation to correct the situation. If soils subject to erosion are protected, it is likely that forages will be grown

on them. Then the most logical use for the grasses and legumes is for animal feed. Ruminants should consume the bulk of it. This could very well result in more feed for sheep, especially since it is possible to produce market lambs with very little grain.

4. Other factors favorable for growth of the sheep industry were discussed by Parker and Pope (1983) in an excellent review of the changes and challenges of the industry. Some of these are:

a. The purebred sheep industry has actually prospered during the decline in commercial sheep numbers. Numbers registered were 176,723 in 1981 in some 20 breed associations. This was a 52.5% increase over the 1960–1964 period. The Suffolk breed, with two associations, registered 50% of all animals. Dorsets increased more (186%) than any other breed from 1963 to 1983. Others showing much growth were the Rambouillets (50%) and Columbias (45%) for the same period. Less popular were Southdowns and Corriedales, which lost 50%. During the same period, there were record-breaking sale prices of purebred rams.
b. New breeds have been developed in the United States and New Zealand. The Polypay breed promises to be a valuable addition to the industry, while the Coopworth and Perendale from New Zealand and the Cormo from Tasmania may also be useful in the United States. The Boorola Merino from Australia, an exceptionally fertile strain of fine wool sheep, could make an important contribution to wool producers in the world, and particularly in the southwest United States.
c. Sheep farmers have always been quite conservative and thus slower to adopt new practices than other producers. Poultry, dairy, swine, and beef cattle breeders have made extraordinary progress in improvement through genetics, with strong emphasis on selection for traits of economic importance. They have conducted herd and flock performance testing with extraordinary results. Only in foreign countries such as Australia, New Zealand, and the United Kingdom has the sheep industry put much emphasis on production testing. New Zealand, with its Sheepplan, has the cooperation of thousands of breeders in keeping records on the performance of their flocks. It seems certain they will make rapid advances in the future. Production testing, although late in coming, is underway in the United States. There were on-the-farm testing programs in at least 23 states in 1980. According to Allen (1980), there were 662 purebred flocks enrolled with 12,772 ewes in the program. Although it represents only a small percentage of the total population, it indicates a trend toward improvement. Production testing should be conducive to rapid improvement in the seed stock of the nation. Breeders who adopt the program and promote it may well be leaders in the future.
d. Progress has been made in the control of reproduction of sheep. Manipulation of light and temperature has made it possible to breed in any season. Use of hormones to synchronize estrus and to increase ovulation has been combined with environmental control with excellent potential. These encouraging results with selection for prolificacy and a long breeding season should ensure much improved reproductive performance in the future. Still to be solved are the problems involved in storing ram semen and artificial insemination of ewes. Nevertheless, even these difficult problems are closer to solution than they were 10 years ago.
e. The outlook is much brighter for the future health of sheep. Screwworms and scabies appear to be eradicated for all practical purposes. Selenium supplementation

has much reduced the incidence of white muscle disease in lambs. There are new vaccines for preventing prenatal lamb losses from vibriosis and enzootic abortion. More recently, a vaccine has been developed for ram epididymitis. New treatments for foot rot, including vaccines developed in Australia, hold promise that this disease may soon be controllable. Internal parasites are still one of the hazards of grazing sheep, but timely drenching to prevent the lactational surge of reproduction of stomach worms in ewes, use of clean ground for lambs, and proper rotational grazing can be combined to reduce the risk from the pests. New, more potent anthelmintics are on the horizon.

f. The figures quoted previously on predator losses are discouraging, but in that battle too there are indications of progress. Exclusion by use of electric fences has been successful in farm states, confinement rearing largely eliminates the problem of predation, use of guard dogs holds promise for the range, and concerted action by sheep producers has been effective. There is some reason to hope that studies of coyotes being conducted by the USDA in cooperation with the Idaho and California experiment stations will assist in developing means of control.

g. Marketing sheep and lambs is at present in a state of transition. Just how it will be organized in the future will depend to a large extent on the leadership of slaughtering and processing interests and the producers. If the two groups are truly interested in maintaining and growing the industry, systems that bring fair returns to both sides will be worked out. Otherwise, it may be necessary for producers to organize co-ops to handle selling live lambs, or it may even be feasible for co-op packing and distributing organizations to be created. Electronic marketing (teleauction) is beyond the experimental stage and offers a fair potential. Where it has been operated correctly, competition in bidding has been increased, with a consequent improvement in prices paid for lambs. There is room for new ideas in that phase of the industry.

h. Much new technology is being developed relative to slaughtering and processing meat. To mention only a few innovations that should be beneficial, the stigma attached to the slaughter and processing of ram lambs has been shown to be unjustified. Carcass weights have been increasing steadily since 1940, with a corresponding decrease in processing cost per pound of lamb. Centralized cutting of carcasses, vacuum packaging, radurization (ionizing radiation), restructuring meat into roasts and chops, hot boning by mechanical means, and merchandising frozen lamb are all possible means of improving the product or implementing the handling or sale of it.

i. In the field of management, there is greater emphasis on efficiency of production. Innovations such as accelerated lambing and confinement rearing are two that are valuable. Others are being studied. While it is true that sheep have not made the extraordinary progress of the dairy, poultry, and swine industries in the technology of production, Terrill (1980) noted that meat production per ewe doubled from 1908 to 1958. Since 1958, the index of production per ewe remained static until 1976 and then increased each year to 1981. This increase amounted to 13.5%, which was only one-third of the improvement of the whole livestock industry for the same period. This could be discouraging until it is remembered that much of the improvement in some species was made using an abundance of grain and relatively cheap fossil energy. Few people believe such conditions are likely to be repeated soon. There has been progress in management such as early weaning and artificial rearing of lambs. Only one-third of the potential productivity of forage acreages is presently in use in the north central and humid south regions. This land could and should be used to produce food and fiber for a growing human population. It has been estimated by Parker and Pope (1983) that the application of available technologies has the potential of more than doubling the present levels of lamb production per ewe. It seems logical to assume that adoption of improved technologies in all segments of the industry should result in a vast improvement in the future.

j. Recently, there has been some emphasis on programs for small farms. Sheep are very well adapted to small units.

1.9 FUTURE OF THE SHEEP INDUSTRY

How well the sheep industry will fare in the rough and tumble arena of economic competition, both foreign and domestic, will to a large extent be decided by industry leaders of the future. The elements exist for a glowing future or a dismal failure. Whether the determination and dedication of all the people involved will be great enough to overcome the considerable obstacles and exploit the favorable factors remains to be seen.

Technology has been a great boon to people, who have cleverly fabricated textile fibers from nonagriculture products and meatlike food from plant proteins, but the substitutes have not replaced natural products. Synthetic textiles have not been as satisfactory as wool. Meat is still the preferred protein food by the American public. Although the per capita consumption is low, all the lamb slaughtered is consumed. Tests have shown more would be purchased by the consumer if it were available.

To answer the questions asked by agricultural leaders relative to the continued existence of the sheep industry would require a crystal ball not available to this author. However, it can be said with certainty that the favorable factors outweigh the constraints. The great accumulation of knowledge now available for application is a priceless asset. The wise use of natural resources will ultimately be favorable to the ovine species. The opportunities for innovation in production, marketing, and processing appear to be excellent.

REFERENCES

Allen, G. A., "What's Right and Wrong with Our Performance Testing Programs," *Proc. Natl. Purebred Sheep Symp.*, SID, Denver, Colo., 1980, p. 96.

CAST, "The U.S. Sheep and Goat Industry: Products, Opportunities and Limitations," *Council Agr. Sci. Tech.* (Rep. 94), 1982.

Coffey, W. C., *Productive Sheep Husbandry.* New York: J. B. Lippincott Company, 1937.

Ensminger, M. E., *Sheep and Wool Science.* Danville, Ill.: Interstate Printers and Publishers, Inc., 1970.

Guthrie, J. F., *A World History of Sheep and Wool.* Australia: McCarron Bird Pty. Ltd., 1957.

Parker, C. F., and A. L. Pope, "The Sheep Industry: Changes and Challenges," *J. Anim. Sci.*, 57 (Suppl. 2), 1983, p. 75.

Ponting, Kenneth G., *Sheep of the World,* Poole, England: Blandford Press, 1980.

Terrill, C. E., "More Food from Efficient Sheep," *IV World Conf. Animal Production*, Buenos Aires, Argentina, 1980.

______, "Trends of Predator Losses of Sheep and Lambs Calculated from U.S. Department of Agriculture Mortality Statistics," E. W. Schafer, Jr., and C. W. Walker (eds.), "Vertebrate Pest Control and Management Materials," Third Conference ASTMSTP 752, American Society for Testing and Materials, Philadelphia, 1981, pp. 135–146.

Towne, C. W., and E. N. Wentworth, *Shepherd's Empire.* Norman, Okla.: University of Oklahoma Press, 1945.

Wentworth, E. N., *America's Sheep Trails.* Ames, Iowa: Iowa State College Press, 1945.

Youatt, William, *Sheep: Breeds, Management and Disease.* London: Baldwin and Craddock, 1837.

QUESTIONS FOR STUDY AND DISCUSSION

1. Why is it so difficult to determine the origin and domestication of sheep?
2. Justify the statement that sheep domestication was related to civilization of people.
3. What are the reasons for the belief that domesticated sheep originated in the Middle East?
4. Trace the spread of sheep from the ancient Hebrews to the United States.
5. What is the connection between the wool industry in Spain and the discovery of the West Indies?
6. What was the reason for the Merino craze in 1812?
7. Why did the center of sheep population continually move westward from 1812 to 1884?
8. Why was wool an important commodity in the early 1800s?
9. What changes occurred in the sheep industry as a result of the invention of the refrigerator rail car?
10. What purpose was served by the first feedlots for finishing lambs?
11. List five causes for the decline in the sheep industry after World War II.
12. Discuss reasons for optimism for the future in the sheep industry.

2 Characteristics of Sheep

2.1 SCIENTIFIC CLASSIFICATION OF SHEEP

Sheep are living organisms. They are classified scientifically as follows:

Kingdom	Animalia	Animal
Phylum	Chordata	Vertebrates
Class	Mammalia	Hairy and suckle their young
Order	Ungulata	Hoofed mammals
Suborder	Artiodactyla	Even-toed ungulates
Section	Pecora	Typical ruminants
Family	Bovidae	Hollow-Horned ruminants
Subfamily	Caprinae	Sheep and goats
Genus	Ovis	Sheep
Species	Ovis aries	Domesticated sheep

There are many breeds and types of sheep, including wild species that are similar to their domesticated relatives. It has been estimated that there are as many as 800 breeds in the world. New breeds are being developed continually. Thus it is difficult to estimate the total number.

Sheep types and breeds differ in physical and other characteristics. This chapter will be concerned primarily with those used domestically in the United States.

2.2 EXTERNAL CHARACTERISTICS

Breeds of sheep may usually be distinguished by inspecting their external characteristics. These differences in conformation, covering, whether horned or polled, color markings, size, and so on, are known as breed character. Variations of external features that may be observed are discussed next.

2.2.1. Heads of sheep Heads vary in size from very small and refined to large and coarse and in width from wide to narrow and in length from short to long. The nose may be straight to Roman. Most domestic sheep are polled, but some breeds in Africa, Iceland, and the Middle East have as many as eight horns. In some breeds, such as the Dorset, both rams and ewes are horned. In others, rams may be horned and the ewes polled, as in some strains of Merino and related breeds. Face color may vary from black to white or be mottled.

2.2.2. Body size Size varies from as small as 80 lb (30kg) in some tropical breeds to more than 400 lb (182 kg) in rams in the Suffolk and some of the long-wooled breeds. Rams are almost invariably larger than ewes of the same breed or genotype.

2.2.3. Body covering Wool is the usual covering of sheep of the temperate zones of the world. Sheep of the subtropics and the tropics are generally covered with hair or coarse wool. Karakul adults have coarse long wool, but the coats of their lambs are a highly prized fur used in manufacturing Persian lamb coats. Wool varies from a length of 1.5 to 18 in. (3.8 to 45.7 cm). Color varies from black to white. Diameter of fibers ranges from 0.0008 to 0.002 in. (20 to 500 microns). Finer fibers tend to be shorter with more waves, called crimp, while the coarser ones may be longer and straighter. Yolk, which consists of fatty secretions plus sweat (suint), varies in wools, with more in the finer fleeces and less in those with a coarser texture.

Shrinkage in wool is that amount of yolk and dirt that is scoured out in processing. It tends to be greater in fine wools and less in those of coarser texture, but it varies with the amount of dust and sand carried by the wind.

2.2.4. Classification of breeds of sheep by wool or hair Classification by wool or hair is as follows:

Wool Class	Breeds
Fine	Merino Rambouillet Debouillet
Medium	English breeds: Cheviot, Dorset, Hampshire, Shropshire, Southdown, Suffolk Crossbred breeds: Columbia, Corriedale, Montadale, Panama, Polypay, and Targhee
Long	Lincoln, Cotswold, Romney, Border Leicester, Navajo
Hair	Barbados, St. Lucia, and other tropical breeds

2.2.5. Skin features Skin on most white-faced sheep is thin and pink. Black sheep may have pink to black skin. Thickness varies from very thin to thick. Generally, pink skin is thinner than black. Skin variations range from smooth to very wrinkled.

2.2.6. Glands of a sheep Sheep have oil (sebaceous) and sweat (sudiferous) glands distributed throughout the surface of the skin. Sebaceous glands secrete the oily material called wool fat, which is necessary for the health and well-being of wool and hair. Finer-wooled sheep tend to have more sebaceous glands than those with coarser fleeces or hair. Sweat glands secrete water and some potassium salts, called suint, which are an aid in cooling the sheep. The mixture of wool fat and sweat is called yolk. Sheep also have other glands that open to the surface. These are:

- *Interdigital:* Located on each foot above the split between the toes.
- *Suborbital:* Located just below the eyes of most sheep.
- *Groin:* Located on the inside of the hind legs.

2.2.7. Points of sheep Figure 2.1 shows the points of sheep. Any student of sheep should study the illustration and learn the names of each part of the animal. Using correct terminology is necessary for communicating with sheep producers and others interested in the species.

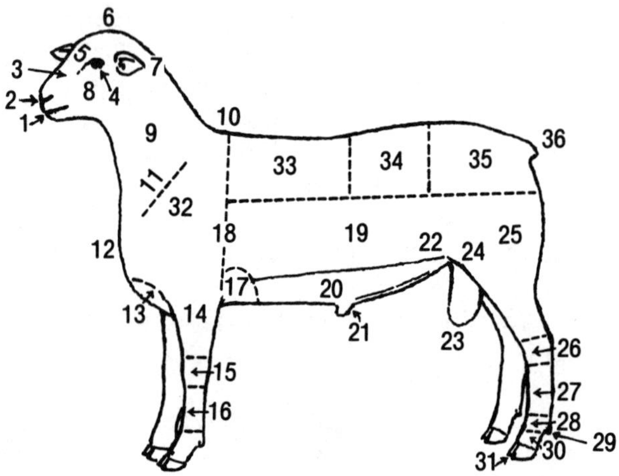

1. Mouth
2. Nostril
3. Nose
4. Eye
5. Forehead
6. Poll
7. Ear
8. Face
9. Neck
10. Top of Shoulder
11. Shoulder Vein
12. Breast
13. Brisket
14. Forearm
15. Knee
16. Fore Shank
17. Fore Flank
18. Heart Girth
19. Side
20. Belly
21. Pizzle
22. Hind Flank
23. Scrotum
24. Stifle
25. Leg
26. Hock
27. Hind Shank
28. Ankle
29. Dew Claw
30. Pastern
31. Toe of Hoof
32. Shoulder
33. Back or Ribs
34. Loin
35. Rump
36. Dock

Figure 2.1 The points of a sheep.

2.3 INTERNAL CHARACTERISTICS

2.3.1. Sheep skeleton The bony structure of a sheep is similar to other vertebrates and includes the neck, back, loin, rump, and tail. The number of vertebrae in each part are:

Part of body	Common name	Number of vertebrae
Cervical	Neck	7
Thoracic	Back	13
Lumbar	Loin	6–7
Sacral	Rump	4
Coccygeal	Tail	4–20

Most sheep have 13 pairs of ribs but some have 14 pairs. The 8 front pairs of ribs (true ribs) are attached to the sternum by cartilages. Four pairs, called asternal or false ribs, are not attached to the sternum but to each other. The thirteenth pair are called floating ribs and not attached to the sternum. The front legs do not have any rigid or flexible skeletal connection with the rest of the bony parts of the sheep. The ribs and vertebrae are suspended between the shoulder blades by strong muscles and tendon ligaments. The width of the chest depends on the size of the muscles and amount of fat. This explains why the width of chest can change as an animal changes condition. Much can be learned by a study of a sheep skeleton. The bony structure determines to a large extent the conformation of the animal. There are differences in the length, shape, and size of various bones. These affect the shape and size of the animal. The shape and size of the various parts determine the conformation and breed character of the animals.

2.3.2. Mouth and teeth Sheep have a mouth uniquely adapted to discriminate grazing. They have 20 temporary teeth and 32 permanent ones at maturity. Lambs may be born with one or two incisors or none at all. The four pairs of incisor teeth erupt within the first week of life. Sheep have incisor teeth only in the lower jaw. The front part of the upper jaw has no teeth, but there is a cartilaginous pad (called dental pad) against which the lower incisors press when grazing occurs. In grazing, the sheep grasps the forage between the dental pad and the incisors and severs it by quick jerking movements of the head. This is the reason that a jaw that is too long or too short hampers grazing and is a serious defect.

A lower jaw that is too long, causing the incisors to miss the dental pad, is called an undershot jaw or monkey mouth. One that is too short is called an overshot jaw or parrot mouth. Both are serious defects and are inherited.

Teeth may be used to estimate the age of sheep up to 4 years of age because the temporary incisors are replaced systematically as the sheep matures. Figure 2.2 shows incisors of various-aged sheep. The following is the sequence of changes:

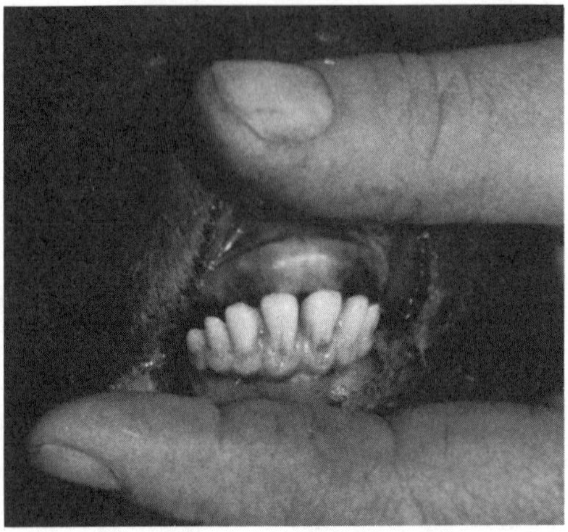

Lamb—all temporary incisors

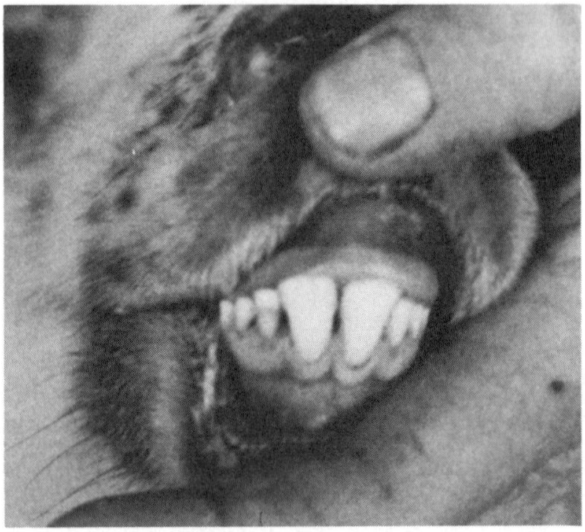

Yearling—one pair permanent incisors

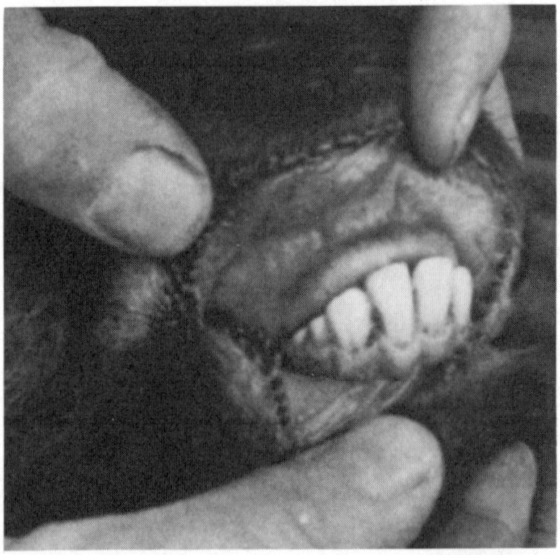

Two-year-old—two pairs permanent incisors

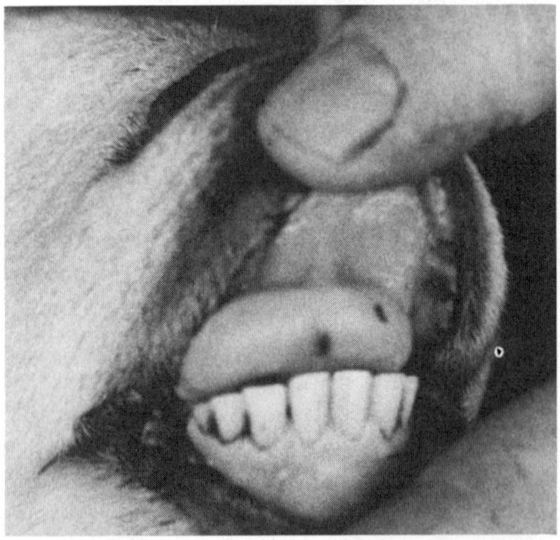

Three-year-old—three pairs permanent incisors

Figure 2.2 Estimating age by inspecting incisor teeth.

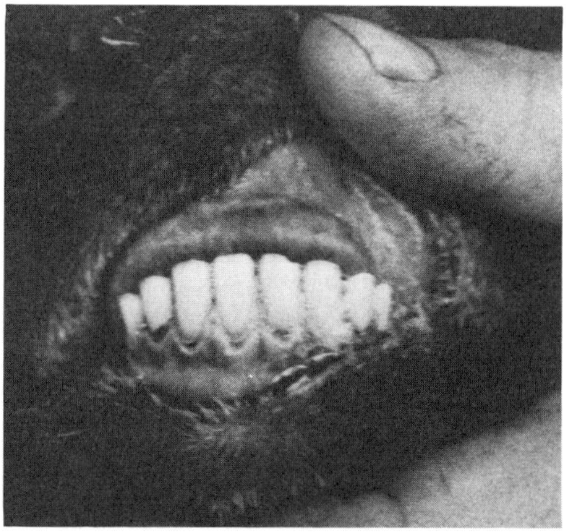

Four-year-old—all incisors permanent

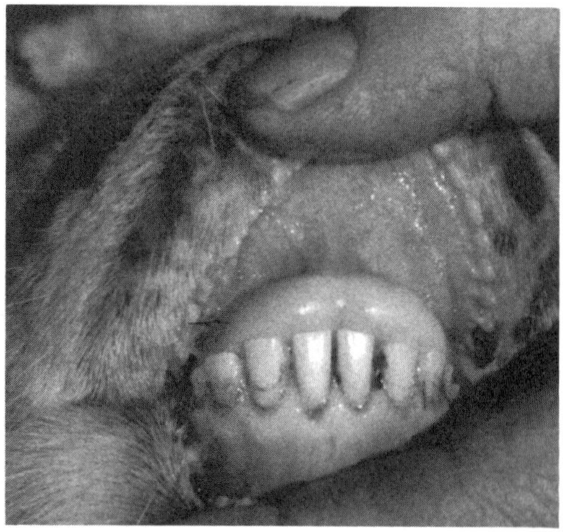

Solid mouth—incisors permanent and close together

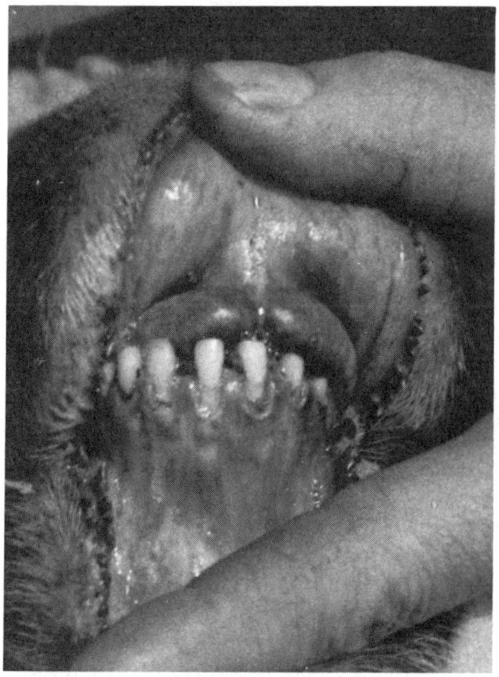

Spreader—incisors small and wide apart

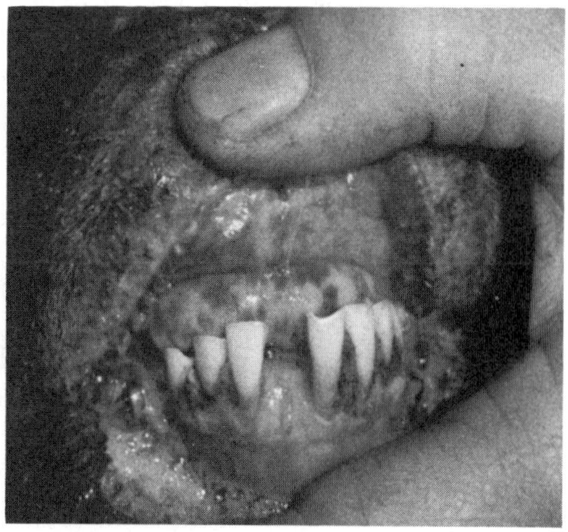

Broken mouth—some incisors broken

Figure 2.2 Continued

Sequence of replacing incisor teeth	Age, years (months)
Middle pair or pincers	1 (10–14)
First intermediate pair	2 (18–30)
Second intermediate pair	3 (28–40)
Corner pair	4 (38–48)

From 4 years of age onward, changes occur but are not so easily observed. The following applies generally:

Description of teeth	Estimated age (years)
All incisors permanent and all teeth close together (called solid mouth)	5
Incisor teeth appear wider apart (called spreaders)	6
Some incisor teeth broken off (called broken mouth)	7–8
All permanent incisors missing (called gummers)	10–12

The permanent incisors are wedge shaped and thus, when they wear with age, they are narrower and appear to be spreading. In areas in the West where windblown sand and dirt are common, wear on teeth is rapid. Not all sheep follow the schedule of replacing temporary teeth. Wide variations may occur for no apparent reason.

It is not always possible to differentiate between temporary and permanent teeth. Typically, temporary teeth are white, small, and rectangular, while permanent ones are larger, yellow, and triangular. The mouth of sheep is not as large as cattle. The muzzle is narrower, the lips thinner and more mobile, and each side can be moved independently of the other. This is because the lip is cleft. The tongue of a sheep is narrow and extremely dextrous. The narrow muzzle, mobile lips, and dextrous tongue enable the sheep to eat the leaves of a thornbush or pick out soybean pellets from a mixture with shelled corn. Unlike the cow, the sheep has a much smaller, shorter, and smoother tongue.

2.3.3. Digestive system The digestive system is composed of the mouth, esophogus (gullet), a four-compartment stomach, the small intestines, the large intestines or colon, and the anus.

A sheep has a large capacity in order to consume large quantities of forage. The following is the approximate capacity of the digestive system of a typical, medium-sized mature sheep:

The small intestine may vary from 85 to 90 ft (26 to 27 m) in length, with 22 ft (6.7 m) for the colon.

2.3.4. Feces and urine of sheep Feces of sheep are typically dryer than those of cattle. The feces are usually pelleted unless the feed is high in moisture. A sheep can control the amount of moisture in the offal as a means of conserving water. Urine is usually more concentrated than for other animals and may be extremely so under

Part of the digestive system	Capacity	
	Quarts	Liters
Rumen	25	23.65
Reticulum	2	1.89
Omasum	1	0.95
Abomasum (true stomach)	3	2.81
Small intestine	9.5	9.0
Large intestine (colon)	2.5	2.37

dry conditions. Urine production is more copious when sheep have adequate water or are grazing lush forage. Under those conditions, the concentration will be much less.

2.4 BODY TEMPERATURE OF SHEEP

Body temperature may vary according to ambient temperature, humidity, wool covering, degree of finish, stress, movement of air, irradiation by the sun, and infections. Normal temperature ranges from 100.8° to 103.8°F (38° to 40°C).

Optimum environmental temperature varies from 50° to 80°F (10° to 26.5°C). Above 80°F (26.5°C), most sheep must make some effort to control their body heat. The following mechanisms are commonly used:

1. Panting
2. Drinking and urinating
3. Transpiration (sweating)
4. Seeking shade and a cool breeze

Homeostasis, the ability to adapt to environmental temperature, does not develop in lambs until approximately 72 hours postpartum. Thus newborn lambs need protection, especially from cold.

2.5 RESPIRATION AND HEART RATE

Breathing and heart action are both more rapid in younger animals and gradually decrease with maturity. Under comfortable resting conditions they are:

- Respiration rate for mature sheep, 12 to 20/min
- Heart rate for mature sheep, 70 to 80/min

Both respiration and heart rate can be increased by exercise, stress, and high temperatures.

2.6 BREEDING HABITS OF EWES AND RAMS

Most sheep are seasonally polyestrus. This means they have a definite breeding season when estrus occurs, and they have a number of cycles within that season. The natural breeding season is in the fall and winter, but the season varies among breeds of sheep.

The Merino and Rambouillet breeds have a longer season than most of the medium- and long-wooled types. Exceptions are the Dorset and breeds that originated near the equator, which have extended periods. Altitude and latitude both affect the tendency to cycle. Other factors are change in length of day, temperature, humidity, and nutrition. Rams also are affected by environmental conditions, but are less sensitive than ewes.

2.6.1. Puberty Puberty is the stage of sexual development when an animal is capable of reproduction. It occurs at a critical age and weight, usually at 8 months in ewes and when they weigh 80 to 100 lb. Rams are sexually mature, capable of siring offspring, at 7 to 8 months and at somewhat heavier weights than ewes.

2.6.2. Estrus When the ewe is physiologically and psychologically ready to mate, she is in estrus. Duration of estrus may vary from 3 to 73 hours with an average of 29 to 30 hours. Ovulation occurs near the end of estrus. The estrous cycle extends from 14 to 19 days, averaging 17 days.

2.6.3. Gestation The period of development of a fetus in the uterus of the ewe is called gestation. The average period varies from 147 to 150 days but is affected by the following:

1. Single or multiple births: The period for singles is usually longer than for multiple births.
2. Larger lambs have a longer period than smaller.
3. Breed: Some breeds, such as the Merino, have longer gestation (149 days) compared with the Finnsheep (140 days).

2.6.4. Prolificacy This relates to the number of young produced per litter. Since this characteristic is affected by heredity, some breeds or strains are more prolific than others. The fine-wool sheep have fewer multiple births than most other types. An exception is the Booroola Merino of Australia, which is exceptionally prolific. Other breeds known for their prolificacy are Finnsheep, Dorset, Texel, East Fresian, Romanov, and Dorper.

Another factor affecting prolificacy is maturity; ewes lambing at 1 year of age usually have singles. Ewes tend to increase multiple births until the age of 6 or 7. According to records of the USDA for 55,000 births, the following was the incidence of single and multiple births among many breeds of sheep studied at their experiment stations:

Singles	52%
Twins	45%
Triplets	2.3%
Quadruplets	0.1%
Quintuplets	0.009%

2.6.5. Birthweights Size of lambs at birth varies from 3 to 25 lb (1.36 to 11.4 kg), with an average of 8 to 10 lb (3.6 to 4.5 kg). Factors affecting birthweight in lambs are:

1. Size of parents.
2. Number per litter: lambs in litters tend to be smaller than singles.
3. Age of the ewe: older ewes tend to have larger lambs.
4. Sex of the lamb: ram lambs are usually larger than ewes.
5. Nutrition: inadequate levels of essential nutrients may reduce size at birth.
6. Environmental temperature: lambs born in the fall tend to be smaller than the ones that arrive in the spring.

2.7 SENSES

Most wild sheep have excellent distance vision. Sheep apparently lost this trait when they were domesticated. Most appear to be rather nearsighted. Smell and hearing are very acute.

2.8 GROWTH AND DEVELOPMENT

There is some variation in breeds and types of sheep as to age of maturity. Small breeds tend to mature more rapidly than those of larger stature. In general, sheep are 80% mature at 1 year and 100% at 2 years. The skeleton tends to mature more rapidly than the soft tissues, with the extremities first and the rest later. Fat deposition increases as an animal matures.

Sheep may live to as old as 16-18 years, but their useful life is rarely more than 7 to 8 years. Fine-wool sheep tend to have greater longevity than other breeds.

2.9 SHEEP PSYCHOLOGY

Sheep are not rated high in intelligence compared with dogs, horses, cats, and swine. They, however, do have the ability to learn and can be trained to do many things. For example, ewes that are being fed in stalls learn to find them. Some sheep can open intricate gate latches or hooks. They are very responsive to people and do not soon forget mistreatment. They respond favorably to the following conditions:

1. Quiet handling
2. Routine feeding
3. Dry quarters protected from rain, snow, and wind

Sheep do not usually thrive in the following environments:

1. Wet, damp, poorly ventilated quarters
2. Closed buildings
3. Wet, poorly drained yards and pastures
4. Being approached and handled by strangers
5. Exposure to loud noises or strange dogs

2.9.1. Sheep behavior There is a saying "a sick sheep is a dead sheep." Another is "you can never tell by looking at a sheep if it is sick." Neither of these statements is true. Sheep respond to treatment the same as other animals. However, if the shepherd does not observe the flock well enough to know there is something wrong with a sheep until it can no longer stand, it is probably too late to treat it successfully. A sheep indicates by the carriage of its head, the brightness of eye, the character of its wool, the way it eats or doesn't eat, and quite often by separating itself from the rest of the flock that all is not well with it. Early in the morning, when the sheep leaves its bed, is one of the best times to observe its health. Gregariousness, the instinct to flock together, is stronger in some breeds than others. The Merino and its related breeds can be handled in large flocks or bands and remain together. Cheviots, for example, do not move in a group but as individuals. Other of the European breeds tend to flock together only when they feel threatened by a predator or are under stress. Then they form a cluster. Very seldom will they attempt to take offensive action except in the case of a ewe protecting her lamb. Even then they are ineffective.

When sheep are being driven, sometimes one will make a break for freedom. If it is successful, others will follow. Then it is best to allow the group to get together and start over. Allowing a single sheep to be isolated may cause it to become berserk, and it may run blindly into a fence or wall and injure itself. If sheep escape, it is almost always the result of trying to move them too fast or not using proper equipment to assist in the job.

Sheep are natural browsers and can pick and choose the most palatable, nutritious, otherwise unusable forage. They like short, tender grass and tend to overgraze certain areas. On rough land, they prefer to graze uphill and at an angle. Sheep do not like to graze where others have urinated or defecated, even though those spots may appear greener and more lush than the rest of the field. They prefer to graze early and late when it is cool and spend the middle of the day under shade. Sheep prefer to bed down at night, in a group, on the highest point of the pasture. Ruckebusch (1972) observed that sheep fed in stalls were wide awake approximately 16 hours a day, drowsing for 4 hours and sleeping for 4 hours. Under pasture conditions, grazing time is related to the quality of forage. If the forage is succulent and plentiful, grazing time may be expected to be no more than 7 hours daily. If the forage is sparse and the quality low, lactating ewes may graze 11.5 hours daily.

Sheep can be fed grain, but they must be accustomed to it gradually; otherwise, they will tend to gorge themselves. They eat more if group fed than individually. If it is possible, they will eat selectively the most palatable feed. They do not like damp, moldy grain or hay. They do not require that most grains be ground because they masticate so thoroughly.

REFERENCES

Coffey, W. C., *Productive Sheep Husbandry.* New York: J. B. Lippincott Co., 1937.

Craig, J. V., *Domestic Animal Behavior: Causes and Implications for Animal Care.* Englewood Cliffs, N.J.: Prentice-Hall, Inc., 1981.

Devendra, C., and I. E. Coop. *Sheep and Goat Production.* New York: Elsevier Scientific Publishing Co., 1982.

Ensminger, M. E., *Sheep and Wool Science.* Danville, Ill.: Interstate Printers and Publishers, Inc., 1970.

Fowler, D. G., and L. D. Jenkins. "The Effects of Dominance and Infertility of Rams on Reproductive Performance," *Appl. Animal Ethol.,* 2, 1976, pp. 327–337.

Hulet, C. V., G. Alexander, and E. S. E. Hafez. *The Behavior of Domestic Animals,* 3rd ed. Baltimore: Williams and Wilkins Co.

Kilgour, R., and Clive Dalton. *Livestock Behavior* Boulder, Colo.: Westview Press, 1984.

Lindsay, D. R., D. G. Dunsmore, and G. J. Syme. "Audience Effects on Behavior of Rams," *Anim. Behavior,* 24, 1976, pp. 818–821.

May, Neil D. S., *The Anatomy of the Sheep.* St. Lucia, Queensland, Australia: University of Queensland Press, 1970.

Ruckebusch, Y., "The Relevance of Drowsiness in the Circadian Cycle of Farm Animals," *Anim. Behavior,* 20, 1972, pp. 637–643.

QUESTIONS FOR STUDY AND DISCUSSION

1. How is the classification of sheep and cattle similar?
2. Why are there so many breeds of sheep in the world?
3. What factors are used in distinguishing the various breeds of sheep?
4. The appearance of breeds of sheep has changed throughout the years. Why?
5. How do fine- and long-wool fibers differ?
6. Most sheep of the tropics are covered with hair instead of wool. Explain the reason for this.
7. Why should fine-wool sheep have more yolk than other breeds?
8. Where are the floating ribs located on a sheep? Why call them "floating"?
9. Explain why a sheep can appear to vary widely in the width of chest.
10. How can you differentiate between temporary and permanent teeth?
11. As a sheep ages, their teeth appear to spread. What causes this?
12. Why should monkey and parrot mouths be avoided in breeding animals?
13. Explain why sheep can graze more discriminately than cattle.
14. In what way can a sheep adapt to droughts? To very warm weather?
15. What factors affect the size of lambs at birth?
16. Explain why many people believe sick sheep do not respond to medical treatment.
17. Why should the likes and dislikes of sheep affect their management?
18. How does feeding a sheep and a hog differ?

3 Types and Breeds of Sheep

Type may be defined as those visual characteristics of form and structure that are ideal for the purpose for which the animal is produced. A breed is a group of animals of common origin that have been selected and bred for certain distinguishing characteristics and are capable of transmitting them to their offspring.

Historically, sheep have been classed according to their production; thus, some were called wool sheep, and others emphasized meat production more than wool and were called mutton sheep. But modern sheep do not fit into such a simple organization. Many breeds of sheep not only produce a heavy wool clip but also raise highly marketable lambs. Although the sales of wool from a flock or band of sheep may not represent as high a proportion of the income as it did in past years, it is still important to the enterprise. The serious student should appreciate the differences in the types and characteristics of the economically important breeds of sheep.

3.1 CLASSIFICATION OF SHEEP BY BODY COVERING

Table 3.1 shows the division of sheep breeds by body covering. Wool is classed by diameter and length of fibers. Other covering includes hair and fur.

3.2 CLASSIFICATION BY COMMERCIAL USE

A committee of scientists of the Sheep Industry Development Program, Inc., sponsored by the American Sheep Producers Council, classified sheep breeds according to their commercial use. These are described next.

TABLE 3.1 CLASSIFICATION OF SHEEP BREEDS BY BODY COVERING

Fine wool	Medium wool	Long wool	Crossbred wool	Hair	Fur
Debouillet	Cheviot Dorset Finnsheep Hampshire Montadale Oxford	Border Leicester Cotswold Lincoln	Columbia Coopworth Corriedale Panama		
Delaine Merino	Shropshire		Perendale		
Merino	Southdown	Navajo	Polypay	Barbados	Karakul
Rambouillet	Suffolk	Romney	Targhee	St. Croix	

3.2.1. Ewe breeds Ewe breeds (see Table 3.2) have characteristics that are most desirable in ewes for producing commercial lambs. They were selected for adaptability to a wide range of environmental conditions. The traits emphasized were reproductive efficiency, wool production, body size, milking and mothering ability, and longevity. Included are ewes from the various wool types, but it should be noted that all the breeds are white faced, which some authorities associate with mothering ability.

TABLE 3.2

		Mature weight			
		Ram		Ewe	
Ewe breeds	Wool type	lb	kg	lb	kg
Border Leicester	Long	175–250	80–114	140–180	64–82
Corriedale	Crossbred	180–275	82–125	130–180	59–82
Debouillet	Fine	180–220	82–100	130–150	59–68
Delaine Merino	Fine	180–220	82–100	120–140	54–64
Finnish Landrace	Medium	175–225	80–102	120–160	54–73
Rambouillet	Fine	200–250	91–114	140–180	64–82
Targhee	Crossbred	220–280	100–127	150–200	68–91

3.2.2. Ram breeds Ram breeds (see Table 3.3) have been chosen for their contribution of superior carcass quality and growth rate to crossbred lambs. They are known as mutton breeds since they have been selected for superior muscling for many years. Their wool is generally inferior to that of the ewe breeds.

TABLE 3.3

Ram breeds	Wool type	Mature weight: Ram lb	Ram kg	Ewe lb	Ewe kg
Cheviot	Medium	160–200	73–91	120–150	54–68
Hampshire	Medium	250–300	114–136	175–225	80–102
Montadale	Medium	175–250	80–114	140–180	64–82
Oxford	Medium	200–300	91–136	175–200	80–91
Shropshire	Medium	200–250	91–114	160–180	73–82
Suffolk	Medium	250–350	114–159	190–240	86–109
Southdown	Medium	175–225	90–102	130–160	59–73

3.2.3. Dual-purpose breeds Dual-purpose breeds (see Table 3.4) may be used either as ewe or ram breeds depending on the traits needed in the lambs. Dorsets are now used for producing replacement ewes in order to increase prolificacy and lengthen the breeding season. They also may be used as terminal sires if they are growthy and well muscled. Columbias are most often used as ewe breeds, but sometimes the rams are used as terminal sires. The Romney is used extensively in the Northwest coastal region of the United States. The Lincoln breed is used very little at present, but has contributed to development of ewe breeds, and rams are sometimes used as terminal sires.

TABLE 3.4

Dual-purpose breeds	Wool type	Mature weight: Ram lb	Ram kg	Ewe lb	Ewe kg
Dorset	Medium	175–225	80–102	130–160	59–73
Columbia	Crossbred	225–300	102–136	150–200	68–91
Romney	Long	200–275	91–125	150–200	68–91
Lincoln	Long	225–315	102–143	150–215	68–98

3.3 HISTORY AND CHARACTERISTICS OF EWE BREEDS

3.3.1. Merino The Merino originated in Spain and was bred for weight and quality of wool. It is reputed to be the oldest breed with the greatest numbers in the world. All fine-wool breeds of sheep in the world trace their origin to the original Spanish Merino. The breed was produced in large numbers in Spain by the Crown, nobility, and the Roman Catholic church. The sheep were grazed in the lowlands during part of the year and then were trailed to the mountains during the spring and summer. They were kept in bands of thousands. Their wool was highly prized and formed the basis of much of the foreign trade of Spain.

The exact origin of the breed is not clear. Some authorities believe the original seed stock used for the foundation of Merinos was brought from Italy and North Africa by the Romans. Regardless of the ingredients used to form the breed, it appears to be an acknowledged fact that they and their relatives were far superior to other breeds in the world both in quantity and quality of fleeces produced. The first importation of Merinos to the United States occurred in the early 1800s.

William Jarris of the Diplomatic Corps was responsible for bringing at least 3500 Merinos to the colonies that later became the United States. British embargoes on wool and wool clothing just prior to the War of 1812 stimulated great interest in sheep and the wool industry. It has been estimated that almost 20,000 Merinos were brought in during 1810–1811. During that period, prices for Merinos reached unprecedented heights, and it was known as the "Merino Craze." After the war, Britain lowered its prices on woolen materials to such low levels that it ruined many sheep raisers. However, the wool industry continued to develop in the United States. In 1824, a protective tariff was imposed and interest revived. Breeders in Vermont developed a special strain of Merinos that was much in demand. From the initial importations, Merinos spread throughout the United States. Various breeders selected for different characteristics and thus types were developed. Some strains were small, angular, and had an exceptional number of wrinkles and much yolk in the wool (type A); others had smoother skin, more muscling, and moderate yolk (type C). Conformation varied with the strain. The C type Merino had few wrinkles, greater size, and much improved mutton qualities. The Delaine was developed in Ohio and was larger and smoother skinned than any of the others. This strain is the most popular in the United States today.

Merinos have white markings and pale skin. Most of the rams are horned, but polled strains exist; the ewes are usually polled or have very small horns. They have strong constitutions and can exist under hard conditions of drought, sparse range, and heat better than most other breeds. They have strong flocking instinct and thus are usually found in one group. Merinos for many years were the mainstay of the Southwest sheep industry because of their ability to withstand the climatic conditions. Merinos are generally small and lack muscling, but when the ewes are crossed with mutton rams they produce satisfactory market lambs. Generally, they are the least prolific of American breeds, probably because range breeders have not selected for twinning. The ewes lamb easily and are good mothers. In recent years, a strain called Booroola has been discovered in Australia. These sheep are much more prolific than most other breeds except for the Finn. According to Piper and Bindon (1982), litter size is controlled by a single dominant gene. This trait, if introduced into the American Merinos, could change the usefulness of the breed in the future. There is also a polled strain that is gaining popularity.

Merinos can be used successfully as ewe breeds under arid conditions, but they also do well on farms in the Midwest.

3.3.2. Rambouillet · The Rambouillet breed originated in France and was the result of improving Merinos by selection. They are popular in the United States, particularly the western ranges. It has been estimated that 50% of the blood of all range sheep is Rambouillet. Generally, the characteristics are the same as Merinos with regard to fleece, color markings, color of skin, horns on rams and not on ewes, flocking instinct, and behavior. They are larger and have straighter lines and more muscling than the Merino. They also appear to be somewhat more prolific. Rambouillets have the ability to thrive on the tough dry southwestern ranges, but also do well farther north. They have a very long, productive life, cross very well with mutton rams, and are good mothers.

Modern Rambouillets have smooth skin and are well covered with wool, including the legs. Excessive wool covering on the face to the extent of wool blindness is considered a defect. On range sheep, wool blindness was shown by Terrill (1950) to result in a reduction of 10% in fleece weights and 10% of the weight of lamb produced. Fleece grades should be fine with no discolored fibers. Skin folds and wool below the eyes is discriminated against. In the last 15 years, breeders have made a great effort to improve the utility of the breed. They have improved the muscling, reduced wrinkles, and increased the staple length of the fleece. On their official scorecard, 60% is devoted to conformation and 40% to wool. They have also developed polled strains of the breed. Rambouillets have a longer breeding season than most other breeds. Many ewes will breed in the spring, but, according to Hulet et al. (1974), this trait is strongly influenced by environmental conditions. In cooperative research work, ewes reared in Texas cycled much longer throughout the year than they did when moved to the Idaho Experiment Station.

3.3.3. Debouillet The Amos D. Jones ranches in New Mexico developed the breed from crosses of Delaine Merinos and Rambouillets. The Debouillet Sheep Breeders Association was started in 1954. All sheep must pass visual and performance tests prior to registration. The breed was developed under practical range conditions to produce acceptable fleeces and lambs under the arid conditions of the southwest ranges.

Debouillet sheep have white open faces and smooth bodies. They have wool on the legs. Rams may be horned or polled, but polled is preferred. They should have uniform fleeces that grade fine, with spinning count of 64's to 80's, with a staple of at least 3 in. (7.7 cm) and a ewe fleece weight of 10 to 18 lb (4.5 to 8.2 kg) on a grease basis of 6 lb (2.7 kg) clean. The ewes lamb easily and suckle well, are gregarious, and handle well.

3.3.4. Corriedale James Little, manager of the Corriedale state in North Otago of the South Island of New Zealand, is given credit for being the architect of the Corriedale breed of sheep. It was developed for the drier hill country in the South Island, and they are still important in that area. In 1980, 5 million of the 63 million head in the country were Corriedales.

The breed was formed primarily by mating Merino ewes with Lincoln and English Leicester rams, but Romney and Border Leicester rams were also used in some of the early crosses. From the original crossbreds between the fine- and long-wool breeds, the progeny were interbred and selected to establish the type. The breed was officially recognized as an "inbred halfbred" in the New Zealand Sheep Breeders Association Flock Book in 1906 and established their own flock book in 1924. First importations to the United States were in 1914 by the USDA. Corriedales are known as dual-purpose sheep, with equal emphasis on wool and carcasses.

Corriedale rams may shear from 15 to 25 lb (6.8 to 11.4 kg) of wool and ewes 12 to 20 lb (5.4 to 9.1 kg), depending on level of feeding. The fleece grades medium to fine or an average spinning count of 56's with a light shrinkage. Markings are white, the skin is pink, and they are wooled to the ground. Wool below the eyes, wrinkles, and white hooves are undesirable. Compared to fine-wool breeds, the Corriedales have straight lines, more muscling, and are deeper and thicker. However, they do not equal the mutton breeds in this respect. The ewes are good mothers and more prolific than fine-wool breeds. Some flocks of purebreds have 150% lamb crops although the average may be nearer 135%. They are docile, easily handled sheep and have retained the flocking instinct from their fine-wooled ancestors. The ewes have about the average length of breeding season, but some flocks have a high percentage of ewes that breed for fall lambs. Generally, the ewes have a long, productive life.

3.3.5. Targhee The Targhee breed was developed at the United States Experiment Station at DuBois, Idaho. They were named for the Targhee National Forest, which is used for summer range by the USDA Sheep Experiment Station. The development of the breed began in 1926 with carefully selected Rambouillet rams bred to selected ewes out of a cross between Corriedale rams on Lincoln × Rambouillet crossbred ewes. The crossbreds were then inbred, tested, and selected to produce a range sheep adapted to northern intermountain range conditions where more compact finer fleeces are required to protect the sheep from severe weather conditions. Authorities estimate the percentage of blood in the breed to be one-fourth long-wool and three-fourths fine-wool breeding. Targhees appear to be somewhat smaller than Columbia sheep.

Mature rams and ewes in range condition weigh in excess of 200 lb (91 kg) and 130 lb (59 kg), respectively. They are polled and have white, open faces, smooth skin, and white markings. They may or may not have wool on the legs. They shear 11 lb or more wool that grades one-half blood (58's to 62's spinning count), with a staple length of 3 in. (7.6 cm). The sheep are docile and easily handled. They have a well-developed flocking instinct. The ewes lamb easily and are good mothers. In research by Bradley et al. of Wisconsin (1972), Targhee ewes weaned 1.6 lambs per ewe exposed during a 2-year trial. They were superior in reproductive performance to Suffolk and Shropshire ewes. It seems reasonable to say that Targhees are probably more prolific than Corriedales and Columbias. They not only do better than most other breeds on hard range conditions, but they are superior under farm conditions.

3.3.6. Finnsheep (Finnish Landrace) The Finnsheep originated in Finland in a region of rugged hill pastures and severe winters. This breed is known as Finnish Landrace in its native land. Although it is undoubtedly an ancient breed, the Finnsheep Breed Society was not established until 1918. Through careful selection and production testing, the breed became widely known for its extraordinary prolificacy. It also has a reputation for hardiness of lambs and mothering ability of ewes. Finnsheep were first brought to the United States in 1968 as a result of the demands of sheep breeders for more prolific sheep. The Finnsheep Breeders Association, Inc., was founded in 1972. Although the breed is new to the United States, it has been used extensively in crossbreeding research in many experiment stations. It has been used in forming synthetic breeds for modern systems of intensive lamb production. Finnsheep are polled, with no wool on the face and legs and pink skin. Rams may shear 7 to 8 lb (3.2 to 3.6 kg) and ewes 5 to 6 lb (2.3 to 2.7 kg). The wool grades one-fourth to one-half blood (48's to 62's). The conformation of the Finn is generally lacking in muscling, with uneven top lines and narrow bodies. Pure Finnsheep may produce litters of as many as five or six lambs, although that is by no means the average for the breed. They lamb easily, mother their lambs well, and have a heavy milk flow. Although the lambs are small at birth, they are alert and have a strong desire to live. The breeding season appears to be somewhat different from domestic sheep. Apparently, they continue to cycle later in the winter than American sheep. It appears that from one-fourth to one-half Finnsheep blood in crossbred ewes will increase lamb production appreciably. The crossbreds bred to meaty terminal sires produce very acceptable market lambs. Some authorities now recommend 25% to 50% Finnsheep blood for farm flocks and less than 25% for range sheep.

3.3.7. Border Leicester The Leicester breed was developed by Robert Bakewell and had much to do with the formation of other breeds, including the Border Leicester. The English Leicester is different from the Border Leicester in appearance and is not as popular. According to historians, there is some Cheviot blood in the Border Leicester breed. The head and the legs are free of wool. The ears are held erect, but not as much as the Cheviot. They also tend to be considerably larger and far more docile and less active than Cheviots.

Besides the clean head and legs, the most striking feature of the Border Leicester is the facial profile, which is usually Roman. The markings are white, the skin pink, the nostrils black, and they are polled. The wool grades low quarter-blood to braid (36's to 50's) and ewe fleeces weigh 8 to 10 lb (3.6 to 4.5 kg). The staple length is 6 to 10 in. (15.2 to 25.4 cm). Border Leicester ewes are noted for high fertility, mothering, and nursing ability. The rams have been used effectively in crossing with Romney ewes to improve the lambing percentage.

Excellent type rams and ewes of the ewe breeds described in the above paragraphs are shown in Figure 3.1.

3.4 HISTORY AND CHARACTERISTICS OF RAM BREEDS

The ram breeds are predominently European in origin. These breeds are also known as meat-type sheep since the emphasis in breeding programs has been more on muscling than wool. The primary contributions of the ram breeds in a crossbreeding program (i.e., the terminal cross) for commercial lamb production are to provide the genes for rapid growth and improved carcass quality. The rams are also expected to be virile and aggressive breeders.

3.4.1. Suffolk The Suffolk breed originated in southeastern England in Suffolk country, a low, flat, fertile area not far from London. Like the Hampshire breed, it is said to be the result of crossing Southdown rams on a native strain of ewe. The native strain was horned and called the Norfolk. From the original cross, descendants

DELAINE MERINO

Fall ram lamb
Courtesy *Sheep Breeder and Sheepman*

Yearling ewe
Courtesy *Sheep Breeder and Sheepman*

BOOROLA MERINO

Mature ram
Courtesy USDA

Mature ewe
Courtesy USDA

RAMBOUILLET

Senior ram lamb
Courtesy *Sheep Breeder and Sheepman*

Yearling ewe
Courtesy *Sheep Breeder and Sheepman*

Figure 3.1 Typical rams and ewes of the ewe breeds.

CORRIEDALE

Yearling ram
Courtesy *Sheep Breeder and Sheepman*

Yearling ewe
Courtesy *Sheep Breeder and Sheepman*

TARGHEE

Yearling ram
Courtesy *Sheep Breeder and Sheepman*

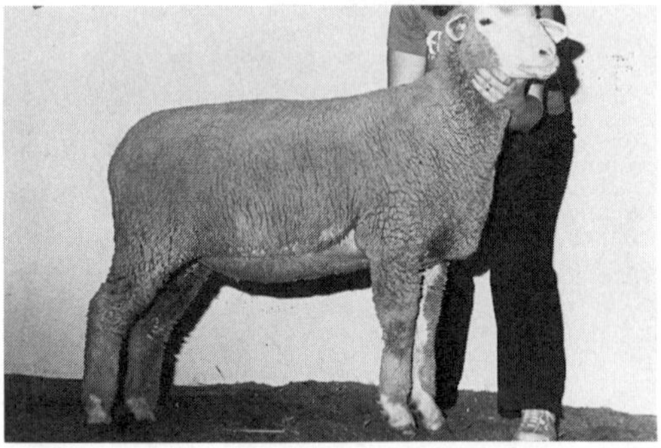

Yearling ewe
Courtesy *Sheep Breeder and Sheepman*

FINNSHEEP

Yearling ram
Courtesy Finnsheep Breeders Association

Ewes and lambs
Courtesy Finnsheep Breeders Association

Figure 3.1 Continued

BORDER LEICESTER

Mature ram
Courtesy American Border Leicester Assoc.

Mature ewes
Courtesy American Border Leicester Assoc.

Figure 3.1 Continued

were inbred and selected to obtain a large sheep with superior mutton characteristics. Suffolks were brought to the United States in 1888. The Suffolk breed has two breed societies, the American Suffolk Sheep Society established in 1929 and the National Suffolk Sheep Association established in 1934. The National Suffolk Sheep Association is the larger of the two.

The Suffolk is a large, rugged, meaty breed of sheep. Their black head, ears, and legs contrasted with their white wool cause them to be distinctive in appearance. They are polled, with no wool on their head in front of their ears. The face and legs should be free of wool and covered with fine black hair. The head should be strong but not coarse. Scurs are sometimes found on rams. The ears should be fine, but long, and held horizontally or slightly drooped. Skin color may be pink or dark, with pink preferred. Suffolks are not known for their wool. Fleece weights of 6 to 7 lb (2.7 to 3.2 kg) for ewes and 8 to 9 lb (3.6 to 4.1 kg) for rams are average. Black fiber is a problem with Suffolks. Staple length ranges from 2 to 2½ in. (5.1 to 6.4 cm). Grade of wool is three-eighths blood (56's). Many rams and some ewes may be bare in the armpits and on the belly.

Modern Suffolks have long bodies and stand taller than most sheep of their class. But although legs and necks may look long, they are a well-muscled breed with strong backs and superior loins and hindquarters.

Suffolk ewes have a long breeding season and are generally conceded to be equal or even superior to any other meat-type breed, with lambing percentages of 150% to 175% percent or more. Ewes have little difficulty lambing and are good mothers with an abundant milk supply. The lambs grow fast and efficiently and have done well in interbreed carcass shows. Suffolk rams have been used extensively on white-faced range ewes to sire market lambs and replacement ewes. Their progeny is superior in rate of gain to other breeds. The breed has increased in popularity in recent years as greater emphasis has been placed on percentage of lean meat in carcasses. For commercial production of quality lambs, Suffolks must be considered a leading terminal sire breed.

3.4.2. Hampshire The breed originated in southern England in an area of rolling hills and fertile soil. It is not completely clear how the breed originated, but most authorities agree that at least two native strains, Wiltshires and Berkshire Knots, and the Southdown breed were used in the initial crosses. Hampshires were recognized as a breed in 1857 by the Royal Agricultural Society. They were brought to the United States before the American Civil War, but the flocks disappeared during the war. Importations in 1880 and 1881 were the start of the breed in the United States. Registration of Hampshires began in 1889. The breed is widely distributed from the East Coast to the West. The sheep do well under farm and range conditions. They are polled and have dark brown or black markings. They have strong heads with large, well-shaped ears that are carried well down on the side of their head either horizontally or with a slight droop. Rams usually have bold strong heads and may have a Roman nose. Ears free of wool are preferred. Legs may be covered with dark or colored wool. Ewes shear approximately 8 lb (3.6 kg) of one-fourth to three-eighths (48's to 56's) wool and rams 10 lb (4.5 kg). Staple length may be 3 in. (7.7 cm). The skin of most Hampshires is dark, but some are pink. Hampshires have a quiet disposition and are easily handled. Ewes are slightly above average in prolificacy (135% to 165%) and have a long breeding season. Rams also appear to be fertile over a longer period than most of the other medium-wool breeds. The ewes are good mothers and milk well, and the lambs make rapid gains and are usually sufficiently finished when they reach market weight. Hampshire rams stamp their characteristics on their offspring. Thus, they are very well regarded as sires by commercial sheep raisers with fine-wool or crossbred wool types of ewes either in farm or range states. Not only do the lambs look like Hampshires, but they grow rapidly and have desirable carcasses. Some Hampshires have exceptionally large heads and shoulders, causing ewes to have difficulty in lambing. The breed is also criticized for black fiber in the wool, and sometimes they have crooked legs.

3.4.3. Shropshire Shropshires originated in central England in the counties of Shropshire and Staffordshire, a land of rolling hills and fertile pasture land. The breed originated from a blending of native strains with Southdown, Leicester, and Cotswold breeds, followed by linebreeding and careful selection. The breed was well established in England by 1859. There is some evidence that Shropshires were brought to the United States in 1855, but not much detail is available. Later, in 1860, importation began and increased in intensity for the next 20 to 30 years. The American Shropshire Registry Association, Inc., was founded in 1884 at Lafayette, Indiana. In 1920, a federal census of sheep breeds showed that Shropshires were the most popular breed in the United States. The breed was favored by small farmers for its production of lambs as well as wool. In fact, the motto of the breed was "Wool from the tip of the nose to the tip of the toes." Shropshire lambs made rapid growth and were well finished when they reached market weight. Only the Southdown had a better record for winning carcass shows. Shropshire rams were used successfully on fine-wool ewes for market lamb production. There were many flocks of grade Shropshire ewes in the farming states. But the Shropshire breeders began emphasizing wool and blocky conformation with a consequent loss of size and an increase of wool on the face to the extent that many became wool blind. Within a very few years, this tremendously useful breed of sheep lost favor with farmers and soon was no longer a competitor with such breeds as the Hampshire or Suffolk. In 1946, some English Shropshires were imported, but it was only after much controversy that the larger, more useful sheep were accepted. The modern Shropshire is a much more useful sheep than the wool-blind pony type that was popular in the 1940s and 1950s. Color markings are dark, ranging from brown to very dark brown but not jet black. The ears are medium in size and are carried slightly erect or level with the base. They should be wooled over the poll and open around the eyes with little, if any, wool below the eyes. They should have solid, dark-colored faces, except that some may have gray noses and gray around the nostrils. Many modern Shropshires have considerable dark wool on legs and around the feet. They should be well muscled with strong tops and with heavy muscling in the leg. Most breeders are attempting to increase the length of body and the scale of the breed. Ewes should shear 10 lb (4.5 kg) of three-eights blood (56's) grade wool with a staple of 2.5 in. (6.4 cm). The breed is polled with a pink skin. Ewes are prolific, with a 150% lamb crop or better. The ewes have little difficulty lambing, mother their lambs well, and are good milkers. Their lambs grow at an intermediate rate and finish without excessive amounts of grain. Carcasses of lambs out of commercial fine-wool ewes and Shropshire rams are quite satisfactory. Shropshires appear to have shorter breeding seasons than some of the other medium-wool breeds, and rams may not be as fertile early in the breeding season as some other breeds. If breeders continue to produce larger, meatier, open-faced rams, they should find a ready market as terminal sires for commercial sheep operations.

3.4.4. Oxford The Oxford originated in the southern third of England in an area well adapted to grain and grass production. The breed was formed by crossing the original Hampshire and Cotswold strains of sheep in about 1833. The Oxford was recognized as a breed in 1851. It was imported to the United States in the early 1850s. The American Oxford Down Record Association was organized in 1882. The Oxford is one of the largest breeds in the United States. They are polled with pink skin and dark markings. Fitted sheep are easily distinguished by the long wool on the top of the head. The color of the face, ears, and legs varies from brown to dark gray. The ears are shorter and wider than Hampshires and carried horizontally. The ears may be covered with wool. The face may be wooled, but not to the extent of wool blindness. Legs usually have some brown wool on them. Oxfords shear more than the other breeds, with average fleeces of ewes 10 to 12 lb (4.5 to 5.4 kg) and rams up to 15 lb (6.8 kg). The wool grades one-fourth (48's), the staple is 3.5 to 5 in. (9 to 12.8 cm), and the shrink is low. The ewes are prolific, with 150% lamb crops common in most flocks. The length of the breeding season is average to shorter than for most mutton breeds. The ewes lamb easily and suckle their lambs well. The lambs grow fast, but must be above average in weight to reach suitable market finish. Carcasses of the lambs tend to have too much finish on the outside and not enough within the meat. The breed does well on very good pasture or with a bountiful feed supply. They are not good rustlers and do not do well under tough conditions. Oxford rams have found favor for crossbreeding with small fine-wool ewes to improve the size and muscling of lambs and the wool production of the crossbred replacement ewes. Large, meaty Oxford rams are suitable as terminal sires.

3.4.5. Southdown The breed originated in the county of Sussex in Southern England in an area of hilly pastures and mild climate where the pastures are short but of high quality; sheep thrive there. The breed was developed by selection and close breeding of the native sheep of Sussex. They became famous for their carcasses and won many interbreed carcass shows. The Southdown Sheep Society of England was founded in 1891, which was actually 9 years later than the establishment of the American Southdown Sheep Breeders Association in 1882. The Southdown is known for its excellent muscling and superior finishing ability. Although it is small in stature, it has been used extensively for crossbreeding on commercial ewes. It has the ability to stamp breed character on crossbred lambs. The Southdown breed has been criticized for lack of size and has not attained the popularity it could have if it had more length, height, and weight. It is interesting that New Zealand breeders have selected for more scale, and their Southdowns are much larger than the American strain.

The Southdown is not easily mistaken for any other breed. The small, erect ears, short, wide head, and prominent eyes and nostrils are notable. Color markings of the face and legs varies from mouse colored to brown. White or very dark color approaching black is not desirable. Modern Southdowns have less wool about the face than the winners of shows in the 1940s and 1950s. The breed is polled with pink skin, and has the finest fleece of any of the medium-wool breeds. The fleece has a short staple, usually less than 2 in. (5.10 cm) and a low shrinkage. Ewes may shear 5 to 6 lb (2.3 to 2.7 kg) and rams 7 to 8 lb (3.2 to 3.6 kg). The wool grades three-eighths to one-half blood (56's to 62's spinning count). Southdowns are moderately prolific (140% to 150% lamb crop) and somewhat below average in length of breeding season. Some ewes have difficulty in lambing, but the lambs are strong and active and the ewes mother them well. Milk production is average, but lambs tend to grow slower than other medium-wool breeds. Thus their weight is less, although the carcass quality is high. Southdowns suffer in comparison with the larger, longer breeds presently in favor with commercial producers. If Southdown breeders are successful in changing their type to longer, taller animals without sacrificing carcass quality, this breed may increase in popularity as a terminal sire on commercial ewes.

3.4.6. Montadale The Montadale breed originated in Missouri from crosses of the Columbia and Cheviot breeds in 1945. The association was formed in 1947. The flock book was closed in 1954. The breed is polled with white markings and pink skin. The head and face are free of wool, and the legs should be free of wool below the knees and the hock. The ears are medium in size and carried erect, although not so much as the Cheviot. The ewes should shear 10 to 12 lb (4.5 to 5.4 kg) of one-fourth to three-eighths blood (50's to 56's) wool with a staple length from 2 to 2.5 in. (5.1 to 6.4 cm), and the rams 12 to 14 lb (5.4 to 6.4 kg) with somewhat coarser fleece. The breed association has emphasized body conformation, and thus improvement has been made. In recent years, they have won some major interbreed carcass shows. Montadale ewes have an average-length breeding season and 135% to 150% lamb crops. They lamb easily, are good mothers, and are reputed to produce a good flow of milk. Montadale lambs grow somewhat slower than some breeds, but are finished at market weights without excessive grain. The breed appears to be gaining acceptance on farms in the Midwest. They have an active breed association that has advertised extensively. Montadale rams have been used for crossbreeding with satisfactory results. Large, meaty Montadale rams are very acceptable as terminal sires.

3.4.7. Cheviot The Cheviot breed originated near the border between Scotland and England. This area borders on the North Sea and is called the Cheviot Hills. Climatic conditions are severe with high moisture and cold weather. The grass is short, but nutritious. Most Cheviots are very active and able to exist under hard conditions.

According to breed history, Cheviots were first brought to the United States from Canada in 1838. The American Cheviot Sheep Breeders Association was formed in 1891. Cheviots are distributed throughout the east and north-central states. Cheviot markings are white and the skin pink. They should have no wool on the head or face in front of the ears. The legs are also bare of wool. Perhaps the most striking breed characteristics are the small, very erect ears and the erect carriage of the head. Nostrils and hooves should be black. The best of the breed have well-muscled bodies and sound feet and legs. Cheviot wool generally grades one-fourth blood (48's to 50's) and has a staple of 4 to 5 in. (10.2 to 12.8 cm). Yolk content of the wool is very low and thus they appear to be whiter than other breeds. Ewes' fleeces average 5 to 7 lb (2.3 to 3.2 kg) and rams' 7 to 9 lb (3.2 to 4.1 kg).

Cheviot ewes lamb easily, are good mothers, and milk well. They are excellent grazers and can stand inclement weather. They are average for prolificacy and length of breeding season. The greatest objection to Cheviots is their lack of size. Some sheepraisers also object to their active nature and lightweight fleeces. In years past, the breed was not highly regarded for their carcasses, but their association has been emphasizing carcass quality, which has improved to the extent that they have won some major carcass contests. It is also noted that efforts are being made to increase the size of the breed.

Figure 3.2 shows typical rams and ewes of the ram breeds of the sheep described in the above paragraphs.

3.5 HISTORY AND CHARACTERISTICS OF THE DUAL-PURPOSE BREEDS

It is difficult to draw the line on classification of sheep by use. There are many commercial producers who are very well pleased with crossbred ewes out of fine- or crossbred-wool types of ewes and sired by Suffolk, Hampshire, Shropshire, and other ram breeds. Thus the classification of breeds as dual purpose is not precise. However, in practice, the breeds in this classification seem to have advocates for using them either as ewe or ram breeds, depending on the traits they expect them to impart to their offspring.

3.5.1. Dorset The breed originated in southern England in a good agriculture area of rolling hills and fertile soil. The climate is mild, but moist, and little housing

SUFFOLK

Yearling ram
Courtesy *Sheep Breeder and Sheepman*

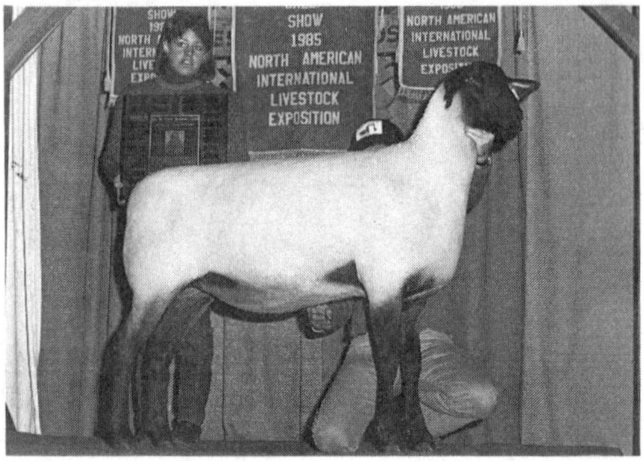

Yearling ewe
Courtesy *Sheep Breeder and Sheepman*

HAMPSHIRE

Yearling ram
Courtesy *Sheep Breeder and Sheepman*

Yearling ewe
Courtesy *Sheep Breeder and Sheepman*

SHROPSHIRE

Senior ram lamb
Courtesy *Sheep Breeder and Sheepman*

Junior ewe lamb
Courtesy *Sheep Breeder and Sheepman*

Figure 3.2 Typical rams and ewes of the ram breeds.

OXFORD

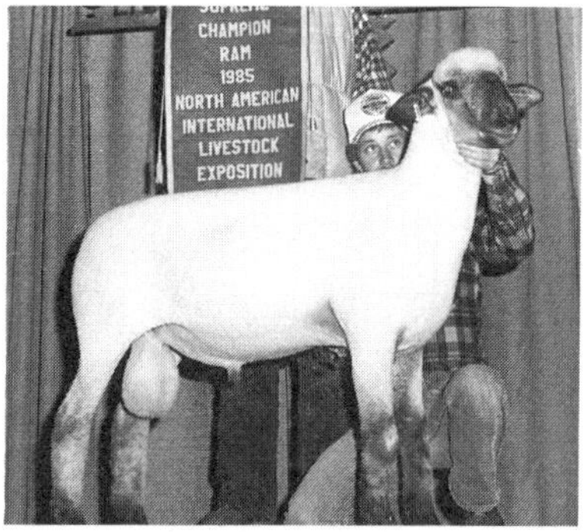

Ram lamb
Courtesy *Sheep Breeder and Sheepman*

Yearling ewe
Courtesy *Sheep Breeder and Sheepman*

SOUTHDOWN

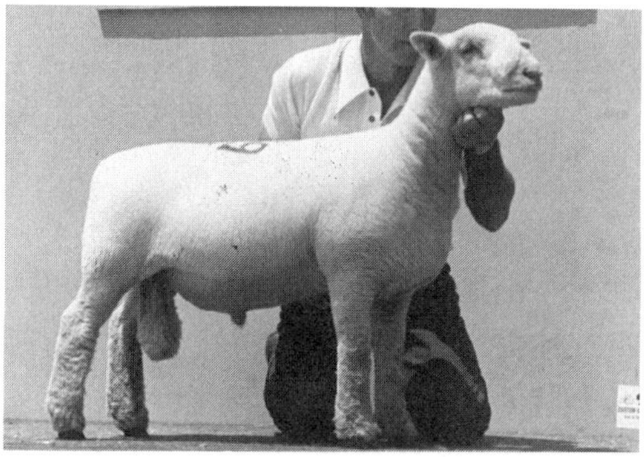

Junior ram lamb
Courtesy *Sheep Breeder and Sheepman*

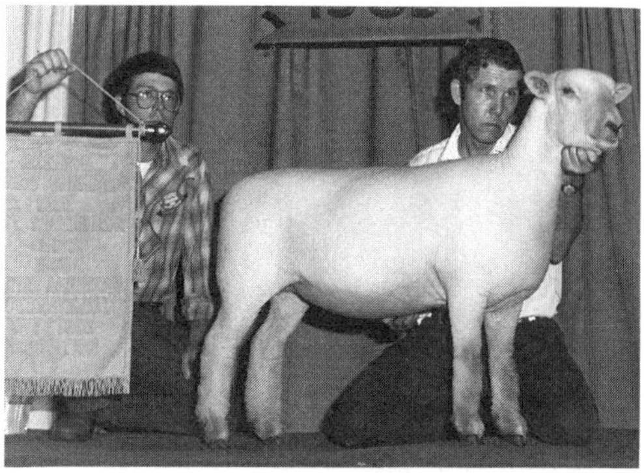

Yearling ewe
Courtesy *Sheep Breeder and Sheepman*

MONTADALE

Junior ram lamb
Courtesy *Sheep Breeder and Sheepman*

Yearling ewe
Courtesy *The Suffolk Banner,* Cuba, Ill.

Figure 3.2 Continued

CHEVIOT

Junior ram lamb
Courtesy *Sheep Breeder and Sheepman*

Yearling ewe
Courtesy *Sheep Breeder and Sheepman*

Figure 3.2 Continued

is used for the sheep. Dorsets have long been known for their prolificacy and early lamb production. Dorsets, known as the Dorset Horn, were brought to the United States in 1860. The breed is well known in the eastern United States for their production of hot-house lambs (lambs weighing 30 to 60 lb when slaughtered from December to April). More recently, it has increased in popularity in the Midwest, Southwest, and western United States.

Dorsets have strains that are horned and polled. Both sexes are horned in the horned strain. Color markings of Dorsets are white with pink skin. Wool is not generally found on the ears or in front or below the eyes. Legs may or may not be covered with wool. Dorsets are not known for wool production. Their fleeces grade one-fourth to three-eighths (48's to 56's), with a staple length that may exceed 3 in. (7.7 cm). It has a very low yolk content and thus is quite white. Ewes may shear 8 lb (3.6 kg) and rams 10 lb (4.5 kg).

Ewes have a very long breeding season, much longer than other medium-wool breeds. They are widely known for their prolificacy and milking ability. Dorset ewes are good mothers and lamb easily. As a breed, they are docile and easily managed, and they have increased in popularity as interest in accelerated lambing and increasing lambing percentage by crossbreeding have developed. A polled strain was developed from a mutation that occurred on the campus of North Carolina State University. Polled Dorsets are found in much greater numbers than the horned strain today; however, performance relative to breeding out of season or prolificacy has declined with the introduction of the polled strains. Crossbred ewes out of Dorset rams and either fine- or medium-wool range ewes have proved to be quite productive on Midwest farms. Large, meaty Dorset rams are very satisfactory as terminal crosses on ewe breeds of sheep.

3.5.2. Columbia The breed originated in the United States through the efforts of scientists in the USDA. The original crosses were made on the King ranch at

Laramie, Wyoming, in a cooperative project between the ranch and the USDA. After a number of crosses between long-wool rams and Rambouillet ewes, it was decided that the Lincoln × Rambouillet cross was superior and that became the foundation stock for the breed. The scientists used a systematic breeding and testing program after they began interbreeding the original crossbred progeny. Outstanding individuals were mated each generation to establish the type. The Columbia Sheep Breeders Association of America was established in 1941.

Columbia sheep are one of the larger, if not largest, breeds. Color markings are white, the skin is pink, the face is open, and there may or may not be wool on the legs. The wool grades one-fourth to three-eights blood with a spinning count of 48's to 58's. Under range conditions, ewes shear 11 to 12 lb (5 to 5.4 kg) of wool and rams more. The wool has a light shrinkage. They have well-developed flocking instinct. Their lambs are large at birth and grow fast. The ewes are good mothers, lamb easily, and suckle lambs well. Prolificacy is about the same as for Corriedales, but higher than for fine-wool breeds. Columbia ewes mated to mutton rams produce excellent market lambs. Although the breed was developed for the ranges in the mountain states, Columbias do quite well under farm conditions.

Because lambs out of Columbia rams have the ability to grow rapidly and have carcasses superior to progeny from Merino and Rambouillet rams, they are often used as terminal sires for market lambs on the range and thus their classification as a dual-purpose breed is to some extent justified. However, it seems to this writer that they are more nearly a ewe than a ram breed.

3.5.3. Romney The Romney breed originated in southeastern England in an area of high rainfall and wet ground conditions. It is reputed to have a certain amount of immunity to foot rot and internal parasites. Although it is not raised in large numbers in the United States, Romneys appear to be adapted to climatic conditions in the northwest coastal states, particularly in Oregon. The breed is very popular in New Zealand, where they are said to number 29 million head. The breed association was established in England in 1894 and in the United States in 1911.

The Romney has a white face that is clear of wool around the eyes, white legs covered with wool, and black nostrils. The wool is finer and not as long as in other wool breeds. It grades one-fourth to braid (36's to 48's), with a staple of at least 7 in. (17.8 cm) and weights of 10 to 13 lb (4.5 to 5.9 kg). The fleece on Romney sheep is considerably more compact and has more crimp and luster than do other long-wool breeds. Romney ewes are only medium in prolificacy, but are good mothers and produce such a good flow of milk that lambs gain rapidly and have satisfactory finish at the low market weights desired in New Zealand. They appear to do better under very favorable conditions than in areas that are dry with poorer range or pasture conditions.

3.5.4. Lincoln The Lincoln originated in England in a fertile area called Lincolnshire. They were brought to the United States early in the 1800s, but never became very popular. The National Lincoln Sheep Breeders Association was founded in 1891. Very few are registered each year. The Lincoln is a polled, white-faced, and pink-skinned breed. Modern Lincolns have little wool on their faces and legs. The

DORSET

Polled ram lamb
Courtesy *Sheep Breeder and Sheepman*

Polled ewe lamb
Courtesy *Sheep Breeder and Sheepman*

COLUMBIA

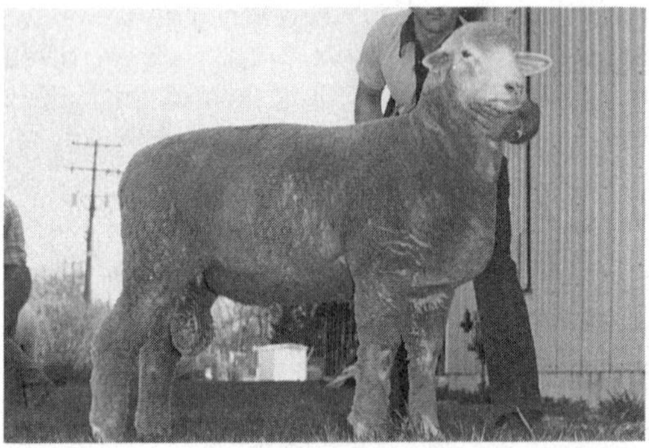

Yearling ram
Courtesy *Sheep Breeder and Sheepman*

Yearling ewe
Courtesy *Sheep Breeder and Sheepman*

Figure 3.3 Typical rams and ewes of the dual-purpose breeds.

wool is coarse, long, and lustrous and hangs in ringlets. Fleeces of ewes may weigh 10 to 15 lb (4.5 to 6.8 kg) and rams more than 20 (9.1 kg). The fleece grades low quarter-blood to braid (44's to 46's) with a staple of 10 to 15 in. (25.6 to 38.5 cm). Lincolns have been used in the development of the crossbred types of sheep to increase body size and length of fleece. The ewes are medium in prolificacy, lamb easily, and are low to medium in milking ability. Lambs may lack some in hardiness and develop rather late. Some lambs are entirely too large when they are fat enough for slaughter. As terminal sires, rams can be used to provide size and muscling to lambs.

Excellent type rams and ewes of the dual-purpose breeds are shown in Fig. 3.3.

3.6 AVAILABLE NEW BREEDS MAY HELP THE INDUSTRY

New breeds that may be useful to the sheep industry are being developed. Some that are now in use are described next.

ROMNEY

Ram lamb
Courtesy *Sheep Breeder and Sheepman*

Ewe lamb
Courtesy *Sheep Breeder and Sheepman*

LINCOLN

Yearling lamb
Courtesy *Sheep Breeder and Sheepman*

Yearling ewe
Courtesy *Sheep Breeder and Sheepman*

Figure 3.3 Continued

3.6.1. Polypay The Polypay is one of the newest breeds to be introduced. It was developed by the U.S. Sheep Experiment Station at DuBois, Idaho, with initial crosses made in 1967. The breed originated from reciprocal crosses of (Dorset × Targhee) (Finnsheep × Rambouillet). The original objectives of the scientists were to develop a range breed that would have outstanding productivity, with twins at 1 year of age and each year thereafter two sets of rapidly growing twins with good carcass quality. Although to achieve such lofty objectives the breed would be little short of miraculous, the fact is that the Polypay appears to be outstanding in lamb production. In comparisons with the four parent breeds, they invariably produce more (except the Finnsheep) and heavier market lambs. There is, however, less wool produced by Polypay sheep than the Targhee or Rambouillet ancestors. The Polypay is superior to other breeds in length of breeding season, but still not capable of producing twins twice each year. They appear to be well adapted to the accelerated lambing program, with lamb crops at 8-month intervals. Polypay sheep are polled and have open, white faces, pink skin, and little wool on their legs. They shear 8 lb (3.6 kg) of wool, compared with 10 lb (4.5 kg) for Targhees. The wool grades one-fourth to three-eighths blood (48's to 58's), with a staple length somewhat shorter than the Targhee breed. The American Polypay Association was organized in 1980. The association has performance criteria that must be met to register sheep.

3.6.2. Coopworth During the 1960s, Ian Coop of Lincoln College in Canterbury, N.Z., in experiments to increase productivity of ewes, developed a strain of sheep by crossing the Border Leicester and Romney breeds and interbreeding the crossbreds. The breed society was established in 1968 and has become quite popular, with estimates of 11 million head in New Zealand in 1980. It is now being exported to foreign countries, including the United States. Coopworth ewes may weigh 135 to 155 lb (61.4 to 70.4 kg) and rams 200 to 225 lb. (90.9 to 102.3 kg). The breed is polled with an open face and little wool on their legs. Markings are white and the skin is pink. The wool grades one-fourth blood or (44's to 50's), with a staple length of 4.9 to 6.9 in. (12.4 to 17.5 cm) and weights of 11 lb (5 kg) for ewes and 13 lb (5.9 kg) for rams. The ewes are average for length of breeding season and give birth to 140% to 150% lamb crops. The ewes lamb easily and are excellent mothers and good milkers.

One reason Americans may be interested in the Coopworth breed is the activity of the breed association to ensure standards of performance in order to register sheep. Breeders must record performance and be members of the National Sheep Improvement Program, Sheepplan, in order to register sheep. Before registering, the owner must prove that the sheep has been bred from highly selected parents, has been fully recorded, and meets entry standards. It must also pass a visual inspection for physical soundness and freedom from faults in wool and carcass. Two examples of minimum requirements for registration are: A stud sire must be from a multiple birth, out of a ewe with 200% lambing record (lambs weaned to number of lambings), and the sire must have an above-average record for adjusted weaning weight and fleece weight. The ewe must be culled if she has not reared a set of her own twins at the end of the second lambing or if on any occasion she is barren or assisted in lambing. The society prohibits competitive showing of Coopworth sheep. The Coopworth appears to be best adapted to more fertile soils with better quality pastures.

3.6.3. Perendale The Perendale breed was developed in New Zealand by Sir Goeffrey Peren of Massey University. Perin crossed the Cheviot and Romney breeds and practiced close selection to produce a type of sheep adapted to the less fertile, rough, hill country pastures. The breed association was established in 1960. In 1980, it was estimated there were 11 million Perendales in New Zealand. The breed is only exceeded in numbers by the Romney and the Coopworth. Perendale breeders actively participate in production testing, and all sheep must meet minimum standards to be registered. The association puts equal emphasis on production of meat and wool.

Perendales are rather distinctive in appearance with white, wool-free faces and white, clean legs. The ears are erect, although less than the Cheviot. Some wool is allowed on the poll. The nostrils are black.

POLYPAY

Yearling ram

Mature ewes

COOPWORTH

Mature ewes

PERENDALE

Mature ewes

Figure 3.4 Rams and ewes of new and exotic breeds that may be useful to the sheep industry.

The Perendale breed is smaller than Coopworths and more active. Ewes may weigh 120 to 145 lb (54.5 to 65.9 kg) and rams 180 to 220 lb (81.8 to 100 kg). Fleeces grade one-fourth to three-eighths (46's to 56's), with a staple of 3.9 to 4.7 in. (10 to 12 cm), and weigh 8 to 11 lb (3.6 to 5 kg).

Perendale ewes are good foragers, are easily handled, lamb easily, are good mothers, and suckle well. They are more prolific (100% to 120%) than the Romney breed. This breed could possibly be used for crossing to produce commercial ewes.

Figure 3.4 shows typical sheep of the breeds described in the paragraphs above.

3.7 *SHEEP BREEDS IN THE WORLD POTENTIALLY IMPORTANT TO THE SHEEP INDUSTRY IN THE UNITED STATES*

There are a number of breeds that could provide genetic material useful to the American sheep industry. At the present time, because of danger of disease, it is legal to import sheep from only a few countries in the world. If it becomes possible, by using quarantine stations, to import sheep from other countries, some of the following breeds will surely be of value to the breeder who is attempting to develop the type of sheep the industry needs for the future.

3.7.1. Texel The Texel breed originated on the Texel Island near Holland and is widely known for its prolificacy, milking and mothering ability, and quality of lamb carcasses. It is a large, well-muscled sheep that is polled and has white markings and pink skin. Ewes shear 10 to 12 lb (4.5 to 5.4 kg) of one-fourth to three-eighths blood (48's to 58's fleeces). If importation becomes possible, the breed should be useful in crossing for commercial ewe production, or purebred rams could be used as terminal sires for lamb production.

3.7.2. East Friesian The East Friesian breed is a native of Germany. It is also called the Westphalian rat-tail. This breed has a mutton-type conformation similar to the Texel. It is noted for prolificacy and high milk production. In fact, they are used in dairies in Germany in certain localities. General characteristics are similar to the Texel. This breed should have a similar usefulness as the Texel if it becomes possible to import them.

3.7.3. Ile de France This is a French mutton breed that is similar in length of breeding season and prolificacy to the Dorset. It is a well-muscled, medium-sized breed with white markings and medium wool of one-fourth to three-eights blood (48's to 56's). Ewe fleeces may weigh 9 to 11 lb (4.1 to 5 kg). This breed should be useful in producing crossbred commercial ewes.

3.7.4. Romanov This breed originated in Russia and was developed from the Russian Northern Short-tailed sheep. They have very coarse wool, similar to that of the Karakul, white and black blazed or mottled faces, and dark skin. They are inclined to be narrow and light muscled. The breed is of possible interest because of their ability to breed out of season and their high prolificacy.

Figure 3.5 shows breeds of sheep that have the potential to contribute to the sheep industry of the United States.

3.7.5. Other breeds with potential The Dorper from South Africa, the Awassi from Asia and Africa, the Hangyang from Northern China, and the Barbados and St. Croix breeds from the Caribbean all are quite prolific, can breed out of season, and are resistant to parasites. None have satisfactory fleeces and all have medium to poor conformation, but they may be useful in crossbreeding for producing highly efficient commercial ewes because of their superior reproductive performance.

REFERENCES

Allison, Jock, "Booroolas, More Lambs More Profits," *Sheepbreeder Magazine,* 105 No. 3, 1985, pp. 16–20.

Bradley, B. P., and others, "Two- and Three-Way Crosses Estimating the Combining Ability of Suffolk, Targhee and Shropshire Breeds of Sheep," *Anim. Sci.*, 34, 1972, p. 541.

TEXEL

Mature rams
Courtesy USDA

Mature ewe
Courtesy USDA

BARBADOS AND ST. CROIX

Yearling ewes: Barbados in foreground
and St. Croix in background
Courtesy Animal Science Department,
Ohio Agricultural
Research and Development Center,
The Ohio State University, Wooster

ROMANOV

Mature ram
Courtesy Canadian Department of Agriculture

Ewes with lambs
Courtesy Canadian Department of Agriculture

Figure 3.5 Breeds with the potential to improve the sheep industry.

Briggs, H. M., *Modern Breeds of Livestock*. New York: Macmillan, Inc., 1969.

Coffey, W. C., *Productive Sheep Husbandry*. Philadelphia: J. B. Lippincott Co., 1937.

Collins, S. B., *Profitable Sheep*. New York: Macmillan, Inc., 1956.

Diggins, R. V., and C. E. Bundy, *Sheep Production*. Englewood Cliffs, N.J.: Prentice-Hall, Inc., 1958.

Epstein, H., *Domestic Animals of China*. Bucks, England: Commonwealth Agricultural Bureau, Farnham Royal, 1969.

Good, Lemuel, "The Barbados Blackbelly," *Sheep Magazine*, August 1983.

Hairsign, W., *Sheep Production*. Boston: Butterworths, 1983.

Harrison, Virdon, "Sheep Production: Intensive Systems, Innovative Techniques Boost Yields," *ERS Agr. Economics Report No. 52*, USDA, 1980.

Hulet, C. V., and others, "Effect of Origin and Environment on Reproductive Phenomena in Rambouillet Ewes," *J. Anim. Sci.* 38, 1974, pp. 1218–1224.

Kammlade, W. G., Sr., and W. G. Kammlade, Jr., *Sheep Science*. Philadelphia: J. B. Lippincott Co., 1955.

Lydekker, R., *Wild Oxen, Sheep and Goats of All Lands Living and Extinct*. London: R. Ward, 1912.

Mason, I. L., *The Sheep Breeds of the Mediterranean*. Edinburgh, Great Britain: T. and A. Constable Ltd., 1967.

McKinney, John, *The Sheep Book*. New York: John Wiley & Sons, Inc., 1959.

Morrison, Iain, *The New Zealand Sheep and Their Wool*. Wellington, New Zealand: Growers Services Division, Wool Board, 1980.

Parker, Ron, *The Sheep Book*. New York: Charles Scribner's Sons, 1983.

Piper, I. R., and B. M. Bindon, "Genetic Segregation for Fecundity in Booroola Merino Sheep." *Proceedings of the World Congress of Sheep and Beef Cattle Breeding*, Palmerston North, New Zealand: Dunmore Press, 1982.

Scott, George E., *The Sheepman's Production Handbook*. Denver, Colo.: Sheep Industry Development Program, Inc. 1970.

Terrill, C. E., *National Wool Grower*, 40, 1950, 6.

QUESTIONS FOR STUDY AND DISCUSSION

1. What is the difference between a type and breed of sheep?
2. In general, what procedures are followed in developing a breed of sheep?
3. What general characteristics of conformation are unique to the fine-wool sheep?
4. What characteristics should ewe breeds of sheep have?
5. What characteristics should a sheep breed have to be classed as a ram breed?
6. Compare the Merino and Rambouillet breeds and explain reasons for superiority of one breed over another.
7. What were reasons for developing new breeds by the USDA?
8. How do the requirements of ewe breeds for range and farm production differ?
9. Why can't breeds such as the Texel, Awassi, and Romanov be used in crossbreeding in the United States?
10. How has the Finnsheep been an asset to the American sheep industry?
11. Enumerate reasons why the Suffolk breed is so popular in the United States.

12. Give some reasons why the Coopworth and Perendale breeds of sheep may make very rapid progress in New Zealand.
13. What are the reasons why New Zealand Southdowns are being imported to the United States?
14. What caused the loss of popularity of the Shropshire breed from World War I to date?

4 Entering the Sheep Business

Sheep raising was a satisfying and rewarding activity long before histories were written. Sheep have rewarded the diligent shepherd by producing food and clothing that, in many cases, made the difference in survival of people and animals. Make no mistake about it: the two are dependent on each other for a successful operation. Most of this book will be concerned with sheep and how they should be managed and fed to make a profit. But there should be no question in the mind of the reader that the most important ingredient in a well-managed sheep operation is the man or woman involved. What he or she sees, believes, and does almost invariably determines the degree of success of the endeavor.

4.1 WHAT KIND OF PERSON SHOULD RAISE SHEEP?

There is no set pattern or mold that can be used to sort out the outstanding sheep producer. Physically, they may vary from extreme strength to handicapped. Mentally, they may vary from average to highly intelligent. Other characteristics may vary so that in a crowd it would be impossible to pick out the talented sheep people merely on the basis of how they looked. Then what characteristics mark the persons with the potential for success with sheep? Some of these are:

1. A strong interest in nature and a desire to work with living things. These people enjoy unconfined spaces and derive a special satisfaction from observing the miracle of life. Everything that grows interests them. Life is precious, and they respect it.
2. A love of animals and a desire to be responsible for them. Interest in and respect for sheep are, although not paramount, quite desirable. If you detest the sight, sound, and smell of sheep and only desire to make money out of them, you may well wish to reconsider and work with another species. Life is too short to be saddled with an uninteresting and irritating career.

3. Good shepherds are careful observers not only by sight, but by sound, feel, and even smell. The best operators are always looking for the exception; the telltale signs that things are not right. They hear the lamb in distress and the ewe in trouble. They note the ewe or lamb that is not eating or is beginning to limp. They know when pastures are ready to graze and when sheep should be moved. They are attentive to all aspects.
4. A compulsion to protect and care for the flock and to correct anything endangering or causing stress to animals. A good shepherd will see that the flock is taken care of before thinking of personal comfort.
5. A well-developed imagination. The ability to visualize a situation and to anticipate problems or opportunities is invaluable.
6. An optimistic outlook on life, an even temper, and patience with people and animals.
7. A sense of purpose, a strong desire to create something worthwhile, and the will to work harder than others to achieve goals.
8. An orderly mind and the ability to establish routines and to follow them.
9. An analytical nature. No decision is made without due consideration of the facts, alternatives, and consequences. Snap judgments are made only under exceptional circumstances.

The desirable characteristics of the person who should do well with sheep are not restricted to race, creed, sex, or any other means of categorizing people. It has been proved that the Spanish Basque were able to manage sheep very well in the mountains of the western United States even though they could speak little English. Regardless of where sheep are raised in the world, the same principles of production apply, and those who are most successful invariably follow them closely.

4.2 FACTORS TO BE CONSIDERED IN THE DECISION WHETHER TO RAISE SHEEP

4.2.1. Available feed Perhaps the first detail that must be considered is the feed supply. In the final analysis, nearly everyone raises sheep to make money. Generally, sheep are used to convert a wide assortment of grasses, grains, and roughage into salable products. When a lamb goes to market, it usually represents feed that could not be harvested by any means other than an animal, because the terrain is so rough or the vegetation so sparse. Thus the sheep on rangeland condenses hundreds of units of vegetation into the weight of lamb and wool sold. These have a ready market throughout the world. Under certain conditions, it may be feasible for lambs to be finished in drylot on harvested grains and other concentrates and roughage. Then the practicality of the operation depends on the cost of those feeds plus other expenses versus the price received for the sale of lambs and wool. In any event, a close study must be made of the prospective feed supply. For this reason, western ranchers must constantly gear their sheep operations to the carrying capacity of available rangeland. They do not like to buy feed, because it is invariably higher than what can be raised on the farm or ranch.

Pasture is usually thought of as cheap feed, but it could be too expensive on very fertile tillable soil. Total productivity of such rich soils is invariably greater if used for silage crops, such as corn or sorghum, or for grain than for pasture. However, by confining ewes in a small area and feeding them forages plus concentrates, some sheep producers are making good profits on the basis of marketing their crops through sheep. Generally, in a given area, feed supply dictates the operation. If you have your heart set on any special operation and the feed supply is inadequate, it may then be better to find an area where the feed supply is favorable.

Other considerations are markets, financing, parasites, predators, climatic and weather conditions, zoning, and environmental controls.

4.2.2. Markets and marketing Markets and marketing continue to be a severe problem for producers. Whether you like it or not, the returns cannot be safely put in the bank until the products are sold. Make certain your product has a market available. You should be sure your product is salable. If you want to produce Karakul lambs, find out details of processing and selling the hides before you start in the business. If you wish to be in the purebred business and expect to sell rams, look at the acceptance of the proposed breed before you start out. Lack of marketing knowledge is responsible for many failures.

4.2.3. Financing The interest rate for operating money is a prime consideration. When the money supply is tight, it is of utmost importance to be assured that one piece of bad luck or a common reverse will not cause your operation to be liquidated. A careful study of money needs throughout the year to determine the cash flow is mandatory with most lending agencies today.

4.2.4. Parasites One of the most serious production problems is control of internal parasites. Be prepared to wage a never-ending battle with these pests. In some areas of the United States where there are few killing frosts to destroy eggs and larvae, it may be impossible to intensively graze sheep with any degree of success. At least know the problems before you begin a sheep operation in an area.

4.2.5. Predators One of the principal causes for the demise of the sheep industry in late years has been the ravages of coyotes and dogs. Are coyotes and dogs real problems in the area? What steps are usually taken to prevent losses? Find out what control measures are effective.

4.2.6. Climatic conditions In deciding whether or not to establish a sheep operation, consider climate as a major factor. Find out about the history and status of the area. Are people producing lambs and wool now? Were sheep ever successfully produced there? If they were, why did the people get out of the business? Before setting up a sheep operation in a strange area, it is important to find out in what way climatic conditions affect the feed supply and other conditions (i.e., breeding and lambing seasons).

4.2.7. Weather conditions Frequency of severe winter storms is a major consideration in certain areas. Then a major expense may be in housing the sheep.

Lot conditions can affect feeder lamb gains and death losses. Droughts and very hot weather have a marked effect on breeding sheep. Know what to expect!

4.2.8. Environmental control problems and zoning A grazing operation usually is safe from problems with the Environmental Protection Agency, but a confinement unit either in drylot or on slotted or other floors is subject to restrictions regarding drainage, disposal of manure, flies, odors, and dust. Many feedlots have been forced to close by neighbors who complain that they are a nuisance. Locating near an urban or potential residential area has some advantages, but is subject to the whims of the populace. If the area is not zoned, it is well to ascertain the probabilities that it will be.

4.2.9. Availability of water and electricity It has been predicted that, although the energy situation may be the present most important problem of the country, water availability may be an even greater one within the not too distant future. Livestock must have a constant supply of high-quality water in order to carry out their own individual functions in a farm operation. Rainfall may be expected to fluctuate and thus affect feed conditions, but any uncertainty with regard to availability of livestock water must be viewed as a major hazard. Electricity is most important for the efficient use of labor and equipment. The more animals involved, the more it is needed, especially in confinement. A reliable source of power at reasonable prices may be a prime necessity. Very few areas are without electrical service now, but it is well to know the status and plans of the power source for the future before committing yourself to an operation that may be vulnerable to radical changes in supply and cost.

4.3 KNOW YOUR OBJECTIVES

You should not progress very far in your thinking about sheep production without some very careful thought on the subject of why you want to do it and what you expect to get out of it.

If you expect to get rich quick, it is likely you will be disappointed. It is true that fortunes have been made, but the sheep business is more steady than spectacular. It lends itself more to long-time investment than to a quick turnaround and in and out maneuvers. It has nearly always brought in a steady and good return on money invested. In fact, few livestock enterprises have a better record during the history of the country. Do you expect to farm without working? Then don't choose the sheep enterprise, because there is always a lot to do and it is not easy work. It does not take a lot of strength, but a day spent drenching sheep, trimming feet, crutching or shearing, feeding or hauling manure is not for the person who likes to lie in bed till late in the day. In fact, the opposite is the case. Most good shepherds like to rise early and work the sheep when they are comfortable. Many claim to derive a special dividend from the beauty of early mornings in the open when the flowers smell sweetest, the birds sing their best, the grass is greener, and the sky bluer than the rest of the day. No, it is not a sluggard's life, but it is interesting and rewarding. And if you really like sheep, you can come in at night tired and dirty but satisfied you have done something worthwhile.

Do you expect your sheep farm or ranch to provide a good environment for family life? It depends on what the family is willing to give to it. The quality of life is far more dependent on human factors than material things. A sheep farm or ranch offers many opportunities to give youngsters responsibility. Even a very young child can learn to feed an orphan lamb. As soon as possible, the children should be given responsibility and, if possible, some sheep for their very own. The extra effort taken to beautify the farmstead should be a family project. Everyone should help with trees, shrubs, flowers, and paint to make your home a joy to live in. They should also understand that living in a lovely place is part of the pay for the work done. There are great opportunities for learning on a sheep farm. It won't be necessary to give the children the "birds and bees lecture," because they would find it absurd. They will have seen reproduction since they were old enough to go with you to work with the sheep.

Is one of your objectives to do something creative? If so, there is great opportunity for anyone in sheep production. Here you have the medium of flesh and blood. These you can mold as surely as though you were working with clay or paints or with words. After many centuries of breeding, we still do not have the sheep we need to fit into the new systems of production that are being developed. Sheep breeders have made pitifully poor progress in developing highly productive strains when you compare their status with dairy cattle, poultry, swine, and even beef cattle. The United States is far behind other countries such as New Zealand, Australia, and Russia in the practical application of genetics and physiology. The way is open for the truly innovative breeder to blaze new trails and to leave his or her mark on the industry.

In summary, do plenty of thinking before you attempt to start in the sheep business. Set up realistic goals that you hope to achieve, and then give yourself a better than average chance to succeed by considering the factors that determine success. If you do go into the business, work hard and spend lots of time thinking and planning as you go. Don't repeat mistakes and try to dwell on victories instead of defeats. Hope for good luck and work hard enough to make it happen.

4.4 WHEN SHOULD YOU START IN THE SHEEP BUSINESS?

The late Roscoe Snapp, professor of Animal Science at the University of Illinois, once said in a speech about production that if you wanted to get in when livestock were cheap and out when they were high, never!! But, if you have the knowledge, skill, and finances, now is as good a time as any. Over the years, there will be lows and highs, but if your operation is sound it will weather the storms. This is a sound philosophy, and it applies to sheep as well as cattle.

4.5 SYSTEMS OF COMMERCIAL LAMB PRODUCTION

A number of options are open to the person who is interested in sheep production. Some of these are as follows:

1. Market lamb production
   a. Range operations
   b. Farm states operation, usually in combination with other enterprises

2. Lamb finishing
    a. Pasture alone
    b. Grain on pasture
    c. Grain and hay in drylot

It is important for the sheep enterprise to be compatible with the operator and adapted to the conditions that exist on the farm or ranch. Some of the considerations are the level of knowledge, skill, and experience with sheep and the background that will enable the person to organize and manage the operation. Many people may be very good at the tactics of day to day caring for sheep, but fail because they do not understand the strategy of organization and fitting together the parts so that they work as a whole. A good operator must do both. The crop production of the farm or ranch, the improvements and facilities, the location, the availability and problems of marketing, the availability and quality of labor, the availability of breeding stock, and financing must be taken into account.

The following is a general discussion of the most important features of various kinds of sheep operations.

4.5.1. Market lamb production on fenced ranges in the southwest Range production of market lambs implies that extensive pasture is available and that lambs will be marketed off pasture. Basically, the problems are the same whether the location is in the Southwest or in the mountains of the North. Pastures and ranges must be managed to supply feed throughout the year. Concentrates and hay must be available during droughts and severe winter weather. Time of lambing and methods of handling the sheep are dictated to a large extent by climatic conditions. Most ranges are fenced in the Southwest and much of the West. In the Southwest, if feed conditions are favorable, ewes may be bred to lamb on pasture in the fall. Others may produce winter or spring lambs. The ewes and lambs on a fenced range are handled a bare minimum. The income comes from the sale of wool and slaughter and feeder lambs. Range operators, in general, produce their own replacements. Lambs are docked and castrated at a young age and vaccinated for soremouth and enterotoxemia. In problem areas, ewes are vaccinated for vibriosis, enzootic abortion (EAE), and blue-tongue. Internal parasite control programs are followed. In areas of higher rainfall or during seasons of abnormal precipitation, lambs and ewes are drenched at least before going to pasture and before breeding. Ewes are usually dipped or sprayed for keds and ticks after shearing. The prime essential is that the sheep operation will be carried on with a minimum of expense. A knowledge of range forages and how to manage them is essential. Anyone wishing to manage sheep operations in the Southwest should have experience before embarking on such a venture. Because very little shelter is provided, the occasional blizzards that hit the area are a severe problem requiring knowledge not easily obtained. The frequent droughts, predators, and disease must be listed as problems requiring solutions most years. Even though special knowledge and skills are required, sheep production in the Southwest has been a stable and rewarding business.

4.5.2. Market lamb production on the ranges of the northwest Much of the sheep industry in the Northwest is located in the rugged areas of the Rocky

Mountains. Most of the operators are owners of some land and hold leases on U.S. government land. They use their own or rented land for fall and winter pasture and depend on leased land for spring and summer grazing. The ewes are usually bred to lamb early enough in the spring so that the lambs are old enough to travel to the range. The ewes are handled in bands of 1000 when suckling lambs and 1500 to 2000 when dry. A herder and dogs are with the sheep constantly and are responsible for their safety to see that they are moved as needed to assure that they always have plenty of forage and water. Knowledge and skill are required to manage a band of ewes and lambs. Experience is a necessity for anyone wishing to go into the sheep business in the Northwest. Some further considerations with regard to the operation are: Most of the ewes are lambed in sheds and they may or may not be sheared before they lamb. The ewes may be fed in drylot for a period before they are put out on pasture. Some producers have winter and spring pasture for ewes before they go to summer range. Lambs may be vaccinated for soremouth and enterotoxemia. In areas where problems exist, ewes are vaccinated for vibriosis, EAE, tetanus, type C and D enterotoxemia, and bluetongue.

If ranges are properly managed, not many problems are encountered with internal parasites, but most prudent shepherds drench ewes in the spring before going to pasture and again before breeding. Under special conditions of unusual rainfall, other drenchings may be needed. Ewes are usually dipped or sprayed for lice and keds after shearing.

Lambs are weaned when they come off the range, and the fat and feeder lambs are sold. Replacement ewe lambs are usually reared by the producer. Depending on the market price, they may be kept for breeding or sold as replacements or slaughter lambs. Income comes from sale of wool and lambs. In this operation, expenses must be kept low and thus elaborate buildings, equipment, or fences are avoided.

4.5.3. Fall lamb production on farms In areas of the United States where winter wheat, barley, rye, alfalfa, rape, or turnip pasture is available in the fall and winter, fall lambs are produced efficiently with low expenses for labor and harvested feed. If the pastures are excellent, very little concentrates or roughage are required. The lambs usually do not bring the top prices of the year, because they must compete with older lambs that have been in the feedlot. But they have a tremendous advantage in that they have been produced with low expenses. The key to this operation is the availability of pasture when it is needed. Droughts or early storms with snow cover are a distinct hazard. The operator must always have a source of feed available in case the weather conditions are unfavorable.

Ewes are bred for fall lambs in May and June, a period when most breeds are in anestrus. Thus only certain types of ewes are adapted to the program. Lambing may be either on pasture or in sheds or barns. Lambing on pasture requires less labor and equipment, but may result in a reduction in number saved. Conception of ewes bred for fall lambs is generally lower than it is with those bred at other seasons. When good pastures are not available in fall and winter, lambs may be weaned when they are 7 to 12 weeks old and finished in drylot.

Ewes are normally sheared in the early spring before the breeding season. They are usually dipped, sprayed, or dusted for external parasites after they are sheared.

A program for controlling internal parasites is necessary. It should be designed to meet the conditions of the area, but at least two treatments per year are almost mandatory. Lambs are docked and castrated at 2 to 4 weeks and vaccinated for type C and D enterotoxemia before they are weaned. In areas where there are problems, ewes should be vaccinated for vibriosis, tetanus, type C enterotoxemia, bluetongue, and enzootic abortion (EAE).

Lamb and wool sales are the sources of income from this system of production. Lambs are nearly always finished when sold. Generally, the ewes are purchased for this operation, although there are no particular reasons why they could not be raised on the farm.

Some advantages of the system are:

1. When pastures are good, no program produces lambs with less expense.
2. Internal parasite problems may be less.
3. Ewes are lambed when the weather is usually comfortable and with less labor.
4. Equipment costs are less than for other systems.

Some disadvantages are:

1. In case of drought, it may be necessary to put a lot of expensive feed in the ewes and lambs.
2. Lambing percent is generally less with fall lambs.
3. In certain areas, lambing may interfere with the harvesting of grain.
4. During years when rainfall is exceptionally heavy, internal parasites may be a problem.
5. The markets have a small differential between old crop fed lambs and the high-quality fall lambs, because the volume of fall lambs is too low to establish a special demand for them.

4.5.4. Winter lamb production on farms Most of the winter lambs produced in the United States come from the farm flocks of the Corn Belt and eastern states. Size of flocks vary from 25 and under to over 500 head. Although some fall and winter pastures are available, they are usually sufficient only for satisfying a part of the requirements of pregnant ewes. Ewes are usually pastured from April until November or December, and then fed hay, haylage, or silage plus concentrates until April. Lambs are fed in creeps and may or may not be weaned before they are marketed. Markets in May and June are excellent, and thus this system has been popular. Breeding stock is usually purchased, but more recently there has been a trend toward raising replacements.

Building requirements are more stringent than those for other systems, because the ewes lamb during January and February when the weather is cold. Thus, an enclosed lambing area is almost a necessity. Feed requirements are greater for grain, protein supplement, and high-quality roughage. Feed expenses may be higher with this system than any of the others. Since ewes are lambed when weather is severe, a high degree of skill is required to save lambs from chilling.

Ewes are bred in late July and August and are usually flushed before breeding. Housing requirements for pregnant ewes consists of a windbreak and a roof so sheep can get out of rain and wet snow. Feeding facilities may be simple feed bunks or, in more sophisticated setups, fence-line bunks or self-feeders. Ewes may be fed daily or on alternate days.

Ewes are put in lambing pens just before or just after they lamb and held there for 1 to 3 days depending on how the ewe and lamb are doing. Efforts are made to foster lambs on ewes that have lost lambs or have only a single. Lambs are docked and may or may not be castrated. Some producers who sell their lambs in early June do not castrate. Lambs are vaccinated for soremouth and enterotoxemia type D at 6 weeks of age. In areas where problems occur, ewes are vaccinated for enterotoxemia type C, soremouth, tetanus, vibriosis, and bluetongue.

Most of the better producers never put winter lambs on pasture. This is to prevent internal parasite infection. Ewes are routinely drenched before going on pasture in April or May, again about July 1, and after the first hard freeze in the fall. If rains are frequent and pastures are lush, they may need to be treated monthly or more often. If lice or keds are present, ewes are dipped, sprayed, or dusted after shearing.

Shearing may vary with the type of ewes. Some very heavy fleeced ewes are sheared before breeding and again before lambing. Many producers who have good shelter and shear only once a year do it before the ewes lamb in January or February.

Nearly all lambs are fed concentrates in creeps beginning at 2 weeks of age. Some winter lamb producers allow their lambs to run with ewes until they are marketed. More recently, many enlightened producers wean lambs at 8 weeks and finish them in drylot.

The advantages of winter lambs are:

1. Market prices are generally the highest when they are sold.
2. Ewes lamb when the farm workload is light.
3. Lambs are usually free of parasites.
4. Ewes and lambs furnish an excellent market for homegrown feeds.
5. Since lambs are confined, predator problems are not usually serious.

The disadvantages are:

1. More skill is required on this system than others.
2. More labor is required.
3. Feed requirements are greater than for some other systems. Feed may cost more per unit.
4. More out of pocket expenses may be incurred.
5. Many producers find it distasteful or more difficult to lamb out ewes in the dead of winter.

4.5.5. Spring (late) lamb production on farms Some sheep owners in the Corn Belt prefer to lamb out ewes when the weather is mild and pastures are good. These are called spring lambs by some and late lambs by others. In this system of production, forages furnish most of the feed. Ewes are turned to pasture in April and May and allowed to graze through the fall and winter as much as possible. Roughage is fed during periods when ewes cannot graze and concentrates only during the last 4 to 6 weeks of gestation. Lambs may or may not be fed in creeps, depending on how they are to be handled. Two options are (1) to creep feed the lambs and finish out in drylot and (2) to graze the lambs all summer, sell the grass-fat lambs in September, and finish out the feeders in drylot after grazing on fall pastures.

Almost any type of ewes can be used. They are bred after October 15 when ewes are at the height of their fertility. They lamb after March 15. Quite often, it is planned that they will lamb on pasture. Shelter requirements are simpler than for winter lambs. Generally, a shed open to the south is sufficient, but this will depend on when they lamb. If the date is before the end of frosts, it may be necessary to provide for protection of the ewe and lamb.

The method of handling lambs will depend on feed supplies and costs and the rest of the farm operation. If very good pastures are available for much of the year, costs of production can be very low. Labor and machinery costs may be lower than other systems where sheep are fed for long periods in drylot. On the other hand, if ewes are lambed in April and May when field crops are being planted, there may be a conflict for labor. There may also be a conflict in the fall when grain is being harvested.

Spring lambs that are allowed to graze with the ewes need much more treatment for internal parasites than winter lambs that are kept in drylot. They are also more vulnerable to external parasites and to predators. Droughts and very hot weather in the summer may reduce the quality and quantity of available feed to the extent that supplementary hay and concentrates must be fed in order to obtain growth and development by the lambs.

Spring lambs may be run with ewes all summer, but there is little advantage and many disadvantages in the practice. Usually lambs are weaned by the first of July or when they are from 50 to 90 days old. Ewes may get too fat on high-quality pastures, and they continue to act as a source of internal parasite infection for the lambs.

Shearing ewes before they lamb in the spring is a usual practice. Lambs are usually sheared at least by the first of July. In many areas, it is the usual practice to clip lambs in May and June. Wooled lambs gain more slowly and may be infested with maggots during the summer. In the better run operations, the lambs are docked and castrated from 1 to 2 weeks of age and vaccinated for soremouth and enterotoxemia type D when they are 6 weeks old. Ewes may need to be vaccinated for vibriosis, type C enterotoxemia, tetanus and bluetongue, depending on the conditions in the area. Agriculture extension agents or the local veterinarian can furnish this information if it is requested. A drenching program must be followed carefully to control internal parasites. Ewes should be dipped, sprayed, or dusted for external parasites after they are sheared.

Some advantages of spring lambs are:

1. Since the ewes are bred when conditions are most favorable, the lambing season is usually short and the fertility of the ewes is excellent.

2. Out of pocket expenses should be quite low since most of the feed can be grown on the farm.
3. Ewes lamb when weather conditions are mild, and it is possible to work comfortably with the sheep.
4. Since most of the lambs are finished in the fall, feed prices are usually lower. In the Corn Belt, lambs can be finished on the new crop of corn.
5. Lamb weights when sold are greater than those marketed in the spring.
6. Building and equipment needs are less than for other systems.

Some disadvantages of spring lambs may be:

1. Greater problems with parasites and predators than other systems.
2. Some conflict with crop production during lambing and in the harvest season.
3. Lambs are sold when prices are lower in the fall and winter.
4. Droughts may cause feeding problems when pastures dry up.

4.5.6. Accelerated lambing on farms Researchers have studied methods of increasing the volume and efficiency of lamb production and concluded that it is possible to breed ewes more often than once each year if a number of factors are favorable. These are that ewes must be capable of breeding in the spring as well as the fall and winter, a year-round feed supply must be available, adequate buildings and equipment must be on hand to handle ewes and lambs during the whole year, and the manager must know sheep and how to care for them and be willing to spend time and thought to solve special problems associated with the system. A number of plans have been tried including the following:

1. Continuous lambing: ewes are run with rams throughout the year.
2. Ewes are exposed at 4-month intervals.
3. Ewes are exposed at 7- and 10-month intervals.
4. Ewes are exposed at 8-month intervals.

Experiment stations at Illinois, Virginia, and Indiana (Purdue) were some of the first to investigate the possibilities of accelerating lamb production by producing three lamb crops in 2 years. They were successful in combining winter, fall, and spring lamb production into a system. Others have attempted to duplicate their efforts with success so that now producers in many parts of the United States are using accelerated lambing.

Ewes must be capable of cycling in the spring, fall, and winter to maintain such a schedule. The number of choices of types of ewes is rather narrow at present, but much work is being done on the problem, and the future holds promise that more options will be available.

Feed supplies needed for accelerated lambing are more critical than for most systems because the ewes will be on concentrates and roughage more months each year. There will also be more lambs to feed at different times during the year. Generally, two approaches are followed. The ewes may be grazed during the forage growing season when it fits the breeding and lambing schedule and then put in drylot or a confinement building for the last part of gestation and lactation. Or they may be confined in drylot or a building throughout the year. Feeding the ewes is critical, because they must be ready to rebreed sooner than in a single lamb crop system. Maintaining condition on the ewes is most important for success. Lambs should be creep fed so that they can be weaned when they are 6 to 8 weeks old. It is also vital to the system that fertile rams be available when ewes are to be bred. Thus, rams must be sheared more often than once each year and protected against heat and other stress.

Recent research indicates the value of light control as an aid in improving ram fertility. Ewes must be very carefully managed during the spring breeding season to assure that they cycle and to protect conception. Multiple births are fewer, and lambs are usually smaller in the fall.

Although ewes may be lambed on pasture in the fall, most producers prefer to lamb them in sheds to increase the number saved. Winter and spring lambs are also handled in lambing quarters not only to save lambs, but also to assure close observation of the ewes.

Some authorities recommend shearing in the fall and again in the spring to improve reproductive performance. If the ewes have heavy fleeces, this practice is advisable. Lambs should also be sheared soon after weaning in the spring.

Lambs should be docked and castrated at 1 to 2 weeks of age and vaccinated for soremouth and enterotoxemia type D when they are 6 weeks old. In areas where problems exist, it is routine to vaccinate for vibriosis, tetanus, soremouth, bluetongue, and enterotoxemia type C.

Research at the Missouri Experiment Station (Thompson, 1980) has shown that attrition rates are higher in ewes on accelerated lambing than winter and late lambing systems. Mastitis appears to be the most common cause for culling ewes. Control of parasites and diseases is similar to those described in late lamb production.

The following are advantages of an accelerated system:

1. Productivity per ewe can be increased.
2. Ram and ewe costs per lamb raised are decreased.
3. Labor costs per lamb raised are reduced.
4. Buildings and equipment are more efficiently used.
5. Volume of business may be increased with very little increase in investment.
6. Home-grown feeds can be marketed through ewes and lambs.

The disadvantages of accelerated lambing are:

1. Level of competence of the shepherd must be high.
2. Investment in building and equipment is high.
3. It is difficult to obtain ewes that will breed out of season.
4. Since ewes are lambing during three seasons of the year, the system is not compatible with heavy involvement in crop production.
5. Attrition rate in ewes may be high.

6. Out of pocket expenses may be much increased over some other systems of production.

4.5.7. Confinement lamb production on farms It has been demonstrated that lambs can be produced either in complete or semiconfinement. In areas where land is quite productive and expensive, some lamb producers are confining ewes and lambs to drylot or to buildings with steel mesh or slotted floors. The ewes may remain in confinement throughout the whole year, or they may be grazed during the period from the time lambs are weaned until they are bred or even most of the gestation period. The amount of time on pasture is usually dependent on the availability of forage, fences, and other facilities. On expensive land that is capable of high tonnage of corn silage or a similar crop, an acre of such land can carry a large number of ewes in confinement. Whether or not such an operation is feasible will depend on a number of factors. Some of these are equipment, buildings, managerial ability of the operator, kind and quality of sheep available to stock the operation, and climatic and weather conditions.

A confinement system requires that feed be brought to the sheep. Self-unloading wagons with fence-line bunks or an auger system may save much labor in feeding. Also, it is important to have feed in a form that can be handled mechanically. Long hay would not usually fit into a mechanized system. Sheep in confinement need shelter from rain, snow, and wind in winter, but also must have shade in the summer. Watering facilities are needed for both winter and summer. Also, if large numbers are to be held, handling facilities are needed. Legally, it is necessary to provide for handling manure and urine so that they will not pollute streams or cause a nuisance with odors or flies. There must also be space and facilities for storage of feeds and feeding equipment. Quality of management must be high in a confinement operation. The manager should know sheep and have had some experience with confinement rearing if possible. There are some problems that are different from an extensive operation, particularly in breeding ewes in confinement.

Most confinement operations are designed to use accelerated lambing because the productivity per ewe is higher and, with volume, expenses per unit of production can be reduced appreciably. If accelerated lambing is to be used, the choice of breeding stock will be restricted to Rambouillet or Merino ewes or crosses between the fine-wool breeds and Dorset or Dorset plus Finn crossbreds.

If mechanized feeding is used, it may be desirable to feed ewes less frequently than daily. Research at the Minnesota and Cornell Experiment Stations among others indicate that feeding pregnant ewes three times a week gave satisfactory results and saved labor and machinery costs. In some confinement operations, ewes are lambed on slotted or similar floors, but there is some doubt whether this is a good practice since one of the most serious problems with ewes in confinement is mastitis. There is reason to believe slotted or mesh floors may contribute to udder infections. Lambs should be creep fed and weaned at from 5 to 7 weeks of age or at least 20 lb in weight and eating well. Space requirements for ewes in confinement vary from 12 to 16 ft^2/ewe. The space will depend on the size of ewe and whether or not she is suckling twins. Lambs require at least 4 ft^2/head.

Shearing of ewes should be done twice yearly, usually in the spring and fall. Spring lambs should be sheared by July 1.

All ewes should be vaccinated for soremouth and enterotoxemia types C and D. Vaccination for other diseases such as tetanus, vibriosis, and bluetongue should be done if there is a problem in the area. External parasites may be controlled by dipping, spraying, or dusting after shearing.

The advantages of confinement rearing are:

1. It is possible to produce sheep on highly productive, expensive land.
2. Feed requirements may be reduced as much as 38% by feeding in confinement.
3. Internal parasites are not a serious problem.
4. Predator losses may be eliminated.
5. Confinement operations can be automated to reduce labor.
6. Ewes can be handled, in uniform groups, more efficiently in confinement.

The disadvantages of confinement rearing are:

1. Requires a more skillful manager.
2. Equipment and building costs are greater in confinement.
3. Pollution, flies, and odors must be controlled.
4. Attrition rate is higher in confined ewes.
5. Ewes are more difficult to breed in confinement.
6. Pneumonia and coccidiosis may be a problem where sheep are closely confined.

4.6 PUREBRED SHEEP PRODUCTION

The availability of superior breeding stock for the American sheep industry depends on purebred sheep breeders. The ordinary commercial producer has neither the time nor the skill to develop the type of sheep required by efficient production systems of the future.

Research has demonstrated the advantages of increased livability, growth rate, and increased production (20% to 25%) that accrues from selective crossing of breeds and strains of sheep. These serve to emphasize the importance of enlightened progressive breeders of purebreds, who can provide the genetic material needed by the industry. If crossbreeding is generally adopted by producers, the opportunities for the purebred breeder may well be greater than they ever were in the past. Not only will sires be needed to provide the terminal crosses with superior carcass characteristics, but rams on the ewe side must transmit traits required in the ewes that give birth to increased numbers and quality of offspring. Purebred sheep require more skillful management than crossbreds, because they lack the vitality of hybrids. The purebred breeder must understand genetics and breeding systems in order to develop a desirable flock. Detailed records must be kept and used in a program of improvement. The purebred producer must also understand and use special methods of advertising and merchandizing in order to realize their true growth. The most successful breeders of the future will be able to demonstrate that their rams give superior performance, and they must be rewarded for their efforts. It should be emphasized that a purebred operation is invariably more expensive to conduct than a commercial one, and the value per head of animals is greater than for grades or crossbreds. Thus the ability to maintain a high level of production is quite important. It, therefore, is mandatory that more intensive care be given to the sheep. The records of purebred flocks will surely be most important to the prospective purchaser as a means of judging the value of genetic material purchased.

Young people going into the sheep business may be attracted to the romance of purebred shows and sales and not realize that the most important knowledge they need is how to do a good job of raising sheep. It is for this reason that some successful purebred breeders encourage beginners to start out raising commercial lambs so that they will understand the basic fundamentals of sheep husbandry. When they have shown that they are proficient in producing for the market, they are ready to engage in the more technical aspects of sheep breeding called for in purebreds.

4.6.1. What breed of sheep should you choose? The breed of sheep to produce will depend on a number of factors. Some of these are as follows: your personal preference, the breeds in the area, the market demand, and the availability of breeding stock. Personal preference for a breed may not assure success, but it could well foretell failure if the prospective breeder has a distinct dislike for the beasts. Successful breeders must be proud of their livestock and convinced they are superior, or they will never have the enthusiasm and desire necessary to do an outstanding job with them. Choosing a breed already popular in an area may or may not be an advantage. On the one hand, the prospective customers already know about them, but it may well be that the competition is quite keen. Then the young breeder may be up against a group of established breeders who more or less control the market. It may take time to establish a reputation for breeding superior stock.

4.6.2. Purebreds must be merchandized Some experience in merchandizing breeding stock should be of advantage. A knowledge of techniques of obtaining gain and efficiency of gain data is most important. It seems reasonable to expect that the commercial producer of the future will place a much greater emphasis on efficiency of production and thus will be searching for breeding stock that will provide those traits. Breeders who can show that their sheep give consistently higher performance than others should have a ready sale for their breeding stock. Test station results with carcass cut-out information on half-sibs provides valuable information to a prospective purchaser. Performance records are the basis for sale of swine, dairy cattle, beef cattle, and poultry. It seems reasonable to assume sheep breeders will see their value, too.

Show rings have served a useful purpose for breeders in the past, and there are still many who believe the publicity achieved by winning shows is their best advertising medium. Whether or not the winners of shows demonstrate what is needed by the industry is debatable. Too often the popular type represents only the ideas of a group of judges rather than values based on performance.

4.6.3. What size unit should you have? Size of the unit is an important consideration. One-ram flocks very much limit the scope of operation of the breeder. How big the operation should be depends on the resources of the person in the business, but it should be pointed out that the facilities, record keeping, merchandising, and genetic improvement of the purebred sheep operation favors one of 300-500 ewes or more. While this may not be immediately obtainable, it should probably be the objective of the breeder. Then the flock can be built gradually, while at the same time the breeder continually improves facilities. Purebred breeders have traditionally taken pride in the appearance of their buildings and equipment. Even though these do not contribute to the breeding value of the livestock, it is associated in the minds of buyers with superior livestock. The farmstead need not be ostentatious, but it should be neat, clean, and in good repair. Judicious planting of trees and shrubs helps the appearance of any grounds, and they are not expensive.

4.6.4. Purebreds require special care and feeding Feeding purebred sheep may differ very little from commercial operations except that more feed may be needed to achieve the rates of gain and the mature size needed to demonstrate their traits. However, if they are to be fitted for show, special skills are required to avoid overfinishing.

4.6.5. Special skill and equipment needed for handling purebreds Since most purebred lambs are born in the winter, dry, draft-free lambing quarters should be provided. Experienced personnel should lamb out the ewes in order to keep losses to a minimum. It is desirable to foster orphan lambs or one of triplets on ewes that have lost their lamb or that have a single. If no satisfactory foster mothers are available, lambs may be reared artificially. It is important to identify all lambs at birth. Lambs should be docked when they are 7 to 10 days old. Generally, it is advisable to wait until weaning to castrate rams not kept for sale as breeding stock in order to obtain performance data.

Weaning of purebreds is usually later than commercial sheep. The age may depend on the system of production, but it is doubtful if they should be more than 60 to 80 days old when removed from the ewes. Shearing may be done once or twice annually depending on the weight of fleece. If twice a year shearing is used, it is recommended prior to breeding and just before ewes lamb. If only once per year, there is a distinct advantage to doing it before they lamb. Lambs held over the summer, except those to be shown, should always be sheared before the first of July. Depending on the system of production, rams may be sheared once or twice yearly. On an accelerated system, it may be preferable to shear them in the spring and fall or, otherwise, 4 to 6 weeks before the breeding season.

Ewes should be vaccinated for enterotoxemia type D prior to lambing. In problem areas, it is advisable to vaccinate for enterotoxemia type C, vibriosis, tetanus, and bluetongue. Lambs should be vaccinated for enterotoxemia and soremouth at 6 weeks of age. Rams should be vaccinated for epididymitis in areas where the disease is a problem. An internal parasite control program should be developed for the farm. All ewes should be drenched before going to pasture and at least once monthly during the summer. They should have a final treatment after the first hard freeze in the fall. Ewes and lambs should be dipped, sprayed, or dusted for external parasites after shearing.

4.6.6. Record keeping is important in purebred operations Decide what records are needed and establish a routine to obtain them. Review your records each year and change them as needed. It requires about the same amount of time to enter records in a computer data bank as it does to record them in a book. One advantage of the computer is the speed of recall of the data, and another is the calculations that can be done rapidly and accurately using the software programs now available. Then use the information to improve your operation.

The advantages of a purebred operation are:

1. Financial returns may be exceptional for the outstanding purebred breeder.
2. There is much prestige and satisfaction for the creative breeder of purebreds.

The disadvantages are:

1. Much skill and knowledge is required.
2. Risks are greater because animals are more valuable.
3. Excellent breeding animals are usually very expensive.
4. Investment in buildings and equipment is greater.
5. Successful purebred production and merchandising requires a much higher degree of knowledge, dedication, and effort than commercial production.

4.7 LAMB FINISHING

Lamb finishing differs from the other types of operations previously discussed in that the starting point is not a flock of ewes and a ram, but lambs that need more growth and finish to satisfy market requirements. The lamb finisher must buy the feeders at favorable prices, put on gains at a low enough figure, and sell the finished lambs high enough to make a profit.

Lamb finishing may be a major or minor enterprise. Some larger operators handle thousands of lambs on a year-round basis. Others may only feed small numbers for part of the year. It may be either a full- or part-time endeavor.

Successful lamb feeders need a wide range of knowledge and skills in order to exist in a highly competitive business. Some of the tools needed by the operator are discussed in general in this section. More specific information is given in Chapter 12.

4.7.1. Knowledge needed by a lamb finisher

1. Nutritional requirements and how to meet them economically.
2. Building and equipment requirements.
3. Kinds of feeder lambs and their capabilities.
4. Availability of feeds and relative prices.
5. Availability of financing and how to use it effectively.

6. Hazards to the health of lambs and how to control them effectively.
7. Understanding of the principles of management.
8. Understanding of marketing lambs and wool.

4.7.2. Skills needed by a lamb finisher

1. Handling lambs according to their needs.
2. Setting up valid management routines.
3. Training personnel to handle lambs properly.
4. Bargaining for feeder lambs.
5. Evaluating and purchasing feeds and supplies.
6. Recognition of health problems and solving them.
7. Negotiating sales of lambs and wool.
8. Operating machinery and equipment effectively.
9. Setting up valid budgets.
10. Analyzing results and planning for the future.

4.7.3. Experience is essential Regardless of how well trained a student may be, some practical experience is needed to recognize problems of lamb finishing and how to solve them. If it is not possible to work with an experienced operator, it is desirable to start out on a small scale and expand gradually.

4.7.4. Rewards of lamb finishing Lamb feeders seem to like the dynamic aspects of the business. There is much buying and selling of lambs, wool, and feed, which provide an opportunity for the use of skill and knowledge by the operator.

Lamb finishing is much more speculative than any other operations discussed in this chapter. There are too many factors that the operator cannot control for it to be a stable business. Most of the best operators try to figure out some means of reducing the risks in lamb finishing. Some have been successful and made a success of lamb finishing. It is a challenge for them.

4.8 HOW TO BECOME A SHEEP PRODUCER

If, after observation, careful study and much thought you know that you want to become a sheep producer, then how do you go about deciding the kind of operation that appeals to you and how do you go about achieving your goal? While it is true that being born and reared on a sheep farm or ranch is a tremendous advantage, not everyone can be so fortunate. However, if the desire is strong enough and you are willing to make the effort, it is possible to become proficient in the management of sheep. Some of the most successful producers had their first experience with sheep when they were grown men and women, but they had an unusually strong desire and were willing to learn the business from the ground up.

4.8.1. Learn the fundamentals of sheep husbandry In almost any worthwhile endeavor, the first requisite for success is to know the basic fundamentals. These may be learned by observation and work on the farm or ranch, by 4-H or FFA programs, or in classrooms and laboratories in high school or college. Many practices may be learned by reading about them, but to truly know how to handle sheep, it is necessary to have hands-on experience.

When the writer was growing up out in the country, we used to hear it said by livestock people that someone was a "city farmer" or a "book farmer." These were not complimentary terms. It meant that the "city" or "book" farmers may have had a lot of information from reading magazines or books, but they were short on practical know-how. They did not really know the basic fundamentals or have the skills to do a satisfactory job with livestock. Such people are prone to make costly mistakes and may soon become discouraged and leave the business. Many of those people could have become successful if they had been properly prepared for the endeavor before they started it.

While it is true that 4-H and FFA projects and classes in vocational agriculture and in college may be a means of learning the basic fundamentals, it is usually necessary for a person who wants to specialize in sheep production to have considerable practice to develop the skills and judgment needed to run a sheep operation. If you fit into any of the categories mentioned, you have probably asked yourself how you can obtain the training you need.

4.8.2. How to get practical experience with sheep If you are a high school student, it may be possible to work on a farm of a classmate or for a sheep farmer in the community. Colleges of agriculture usually have flocks that must be cared for. If a student has a sincere desire to learn and is willing to work, hands-on experience can be gained on the campus. Some colleges have courses or will make arrangements for such training. For others who want and need the training, it may be a bit more difficult, but not impossible. The most important thing to remember is that you must ask someone if you expect to get help.

There are always exceptions to the rule, but, generally, livestock people are sympathetic to persons, especially the young ones, who want to learn how to care for animals. Pick a successful operation and go and visit the manager or owner and explain that your objective is to gain experience. Don't expect to be paid well or at all, because you won't be worth much in the beginning. If you are accepted, work hard and try to make a contribution. Anticipate what is to be done and make yourself useful. If you save your teacher time and work, then he or she will be justified in spending time with you. If you are pleasant, polite, and demonstrate that you are interested and want to learn, it will be a pleasure to work with you and a feather in the cap of the person who trained you.

4.8.3. How to be a good trainee Being a good trainee requires some doing on your part. Be dependable. Just because you are not receiving a salary or wage does not excuse laxness. If you have a schedule, follow it. Be on time or a little early and ready to do the job. Don't avoid something because it is disagreeable, difficult, or menial. Be able to say you have run the whole course.

It is interesting to visit the best livestock people and to note how few are talkative. It seems to be a trait that goes with the job. The best ones are happy to talk at length about problems of the industry or even their own operation, but are uncomfortable in a trivial conversation. You should ask questions and remember the answers. If you don't understand something, find out about it. Be eager to do the jobs that require skill. Learn how to do the jobs the easy way, but don't get careless. Treat the sheep as though they were your own.

One final note relative to getting the most out of your training period is perhaps the hardest thing to do. That is to say "thank you." Not many people are self-made. Most everyone who is at all successful owes a lot to many people; people who took the time and made the special effort to help someone who needed it. Show them that you appreciate what they have done.

4.8.4. How much training do you need? Don't expect to learn all there is to know in your lifetime. You won't. It may seem trite to hear the "old timers" say they are still learning, but it is true. When you close your mind and no longer try to see or hear something new, your growth is ended and your usefulness will be limited. Your training should never be complete, but there must be an end to the learning of the fundamentals. If the person with whom you are working is perceptive, he or she will tell you when you are ready to go on your own. You should ask from time to time for their evaluation if it is not volunteered.

4.8.5. Value of formal education The amount of knowledge you need to be a success will depend on your objectives. How far do you wish to go? What do you want to achieve? These are questions you must answer in order to determine the level of training you should strive for.

It is well documented that there is a close relationship between the level of schooling and lifetime earnings. There are, of course, exceptions, but the facts show that high school graduates have higher achievement than those who complete only grade school. College graduates have a decided advantage over those with a high school diploma.

Should you obtain a college degree? That is a question you must answer. It entails more than the professional preparation you will receive in college. Much of the value of attending college is designed to improve your ability to communicate and to relate to the world you live in. The basic sciences provide the background for an understanding of life. They will become more important with the increasing complexity of all biological systems. Sheep production is, in essence, applied biology.

An understanding of the principles of economics and management provides the serious student with the tools to compete in the business world. Courses in agronomy and agricultural engineering may be fully as important as those in animal science, because they are so closely related to the management of the natural resources that provide the feed supply.

It is virtually impossible for a young person to appreciate the amount of information being generated by scientists today. The classroom is still the most efficient way to learn. Courses in genetics, nutrition, physiology, livestock evaluation, production, and management provide the technical training for the future leaders in livestock production. For those who have special interests, internship programs are available in most universities. These are for students who have completed much of their training, but need experience on a higher level to fit them for management of an operation. Internships are done under the supervision of persons who have achieved success in their field. The student works in the operation and has constant contact with the manager. In a good internship, the student will learn how the operation is organized and controlled, the problems encountered, the decisions required, and how they are made. After the internship, most students reevaluate the remainder of their coursework and may alter their program in order to bolster up any weakness discovered while serving on the job. The other possibility exists that the experience gained on the internship will cause the student to change objectives entirely.

4.8.6. Practical experience and some financial assets needed Ideally, persons who hope to own and manage a sheep farm or ranch should work for a while for a first-class operation and then be ready to go on their own. But this is not always possible. Land, equipment, and livestock are so very expensive that it requires a large investment to get started. Banks or other lending agencies are reluctant to finance anyone who has no financial assets or practical experience. It may be very discouraging at first, but it has been this writer's observation that a well-trained, determined person somehow finds a way to get started. This, to a large extent, is the justification for bankers or other lenders to give worthy people an opportunity.

Throughout the United States, much of the land is in the hands of absentee landowners. They own land for a variety of reasons, including a hedge against inflation. Many would like greater returns on their investment than they are receiving. Some of them would be receptive to entering a sheep operation if they could find the right manager. These contacts are made in a variety of ways, but they usually hinge on the recommendation of a successful sheep producer, banker, extension agent, teacher, or other influential person.

4.8.7. A good reputation is your greatest asset If you are well trained and struggling to establish a career, remember your reputation is your most priceless asset. Out in the country, farm people make a special effort to know their neighbors. If you work for someone, your neighbors form an opinon of you. They soon know if you are pleasant, intelligent, and a hard worker. They don't, at first, know about your integrity, but they will be extremely observant of what you say or do. If there is any doubt about your character, everyone will soon know about it.

Whatever you do, do it well. While it is true good news seems to travel slower than bad, it doesn't take many years to establish a reputation. Use that as your steppingstone to bigger things. It doesn't hurt to let people know you are ambitious and want to run your own spread. Sometimes, if you have demonstrated a very great interest in the operation and it is obvious that your efforts have contributed to the success of it, an intelligent owner will make it possible for you to own stock in the company or become a partner. Many successful livestock people who have devoted their lives to building up an operation want to see it carried on after they retire. Such an opportunity should serve as a powerful incentive for continued good work.

4.8.8. Sheep producers should help others get started in the business Not many persons who have achieved success can claim to have done it alone. Quite the contrary, most of them are indebted to the many people who unselfishly assisted them in their career. This idea brings to mind the story of the fig tree, author unknown, which illustrates a good point:

THE FIG TREE

A nobleman riding through
the countryside saw a very old
man working in his garden, planting
a small fig tree.

The nobleman called out,
"Why are you planting a fig tree
that will take years and years to
bear fruit? You will never eat
figs from that tree!"

"Tell me," said the old man,
"Did you eat figs when you were a
boy?" "Certainly, Why?"
"Who planted the tree from
which those gifts were picked?"

The nobleman hesitated, "I
don't know," he said.

"You see, Sir, our forefathers
planted trees for us to enjoy, and I
am doing the same for those who
will come after me. How else can
I repay my debt to those who came
before me?"

When this writer was an undergraduate at the University of Missouri in 1934, E. L. Morgan, of the Rural Sociology Department, was fond of repeating, time after time, this sentence: "An organization or group will progress no further or faster than their leaders see and believe." When all the people involved in any way with sheep combine their efforts to develop leaders and support improvement programs, the sheep industry should recover from the long-term, downward trend it has been in since the Korean War. When you become successful, remember how you got your start and repay the ones who helped you by helping others.

REFERENCES

Coffey, W. C., *Productive Sheep Husbandry.* Philadelphia: J. B. Lippincott Co., 1957.

Collins, S. B., *Profitable Sheep.* New York: Macmillan, Inc., 1956.

Diggins, R. V., and C. E. Bundy, *Sheep Production.* Englewood Cliffs, N.J.: Prentice-Hall, Inc., 1958.

Ensminger, M.E., *Sheep and Goat Science.* Danville, Ill.: Interstate Printers and Publishers, Inc., 1986.

Harrison, Virdon, "Sheep Production: Intensive Systems Innovative Techniques Boost Yield," *ERS Agr. Economics Report No. 452,* USDA, 1980.

Kammlade, W. G., Sr., and W. G. Kammlade, Jr., *Sheep Science.* Philadelphia: J. B. Lippincott Co., 1955.

McKinney, John, *The Sheep Book.* New York: John Wiley & Sons, Inc., 1959.

Scott, George E., *The Sheepman's Production Handbook.* Denver, Colo.: Sheep Industry Development Program, Inc., 1982.

Thompson, J. M., *Comparison of Four Systems of Lamb Production.* Ph.D. thesis, University of Missouri, Columbia, 1980.

QUESTIONS FOR STUDY AND DISCUSSION

1. What characteristics are desirable in a prospective shepherd?
2. Why should feed supply be a prime consideration in deciding whether or not to enter the sheep business?
3. Why is financing such an important consideration in deciding whether or not to start a sheep operation?
4. Discuss values other than profit that may be obtained from managing a sheep operation.
5. How do range and farm sheep operations differ?
6. What cropping system fits well with a fall lambing operation?
7. What are the advantages of a winter lambing operation over spring lambing?
8. Why should accelerated lambing be more efficient than winter or spring lambing?
9. What are the special problems of an accelerated lambing program?
10. Under what circumstances should a confinement operation be especially attractive?
11. What system of production should be most adaptable to confinement?
12. What special talents are required by a purebred breeder of sheep?
13. Why is it important for a prospective purebred breeder to understand the fundamentals of commercial sheep production?
14. Discuss the importance of record keeping in a purebred operation.
15. How do the skills required for purebred sheep production differ from those of a lamb feeder?
16. What are the reasons for finishing lambs?
17. Explain why lamb finishing is considered to be "risky."
18. What are the greatest hazards of finishing lambs on pasture?
19. What conditions are favorable to finishing lambs by hand feeding self-feeding ground mixed or pelleted rations?
20. What are the advantages of finishing lambs on pasture versus drylot?
21. Is it true that you must be born and reared on a farm or ranch in order to be successful with livestock? Explain.
22. What is meant by hands-on experience?
23. What methods can you use to learn the fundamentals of sheep husbandry?
24. How can a sheep producer justify a college education?
25. What should a student receive from an internship?

5 Genetics of Sheep Breeding

5.1 INTRODUCTION TO SHEEP GENETICS

All our present breeds of sheep resulted from programs of selective mating. Whether or not the members of the royal family, the leaders of the Catholic church, or the wealthy landowners in Spain who developed the Merino breed of sheep knew anything about genetics, they did succeed in producing a most unique and useful animal. Undoubtedly, rams must have been selected with the most care in order to produce the extraordinary wool, the tremendous hardiness, and the flocking habits of the Merino. Other examples of livestock development, such as the Arabian horse and the greyhound, also are proof that some understanding of breeding existed before modern times. In fact, the very existence of definite breeds and types is substantial proof that either instinctively or otherwise people have been able to improve animals by selection.

It is generally conceded that the pioneering work of Gregor Mendel from 1857 to 1865 with peas was the beginning of the understanding of genetics. He demonstrated that inheritance was logical and predictable and accomplished by genes from the two parents. He showed that segregation occurred, and there was an independent assortment of genes. Even though much more is now known about the mode of action of genes, the principles discovered by Mendel are still the basis for the science. Genetics, then, is the science that explains the similarities and differences in related organisms. By manipulating genes, the animal breeder seeks to obtain superior breeding stock. Unfortunately, the ability of a gene to affect the development of an animal cannot be measured by inspection and thus must be discovered by indirect means. Genes are located on the chromosomes of sheep. The chromosomes are part of the nucleus of all body cells. Chromosomes are very small and are characteristic of a species. Sheep normally have 54, but some have 56. The 54 chromosomes of a sheep are paired into identical chromosomes. One of the 27 pairs in sheep is called the sex chromosomes, because they are responsible for determining the sex of the offspring.

There are two kinds of sex chromosomes, X and Y chromosomes. The female has a pair of X chromosomes and the male an X and Y. When the chromosome pairs divide into the haploid state in the ova and sperm, the ova has only X chromosomes but sperm can have either an X or Y. When fertilization occurs, the sex of the offspring is determined by the sex chromosome carried by the sperm. If it is X, the lamb will be female (XX). If the sperm carries a Y chromosome, the lamb will be a male with an X and a Y chromosome. What the individual will be like is dependent on the assortment of genes carried by the chromosomes from the female and the male. If the genes are the same on each of the chromosomes received from the parents, the trait should be expressed in the same way as the parents. Thus, if the parents had many similar genes, the offspring should closely resemble the parents. The degree to which the offspring varies from the parents should indicate the difference in the genetic makeup of the parents.

5.2 GENES DETERMINE INHERITANCE

The gene is the unit of inheritance. It is a component of the DNA molecule carried on the chromosome. Genes are responsible for controlling all biochemical reactions in the animal body. When an egg is fertilized and the chromosomes from one parent are paired with the same from the other, the genetic code for the new individual is established and becomes a part of all cells in the body. The program for the development and function of the organism is followed throughout life and is only modified by the environment under which it exists.

Much is known about genes and the way they act. Plant and animal geneticists are making much progress in understanding their structure and function and using this knowledge to manipulate genes within the chromosomes. In recent years, molecular genetics has become a most intriguing and stimulating field of study. It will undoubtedly someday have a great impact on the livestock industry. That, however, is for the future and should provide much opportunity for a new crop of scientists. In this discussion, emphasis will be on how genes express themselves phenotypically and how individuals possessing superior genes can be identified and exploited.

5.3 SEGREGATION AND RECOMBINATION OF GENES

Genes are carried on chromosomes, normally in pairs, in body cells, and in half-pairs in the gametes (sperm and ova). A single pair is carried at the same location (called locus) on a pair of chromosomes, which are called homologous chromosomes because they are similar. Then, if only one pair of genes is to be considered, one of the pair would come from each parent, and when the individual reproduced, one of each pair would be transmitted to the offspring. If the genes were different for the trait, the offspring of the individual could receive either of two genes. If two offspring were born and each got a different gene, the two should be different for a trait. Multiply this one pair of genes by enumerable others, and the possible combinations tend to boggle the mind. A simple illustration should be helpful. Supposing there were two genes carried at the same location on homologous chromosomes, for example, color of the coat with white dominant and black recessive. We will represent white by a large W and black by a small w. Then if one parent was pure (homozygous) for white, it would have two large W genes or WW. The black would also be homozygous with two small ww's. Mating the two would result in a large W and small w (Ww) for all offspring. All would be white because white is dominant but all would carry the gene for black and be heterozygous. If 2 Ww's were mated, the possible combinations of genes would be WW, ww, and Ww. In this case, there could be pure white (two large WW's), pure black (two small ww's), and heterozygous white (one large W and one small w; Ww). That is the genetic makeup of the individuals that could result from mating the Ww individuals. Would there be an equal number of each? A simple table would demonstrate the possible combinations and ratios. Remember each individual carries either a W or w gene in the haploid state.

Genetic combination of offspring of two Ww parents

	W ♀	w ♀
W ♂	WW	Ww
w ♂	Ww	ww

The possible combinations indicate that there would be the following ratio of genotypes:

	Genotypes	Phenotypes
2	Ww	Impure white
1	WW	Pure white
1	ww	Pure black

There would be a ratio of two heterozygous white (Ww), one homozygous white (WW), and one homozygous black (ww). There would be distribution of three white and one black individuals. This is called the phenotype. Then, what is the difference between genotype and phenotype? The genotype refers to the genetic makeup of the animal and indicates the genes that it carries. The phenotype refers to characteristics that may be observed or measured. Thus, it is often impossible to discriminate between the two.

The problem of inheritance is not simple. A further illustration of gene segregation and recombination is of interest. In most cases, the polled (P) trait is dominant over horned (p). If a pure-white polled ram were mated to a pure-black horned ewe, the pure-white polled ram would carry WWPP genes, and the pure-black horned ewe wwpp. The first generation would all be WwPp or heterozygous white and polled.

WP ♂ × wp ♀ = WwPp, the genotype of offspring

Then, if the offspring of the mating were mated (i.e., WwPp × WwPp), the gene segregation in the offspring would be as follows:

Possible combination of genes from mating two WwPp individuals				
	Wp ♀	WP ♀	wP ♀	wp ♀
Wp ♂	WWpp	WWPp	WwpP	Wwpp
WP ♂	WWPp	WWPP	WwPP	WwPp
wP ♂	wWPp	WwPP	wwPP	wwPp
wp ♂	wWpp	wWPp	wwPp	wwpp

Distribution of the genotypes and the phenotypes is as follows:

	Expected distribution			Expected distribution	
Genotype	Ratio	%	Phenotype	Ratio	%
WWpp	1	6.25	White, horned	3	18.75
Wwpp	2	12.50	White, horned		
WWPp	2	12.50	White, polled	9	56.25
WwPP	2	12.50	White, polled		
WwPp	4	25.00	White, polled		
WWPP	1	6.25	White, polled		
wwPp	2	12.50	Black, polled	3	18.75
wwPP	1	6.25	Black, polled		
wwpp	1	6.25	Black, horned	1	6.25

This illustration is used merely to demonstrate the many possible combinations that can occur. Actually, neither color nor horns are always inherited so simply. For example, black in Karakul sheep is dominant to white, but in most white breeds the black allele is recessive. Horns are recessive to polled, but may be modified by other genes; for example, in Merinos and Rambouillets, horns are found on rams that are homozygous pp but not for ewes because of another pair of genes that suppress the horns. Dorset Horns have the same genes for horns as Merinos, but do not have the other gene that suppresses horns in females.

There are very few traits in sheep that are controlled by a single pair of genes. Perhaps the most important ones are those that cause congenital defects.

5.4 INHERITED TRAITS IN SHEEP

There are two kinds of inheritance that affect traits in sheep. These are known as qualitative and quantitative inheritance. Qualitative inheritance is characterized by easily distinguishable difference between phenotypes and is little affected by environment. It is controlled by a single or limited number of pairs of genes. Examples are inheritance of horns, color of coat, or lethals. Quantitative inheritance involves traits where there is no sharp distinction between phenotypes. The traits are very sensitive to environment and are controlled by many pairs of genes. Examples are body size, loin eye area, staple length, and clean weight of fleece. Qualitative inheritance will be discussed first.

5.5 QUALITATIVELY INHERITED TRAITS

5.5.1. Fleece color The color of the fleece is controlled by only one pair of genes. White is usually dominant, but there is a dominant black found in Karakul and Black Welsh Mountain sheep and a recessive black is carried in most white breeds.

5.5.2. Face and leg color Face and leg color varies from black to white with many variations in colors and patterns. These are controlled by two or only a few pairs of genes.

5.5.3. Multiple teats Sheep usually have only one pair of teats, but some may have an extra pair of supernumerary ones. Sometimes the extra teats are functional, but usually they are not. There appears to be no advantage in milk production or any other desirable trait to the extra teats. The trait is controlled by one or more pairs of genes.

5.5.4. Birth coat of lambs The character of the coat of a newborn lamb is controlled by only a few pairs of genes. Hairiness, a recessive gene, appears to be fairly easily eliminated by selection.

5.5.5. Horns Horns were undoubtedly natural to the original breeds of sheep. Mutations were probably responsible for the polled condition. There are horned and polled breeds in our domestic sheep. Both sexes are horned in the Dorset Horn and Scottish Blackface breeds, but only the rams have horns in the Merino and Rambouillet breeds. Horns are controlled by one pair of genes, but a second pair of genes can modify the horned genes, as in Merino and Rambouillet ewes.

5.5.6. Blood groups Blood types differ in sheep, and they are controlled by only a few genes. The practical significance of this phenomenon is yet to be determined.

5.6 INHERITANCE OF LETHALS AND OTHER DEFECTS

Many lethals and other defects are either recessive or inherited in a simple way with only a few pairs of genes. The following list includes defects and lethals that have been described in the literature. Many of the conditions were discovered as a result of inbreeding.

1. *Amputated.* This is a lethal condition in which there is no limb below the fetlock. The exact mechanism is unknown.
2. *Ataxia.* A recessive lethal in which lambs are born with a nervous incoordination and die.
3. *Atresia ani.* Lambs are born without an anus. The condition is lethal in males and prevents breeding in females. It is recessive.
4. *Blindness.* Lambs are sightless at birth. The condition is recessive.
5. *Brachygnathia.* The lower jaw is too short for the lamb to obtain nourishment. This is lethal. Mode of inheritance has not been proven.
6. *Cryptorchidism.* One or more of the testicles of ram lambs are retained in the abdomen. The condition is recessive.
7. *Daft lambs.* Lambs are unable to walk due to poor balance, or they are paralyzed. The condition is recessive.
8. *Dwarfism.* A recessive, lethal condition in which lambs are born with abnormally short legs, thyroid disfunction, and sometimes a short underjaw.
9. *Earless and cleft palate.* A recessive, lethal condition. Lambs are born alive, but soon die. Some lambs have extra toes.
10. *Earless.* Lambs are born without ears when the genes are homozygous. Short ears occur in the heterozygous condition. It is inherited as an incomplete dominant.
11. *Entropion.* Eyelids turned in, causing blindness if not corrected. Mode of inheritance not determined.
12. *Freemartinism.* Females incompletely sexed. This condition not caused by twins of different sexes, as in cattle. Mode of inheritance due to chromosome aberration (Gerneke, 1965; Moore, 1958).
13. *Goiter.* Lambs born with goiter not associated with iodine deficiency or goitergenic substances. Lethal in some lambs. Mode of inheritance not established.
14. *Lethal gray.* Occurs in Karakul sheep. A recessive gene in which pure gray lambs die at 1 to 48 months of age.
15. *Inguinal hernia.* A lethal in which lambs are born alive, but are unable to breathe. Inheritance not positive, but is apparently a recessive gene.
16. *Muscle contracture.* A recessive lethal. Lambs are usually stillborn with stiff joints in the lower legs.
17. *Naked.* Lambs are born with a shiny skin and little or no hair or wool covering and thus are very sensitive to temperature. Inheritance simple recessive.
18. *Occlusion.* Failure of incisor teeth to make contact with the dental pad either because the underjaw is too short or too long. Inherited recessively.
19. *Paralysis.* A recessive, lethal gene is the cause of paralysis in the hindlegs of lambs. Lambs soon die.
20. *Photosensitivity.* Lambs born with bile products in the blood causing sensitivity to sunlight with dermatitis. This is a semilethal condition caused by a single

autosomal recessive gene. The condition may also be caused by ingestion of noxious plants by the ewe or lamb.

21. *Rigid fetlocks.* A recessive lethal. The lambs are born with legs bent and rigid fetlocks. The body is short and thick, and the claws are deformed.
22. *Scurs.* Horny growth attached to the skin in hornless sheep. The cause is a recessive gene.
23. *Skeletal defects.* Deformed skeleton, includes the skull. The condition is sometimes accompanied by hernia. Caused by a recessive gene.
24. *Wattles.* Wattles are pendulous, cylindrical growths of skin attached at the throat. They are inherited recessively.
25. *Yellow fat.* Carcass fat is yellow. Caused by a recessive gene.
26. *Rectal prolapse.* A defect that appears to have some hereditary basis. The condition usually occurs in lambs weighing 70 to 100 lb. The rectum pulls away from the attachments holding it inside the body cavity, and it is ejected. The mode of inheritance has not been established. The condition also may be caused by dusty feed and docking lambs too short.
27. *Spider Lamb Syndrome.* The condition is characterized by malformation of the skeleton, including kyphoscoliosis (crooked back involving the thoracic vertebrae) and extreme height with long legs that have angular deformities causing splayed legs. Also, lambs may have malformed sternums and/or rib cages and fragile bones that heal slowly. Lambs may die from birth to a few weeks in age. Apparently, the condition is caused by a single autosomal recessive gene.

5.7 AVOIDING INHERITED DEFECTS

Since most defects in sheep are recessive or controlled by only a few genes, a recessive gene may be carried with a flock for many generations and never appear simply because the breeder has always gone outside the flock for rams. If outbreeding is followed, it may be that none of the rams used carried the recessive gene, and thus it could not appear. Inbreeding tends to pair recessive genes, and then the condition can be recognized phenotypically. If the dominant gene is shown as DD and the recessive gene is dd, the only way to discover that an undesirable recessive is present in the flock is to use a ram that carries the recessive gene. If the gene is present, it must be carried by ewes heterozygous for the trait, or Dd. For example, if a ram carrying dd genes is bred to the flock, any ewe producing a lamb with the recessive condition must be heterozygous for the condition and she should be eliminated. A progeny test of a ram can be conducted to determine if he is carrying a recessive gene. If he is mated with a group of known recessive ewes and one recessive offspring occurs, the ram is carrying the recessive gene. Since recessive ewes may be difficult to obtain, how many recessive ewes should be used to be confident that he is not a carrier? If five recessive ewes are bred to him with no recessive offspring, the confidence level is 95%. If seven recessive (dd) ewes are used with no recessive offspring, the confidence level increases to 99%. Another alternative is to use known heterozygous females (Dd). Proof that the ewes are carriers of the gene are those ewes that give birth to a homozygous recessive lamb or that are offspring of a homozygous recessive parent. If 11 heterozygous (Dd) ewes are bred to the ram and no homozygous recessive appears, the confidence level of freedom from the recessive gene is 95%. The confidence level is 99% if 16 ewes are mated without the appearance of the recessive.

Mating the ram to his unselected daughters is a drastic procedure because of the lack of vigor in the offspring, but it is a test for recessive genes. If the progeny of 23 of his daughters show no recessive genes, the confidence level that none are present is 95%. The level increases to 99% with 35 ewes. It should be pointed out that progeny tests are very expensive and time consuming. They should be used only under extreme circumstances.

5.8 VARIATIONS OF TRAITS OF SHEEP

5.8.1. Mutations Mutations are changes in genes that produce new phenotypes. This phenomenon occurs rarely; it results from a change in the genetic code, which is transmitted via the RNA molecule to the cytoplasm. This results in the synthesis of a different protein in place of the original one. If mutations occur in genes carried on the autosomes or in those carried in the sex chromosomes, they are inherited. The appearance of a polled lamb in a Dorset Horn flock at North Carolina State University is an example of a mutation. Most mutations are recessive and are lethal or detrimental to the proper function of the lamb and, therefore, are discarded. It is probable that nearly all flocks of sheep carry undesirable recessive genes that never are expressed. However, a program of inbreeding may cause the appearance of the defect. The undesirable animals may be discarded, but it is seldom that the breeder will be willing to take the radical steps necessary to eliminate the trait from the flock.

Before leaving the topic of mutations, it should be pointed out that desirable ones occur sometimes, and occasionally they may provide the means of extraordinary improvement of a trait. An example that may have resulted from a mutation is the Booroola Merino of Australia. This strain, in contrast to the typical Merino, is quite prolific. The trait appears to be controlled by a single pair of dominant genes, easily transmissible to other Merinos and potentially of great importance to the wool industry of the world.

5.8.2. Diversity in the sheep population Sheep, because of their wide distribution from the Arctic to the Anarctic, from desert wastes to the humid tropics, have a great diversity of traits. They have been selected for different purposes depending on the culture and needs of the people using them. The variability that exists is a distinct advantage to the sheep breeder who can select genes from a much wider pool. Progress can be increasingly rapid and precise as new traits are introduced and tested to determine their practical use.

Sheep of the future are likely to vary even more than in the past as they are developed for use on the range and farms versus in confinement and improved systems of production. Sheep breeds and crosses will surely be tailored to the needs of specialized operations. To date, there have been only a few studies of genetic/environmental interactions. Although the research has been limited, it has indicated the importance of this interesting field. More work must be done in the future.

5.9 QUANTITATIVELY INHERITED TRAITS

5.9.1. Additive genetic action Most traits of sheep that are of economic importance are inherited by additive genetic action. This means that the traits are controlled by many pairs of genes. In additive gene action, the phenotypic effects of one gene add to the phenotypic effects of others to generally change the phenotype. Since the phenotype can be measured, it is possible to improve the trait by identifying the best and mating them. The improvement is usually not spectacular, but it is continuous. The rate at which improvement can be made is affected by genetics and environment.

5.9.2. Heritability By the study of traits made by geneticists, it has been possible to estimate the proportion of the total phenotypic variation that is due to heredity. This is defined as the *heritability estimate.* Relatives should resemble each other more than nonrelatives, because they should have more genes in common. Many estimates of heritability have been made throughout the world. They differ between investigators and thus at this date are not precise. However, averages are generally accepted as guides and are useful to the breeder. Table 5.1 shows averages for traits of sheep.

5.9.3. Relative importance of heritability estimates The values shown in Table 5.1 are average and do not reflect the extreme ranges reported in the literature. Like almost all averages, they tend to make the estimates conservative. These data are being changed continuously as more information becomes available.

TABLE 5.1 HERITABILITY ESTIMATES FOR CERTAIN TRAITS IN SHEEP

	Percentage heritable
Economic traits	
Number of lambs born	10–15
Birth weight	30–35
Weaning weight	30–35
Yearling body weight	40–45
Postweaning daily gain	40–45
Efficiency of gains	20–25
Type score (weaning)	10–20
Type score (yearling)	20–40
Condition score	10–15
Slaughter grade (live)	20–25
Carcass characteristics	
Fat covering over the loin	20–25
Loin eye area	40–45
Marbling	20–25
Tenderness	30–35
Fat weight	35–40
Percent lean cuts	30–35
Carcass grade	15–20
Carcass length	30–35

TABLE 5.1 CONTINUED

	Percentage heritable
Wool characteristics	
Face cover	40–45
Neck folds	25–30
Skin folds on body	35–40
Grease fleece weights	45–50
Clean fleece weight	45–50
Staple length (yearling)	35–45
Fiber diameter	40–45
Crimp per inch	40–45
Fleece grade	30–40

Since the heritability estimate is an indication of the proportions of the expression of the trait that are governed by heredity and environment, the values provide a basis for estimating the progress made possible by a selection program. It should be expected that the higher the heritability, the more rapid the progress. In general, heritabilities of 40% or above are considered high, 20 to 40% medium, and less than 20% low. The progress, however, is not only limited by the heritability but by two other very important factors. These are (1) how accurately it is possible to measure the trait, and (2) how much emphasis will be placed on the trait in the breeding program. The breeder must decide on objectives and adhere to them if improvement is expected.

5.10 ECONOMIC TRAITS AND METHODS OF IMPROVEMENT

5.10.1. Prolificacy Few factors are more important in a sheep operation than the ability of a ewe to produce lambs. The production of more than one lamb by a ewe is much more efficient than a single. Sheepbreeders throughout the world are now very much aware of the importance of regular breeding, prolific ewes. The natural breeding season of most sheep is in the fall. The ewe lambs in the spring when forage is available so that she can suckle her young (January to April in the northern hemisphere and September to December in the southern hemisphere). Most of the lambs in the world are born in the spring. The average litter of lambs dropped in the United States is only slightly more than one lamb, and the weaning rate is barely one. There has been very little improvement in prolificacy throughout the history of the American sheep industry. Very little effort has been made to improve the trait, yet the existence of breeds such as the Finnsheep, Dorset Horn, and Romanov indicate that it can be improved.

Because heritability is low, many breeders are reluctant to emphasize twinning in their selection program. In fact, purebred producers have tended to do the opposite in selecting their stud rams. Quite often, they choose the larger, more developed ram that was born and reared as a single, thus increasing the chances that the prolificacy will not improve. Research has shown that the twinning percentage can be increased by as much as 5% to 10% in one generation by selection. In long-term experiments, it has been concluded that a 3% increase per generation can be expected.

It should be mentioned that ewes naturally tend to increase prolificacy until they are 6 years old. Ewes bred as lambs produce very few twins, but tend to produce more than their counterparts that fail to breed the first year of life. Some breeders evaluate the potential prolificacy of replacement ewes by using vasectomized rams to determine which ewe lambs cycle early in life. Others, with the skill and equipment, perform laparotomy, a surgical operation to expose the ovaries and determine the ovulation rate by inspection.

Research work with rams has demonstrated that testicular circumference is an accurate indication of fertility. It also appears to be an excellent indication of fertility of sisters of the ram.

Program to Improve Prolificacy in Ewes

1. Individually identify all ewe lambs at birth.
2. Record numbers of ewe lambs born as twins and triplets and save for breeding. If singles must be used as replacements, choose from the offspring of young ewes.
3. Use rams from sets of twins or triplets from highly prolific ewes. Measure testicular circumference and require 32 cm or more at 1 year of age.
4. Cull ewes that fail to lamb in two consecutive seasons.
5. Consider breeding ewe lambs and eliminate those that fail to produce a lamb.

5.10.2. Birth weight Birth weight is the first indication of the growth potential of the lamb. It is medium in heritability (30%–35%) and easily measured. Birth weight is influenced by sex of the lamb (males heavier), whether single or multiple birth, age and size of the ewe, and breed of sire and dam. Some caution should be urged in emphasizing this trait, because lambs that are too large at birth may be difficult to deliver.

5.10.3. Weaning weight Lambs are weaned at various ages. There is a strong trend toward weaning when lambs are younger. Studies of milk production show there is very little advantage to allowing lambs to nurse much longer than 8 weeks, because the milk flow generally decreases continually after the first 3 weeks. Weaning weights are usually calculated on the basis of 60- or 90-day weaning weights. Weaning weight is medium in heritability (30%–35%) and is easily measured. Weights should be corrected for age of the ewe, sex of the lamb, type of birth (single or multiple), and whether reared as a single or in a litter. Weaning weights not only provide information on the growth rate of the lamb but also on the milking ability of the ewe.

5.10.4. Postweaning gains Gain from weaning to market is highly heritable (40%–45%) and valuable information to the breeder. Test stations uniformly use this period to evaluate rams and ewes. They feed sheep a standard ration for a specified period under specified environmental conditions.

5.10.5. Yearling weight Weights from 1 year of age to maturity are closely related, and yearling weights appear to be highly heritable (40%–50%). They are useful as an indication of the mature weight of the ewe or ram. In the absence of weights at weaning or for a postweaning period, they are the best indication of gaining ability.

Program for Improving Gains of Sheep

1. Either ear tag or tattoo lambs individually.
2. Weigh lambs at birth, at weaning, after a postweaning feeding period, and at 16 months of age.
3. Correct weaning weights for age, age of dam, type of birth, rearing, and sex.

5.10.6. Efficiency of gain Data on inheritance of efficiency of gains indicate the heritability is lower than rate of gain, but there is a high correlation between rate and efficiency of postweaning gains. From the information available, it appears that efficiency of gains is medium in heritability (20%–25%). New equipment such as the Pinpointer or automatic head gates enable test stations or breeders to obtain feed conversion ratios on individual sheep. Ram producers should put emphasis on this trait.

5.10.7. Body conformation Scoring for type or shape of body is difficult with variable results. It appears the heritability is higher with older animals. Conformation cannot be ignored, especially where defects that may affect production are concerned. Unsoundnesses or structural weakness must be eliminated from the flock regardless of emphasis on other traits.

5.10.8. Carcass characteristics Carcass merit is medium to high in heritability. Loin eye area, percent of lean cuts, percent fat, and tenderness are all highly heritable, but they require the animal to be slaughtered in order to obtain direct measurements. It is, however, possible to estimate the loin eye area and fat covering over the loin by the use of ultrasonics. The K^{40} counter, which uses the naturally radioactive potassium in the body to estimate fat content and then lean mass by difference, may be useful to breeders if such equipment is or some other becomes available. It should be possible to improve carcass cutability rapidly. To date, no device has been found to directly estimate tenderness in a live animal. The only method that can be used is to slaughter sibs, half-sibs, or progeny.

It is possible, with training, to learn to estimate cutability of sheep quite accurately by eye appraisal, but it is necessary for the person to check estimates frequently with kill sheets in order to maintain the skill.

5.10.9. Fleece characteristics With only a few exceptions, those traits that have to do with fleece and skin are highly heritable. There are reports indicating that wrinkling is associated with low fertility, and there are negative correlations between fleece and carcass characteristics. Fleece weight and staple length may be negatively correlated to meatiness and positively with carcass fat. The amount of importance to be placed on fleece traits will depend on the type of operation. It seems doubtful that range producers will soon de-emphasize the improvement of wool. On the other hand, the farm flock owner who realizes only 15%–20% of income from wool is apt to put selection pressure on other traits. This is a decision that must be made by the owner or operator based on the individual operation.

Program for Improving Fleece Traits. If it is desirable to improve fleece merit, the following steps are proposed:

1. Weigh individual fleeces and measure length of staple at shearing.
2. Obtain estimated clean fleece weights for individual fleeces.
3. Obtain individual fleece grades.
4. Select stud rams that do not have skin or fleece defects.
5. Choose sheep with heaviest-weight fleeces and the longest staple.

5.11 RELATIONSHIP BETWEEN TRAITS

Research has been conducted to study the effect of traits on others. Results of such studies were summarized by Lasley (1978) and are shown in Table 5.2. The data show that birth weight is related to weights at other ages. Thus, selection for weight at any age should result in improvement in weights at other ages. Of particular interest is the negative correlation between average daily gain on test and feed per pound

TABLE 5.2 SOME GENETIC AND PHENOTYPIC CORRELATIONS BETWEEN ECONOMIC TRAITS OF SHEEP

	Correlations[a]	
Trait	Genotypic	Phenotypic
Birth weight and:		
Weaning weight	0.34	0.29
Gain birth to weaning	0.30	0.30
120-day weight	0.33	0.45
Weaning weight and:		
Average daily gain on test	0.53	0.30
Feed per gain	0.55	-0.15
Grease fleece weight	0.06	0.30
Fleece grade	-0.24	-0.01
Staple length	-0.15	0.01
Average daily gain on test and:		
Feed per gain	-0.73	-0.60
Grease fleece weight	0.17	0.22
Fleece grade	0.16	0.10
Staple length	-0.20	-0.01
Live fat depth and:		
Weaning weight	-0.19	
Average daily gain on test	-0.19	
Grease fleece weight	0.08	
Fleece grade	-0.21	
Staple length	0.11	
Carcass grade and:		
Live fat depth	0.31	
Loin eye area	-0.28	
Percent retail cuts	-0.23	

[a]Average figures from many studies.

of gain. This indicates that more rapidly gaining lambs are more efficient. Then, if sheep are selected for rapid gains, there should also be progress in efficiency of gains. Turner and Young (1969) in Australia have shown that a positive correlation exists between clean wool weight and yield, staple length, fiber numbers per square millimeter, and fiber diameter. They also demonstrated a negative relationship between clean wool weight and crimps per inch of staple and a positive correlation between wrinkle score and body weight and wrinkle score and greasy wool weight. Wrinkle score has a negative relationship with staple length, but is positive with crimps per inch. Crimp number per inch is negatively associated with greasy and clean wool weight, yield, and staple length of fleece.

5.12 PRODUCTION TESTING

Production testing simply stated is nothing more than measuring performance under specified conditions. There are programs for production testing at official test stations, as well as on the farm testing. In the United States, there has been very little production testing of sheep compared with other countries such as New Zealand, Australia, South Africa, and a number in Europe. Breeders of dairy and beef cattle, swine, and poultry have made extraordinary progress in their extensive performance testing programs. There is no ready explanation for the small number of sheep tested each year.

However, efforts are being made to carry on flock improvement programs in such states as Ohio and Wisconsin. There are also ram test stations scattered throughout the United States. A few breed associations have production testing programs, but it is estimated that less than 2% of purebred sheep are in such a program.

5.12.1. Who should performance test? It is difficult to visualize any serious purebred breeder who can justify not doing some performance testing. Whether or not sheep are put on test in an official test station or an on the farm test program should depend on the individual situation. Is the breeder interested in selling stud or commercial rams? Is the volume of business sufficient to justify the expense of testing? Is performance testing crucial to the breeding program? Since growth and feed efficiency traits in sheep are medium to high in heritability, breeders of the ram breeds can obtain much information on their sheep by putting them through a test station. If, in addition, the breeders can use the Sonoray to determine fat thickness and area of the loin eye, they can select rams with lean carcasses, but may also obtain some estimates of the ability to convert feed into gains. Preweaning and postweaning performance of lambs in the feedlot can be much improved by using tested rams. Performance testing of ewes is not usually done because of the expense involved and the potentially detrimental effects of finish on the production and longevity of ewes.

5.12.2. Ram test stations It is not certain how many ram testing stations there are in the United States. The *Progressive Sheep Breeder* for fall of 1981 published details on test stations from 14 states and Canada. It was interesting to observe the great variation in the rules for each station. While it is true that the sheep industry is different in various parts of the country, it seems reasonable that there should be more standardization in the procedures. Surely the date of birth should be standard for similar sheep and localities. Weight requirements vary from none to a definite figure. The figures quoted by the 8 stations that have the requirements varied from 30 to 60 lb. Most stations have an adjustment period, but these varied from 7 to 21 days. Days on test also were much different. The periods ranged from a low of 56 days to a high of 150 days. The rations fed the rams being tested varied from a low of 11% crude protein to a high of 18%. All the stations furnish the sheepbreeder with average daily gains (ADG), and 13 of the 16 provide ADG ratios. All but 3 stations figure weight per day of age (WDA), but only 3 provide WDA ratios. Only 6 of the 16 can measure feed per unit of gain. Twelve of the 16 have some inspection for soundness, but not all are the same. Only 5 stations will do progeny testing, and they vary from 4 to 6 progeny required. The majority of stations sell tested rams.

Some stations provide scores on breed type, muscling, conformation, and wool. Some also measure testicular circumference and estimated carcass merit. Some western stations score rams for face cover and skin folds. The number of ram test stations appears to be increasing in late years. This would indicate that producers are placing some value on test results. The breeder must decide whether or not such a program offers benefits to him or her. Some advantages of central testing are:

1. A breeder can be reasonably certain that the records will be unbiased.
2. Heritability of most traits associated with productivity is highest when measured during the postweaning period.
3. Performance can be measured in such a manner that comparisons can be made between flocks.
4. Many breeders prefer to go outside their flocks for sires. Test stations provide some convenience in locating worthy individuals.
5. In other species, tested sires sell higher than those with no record. In ram tests to date, the same can be said for sheep.

But the disadvantages are:

1. Ram testing is expensive.
2. Only a few rams per flock are tested and their selection by the breeder may be biased.
3. Rams are highly finished at the end of the test. High finish may be detrimental to fertility and longevity.
4. Some breeders manage their rams in an atypical manner so they will make compensatory gains on the test. This provides data that appear good, but is not a true indication of the performance of the ram.

Although it is not possible to control pretest environmental conditions, manipulation of the animals to obtain better test results may be countered by putting rams on when they are younger and feeding for longer periods.

A partial list of central test stations by states is as follows: California, Colorado, Illinois, Indiana, Iowa, Missouri, Ohio, Ontario (Canada), Pennsylvania, South Dakota, Texas, USDA (DuBois, Idaho), Virginia, Washington, West Virginia, Wisconsin, and Wyoming.

5.12.3. Progeny testing of rams Progeny testing of rams may prove useful, although it requires time and is expensive. Too often the ram may be too old by the time sufficient information has been obtained on him. This could very well give information on the ability of the sire to alter the carcass characteristics of his progeny. Theoretically, a combination of performance test data and progeny testing of offspring of the ram should be more valuable in predicting the breeding value of the ram than either alone. It has been estimated that at least 20 progeny are required to estimate the value for lowly inherited traits. Accuracy of selection based on the sire's own record plus his progeny increases with the number of progeny tested, but the increase in precision is greater in traits of lower heritability.

Progeny tests of rams for economic traits are often disappointing because there may be only small differences between the offspring of rams, but the occasional good ram that can transmit his characteristics to his offspring is of great value and he should be used extensively.

Performance and progeny testing have been valuable tools in improving dairy and beef cattle and swine, and they have increased the value of males that have been tested. Heritability of economic traits is as great in sheep as other species and thus the potential is just as great.

5.13 RECOMMENDED SELECTION PROGRAMS FOR SHEEP IMPROVEMENT

A purebred producer can develop an individual program for his or her own flock by putting emphasis on traits desired to be improved. However, most breeders may wish to enter into a general program with the obvious advantages of not having to work out the details, and, in many cases, the computation is part of a service paid for by the breeder. Also, if the program has been standardized, it may have a greater acceptance by other breeders.

In an effort to standardize procedures and to improve communication, the National Extension Sheep Committee proposed terminology and methods for selection programs in sheep breeding (Allen et al., 1968). They emphasized selection for number of lambs born and weaned per ewe exposed and rapid growth rate of lambs. Emphasis on wool production would depend on the area and type of production.

5.13.1. Importance of records The Sheep Committee emphasized the importance of accurate records on all animals and that weighing conditions be the same for all animals compared. They recognized that the great diversity in environmental conditions and levels of management preclude adjustments for such factors and, therefore, recommended that comparisons be limited to animals reared in the same flock or test station.

5.13.2. Records needed for improvement programs Records essential for purebred and farm flock selection systems are:

1. Individual identification of each animal (eartag, tattoo, or brand)
2. Sire number
3. Dam number
4. Dam's age at lambing
5. Birth date of lamb
6. Birth weight of lamb
7. Sex of lamb
8. Type of birth (single, twin, etc.)
9. How reared
10. Dates weighed
11. Corrected weights to a standard age (90, 120, 140 days). Correct weights as follows: if the standard age is 90 days, then 90-day corrected weight = (actual wt. × 90)/age in days.
12. Adjusted weights: corrected weights are adjusted on a single ewe lamb out of a mature ewe basis.

    Optional information

13. Adjusted weight ratios: Relationship between adjusted weight of individual to the adjusted weight of all lambs of same sex in the flock = (adjusted weight of lamb)/(adjusted weight of all lambs) × 100.

Depending on the type program being followed, other information may be recorded, such as:

1. Whether or not the lamb was creep fed
2. Yearling or other weights
3. Type score, conformation grade, live slaughter grade, loin eye area, scores for face cover, wrinkles or scurs or horns for the individual, plus carcass information for sibs
4. Grease fleece weight and staple length
5. Staple length of fleece
6. Defects observed in the lamb

Factors for correcting weaning weights for age of dam, sex of lamb, and single or multiple births are shown in Table 5.3.

Note: The corrected weaning weight is multiplied by the appropriate factor, as shown in the following example.

TABLE 5.3 FACTORS FOR ADJUSTING WEIGHTS OF LAMBS

	Age of Dam		
	3–6 Years	2 Years or over 6	1 Year old
Ewe lamb			
Single	1.0	1.09	1.22
Twin raised as twin	1.11	1.20	1.33
Twin raised as single	1.05	1.14	1.28
Triplet raised as triplet	1.22	1.33	1.46
Triplet raised as twin	1.17	1.28	1.42
Triplet raised as single	1.11	1.21	1.36
Wether			
Single	0.97	1.06	1.19
Twin raised as twin	1.08	1.17	1.30
Twin raised as single	1.02	1.11	1.25
Triplet raised as triplet	1.19	1.30	1.43
Triplet raised as twin	1.14	1.25	1.39
Triplet raised as single	1.08	1.18	1.33
Ram			
Single	0.89	0.98	1.11
Twin raised as twin	1.00	1.09	1.22
Twin raised as single	0.94	1.03	1.17
Triplet raised as triplet	1.11	1.22	1.35
Triplet raised as twin	1.06	1.17	1.31
Triplet raised as single	1.00	1.10	1.25

Problem

To determine the adjusted 90-day weights of twin lambs (a ram and ewe) out of a 2-year-old ewe: Lambs were weighed at 100 days of age, and the ram weighed 105 lb and the ewe 95 lb.

$$\text{corrected 90-day weight of ram} = \frac{105}{100} \times 90 = 94.5$$

$$\text{corrected 90-day weight of ewe} = \frac{95}{100} \times 90 = 85.5$$

$$\text{adjusted weight of ram} = 94.5 \times 1.09 = 103$$

$$\text{adjusted weight of ewe} = 85.5 \times 1.20 = 102.6$$

5.13.3. Lamb indexes The Ohio Sheep Production Testing Program uses an index for lambs. It includes a value for wool production as well as lamb weights.

$$\text{index for single lambs} = \text{adjusted weight of lambs} + (0.6 \times \text{grease fleece weight of ewe})$$

Problem

A ewe produces a single ewe lamb with an adjusted weight of 85 lb and 10 lb of wool.

$$\text{lamb index} = 85 + (10 \times 0.6) = 91$$

The formula for multiple births is

$$\text{lamb index (multiple births)} = \text{adjusted weights} \times 1.1 + (0.6 \times \text{fleece weight of the ewe})$$

Indexes may vary with emphasis on traits; for example, fleece credit could be raised or lowered by changing the fleece factor.

5.13.4. Ewe indexes Progress by selection is most rapid when selection pressure is focused on a restricted number of traits. This however, is seldom practical. It is usually more satisfactory to choose a limited number of important traits and incorporate them into an index that puts a weight on each trait. The formula for the index may then be varied to put more or less emphasis on the traits.

Flock Records. It is recommended that the following essential data be kept on the flock:

1. Ewe identification
2. Sire identification
3. Dam identification
4. Age of the dam
5. Birth date of ewe
6. Type of birth of ewe (single, twin, etc.)
7. Details of rearing (naturally, artificially reared as single, twin, etc.)
8. Date ewe lambed
9. Number of lambs born
10. Number weaned
11. Adjusted weight of lambs weaned
12. Grease fleece weight

Other data frequently recorded are:

1. Shearing dates
2. Staple length
3. Conformation grade and scores for types, face cover, wrinkles, horns, and scurs

Lamb productivity should be calculated for each ewe in the flock.

The Ohio program has the following basic formula for calculating ewe indexes:

1. Ewe giving birth to a single lamb that does not survive = no index.
2. Ewe giving birth to a single lamb that survives = adjusted weight of lamb + (3 × fleece weight).
3. Ewe with multiple births and none survive = no index.
4. Ewe with multiple births and one or more survives:

$$\text{ewe index} = \frac{(\text{sum of adjusted weights} \times 1.17) + (3 \times \text{fleece weight})}{\text{number of lambs}}$$

Note: The factor of 1.17 is the creditation for genetic value of multiple births.

Problem

A ewe rears mixed twins with adjusted weights of 85 lb for the ram and 88 lb for the ewe and shears 9 lb of wool. What is her index?

Solution

$$\text{Ewe index} = \frac{(85 + 88) \times 1.17 + (3 \times 9)}{2}$$

$$= \frac{173 \times 1.17 + 27}{2}$$

$$= \frac{202.4 + 27}{2}$$

$$= 101.2 + 27$$

$$= 128.2$$

Ewe Index Involving Fostering Situations. The true dam's index is calculated as shown in items 2 and 4 except that the fostered lamb's corrected weight will be adjusted (Table 5.3) based on foster ewe's age.

5.13.5. Ewe index ratio This information allows the breeder to compare the index of each ewe with the average of the rest of the flock.

$$\text{avg. index of rest of the flock} = \text{number of ewes} \times \text{avg. index} - \frac{\text{index of the individual ewe}}{\text{number of ewes}}$$

$$\text{ewe index ratio} = \frac{\text{index of individual ewe} \times 100}{\text{avg. index of rest of flock}}$$

For example, if the average index of the flock of 60 ewes is 120, the following would be the computation if a single ewe had an index of 131.18:

$$\text{average index} \times \text{rest of flock} = \frac{(60 \times 120) - 131.18}{59}$$

$$= \frac{7200 - 131.18}{59}$$

$$= \frac{7068.82}{59}$$

$$= 119.81$$

$$\text{Ewe index ratio of individual ewe} = \frac{131.18 \times 100}{119.81}$$

$$= 109.4$$

The index described has been very useful in farm flocks. Range producers will perhaps be interested in indexes that give more weight to wool quantity and quality.

If a special index is needed to achieve certain goals, it would be wise to consult with extension specialists or the experiment station geneticist of the state of the breeder's residence.

Suggestions for a Sheep Improvement Plan.

1. Permanently identify all individuals in the flock using ear tags, ear marks, or tattoos. Record breeding dates. All lambs should be given an individual number when born.
2. Weigh and record weights at birth, weaning, postweaning, and at 16 months.
3. Adjust weaning weights for age, sex, multiple births, and age of dam.
4. Obtain postweaning data under controlled feeding of rams.
5. Weigh individual fleeces and record.
6. Figure lamb and ewe index ratios and use them as a basis for selection. Replace low-indexing ewes with ewe lambs from dams with a high index and with a high lamb index.
7. Cull ewes that are unsound, are erratic breeders, or have poor fertility.
8. Select rams that are twins from a ewe with a record of high prolificacy and, if out of season lambing is important, born in the fall of the year.
9. Select rams that have a high performance record from birth to weaning and, if possible, an outstanding record in a test station or on the farm testing program.
10. If more than one ram is used in the flock, compare progeny records and discard the one with the poorest performance. Replace with one that has a high production record from a highly productive line.

5.14 OTHER TECHNIQUES IN BREEDING

5.14.1. Inbreeding Most of the discussion on improving sheep by breeding has been concerned with outbreeding and selection. Generally, breeders prefer to use unrelated sires because they are not willing to risk the pitfalls of inbreeding. They are aware that such practice may very well expose recessive genes and appreciably decrease the number and vigor of offspring.

Robert Bakewell, who lived in England from 1726 to 1795, was well aware of the taboos against mating close relatives, but he did use inbreeding with extraordinary results. His principles of only looking at the utility of the animal, careful selection, and mating the best to the best regardless of their relationship resulted in remarkable improvement in a comparatively short time.

Bakewell did not practice mating of sire to daughter or siblings except occasionally, but he did mate relatives to concentrate the blood of individuals, usually rams, with outstanding performance. This procedure, called linebreeding, has the advantage of greater numbers of descendants and allows more opportunity to prove merit of the ram by progeny testing.

Reproductive rate appears to be very much affected by inbreeding, not only from the standpoint of fertility but also livability of the young. Inbreeding has been shown to decrease weaning weights of sheep as much as 0.3 lb per 1% of inbreeding.

Inbred lines have been established by both the USDA and the Minnesota Experiment Station. They developed inbred lines of sheep but felt that they could have made more progress by selecting for desirable traits without increasing inbreeding.

Linebreeding can be defined as a form of inbreeding, but by mating individuals not as closely related, such as half-sibs and cousins. It is a useful tool of breeders with such outstanding individuals that it is difficult to obtain a superior sire from unrelated stock. Breeders with a single-ram flock should usually go outside for a stud ram. Then they may have better results by the use of a ram from a linebred flock.

5.15 CROSSBREEDING

Commercial sheep producers have known for years that higher-quality market lambs can be produced by fine-wool ewes if they are bred to mutton-type rams rather than to fine-wool rams. They are well aware that crossbreeding makes it possible to introduce desirable characteristics to the offspring. They also have observed that crossbred lambs have unusual vigor. In late years, research has shown that there are many other advantages of crossbreeding, especially in the performance of crossbred rams and ewes in commercial production.

5.15.1. Heterosis If unrelated lines or breeds of sheep are crossed, the opposite to inbreeding occurs. Where genes become more homozygous in inbreeding, they become more heterozygous in crossbreeding. When heterozygosity occurs, there is a stimulative effect so that the offspring often may exceed either parent. This is known as hybrid vigor, or heterosis. Fertility and survival of offspring are especially affected by heterosis. It has been shown that crossbred rams have greater fertility and more libido than purebreds. In experiments with crossbreeding using two, three, and four breeds of sheep, workers of the USDA have shown that number of lambs produced and weaned and the weaning weight were greater for two-breed crosses than the lambs from either of the purebred breeds that were crossed. There was a further improvement when a third breed was used on the crossbred ewes, and the best results were obtained from breeding the three-way-cross ewes to a fourth breed.

5.15.2. Complementarity By selective crossbreeding, it is possible to introduce characteristics from both parents that are beneficial to the offspring. This is known as complementarity. A good example is the use of Finn × Dorset rams on Rambouillet ewes to produce very productive ewes for commercial lamb production. Finn blood should increase prolificacy, improve milk production, and increase livability. Dorset blood should result in increased prolificacy and milk production and extend the breeding season. The Rambouillet ingredient should result in good wool production, improved mothering, hardiness, size, and extended breeding season. If the three-way-cross crossbred ewes are bred to growthy, meaty rams of the Suffolk, Hampshire, or other similar ram breed, then excellent market lambs should result.

Many crosses are being made to produce better ewes for commercial lamb production. It seems probable that crossing will be used to achieve maximum results. Breeds or strains could very well be developed for crossing, with each contributing a certain group of desirable characteristics.

Ewe breeds should be expected to furnish the following to a terminal cross:

1. High fertility and prolificacy and breed at any time of the year
2. Gentle to handle, lamb easy, and good mothers
3. Heavy flow of milk
4. Heavy, high-quality fleece
5. Long productive life.

Ram breeds should provide the following:

1. Sexual aggressiveness and high fertility
2. Rapid and efficient growth
3. Muscling and high-quality carcasses
4. Maturity of lambs within accepted market weights

There appears to be some relationship between physical characteristics and productivity of ewes. For example, most of the ewe breeds have white faces. This may be the result of the natural selection that occurred in range sheep. Unless the ewe was a good mother, the lamb failed to survive. This was particularly true of the Merinos and Rambouillets from Texas, where most of the ewes lamb on fenced range and receive very little individual attention from shepherds. It is also generally conceded that Northwest range ewes lamb with less trouble than most native ones.

Wool-blind ewes were shown by the USDA to be less productive than ewes with open faces. The difference amounted to as much as 10 lb per lamb per year.

5.15.3. Possible breeds to cross for foundation ewes The following are possible breeds that could be used in developing crossbred ewes: Rambouillet, Merino, Delaine Merino, Targhee, Columbia, Corriedale, and Panama. Ram breeds for crossing to produce crossbred ewes could be some of the following: Dorset, Finnsheep, and Border Leicester. If rams can be obtained from a high-producing flock that has been selected for twinning and fast growth rate, Suffolks, Hampshires, Shropshire, Oxfords, or a number of other breeds could be used. It must be said, however, that if breeds such as the Hampshire or Suffolk are to be used to sire crossbred ewes, then it complicates the choice of breeds of rams for the terminal cross for market lamb production. It is probably better to use them to provide the meatiness and growth rate, as well as heterosis, to market lambs.

Other possibilities could be examined. The Texel breed in Holland and the East Friesian of Germany have many of the qualities desired. Both breeds are very prolific, good milkers, and superior mothers. Both have heavy fleeces, although the wool is coarse textured.

Research is being conducted to determine the possibility of using the St. Croix breed from the Caribbean, which is heat tolerant and prolific and has a long breeding season. They have rather low-quality fleeces, but are reputed to have more resistance to internal parasites than most breeds produced in the United States. Other possible breeds include the Barbados from the Caribbean, a prolific hair sheep with a long breeding season and resistance to parasites, the Hangyang of China, and the Romanov from Russia, which are noted for prolificacy.

There could very well be a place for the Booroola Merino and such new breeds as the Perrindale and Coopworth of New Zealand or the Polypay from the United States. The Perendale and Coopworth are being selected for production and could very well be the kind of rams needed for crossing on Rambouillets or Merinos to produce foundation ewes.

It seems logical that crossbreeding will become more sophisticated in the future. Breeds and strains will be developed for specific needs of the industry. Purebred breeders who have developed flocks with the needed traits should prosper. If crossbreeding does become the rule, sheep producers may find it practical to produce their own replacement ewes. Then they can be assured that the most favorable combinations are made for environmental conditions and that the breeding animals are from stock selected for productivity. Crossbreeding may also be used for grading up, developing new breeds, and corrective breeding, as discussed next.

5.15.4. Grading up Grading up is accomplished by the repeated use of purebred rams of one breed on a flock of ewes. If outstanding rams are used, remarkable improvement can occur in a short time. Changes in germplasm by grading up are illustrated in Figure 5.1. This system was used in cattle to increase the numbers of exotic breeds in the United States in cases where it was very difficult to import animals. It should be emphasized that in the eyes of most established breed associations it would never be possible to register the graded-up flock. For that reason, it should not be undertaken without due consideration as to whether or not the breeder

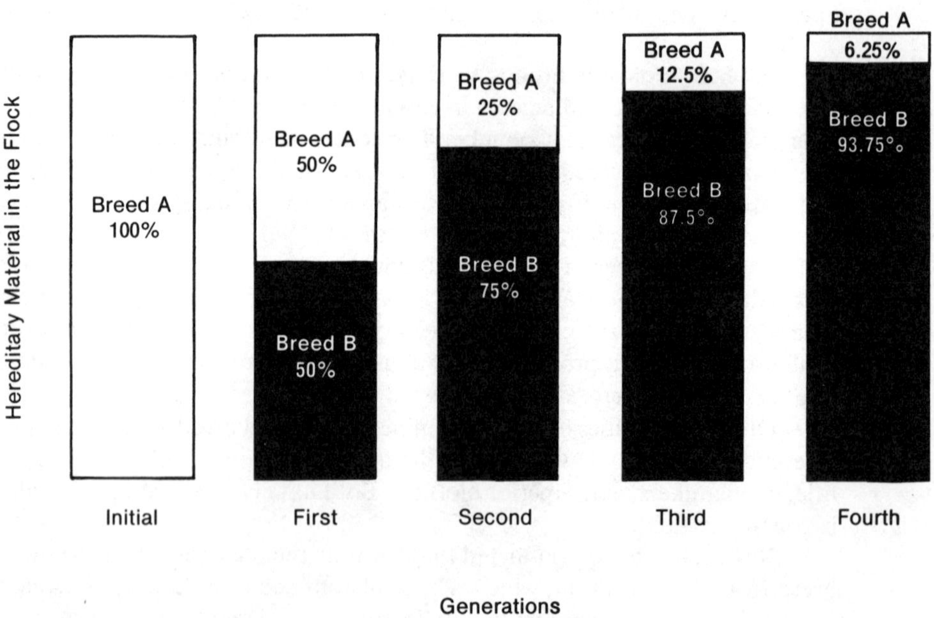

Figure 5.1 Effect of grading up on changes of purebred blood in a flock.

would be better off to start with purebreds initially. Then any improvement made should be acceptable to both the commercial and purebred breeder.

5.15.5 New breed formation New breeds are still being formed. Crossbreeding is a practice that is commonly used. For example, the Columbia, Corriedale, Targhee, Montadale, and Polypay breeds were all synthesized from crossing well-established breeds. Others are in various stages of development. New breeds will undoubtedly be established as the need arises.

One example of new thinking in breed formation was the origin of the Combo 6 sheep at the University of Illinois. This strain of sheep was developed in an effort to provide ewes that would breed in the spring and be very fertile. The basic ewe flock was white-faced Western-type ewes predominantly of Rambouillet breeding. In the spring of 1972, the ewes were inseminated with pooled-mixed semen from rams of each of the following breeds: Border Leicester, Dorset, Finnsheep, Rambouillet, Suffolk, and Targhee. From the original mating, replacements were selected on the basis of fertility in the spring, prolificacy, early puberty, and response to exogenous hormones and artificial insemination. Such a breed should have great commerical usefulness.

It should be pointed out that a great number of sheep are required in any new breed formation in order to have a broad genetic base and to allow rigorous selection pressure.

5.15.6 Corrective breeding Crossbreeding may be the most efficient means of correcting a deficiency in a one-ram flock of sheep. If corrective breeding is to be used, study the situation carefully before taking the step. Make certain undesirable traits are not being introduced. After the first lamb crop, evaluate the results not only to be sure improvement has been made but to be certain other traits haven't been adversely affected. Then discard the progeny if they are not satisfactory.

5.15.7. Crossbreeding and the purebred breeders Some purebred breeders are of the opinion that crossbreeding detracts from the pure breeds. They are right if they object to the introduction of outside blood into a pure breed. Such an act is indefensible and a reflection on the whole breed. A breed is only as strong as the integrity of the breeders who make up the organization. Strict rules should be made and followed.

If crossbreeding is to be the rule for commercial breeders, then they will need many purebreds with fixed traits and as homozygous as possible to do the crossing. The future should be very attractive to the breeder who has the type and quality of genetic material needed by the producer.

REFERENCES

Allen, G. A., and others, "Recommendations for Uniform Sheep Selection Programs," *National Extension Sheep Committee Report,* USDA, FES, ASTM, 223, 1968.

Bogart, Ralph, *Improvement of Livestock.* New York: Macmillan, Inc., 1959.

Botkin, M. P., "Eliminating Inherited Defects in Sheep," *Proc. Symposium Genetic Improvement of Wool and Lamb Production,* Texas A and M University, Livestock and Forage Center, McGregor, Texas, 1968.

Boyd, L. H., C. L. Taylor, and O. Mendoza, "Indicators of Environmental Adaptation in Sheep," *J. Anim. Sci.,* 53, Suppl. 1, 1981, p. 176.

Bradford, G. E., "Principles of Genetic Improvement of Wool and Lamb Production Symposium," American Sheep Producers Council, Texas A and M University, Denver, Colo., 1968.

Carter, R. C., and others, "Genotype–Environmental Interaction in Sheep. I. Ewe Productivity," *J. Anim. Sci.,* 33, 1971, p. 556.

______, and others, "Genotype–Environmental Interaction in Sheep. II. Lamb Performance Traits," *J. Anim. Sci.,* 33, 1971, p. 732.

______, and others, "Genetics–Environmental Interactions in Sheep. III. Lamb Carcass Traits," *J. Anim. Sci.,* 36, 1973, p. 617.

Dennis, S. M., and H. W. Leipold, "Ovine Congenital Defects," *Vet. Bull.,* 49, 1979, p. 233.

Ercanbrack, S. K., and A. D. Knight, "Frequencies of Various Birth Defects of Targhee and Columbia Sheep," *J. Heredity,* 69, 1978, p. 237.

Gerneke, W. H., "Chromosomal Evidence of Freemartin Condition in Sheep, *Ovis aries,*" *J. South Africa Med. Assn.,* 36 (l), 1965, p. 99.

Hohenboken, W. D., and S. E. Clarke, "Genetic–Environmental Interaction Effects on Lamb Survival, Cumulative Lamb Production and Longevity of Crossbred Ewes," *J. Anim. Sci.,* 53, 1981, p. 966.

______, W. H. Rennick, and R. Bogart, "Genetic–Environmental Interaction Effects on Sheep. II. Lamb Growth and Carcass Merit," *J. Anim. Sci.,* 42, 1976, p. 307.

Hulet, C. V., and others, "Effects of Origin and Environment on Reproductive Phenomena in Rambouillet Ewes. I. Breeding Season and Ovulation," *J. Anim. Sci.,* 38, 1974, p. 1210.

Lasley, John F., *Genetics of Livestock Improvement.* Englewood Cliffs, N.J.: Prentice-Hall, Inc., 1978.

Miller, K. P., and D. L. Daily, "A Study of Crossbreeding in Sheep," *J. Anim. Sci.,* 10, 1951, p. 462.

Moore, N. W. and L. E. A. Rawson, "Freemartinism in Sheep," *Nature,* 182, 1958, p. 175.

Morley, F. H. W., "Selection for Economic Characters in Merino Sheep. III. The Effect of Inbreeding," *J. Agr. Res.,* 5, 1954, p. 305.

Ragob, M. T., and A. A. Asker, "Effects of Inbreeding on a Flock of Ossimi Sheep," *J. Heredity,* 45, 1954, p. 89.

Rastogi, Ray, and others, "Crossbreeding in Sheep with Evaluation of Combining Ability, Heterosis, and Recombination Effects for Lamb Growth," *J. Anim. Sci.,* 54, 1982, p. 524.

Saperstein, G., H. W. Leopold, and S. M. Dennis, "Congenital Defects in Sheep," *J. Am. Vet. Med. Assoc.,* 167, 1975, p. 314.

Scott, George, E., *The Sheepman's Production Handbook.* Denver, Colo.: Sheep Industry Development Program, Inc., 1970.

Shelton, M., "Estimation of Genetic Change in a Performance Testing Program for Sheep," *J. Anim. Sci.,* 48, 1979, p. 26.

Sidwell, G. M., and L. R. Miller, "Production in Some Purebreeds of Sheep and Their Crosses. I. Reproductive Efficiency in Ewes," *J. Anim. Sci.,* 32, 1971, p. 1084.

———, and ———, "Production in Some Purebreeds of Sheep and Their Crosses. II. Birth Weights and Weaning Weights of Lambs," *J. Anim. Sci.*, 32, 1971, p. 1090.

———, D. O. Ererson, and C. E. Terrill, "Lamb Weights in Some Purebreeds and Crosses," *J. Anim. Sci.*, 23, 1964, p. 105.

Terrill, C. E., "Sheep Improvement through Selection," *National Wool Grower*, 47, 1957, p. 14.

———, "Fifty Years of Progress in Sheep Breeding," *J. Anim. Sci.*, 17, 1958, p. 944.

———, "Review and Application of Research on Crossbreeding of Sheep in North America," 1st World Congress on Genetics Applied to Livestock Production, No. 1, 1974, p. 765.

The Progressive Sheep Breeder, Special Report, "Ram Performance Testing Is Alive and Well," Fall 1981.

Turner, H. N., "Genetic Improvement in Reproductive Rate in Sheep," *Anim. Breed Abstr.*, 37, 1969, p. 545.

———, and S. Y. Young, *Quantitative Genetics in Sheep*. Ithaca, N.Y.: Cornell University Press, 1969.

QUESTIONS FOR STUDY AND DISCUSSION

1. Why is it important for the sheep breeder to understand the principles of genetics?
2. How many pairs of chromosomes do sheep have? What cells contain chromosomes?
3. Which sex of the sheep controls the sex of the offspring?
4. What is the difference between the genotype and phenotype of an individual?
5. What is the unit of inheritance?
6. What is a mutation?
7. What is the difference between qualitative and quantitative inheritance? Give examples of each.
8. What is a lethal? Give examples of the condition.
9. Why are most lethals recessive?
10. How is it possible for a recessive gene to occur from mating two phenotypically normal parents?
11. What is the difference between heterosis and heritability?
12. Why should relatives resemble each other more than nonrelatives?
13. What is meant by additive gene action?
14. What is meant by a heritability estimate?
15. Which of the following traits are high, medium, or low in heritability?
    a. Number of lambs born
    b. Yearling body weight
    c. Condition score
    d. Loin eye area
    e. Fat weight
    f. Skin folds on the body
    g. Fiber diameter
    h. Staple length
    i. Face cover
16. How is the best way to improve prolificacy in a flock of commercial sheep?

17. What is the importance of testicle size in selecting ram lambs? What is the procedure for measuring testicular size?
18. What method is used to measure efficiency of gains in sheep?
19. What is a performance test?
20. What is a progeny test?
21. What factors are used in a ewe index?
22. Outline a program for improving the prolificacy of a ewe flock.
23. What information can be gained by studying ewe index ratios?
24. What are the dangers of inbreeding?
25. What can be gained by crossbreeding?
26. What is grading up?

6 Establishing the Breeding Flock

6.1 IMPORTANCE OF COMPETENT SELECTION AND PURCHASING OF BREEDING STOCK

Breeding stock represents a major part of the investment in a sheep enterprise. The effort expended in purchasing desirable animals, at a reasonable price, capable of the performance required for a successful operation may well be the difference between the success and failure of the operation. Unlike products in stores that have quality standards, guaranteed weights, and a statement of contents, sheep are not set out on counters for inspection and comparative shopping. A knowledge of the desirable characteristics in breeding animals, sources of supply, mechanisms of purchasing, and skill in bargaining is invaluable in establishing the flock. Too often the neophtye discovers too late that apparent bargains are unsatisfactory or disastrous. While it is true beginners may not be skilled in judging or purchasing sheep, they must learn the essentials and develop the skills promptly. This chapter covers those subjects.

6.2 KIND OF SHEEP TO BUY

The system of production to be followed may require certain types of sheep. For example, the ewes in a fall lambing operation must breed in the spring. Purchasing ewes from the northwest ranges, which are largely of Columbia, Targhee, or Corriedale breeding, would not be satisfactory because few ewes of these breeds mate in the spring. The point is to buy the ewes to fit the operation. If there are options, it is well to learn what they are and the pros and cons of each.

6.3 TYPES OF SHEEP AVAILABLE

Commercial sheep in the United States are seldom purebreds. They may be high grades or a mixture of several breeds. They may be classified by type depending on where they are grown and how they are used. These types are recognized in the industry and considered to differ in their adaptability and thus production capabilities.

6.3.1. Natives Natives are sheep from the midwest farm region of the United States. They may be the result of crossing mutton-type rams on fine- or medium-wooled ewes, or they may be grade Hampshire, Suffolk, or other medium-wooled breeds. Natives may be expected to have medium wool and smooth skin, and they can be either black, mottled, or white faced. They are usually found in flocks of less than 100 ewes.

Strong Points of Natives

1. Prolificacy is usually medium
2. Conformation superior
3. Adapted to midwest environment
4. May be readily available

Weaknesses of Natives

1. May be parasitized
2. May be diseased
3. Some lack thrift and size
4. Sometimes they are not docked
5. Often they lack uniformity in size and conformation
6. Fleeces may be inferior in weight and quality
7. Ewes may be bred when purchased
8. It is difficult to put together a medium to large uniform flock of productive ewes

It should be stated that Natives from flocks that have been carefully bred and managed as those shown in Figure 6.1 may be as productive as any other type, but they are very much in the minority.

6.3.2. Westerns Sheep from the western ranges are called Westerns by local ranchers but Northwesterns by producers in the farm states. They may be called by the state of their origin such as Colorados, Idahos, Montanas, or Wyomings or by an area within the state. Generally, they will be of Columbia, Rambouillet, or Targhee breeding or crosses between the breeds. Most of the ewes will have white faces, but dark or mottled markings of ewes sired by Suffolk or Hampshire rams are common. Most of them will have medium-wool fleeces, but some may grade fine. They usually have open faces. Figure 6.2 shows typical Western ewes.

Strong points of Western Ewes

1. Larger size than other types
2. Uniform in type and wool
3. Shear heavy fleeces

Figure 6.1 High-quality Native ewes.

Figure 6.2 Typical Western ewes.

4. Healthy and free of parasites
5. Hardy, lamb easily, and are excellent mothers
6. Ewes are docile and easily handled
7. Conformation: large frame, medium in muscling, with good feet and legs
8. Prolificacy: medium compared with other types

9. If mated to large meaty mutton rams, they produce superior lambs
10. Ewes are usually available in larger numbers, so uniform groups may be purchased

Weak Points of Western Ewes

1. Yearling ewes are not usually available early in the fall
2. Ewes have a short breeding season
3. Ewes sell at higher prices

6.3.3. Southwesterns The origin of Southwesterns is from states in the southwestern United States. The majority of Southwestern ewes are from Texas, but substantial numbers are available from New Mexico and Arizona. The ewes are also designated by state or even by region in the state. For example, ewes may be called Texas, Del Rio ewes. Breeding of Southwestern ewes is predominently Rambouillet, with many Merinos in Texas. In some of the more mountainous areas, there may be considerable medium-wool blood from the use of Columbia, Targhee, Corriedale, and black-faced (mostly Suffolk) rams. The ewes may be either white or black faced, but most have white faces. Fleeces are usually fine, but may be medium grade. Figure 6.3 shows ewes from southwest Texas.

Strong Points of Southwestern Ewes

1. Hardy, healthy, and adapted to a wide range of climatic and feed conditions
2. Usually have a long, productive life
3. Fleeces are usually heavy and high in quality
4. Long breeding season
5. Lamb easily and are good mothers
6. Available over much of the year
7. Available in large numbers of similar-type ewes
8. Cross well with mutton rams

Weak Points of Southwestern Ewes

1. Lack prolificacy
2. May be small and lack muscling
3. May be wrinkled and wool blind
4. Type may be variable depending on area of origin
5. Ewes from some areas may have heavy infections of internal parasites
6. Sometimes ewes are already bred when purchased
7. Ewes are often wild and not easily handled
8. Teeth may be worn down

Figure 6.3 Typical Southwestern ewes with lambs sired by Targhee rams.

6.4 SOURCES OF BREEDING SHEEP

6.4.1. Commercial ewes Producers usually buy and sell breeding stock by private treaty within their area, but where to purchase satisfactory commercial ewes has always been a problem for the inexperienced buyer. Not only is it difficult to obtain information on the market prices, but it is no small task to locate sheep for sale, negotiate the purchase, and arrange for delivery to the farm or ranch. This is the reason for order buyers, commission firms, and dealers in sheep. Their justification for existence is to provide the expertise so that the buyer receives the kind of sheep needed at a fair market price.

State or area livestock extension specialists and county agents may be the best starting point. The specialist will often be aware of the people or firms to contact. Others who may be helpful are sheep association representatives, sheep and lamb salespeople, and successful producers. Any of these may be able to provide information on a reliable source of ewes or can recommend individuals or organizations who can do the job. Some sources are discussed next.

Sources in the Midwest. Numbers, types, and quality of ewes in the farm states are variable. Unless it is a dispersal sale, it may be impossible to purchase enough Native ewes from any one breeder to establish a flock. If ewes must be obtained in small groups from many sources, there is a danger of putting together a group lacking in uniformity of size, type, and quality. If the buyer is acquainted with the producers in the area and knows their reputation, direct purchases may be quite satisfactory. Other ways of procurement are farm, community, and area auctions. Buyers at such sales are expected to be able to judge the sheep well enough to discriminate against diseased, unsound, and barren ewes. Too often, ewes bought at such sales have been culled because they were unproductive or have some serious defects. There are some exceptions, such as consignment sales, sponsored by sheep associations or other organizations. The ewes offered in consignment sales can be expected to be healthy and sound and to fulfill the conditions of sale; that is, if it is a bred ewe sale, they are pregnant, or if it is an open ewe sale, they are not bred.

There are companies in the Midwest that specialize in buying and selling sheep. These firms purchase ewes from the West and offer them for sale to farmers in the Midwest. Ewes may be bought at the yards or on order. If the firm has a reputation of giving satisfactory service to sheep producers, there are some advantages in dealing with them. These are:

1. The selection may be wider than otherwise possible.
2. The age, health, soundness, and background have been investigated by personnel experienced in handling sheep.
3. Transportation and other costs may be less since they have been handled and shipped in larger groups.
4. They may be purchased in groups of various sizes.
5. The sheep may be available when needed.
6. The buyer knows what the ewes will cost when delivered.

Sources in the West. Western range ewes tend to be marketed from early summer to late fall, after shearing and before breeding. Ewe lambs are usually available in September and October after they come off the ranges in the fall. They may be sold direct or come to a central market and be sold either by auction or by private treaty. Many ewes are sold on contract or through order buyers.

Various organizations are involved in merchandizing ewes. Some states have sheep associations that sell ewes. Independent commission firms and cooperatives have personnel who work with purchasers to locate sheep and negotiate sales. The personnel also make the arrangements for sorting, weighing, watering, feeding, yarding, inspection, and transporation. They are paid on a commission at a designated rate per 100 lb or by the head. There are also order buyers who have, through years of experience, built up clientele in various areas of the country. The order buyer receives instructions from the purchaser and, on that basis, locates the ewes, negotiates the price, provides the services, and ships them. Order buyers also receive a commission for their services.

Some sheep associations in the East and Midwest send a representative to the West to buy ewes. After the agent becomes acquainted with the area, this has been a very satisfactory method of procurement, but it entails considerable travel time and expense, especially during the period the agent is getting acquainted. Unless a considerable volume of ewes is to be purchased, it may be less expensive to use order buyers or dealers. The same can be said for individuals who want to establish contact with ranches. Some ranchers have developed a working relationship with order buyers or dealers and are reluctant to deal directly with individual producers. The inexperienced buyer also is at a disadvantage from the standpoint of knowing the market, trading practices, available services, and transportation.

Southwest. Buying ewes in the Southwest is similar to in Northwest. The inexperienced buyer will usually find it profitable to use the services of reputable commission firms or order buyers until they know the ropes.

6.4.2. Sources of purebred ewes Purchasing purebreds is entirely different from purchasing commercial ewes. There are few public markets where purebreds are sold as foundation stock. Many state breed associations have open and bred ewe sales, but these are few in number and volume for sale is low. Sale by private treaty may be the method of choice for the purchaser. The most important advantage of purchasing purebreds direct is that it is possible to inspect the flock and to observe the environment where the sheep were raised. State and national breed associations can usually provide names and addresses of breeders within the area. Some producer associations maintain lists of available females and are pleased to provide them to prospective buyers. There are breed publications, magazines, and livestock newspapers that carry ads showing breeding sheep for sale. These publications are listed in the appendix. Another possibility is to obtain published reports on production testing association results.

6.4.3. Commercial rams The kind of rams to purchase will depend on the system of production to be used and the breeding of the ewes. If the ram is to be used for siring crossbred commercial ewes, then the source of rams will be entirely different from rams used in terminal crosses for commercial lambs. The producer should have in mind what is expected of the rams and search for that kind.

Some state and county sheep associations hold ewe and ram sales for which the animals have been graded and inspected for soundness and health. Ram test stations usually hold sales at the end of each test period. Sheep associations may also have lists of available rams, which can be requested. Purebred breeders may have suitable rams for sale in the area.

Rams can be purchased at local farm or community sales. Rams are also sold at central markets. These rams may appear to be bargains, but the producer should be cautious about using them. In such sales, the rams may be sterile, diseased, or unsound or lack libido. There is usually a good reason why desirable appearing rams are on the market. Many unwary producers have discovered too late that the "cheap" ram was, in reality, far too expensive when he failed to settle ewes or brought disease to the flock.

It is well to remember that the lamb crop will carry 50% of their genes from the rams used on the flock. If the producer wants improved rate and efficiency of gains and carcass quality, the quickest and most effective way to do it is through wise selection of superior purebred rams. Breeders who carry on intelligent improvement programs should be rewarded for their efforts. Their rams usually cost more, but they may be worth the price. Elsewhere, the value of using crossbred rams has been discussed. If crossbreds are desired, they should be sired by superior purebred rams and out of superior purebred or high-grade ewes.

6.4.4. Rams for the purebred flocks Sources of purebred rams are many, considering the many flocks of sheep in the country. Not so easily answered is the question of where can a superior ram be bought. Most breeders of registered stock have rams for sale. They may allow buyers to inspect the flock and pick rams for purchase, but many breeders plan to sell their rams through a yearly production sale or consign them to ram sales. Some breeders have more than one sale each year.

Other than private treaty and production sales, the following are types of sales where rams can be bought:

1. County, district, state, and national consignment sales: The National Wool Growers national ram sale is the oldest in the nation. They sell both stud and commercial rams at their annual sale at Salt Lake City, Utah.
2. Regional and national stud ram and ewe sales: for example, the largest in the United States is the Midwest Stud Ram Sale at Sedalia, Missouri. Others are the Eastern Stud Ram Sale at Staunton, Virginia, the Keystone Sale in Pennsylvania, the New England Show and Sale, the Black and White Sale in Oklahoma, and the California Ram Sale.
3. Sale of rams from test stations: Most consignment sales include a show so that buyers can compare the sheep in the same class. Information given on the sheep includes the name and number of the sire and dam, birth date, and whether the ram was a single or twin. Sometimes notes include show ring records of the sire or dam and the individual. They may also include information on selling prices of any near relatives. In some sales, rams are weighed and the information given on sale day. Ram test stations usually have sales at the completion of test periods. Breeders have the option of selling or keeping the rams. For those who wish to sell, there is usually an auction. Information on each ram is summarized and given to the buyers, and there is a period for inspection before the sale begins. Information provided usually includes the birth date and weight, weaning weight, beginning and ending weights at the station, and feed required per gain. It may also include loin eye area, carcass, and other data on half-sibs.

6.5 JUDGING SHEEP

6.5.1. Value of judging Many animal scientists believe the practice of judging sheep is time wasted. They point to production testing of ewe flocks, testing rams in official stations, and carcass evaluation as much more meaningful than visual observation. Their arguments have some validity if it is necessary to choose between judging and other means of evaluation. That, however, is not the case. Both methods should be used, and in so doing the whole process can be improved. There are many sheep shows in existence. The bulk of sheep are bought and sold on the basis of visual appraisal of characteristics of individuals. Good sheep breeders still use eye evaluation in their programs.

Some important benefits of judging sheep are:

1. Eye appraisal is the only way to evaluate such characteristics as breed type and character, sex character, balance and symmetry, constitution, structural soundness, and evidence of disease and parasite infections.
2. The person who has access to carcass and other data may, through practice, become quite adept at estimating the carcass cutability of lambs.
3. The person who is trained in judging tends to observe livestock more closely and thus may be a better shepherd.

6.5.2. Judging procedure In practice, judging may be done informally in the barn lot and pasture or formally in the show ring. Lambs or sheep at an auction or central market are bought and sold on the basis of very brief visual appraisal. In shows and judging contests, sheep are usually restrained either by persons or equipment so that they may be examined in detail. It is best to examine the animal systematically. The following is a procedure that should ensure that the judge has thoroughly observed the animals.

1. First look at the sheep from a distance of at least 15 ft. A side view is perhaps best. It is desirable to start either at the front or rear and follow the outline of the sheep. The top line, including the rump, loin, and back should be level. The neck should blend smoothly into the shoulders. Examine the head carefully. Look for soundness of the mouth and eyes and for sex and breed character. The front legs should be straight. Look at the shanks for substance (bone) and length. The chest should be deep. Follow the under line. It should be trim in a market lamb, but fuller in a ewe. Inspect the set of the hind feet and legs and the development of the hind leg. Overall, do the parts appear to fit together? Visualize the sheep with regard to size, length, depth, smoothness, substance, and balance and symmetry. Does the sheep have breed and sex character?

2. Inspect the sheep from the front. Look for width over the shoulders and through the floor of the chest and the distance between the two. Check the spring of ribs; that is, do the ribs arch from the vertebrae like a barrel or tend to be more vertical? Check the straightness of the front legs. Do they set on the corners or on the inside? Observe the head for width through the muzzle strength and breed type and sex character.

3. Examine the sheep from the rear. Look down the back for width of the neck, over the shoulders, back, loin, and rump. Notice the smoothness with which the parts blend together. Inspect the thickness of the thighs at the bulge of the hind leg. Check the depth from the top of the rump to the junction of the two hind legs. This is the twist, and it indicates the amount of finish the sheep is carrying. Inspect the straightness of the legs. The hind legs should also be under the corners of the body. A plumb line dropped from the pinbones should touch the point of the hock and follow down the back of the hind shank to the feet. The toes should point forward.

4. Handle the sheep. Most beginning judges handle sheep more than necessary. Unless the sheep has a heavy fleece, most of the handling should be done to confirm visual observations. Start from the rear to one side of the sheep. First, feel the width of the dock. Use the thumb and index finger. Hold fingers together and use the first joint of the fingers for feeling the back. Use only enough pressure so that the sheep can be felt through the wool. Perhaps the best description is to press firmly, but not hard. Start from the dock and follow the vertebrae to the loin. (Note whether the rump is sloping or level.) At the loin, open the hand and span the loin for width. Apply some downward pressure to determine strength and thickness. Continue feeling down the vertebrae past the back with the hand to determine the levelness and strength of the top line. Move the hand in short steps, pressing firmly on the vertebrae. Also, move the hand from side to side to find out how much the vertebrae are covered with fat. Open the hand at the top of the shoulders to span the shoulders. Move the hand back and forth to determine the smoothness over the shoulders. How do they blend into the back? Continue moving from the top of the shoulders to the neck to see how well they fit together. Open the hand to gauge the width of the neck. Inspect the horn pits, eyes, and mouth. Check the bite and condition of the teeth. At close range, inspect the width of the muzzle and covering on the face. Using both hands, examine the depth of the chest. Check the width of the brisket. Stand beside the shoulder facing forward and using both hands determine the width through the shoulders. Move back slightly and feel the width through the chest and spring of rib. Check width of shoulders, loin, rump, and through the thighs. Feel the size of the leg. Then inspect the testicles of the ram and teats and udder of a ewe. Last, but not least, examine the wool on the outside of the thigh, midway on the side and at the shoulder. To examine the wool, keep fingers together and gently part the fleece at a natural break.

5. Make a last inspection from a distance and reach a decision.

6.6 DESIRABLE CHARACTERISTICS OF EWES

Breeding ewes should be capable of performing the following functions:

1. Reproduce regularly at specified times
2. Be prolific and lamb without difficulty
3. Be a good mother and rear her lambs
4. Produce rapid, efficient gaining lambs that meet market demands
5. Maintain herself on pasture and roughage, except during late gestation and lactation when concentrates are needed
6. Produce a heavy high-quality fleece
7. Have a long, productive life

It is not possible to determine these by inspection, but there are some characteristics that should contribute to the ability of the ewe to do the job expected of her. Close inspection of individuals should be made before purchasing.

Points to observe in individual ewes are discussed next.

Figure 6.4 Contrast between desirable and undesirable ewes. The one on the left is undersized, weak, and unthrifty compared to the larger, healthier, heavier muscled ewe on the right.

6.6.1. Health Satisfactory breeding ewes should be thrifty and vigorous (Figure 6.4). Ewes that are gaunt, stand with head down, and are the "tail enders" when they are driven are probably sick or full of internal parasites. Other signs of parasitism are anemia, "bottle jaw," and scouring. Ewes with inflamed, weeping eyes may have pinkeye. Usually individuals with excessive discharge from the nostrils, chronic cough, or thick breathing are good candidates for respiratory problems. Lame sheep should be inspected for signs of foot rot.

If ewes are being purchased in unfamiliar territory, the buyer should investigate the possibility of problems with blue-tongue, enzootic abortion (EAE), or progressive pneumonia. Sheep that have been through a number of sales and other stress may be exposed to soremouth and shipping fever. If some individuals in a group have a fever and are depressed, it is reason enough not to buy any of them.

6.6.2. Soundness The teats and udder should be examined to determine if they are sound. The ewe must have two normal teats to suckle more than one lamb. Sometimes the end of teats have been clipped off by careless shearers. Such a teat is never functional again. Teats may be too big for the lamb to suckle. The udder of a dry ewe should be well attached, not pendulous and pliable when handled. The presence of lumps or a caked udder indicates the ewe has had an infection (Figures 6.5, 6.6, and 6.7). Small lumps found in the udder of a dry ewe become much enlarged when the ewe lambs and may be infected and not functional. Ewes with pendulous, lumpy, or caked udders should never be selected.

The abdomen should be inspected for sagging or breaking away. If the ewe has such a problem when she is open, it will be multiplied when she is pregnant.

Figure 6.5 Defect of the mammary system—pendulous udder.

Figure 6.6 Defect of the mammary system—greatly enlarged teat.

Figure 6.7 Defect of the mammary system—spoiled udder.

Normal feet and legs are necessary for a ewe to graze and be productive. Deformed toes, exceptionally weak pasterns, or hocks too straight (postlegged) or set too close together with the feet turned out (called cowhocked) are all defects to be avoided. Any swelling or puffiness about joints is an indication of future trouble, and animals with these defects should be sorted off.

The head should be examined to see that the eyes are clear and healthy. The incisor teeth should meet the dental pad. Ewes with overshot or undershot jaws on crooked jaws will not function normally and should be eliminated.

6.6.3. Size and condition Ewes with larger frames tend to have a greater capacity for feed, produce more milk and wool, and lamb easier; their lambs grow faster than fat, lowset, and short ewes. There is also a linear relationship between body weight and twinning up to 155 lb. For each increase of 10 lb of body weight of ewes, twinning tends to increase 6%.

In studies by Joseph (1931) of the Montana Experiment Station, it was observed that each increase of 1 lb of yearling ewe weight above flock average resulted in a ½-lb increase in lamb produced each year of her life. Generally, large ewes can be maintained more efficiently than small ewes since they have proportionately less surface area than small ewes per unit of body weight.

The subject of ewe size is the topic of much debate with very few answers. It appears that neither the very large nor small ewe is the most efficient. Size, when measured by body weight, is misleading unless attention is paid to the amount of fat on the ewe. Except at the end of gestation and in early lactation, ewes should not be fat. In older ewes, degree of fatness is an excellent clue to past productivity. Fat, older ewes are often barren and thus should be viewed with suspicion.

6.6.4. Fleece characteristics The nutritional requirements for wool growth are not great. Production of a heavy, high-quality fleece requires very little more feed and care for the ewe than a lighter, poor-quality fleece. Even though the value of the fleece in late years is much less than lamb, it is still a substantial source of income. Ewes with off-colored, poor-quality and lightweight fleeces should not be selected. If records on the parent flock are available, study them. The fleece weight and quality grades are important information.

6.6.5. Age Age is important because it indicates the extent of the productive life. However, the younger the ewe when she lambs, the more difficult it may be for her to give birth to lambs. Ewe lambs and yearlings have much more lambing problems than older, more mature ewes. Old ewes (beyond 7 years) may also have difficulties as their teeth begin to be a problem. The inexperienced producer should seriously consider buying 2- and 3-year-old ewes rather than starting with ewe lambs or yearlings. Price may very well determine the choice since 2- and 3-year-old ewes are usually more expensive than yearlings. Sometimes 5- and 6-year-old range ewes that ranchers consider to be too old for their purposes can be bought very reasonably and used for several years in farm states. These ewes are often very satisfactory for beginning producers.

6.6.6. Type It is well to stick to one type when purchasing ewes. The crossbred wool types have many good qualities for production of early lambs in the Midwest, but they would not be nearly so well adapted to fall or accelerated lambing systems. Generally, the Northwest ewe does not adapt well to ranges in the Southwest. It has been observed that ewes from warmer climates adapt more easily to colder conditions than the reverse.

6.6.7. Conformation Any commercial ewe should be structurally sound. She should stand on reasonably straight feet and legs, with plenty of bone. Her back should be strong and straight, and she should have some width and depth to her chest and indication of body capacity. While it is true the commerical ewe is not expected to contribute greatly to the muscling of her lambs, she should indicate some development in that characteristic. If the ewes are carrying much wool, handle them to determine their width of the rump, loin, back, and over the shoulders. The size of the leg can be determined by spanning it with two hands at the middle of the leg.

A commercial ewe should be refined and feminine about the head, neck, and shoulders (coarse, staggy ewes are inclined to be poor breeders). The ewe, however, should have a good constitution as indicated by the width of the muzzle, chest, depth of chest, and spring of ribs. Weak heads, narrow bodies, light bones, rough shoulders, weak loins, emaciated, unsound, and diseased animals and poor fleeces are all defects to avoid in purchasing ewes.

Since high-quality, purebred ewes can be expected to sell for much higher prices than those used for commercial purposes, it is desirable to inspect them in greater detail. Ewes of ewe breeds should be handled as described in judging sheep (Section 6.5.5). They should be smooth in their lines, well balanced, adequate in size, correct in the set of their feet and legs, and show strong constitution and be moderately muscled. They should not carry excessive fat.

Purebred ewes of the ram breeds should be expected to possess the same qualities as those of the ewe breeds except that much more emphasis should be placed on muscling in the hindquarters, loin, and shoulders.

6.6.8. Temperament As discussed in the section on ewe types, there is usually a difference in temperament of sheep produced in different regions of the country. Generally, those from Texas or the Southwest have grazed on fenced ranges and been worked by herders on horseback. They are, therefore, inclined to be wild and easily excited when approached by someone on foot. This characteristic is not nearly so prevalent in Western and Native ewes. It should be said that Southwestern ewes, if handled gently and without pressing them, can be tamed and will become tractable. There are marked differences in the temperament of individuals in groups of sheep. Some differences are to be expected and can be tolerated. The exceptional individual merits attention. Extremely nervous ewes are not usually good producers. They also may exert a bad influence on the whole group. One such ewe can actually cause the flock to be hard to handle. This is the one that makes the break when they are being driven and leads others with her. Such exceptional individuals will always be the last one in the pen if ewes are being hauled and the hardest to catch. These ewes should be spotted and eliminated. Temperament is even more important in purebred ewes.

While there is little scientific data to prove nervousness is inherited, experienced shepherds are certain that there is some basis for believing it is. It is much more desirable to have manageable sheep than those with problems. Thus, it is advisable to eliminate problem sheep from the flock as soon as they are recognized.

6.6.9. Uniformity When replacements are purchased, it is desirable to buy sheep that are similar in age, size, type, and quality. This simplifies feeding, management, purchasing sires, and handling wool. It also may result in more uniform lambs when they are marketed. Another factor, which is rather intangible, has to do with the effect of appearance of livestock on the attitude of personnel working with them. Personnel generally take better care of livestock that are pleasing to them.

6.7 DESIRABLE CHARACTERISTICS OF RAMS

The old adage that the sire is half of the flock is as true now as it was when it was coined by ancient sheep breeders. The efforts spent in the selection of rams are probably unequaled in importance by any other activity the producer can perform. In practice, 80% to 90% of the improvement by breeding comes through the ram (Figure 6.8).

Figure 6.8 Influence of the sire on conformation of progeny. Two rams are shown. A typical lamb sired by each is shown below its sire. The ram and the lamb on the left in both pictures are purebreds and the ram and the lamb on the right in both pictures are scrubs. Note the difference in muscling between rams and lambs.

If the production plan has been well conceived and the objectives are clearly in mind, the rams should be selected to do the job. Besides furnishing the genetic material expected by the sire, the ram must be sexually aggressive, fertile, and healthy. A number of factors, discussed next, determine whether a ram can function properly. It is highly desirable to evaluate the ram for these either by inspection or any other means available.

6.7.1. Health Symptoms of disease, such as a fever, respiratory difficulties, bloat, inflamed eyes, soremouth, scours, foot rot, or infections of the reproductive system, are excellent reasons for rejecting a ram. Only select rams that are vigorous, with clear eyes and healthy skin. An enlarged, swollen, or puffy scrotum may be a symptom of epididymitis. Unvaccinated rams may also be suffering from soremouth. A ram with a fever will most probably be sterile for weeks. It pays to be cautious and investigate the flock health before buying a ram.

6.7.2. Soundness Any ram used for breeding should be sound in the reproductive organs. He should have two well-developed testicles that are approximately the same size, carried descended into the scrotum, and of firm, not hard or soft, texture. Enlarged, pulpy testicles are often not functional. Yearling rams should have a minimum testicular circumference of 32 cm. The penis should be normal in size and shape and clear of infection. Look for correct set of feet and legs. Hocks that are too straight are a serious defect in rams. Postlegged rams are prone to knuckle over and be unable to service a ewe. Inspect for strength of pasterns and normal-shaped feet. Rams with enlarged, swollen joints should be rejected. Inspect eyes and mouth. Check to see if the incisors are normal in shape and meet the dental pad properly. Never purchase an unsound stud ram.

6.7.3. Size and condition With the present emphasis on growth rate and efficiency, larger animals have a distinct advantage. Usually, it is desirable to use large rams to increase the size in the offspring. Size as measured by body weight alone is not necessarily a desirable trait. If, in addition to being large, the ram is also lacking in quality and smoothness, he may not be desirable. The ram also may weigh heavy because he is quite fat. Rams for siring commercial lambs should be well grown, with a large frame, and be heavily muscled. A ram that is ready for service should be in medium condition.

Purebred rams for both the ewe and ram breeds should meet the specifications of the breed for size. If the ram has quality, smoothness, and the other desirable traits, few are discriminated against for being too big. On the other hand, nearly all breeds are quite critical of rams that are too small.

6.7.4. Maturity The usefulness of a ram may be severely impaired by immaturity. It is especially important for ram lambs to be well developed before they are used. A ram lamb by 8 to 10 months of age should be mature enough for limited service. At that age, he should be masculine, as shown by the development of the head, neck and shoulders.

6.7.5. Age The optimum age for rams is 2 to 5 years. Ram lambs can be expected to breed a limited number of ewes depending on their size or maturity. Yearling rams should also be expected to service fewer ewes than mature ones. Rams older than 7 years should also be given fewer ewes. Most breeders buy yearling rams because they want more years of service.

It is possible, sometimes, to purchase a proven sire from a breeder who has saved many daughters from an outstanding ram. This may be a means for a young purebred breeder to made extraordinary progress in a very short time.

6.7.6. Type The type of ram to buy depends on the operation and the kind of ewes in the flock. In a commercial ewe-lamb operation, the ram should be of mutton type. In purebred flocks, the ram should be true to the breed type.

6.7.7. Conformation In commercial operations, the meatiness and rate and efficiency of gains of lambs should be inherited from the ram. Therefore, the ram should have adequate size to provide growth rate and should have thick, heavy muscles through the hindquarters and over the rump, loin, back, and shoulders. The ram should be uniform in his lines and have quality in his bones and joints. The ram should not have an excessively large head on coarse shoulders. Ewes may have difficulty delivering lambs out of such rams.

The purebred ram should be outstanding in conformation, depending on the emphasis by the breed association. Regardless of the breed, the ram should be smooth and well balanced, show masculinity about the head and shoulders, stand correctly on legs with adequate substance, and have a strong constitution.

6.7.8. Growth and efficiency Judging sheep may give some indication of the growth potential, but the only way to know for sure what the ram is capable of is to look at records. Birth, weaning, and yearling weights are all useful in evaluating his growth rate and estimating what may be transmitted to his offspring. Ram test station results are invaluable to the purchaser if they are available. On the farm testing records are useful in comparing the ram with others from the same flock and should be studied. It is also helpful if data are available on the sire and dam of the ram being considered.

6.7.9. Prolificacy While it is true that prolificacy has a low heritability, the trait can be improved by selection. Choosing rams that were born twins or triplets from a line of prolific ancestors should result in a marked increase in lambing percent in a flock with a low lambing rate. Progress can be further enhanced if some attention is given the trait in selecting replacement ewes. The decision to only use rams from multiple births is not at all easy to make by the purebred breeder at a consignment sale who observes that the winners of the ram lamb and yearling classes are often from single births. Nevertheless, the breeder who wants to make a real contribution to the sheep industry will do well to follow such a policy. Flock, as well as sire and dam records if they are available, should be studied carefully.

6.7.10. Fertility There are some indications of fertility of the ram. Physical characteristics such as cryptorchidism, swollen, feverish testicles, extremely small testicles, infections of the penis or sheath, or high fever can have a direct effect on fertility. If the rams have been inspected and have health certificates, it may be enough assurance that they are breeders. Most purebred breeders guarantee their rams are fertile and will copulate. Occasionally, breeders will have had semen evaluation made on their rams. Sometimes, yearling rams have been used for breeding as lambs. Such information is helpful but it is no sure guarantee the ram will be a satisfactory breeder. The safest procedure is to make sure the ram is healthy and sound and bought from a reputable breeder who will stand behind his sheep.

6.7.11. Temperament Temperament of ewes was discussed in Section 6.5.8. The same things apply to rams. The ram should be alert, vigorous, and bolder in its actions than ewes. Nervousness, wildness, timidity, and sluggishness are all defects that should detract from the value of the ram.

6.8 WHEN TO BUY SHEEP

If open ewes are to be purchased, it is desirable, if possible, to buy them 4 to 6 weeks before the breeding season. Then they can be wormed, tagged, or sheared, their feet can be trimmed, and they can be flushed. This also has the advantage that they may be less likely to be bred when they are delivered. Unfortunately, the ewes may not be available at the most opportune time for the producer. Under those circumstances, they may be contracted ahead for delivery at a specified time. If older ewes can be inspected before the lambs are weaned, it is an excellent opportunity to evaluate their productivity. If that is impossible, select ewes with the most udder development and those that are suckled down in condition. Dates of ewe and ram sales are usually set well before the breeding season.

Rams should also be procured far enough ahead of time so they can be prepared for breeding.

6.9 HOW TO BUY SHEEP

Buying sheep is as much an art as selling them. There are no deep secrets regarding the way to obtain good sheep worth the money. Anyone with a good eye for livestock and sound judgment who is willing to work at the task can become a competent buyer. The following suggestions may be beneficial:

1. Know what is needed.
2. Talk with dealers, order buyers, other producers, and livestock extension specialists to find out what is available and the asking price.
3. Read livestock papers, magazines, and market reports to get a broader view of sheep markets. The market for breeding ewes is related to the prices of slaughter lambs and ewes.
4. If range ewes are to be purchased, it may be useful to pay a visit to the area. For example, if the production system requires Southwestern ewes, visit the San Angelo, Texas, auction in July or August and watch ewes being sold. If Western

ewes are needed, visit sheep country in Colorado, Wyoming, Idaho, or Montana in summer or early fall. Get acquainted with extension personnel, dealers, order buyers, and others. A lot can be learned about a market that will be useful later. A competent order buyer will help the producer to meet knowledgeable people.

5. Good traders are usually deliberate. The buyer who is too eager, impolite, or "pushy" seldom gets the best prices. Buyers who reveal immediately what price they are prepared to pay will seldom get sheep for less. It is well to leave some room for bargaining. It is usually good strategy to give the salesperson some time to think about any offer that has been made. It is also good tactics when the negotiations are nearing a deal to stop and think about it.

6. Many livestock are bought and sold by word of mouth. It is a point of honor for buyers and sellers to back their word. Anyone who violates those rules will soon find it difficult or impossible to deal with reputable individuals or firms. It is unwise to have any dealings at all with individuals or firms with questionable reputations.

7. Bargains are usually dangerous. If sheep are offered for sale at unusually low prices, it is advisable to proceed very cautiously if at all. The basis for most swindles is to convince the victims that they are getting something for nothing. If in doubt, a consultant who knows the situation should be used. Most sheep people are honest and will not take advantage of a "greenhorn," but among the human flock there are predators who are on the lookout for the unwary. They specialize in shearing the innocent lambs who stumble into their clutches.

8. Buying at auction is different than face to face trading. Part of the psychology of an auction is to create enough excitement that the buyer will get caught up in the action and pay more money for the sheep. Some suggestions for strategy are:

a. Study data on sheep well ahead of time if possible. Know something about the breeding program and the reputation of the seller.

b. Inspect the sheep at the show in detail before the sale.

c. Obtain a sale order, buyer's number, and conditions of the sale.

d. Decide what sheep are suitable and how high to bid. Stay with your decision during the sale.

e. Either talk to the auctioneer or the persons working the ring and indicate your bidding signal. There is an advantage in maintaining anonymity.

f. Don't reveal animals you are interested in buying or how much you would pay.

g. Take your time. You may miss some, but you may also buy others worth the money.

h. Remember that often the best buys are early in the sale.

6.10 HOW MUCH SHOULD BE PAID FOR BREEDING STOCK?

There is no set formula for how much to pay for sheep. A lot will depend on the availability of breeding animals, the live lamb market, the weather conditions in the country, the general price level, and other factors such as international conditions. If war is imminent, the price of sheep usually advances rapidly.

Besides the general factors mentioned previously, such specific factors as the age, size, quality, and breed affect the price. So does the reputation of the breeder, the improvement programs used, and the level of management practiced.

One possible method of estimating the price to pay for yearling ewes is to relate it to the value of market lambs. This may vary from 100% to 200% of the selling price of a choice market lamb. Another rule sometimes used is to pay as much as the ewe produces in lamb and wool the first year. Ewe lambs satisfactory in size and finish may be bought for breeding purposes either through order buyers, dealers, or direct. If the ewe lambs must be sorted from the wethers or if there is a privilege of sorting off any which are undesirable, there may be an extra charge.

If the ewes are 5 to 6 years old, they should produce two-lamb crops, but the attrition rate is very high on aged ewes. For purposes of estimation, assume one-third of them will be gone from the farm within two years. Estimate they will have a salvage value equal to that of slaughter ewes.

Commercial rams are relatively inexpensive compared with the price of ewes, but production-bred rams are very difficult to find and thus would be more expensive. It is probable that a producer would be justified in paying as much for such rams as the value of 8 to 10 ewes. In purebred operations, the breeder may be justified in paying even more for rams than the commercial producer must spend.

REFERENCES

Collins, S. B., *Profitable Sheep*. New York: Macmillan, Inc., 1956.

Ensminger, M. E., *Sheep and Goat Science*. Danville, Ill.: Interstate Publishers and Printers, Inc., 1986.

Joseph, W. E. "Relation of Size of Grade Fine Wool Ewes to Their Production," *Montana Bulletin*, 242, 1931.

Kammlade, W. G., Sr., and W. G. Kammlade, Jr., *Sheep Science*. Philadelphia: J. B. Lippincott Company, 1955.

Parker, Ron, *The Sheep Book*. New York: Charles Scribner's Sons, 1983.

Simmons, Paula, *Raising Sheep the Modern Way*. Charlotte, Vt.: Gardenway Publishing, 1976.

QUESTIONS FOR STUDY AND DISCUSSION

1. What special knowledge and skills are needed to do a superior job of purchasing breeding sheep?
2. Why is there a difference in the kind of ewes required for different systems of production?
3. What characteristics are used to differentiate Native ewes from Westerns?
4. Where are most of the Natives produced in the United States?
5. Rank Native, Western, and Southwestern ewes for the following: prolificacy, mutton conformation, fleece weight, freedom from parasites, and possibility of obtaining a uniform flock of ewes.
6. Ewes with the longest breeding season may be found in which part of the United States?
7. What are the dangers of buying ewes at local sale barns?
8. What are some advantages of purchasing breeding ewes from an established breeder?
9. What contacts are possible in locating breeding sheep?
10. What services can be expected from Western order buyers?
11. Discuss the reasons for using purebred rams on commercial ewes.

12. What special value should there be in purchasing rams from a test station sale?
13. Justify the statement that judging sheep is still useful.
14. What is meant by the following terms: breed type, sex character, balance and symmetry, muscling, unsound feet and legs, caked udder, lacking in constitution, postlegged, undershot jaw?
15. What functions should all commercial ewes be capable of performing?
16. Discuss some advantages of larger size in ewes.
17. What is meant by temperament, and why is it important?
18. What are some advantages of stressing uniformity in putting together a flock of sheep?
19. Why pay a lot of attention to the soundness of feet and legs in a ram?
20. In what ways can a buyer increase the growth and efficiency in a flock of sheep?
21. What are indications a ram may not be fertile?
22. When is the best time to buy ewes?
23. How much should you pay for ewes and rams?

7 Physiology of Reproduction of Sheep

7.1 IMPORTANCE OF REPRODUCTION

No other single factor is more influential to the success of a sheep enterprise than reproduction, not only from the standpoint of regularity but also the rate and persistence over a long lifetime. It provides the base on which a sheep operation begins, and it affects most aspects of the financial returns.

Reproduction is a natural physiological function that appears simple to the uninitiated but is in reality very complex when the various factors controlling the functions are known and understood. It is not the purpose of this book to delve deeply into the science of reproductive physiology, but to provide enough of the principles involved to allow the reader to understand the basics of the process and to be able to use them to advantage in practical production.

It has been wisely said that lambing time is the shepherd's harvest. It is equally true that the breeding season is the time of seeding. "As a man sows, so shall he reap" can be applied as much to a sheep operation as to a grain crop. An understanding of sheep reproduction is a prime requisite for success with the species. Ewes that produce and rear two or more lambs are much more efficient than those that raise one lamb because many basic expenses are the same for the ewe that rears single or multiple births. Feed costs, which represent the greatest single expense of production, are only 30% more for ewes producing twins than singles.

There is a great opportunity to improve the reproductive performance of ewes. The average lambing rate in the United States is less than one lamb per ewe per year. It is higher in the farm states than those of the western ranges, but it is still much too low. If one lamb per ewe yearly is not satisfactory, then what is the optimum? At least one author, Owen of England, has stated that ewes lambing at 8-month intervals should be capable of producing 2.75 lambs per year.

How well you do in producing and saving lambs will, to a great extent, determine your success or failure in the sheep business. Whether you attain exceptional results or merely exceed the average, it all starts with the breeding season.

7.2 MEASURES OF BREEDING EFFICIENCY

Since the ability to produce offspring is basic to the success of a practical sheep operation, it is of great importance for the sheepbreeder to record and carefully analyze individual and flock reproductive performance. This information can then be related to past performance as a measure of improvement, to other similar operations to indicate achievement, and to set standards or objectives for the future. Information on individual ewe performance is an excellent basis for culling the flock.

What records are to be kept will depend on the individual operator and the use made of them. The following are some factors that may be considered in determining reproductive efficiency.

7.2.1. Lambing percentage Lambing percentage may be expressed as lambs born per ewe exposed in the case of a single lamb crop per year or, in the case of multiple lamb crops such as accelerated lambing (8-month intervals), on a yearly basis. It may also be expressed as lambs weaned per ewe exposed.

7.2.2. Prolificacy The number of lambs born per parturition is defined as prolificacy. This may be calculated for a single lamb crop or on the lifetime production of the ewe.

7.2.3. Lambing interval Lambing interval is the period between lamb crops and is also called the breeding frequency. It is important on a flock or individual ewe basis if improvement is to be made in out of season lambing.

7.2.4. Conception rate Conception rate is based on the percentage of ewes exposed that conceive. To date, there does not appear to be a practical means of determining whether or not a ewe has at some time been pregnant except by surgery. However, laparotomy has been used for that purpose on an experimental basis. Also used in research is a technique that measures blood levels of progesterone. Neither procedure is projected to be within the reach of a practical producer within the foreseeable future. It is possible to obtain an estimate of the conception rate on the basis of a pregnancy test 60 days after breeding. But this test does not take into account the ewes that conceived and aborted before they were tested.

7.2.5. Ovulation rate The number of ova shed per ewe may be estimated by laparotomy and examining the ovaries. It has been shown to be valuable in estimating the future productivity of ewe lambs.

7.3 REPRODUCTIVE ORGANS OF THE EWE

The reproductive organs of the ewe are composed of the ovaries, oviducts, uterus, cervix, vagina, and vulva.

7.3.1. Ovaries The ewe has two ovaries located in the abdominal cavity. The ovary is almond shaped with an average weight of 3 to 4 grams, but it varies in size with the stage of the estrous cycle. Cells in the cortex of the ovary are capable of developing into an ova. The ovary has two functions, to generate ova and to provide hormones for the support of reproduction.

7.3.2. Oviducts The oviducts serve the purpose of transporting and maturing the ova from the ovary to the site of fertilization, and they convey the spermatozoa in the other direction to the ova. Then the fertilized egg passes through them to the uterus. The opening into an oviduct, the infundibulum, is funnel shaped and lies near the ovary. When ovulation occurs, it receives the ova. The oviduct is quite convoluted in a sheep. The middle part of the duct is called the ampulla, and it is there that fertilization occurs. The ampulla merges with a constricted section known as the isthmus, which is joined to the horn of the uterus.

7.3.3. Uterus The functions of the uterus are to transport sperm from the vagina to the oviducts, to secrete hormones that maintain the corpus luteum and thus the pregnancy, to support and nurture the fetus, and to initiate and provide a passage for the fetus at parturition.

The uterus consists of two horns attached to the body and a cervix that connects with the vagina. The uterus contracts by peristalsis to move sperm to the oviducts. The contractions are controlled by estrogen and are greatest during the height of estrus.

7.3.4. Cervix The cervix is a sphincterlike structure that protects the uterus except in estrus or parturition, when it is relaxed. The cervix also facilitates the movement of sperm from the vagina. It is a reservoir for sperm, and it apparently functions to prevent transport of nonviable and defective sperm. The cervix is tightly closed and sealed during pregnancy.

7.3.5. Vagina The vagina is the copulatory organ where semen are deposited by the ram. The seminal plasma is not transported into the cervix, but is absorbed or expelled. The environment of the vagina is not favorable for the survival of sperm until a combination of cervical, vaginal, and seminal fluids provide a suitable buffering effect. The vagina serves as an excretory duct for secretions of the reproductive tract and urine and as the birth canal for the fetus.

7.3.6. Vulva The vulva is the external sexual organ of the ewe. It consists of the labia, clitoris, and vestibule. The junction of the vagina and the vestibule is the location of the opening to the urethra. The clitoris corresponds to the penis in a ram and is quite sensitive when the ewe is in heat. The vulva is the external opening of the reproductive tract through which secretions and urine are voided. It is the passage used for entrance of the penis during copulation and the end of the birth canal for the fetus at parturition.

7.4 REPRODUCTIVE ORGANS OF THE RAM

The reproductive organs of the ram consist of the testicles, epididymis, vas deferens, penis, and accessory sex organs.

7.4.1. Testicles The testicles generate the male gamete, spermatozoa, and secrete hormones necessary for reproduction in the male. The male sheep has two testicles that are normally carried in a sac called the scrotum outside the abdominal cavity of the body. The scrotum is muscular and can be retracted and released at the will of the ram. The walls of the scrotum contain many sweat glands and very little fatty tissue. The scrotal temperature is usually 5° to 7°F (2 to 3°C) cooler than the internal body temperature. In cold weather, the testicles can be retracted near the body, and in high temperatures they can be relaxed to provide greater surface area for cooling. Rams that have retained their testicles within their body cavity (cryptorchids) are invariably sterile.

The testicle is made up of numerous tubules separated by interstitial cells. The cells lining the tubules are germinal and capable of developing into the male gamete, spermatozoa, which carry the genetic material of the ram. The tubules connect with the convoluted tube at the posterior end of the testicles. This tube is called the epididymis.

7.4.2. Epididymis This tube functions as a passage for the sperm, and it also provides storage during the 10 to 13 days the sperm remain there. During the storage, they are matured and activated so that they will be motile when ejaculated. The epididymis joins the vas deferens, which leads to the urethra and the penis.

7.4.3. Function of interstitial cells The interstitial cells in the testicle secrete a hormone called testosterone that causes the development of sex organs, the secondary sex characteristics, and the male sex drive or libido.

Size of testicles of a ram relative to stature is greater than the male of any farm species with the possible exception of the Angora goat. They may represent 0.25% of the body weight compared with 0.05% for the bull or 0.1% for a boar. This appears to be the reason a ram can breed a large number of ewes in a short period. Weights of the testicles may vary from 250 to 400 grams. The shape of the testicle is oval and carried upright in the scrotum. Average dimensions according to Sorenson (1979) are 3.9 in. (10 cm) from top to bottom, 2.3 in. (6 cm) wide, and 2 in. (5 cm) thick as viewed from the rear.

7.4.4. Penis The penis is the organ used by the ram to deposit semen in the vagina of the ewe. It is also the passage for urination. It is formed in the shape of an S curve, which is 7.8 to 9.8 in. (20 to 25 cm) in length in the nonerect organ and 2.73 to 3.9 in. (7 to 10 cm) longer in the erect state. Figure 7.1 shows a diagram of the male and female reproductive organs. The S curve in the penis is known as the sigmoid flexure. Muscles allow the penis to straighten and thus thrust outside the body cavity. There are also retractor muscles for returning it to the body cavity. The external part of the penis is known as the glans penis. Figure 7.2 shows the shape of the organ. The glans penis contains erectile tissue that becomes engorged with blood to increase its size and stiffen it for service. The urethral process extends beyond the end of the glans penis by 1.17 to 1.56 in. (3 to 4 cm.) This is known as the filiform appendage or sometimes as the vermiform process because of its resemblance to a worm. The exact function of the filiform appendage is unclear, although some authorities say it allows the ram to deposit semen into the cervix. It is known that rams are fertile after it is removed. The juncture of the appendage and the glans penis is often the site for lodging of kidney stones or gravel in wethers and rams.

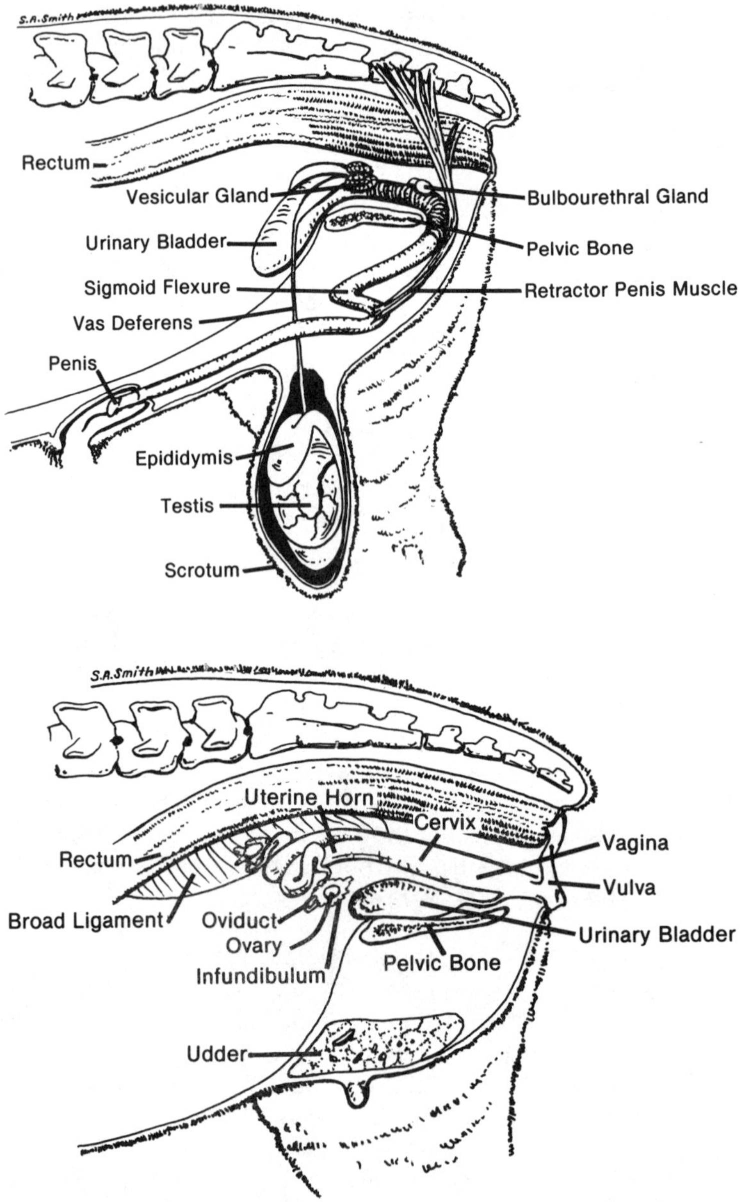

Figure 7.1 Diagram of reproductive organs of a ram and ewe.

7.4.5. Accessory glands The seminal vesicles, prostrate, and bulbo urethra or cowpers glands are the accessory glands. They secrete the fluids that together make up the seminal plasma. The seminal plasma suspends the sperm so it can be transported through the genital tract, and it nourishes and activates it. It also acts on the cells of the vagina to cause buffering to protect the sperm.

Figure 7.2 Diagram of the penis of a ram.

7.5 HORMONAL CONTROL OF REPRODUCTION IN THE EWE

Both male and female reproduction is regulated by hormones controlled by the central nervous system. The hypothalamus, which is part of the brain; the pituitary gland, located at the base of the brain; the ovaries; the adrenals; the thyroid; and the placenta all secrete hormones that have a role in reproduction.

7.5.1. Hormone activity in the open ewe The hypothalamus secretes gonadotrophin releasing hormone (GnRH), which stimulates the pituitary to secrete follicle-stimulating hormones (FSH) and luteinizing hormones (LH), which act on the ovary to cause the development of an ova in a follicle. As the follicle develops, it secretes the hormone estrogen, which causes the development of the sex organs and secondary sex characteristics and estrus or heat.

When the follicle has reached a certain stage of development, the estrogen acts on the brain to cause the ewe to be in estrus. Then the ewe is excited and seeks a ram. Her vulva swells somewhat and is lubricated by vaginal secretions. Estrus or heat extends from 30 to 36 hours depending on factors such as breed, individuality of the ewe, and age. Ewe lambs may be in estrus as much as 10 hours less than mature ewes. Estrogen also suppresses the further secretion of FSH in the pituitary, but causes a high level of LH, which triggers ovulation. When the contents of the follicle, including the ova, pass into the oviduct, it is known as ovulation. Ovulation occurs near the end of the estrus. Exact time varies between 21 to 33 hours after onset of estrus. If the ewe does not become pregnant, progesterone secretion reaches a maximum at 10 days after ovulation, tapers off rapidly through day 12, and on day 14 will have ceased. A hormone, prostaglandin F_2 alpha ($PGF_2\alpha$), which is secreted in the uterus, is responsible for the regression of the corpus luteum. When the levels of progesterone in the blood are low, the hypothalamus secretes GnRH, which stimulates the pituitary to secrete FSH and causes the development of more follicles to start the estrous cycle again. The estrous cycle requires approximately 17 days to complete.

7.5.2. Hormone activity in the pregnant ewe The ruptured follicle develops into the corpus luteum. The corpus luteum forms rapidly and secretes progesterone, the hormone that prevents the secretion of FSH. While the corpus luteum is active, estrus is blocked. The blood levels of progesterone remain until the last 7 to 10 days prior to parturition.

7.5.3. Estrus synchronization The ability of progesterone to block the secretion of FSH is the basis for synchronizing estrus in a flock of ewes. Substances similar in biological activity to progesterone, called progestigens, may be fed, injected, implanted subcutaneously, or implanted as a vaginal pessary for 12 to 14 days and then withdrawn. The ewes will then rapidly form new follicles and be in estrus within 36 to 60 (average 48) hours. This treatment has been remarkably successful in research and under practical conditions when the ewes are cycling naturally. There is some evidence that a combination of the prostaglandins may be superior to other compounds, but that treatment is still being developed. It is also possible to achieve some synchronization by exposing ewes to rams near the end of the anestrous period. It has been shown that ewes do, in fact, ovulate approximately 6 days after sudden contact with a ram but the ewes show none of the symptoms of estrus, and it is known as *silent heat*. However, at the end of a 17-day period, many of the ewes will be in estrus and ovulate.

7.6 HORMONAL CONTROL OF REPRODUCTION IN THE RAM

As in the ewe, the GnRH from the hypothalamus stimulates release of FSH and LH from the pituitary gland, which causes the development and function of the male reproductive system. FSH causes germinal cells in the tubules to develop so that sperm is formed and prepared for ejaculation. LH acts on the interstitial cells of the testicle to cause the secretion of the male sex hormone, testosterone. Testosterone causes the development of sex organs and the secondary sex characteristics that distinguish the male from the female, and last, but by no means least, it is responsible for the sexual aggressiveness, known as libido, of the ram. There is a delicate balance between the hormones of the hypothalamus, pituitary, and the testicles so that testosterone levels influence the amount of FSH and LH secreted. For this reason, injection of testosterone would probably decrease fertility, although it may appear to increase libido.

7.7 FACTORS AFFECTING PUBERTY IN SHEEP

The age of sexual maturity or the age at which an animal is capable of releasing gametes (spermatozoa and ova) and copulating is called puberty. In sheep, it occurs from 6 to 9 months of age, but it may be influenced by such factors as breed, size, crossbreeding, inbreeding, health, nutrition, and season of birth.

7.7.1. Breed Larger breeds tend to reach sexual maturity later than smaller ones. Some estimates have been made on the basis of attainment of mature weight. For example, it has been observed that Romney ewe lambs reach puberty when they attain 40% of their mature body weight, compared with 50% for Suffolks and 63% for the Scottish Blackface.

7.7.2. Growth rate Faster weight gains are related to earlier sexual maturity. Conversely, sheep that gain slowly due to poor nutrition, parasites, or disease reach puberty when they are older.

7.7.3. Crossbreeding Crossbred sheep seem to reach sexual maturity earlier than purebreds. Dickerson and Laster (1975) observed that Finnsheep crossbreds reached puberty at 219 days and 88 lb (40 kg) and Rambouillet crossbreds at 238 days and 97 lb (44 kg), compared with purebred lines at 245 days and 99 lb (45 kg).

7.7.4. Season of birth Another factor is season of birth, especially with breeds that have a short breeding season. For example, fall- and winter-born lambs should mate the next fall. Late spring lambs may be too immature to cycle during the breeding season, and thus may show no sexual activity until the second fall.

It has been observed at the University of Missouri that Finn × Dorset × Rambouillet crossbred ewe lambs settle readily in the late fall if they weigh 80 lb (36.4 kg) and are approximately 8 months old. Purebred Hampshires achieved the same results as the crossbreds, but were 9 months old and weighed 100 lb (45.4 kg).

7.8 EWE AND RAM LAMBS REQUIRE SPECIAL CARE IF USED FOR BREEDING

Although it is possible that ewe lambs may ovulate without showing signs of estrus, they seldom cycle as often or remain in estrus as long as mature ewes. They may only cycle once or twice during their first breeding season. Unlike older ewes, they tend to be timid and fail to approach the ram. This is an excellent reason for breeding ewe lambs separately rather than in a mixed group with older ewes. It is always a good practice to observe breeding activity of rams and make certain that the more aggressive ewes do not monopolize the ram.

Ram lambs are sexually mature at about the same age as their female siblings, but the initial sperm production may be unsatisfactory because of abnormal and dead cells and inadequate motility. From onset of puberty, it requires an additional 2 to 2.5 months for the semen to be of high quality. It is generally agreed that ram lamb usage should be limited to no more than 20 to 30 ewes during a 6-week breeding season. Even then, it may be more satisfactory to hand breed the ewes. This practice consists of using a teaser ram to locate the ewes in standing heat and bring them to the ram one at a time for mating. If large ewes are to be bred to a smaller ram, it may be necessary to stand the ewe in a pit or for the ram to work off of a platform so that contact can be made.

7.9 FACTORS AFFECTING THE REPRODUCTIVE PERFORMANCE OF EWES

Reproductive performance of ewes is influenced by genetic, physiological, environmental, and perhaps even by psychological factors. These will be discussed individually.

7.9.1. Genetic In the chapter on genetics and breeding, it was shown that there is little basis for believing that sterility is inherited. Although twinning rate has a low heritability, it can be improved by selection. Some breeds are more fertile and more prolific than others. Crossbreeding generally improves and inbreeding decreases fertility.

7.9.2. Age Age affects reproduction of ewes. A ewe increases in fertility and prolificacy from her maiden year to approximately 7 years of age.

7.9.3. Nutritional status The nutritional status of the ewe appears to affect reproduction. At least in the initial stage of the breeding season, it is possible to increase conception and improve the twinning rate by feeding the ewe so that she is in a gaining state before and during estrus. This is called flushing and has been credited with improving breeding performance in many areas of the world. It can be accomplished by turning out to an excellent pasture or by feeding concentrates. Conversely, a ewe in a losing condition or at submaintenance levels may not conceive. Ewes require nutrients to maintain their body and to develop the lamb. Deficiencies in protein, energy, minerals, and vitamins may cause the ewe to abort or to produce weak or deformed lambs. These will be discussed thoroughly in the chapter on nutrition.

7.9.4. Disease and parasites Diseases that cause abortion or sterility are very few in sheep, but if a ewe is sick and running a fever during the breeding season she is unlikely to conceive. A heavy load of internal parasites may cause the same results as malnutrition in a ewe.

7.9.5. Pregnancy Pregnancy usually prevents cycling in a ewe and thus further reproduction. There are, however, cases on record of such an occurrence. Whiteman (1982) of the Oklahoma Agricultural Experiment Station in a personal communication has observed that ewes may accept a ram and several days later be serviced again. He noted that ewes carrying twins sometimes give birth to two lambs several days apart. This has led him and other animal breeders to the conclusion that the ewe ovulated at separate times with her two ovaries and conceived with both.

7.9.6. Parturition Parturition usually causes the ewe to be sexually inactive for the period called postpartum anestrus. The process in which the uterus sheds the materials of gestation and returns to normal, ready to conceive another embryo, requires at least 24 days to involute and another 10 to 12 days before conception is possible.

7.9.7. Lactation Lactation generally results in an anestrous period for many breeds of sheep. However, ewes of breeds such as the Rambouillet or Merinos that are suckling lambs in the fall will frequently breed and successfully carry lambs to term.

7.9.8. Environment Day length exerts a strong influence on the reproduction of a ewe. This is called photoperiodism. As the days grow shorter, the hormonal activity relative to reproduction of the ewe increases, possibly through the influence of the pineal gland, to activate the hypothalamus to initiate the estrous cycle. Research has demonstrated that ewes in anestrus can be caused to cycle by subjecting them to 8 hours of light and 16 hours of dark for 8 to 10 weeks.

High environmental temperature causes a reduction in breeding performance of ewes. The mortality rate of embryos is high if the ewe is subjected to high temperatures during the period prior to breeding to 8 days postbreeding, according to Dutt (1959). In trials, ewes subjected to 90 °F temperature and 60% relative humidity lost 91% of embryos. The fetus is less likely to be destroyed later in gestation, but it has been shown that lambs born of ewes subjected to hot weather are generally smaller and weaker than lambs produced by the same ewes during cooler weather. Ewes may be able to maintain pregnancy during high daytime temperatures if nights are cool. Sheared ewes also are less susceptible to heat stress than those carrying a full fleece. Ewes in medium body condition are less affected than those that are fat. Providing shade with good air circulation may be beneficial in counteracting high temperatures. Humidity intensifies the effect of high temperature by reducing the ability of the ewe to dissipate heat. For this reason, greater care must be taken to protect pregnant ewes when both temperature and humidity are high. Solar radiation is capable of raising the body temperature and may cause damage to the fetus. Shade, especially at midday, should be helpful.

7.9.9. Ram fertility and activity Not only is it important for the ram to provide adequate numbers of normal and viable sperm, but his activity affects the reproduction of a ewe. Rams with high libido are more successful than those with little interest in the ewe. Crossbred rams have been shown to have more libido and settle more ewes than straight breds. There appears to be an advantage to rams working in groups versus singly. The ram to ewe ratio can also be a factor.

7.9.10. Natural estrogens in forage Estrogenic substances in pastures, such as red clover, white clover, ladino, bird's-foot trefoil, and subterranean clover, have been observed to reduce numbers of ewes settled. Natural estrogens suppress the activity of the pituitary gland and prevent the formation of FSH, and thus ewes will not ovulate.

7.10 FACTORS AFFECTING THE REPRODUCTIVE PERFORMANCE OF RAMS

Ram performance is affected by genetic, physiological, and environmental factors.

7.10.1. Genetic Although sterility appears to be low in heritability, cryptorchidism may result in sterility or reduced fertility. Rams from more prolific breeds appear to be more fertile than those from breeds that produce fewer multiple births. Rams from breeds with a long breeding season are fertile and have greater libido throughout the year than those from breeds that have more seasonal reproductive habits. Crossbred rams tend to be more fertile and have more libido than purebreds.

7.10.2. Age Age affects the performance of rams. Immature rams may be limited in sperm quality and quantity. Generally, the reproductive ability of a ram improves until he is 2 years old. Little is known about the effect of aging on the quality of sperm of a ram. Healthy, vigorous rams as old as 12 years have successfully sired lambs. Frequently, rams beyond the age of 8 years have problems of mobility, which reduces sexual activity. Rams beyond the age of 7 years should be expected to service only a limited number of ewes. Even though he is healthy and active, 10 to 12 ewes may be his limit. Hand mating could extend the ram's capacity to 15 to 20 ewes.

7.10.3. Nutritional status Adequacy and level of nutrition affect growth and development of young rams and thus puberty. Nutrition also appears to affect older rams, although they have an extraordinary resilience to extremes in nutrient levels. Feeding rams so that they are in a gaining condition, similar to flushing in ewes, before the breeding season appears to be a sound practice, provided the ram does not become obese. Obesity may be expected to decrease semen quality, although some fat rams seem to be unaffected. Fat also may decrease libido. Underfeeding to the extent of thinness apparently has little effect on quality factors of semen, but does reduce volume. A more serious fault is the lack of libido in rams on prolonged low-energy diets. Both protein and phosphorus deficiencies appear to be detrimental to ram performance. A serious vitamin A deficiency may cause sterility.

7.10.4. Testicular size Testicular size is closely related to body development and breeding. Ram lambs, on a high level of nutrition with well-developed testicles, reach puberty earlier than those on inadequate diets. It is interesting that larger testicular weight in ram lambs is positively related to ovulation rate in their female siblings. Testicle size may be estimated by measuring the circumference of the scrotum. Providing the testes are sound, larger ones indicate greater breeding capacity. Very small testicles should be viewed with suspicion.

7.10.5. Environment Rams are sensitive to the environment, particularly light and temperature. They have been shown to respond to photoperiodism, although the response is complex and not easily demonstrated. Experiments have shown a definite improvement in semen quality and libido in off season by subjecting rams to 8 hours of light and 16 hours of darkness for 8 to 10 weeks. Elevated environmental temperatures for prolonged periods can cause sterility. Apparently, temperatures above 80 °F (26.5 °C) are detrimental to sperm development. Other factors contribute to the heat effect. Sheared rams withstand heat much better than those in fleece. Obese rams are less able to dissipate heat than those in medium condition. Irradiation by the sun accentuates the effects of heat, as does high humidity. Exercise or stress has a similar effect. Moving air tends to alleviate some of the effects of heat stress. The author has observed that in hot weather dipping rams in 70 °F (21 °C) water for 10 minutes reduced body temperature 2° to 3 °F (1° to 2 °C), and it remained at that level for 3 to 4 hours. Wool covering on the scrotum can reduce fertility and thus should be removed before the breeding season. Some of the first research on the physiology of reproduction showed that covering the testicles with a woolen bag resulted in sterility.

7.10.6. Disease Diseases, such as epididymitis, may cause partial or complete sterility. Lumps, adhesions, or lesions in the testicles may indicate this condition, which is caused by the organism *Brucella ovis*. Testicles should be symmetrical. Wide differences may indicate injury, disease, or an abnormality. Enlarged, soft testicles indicate degeneration and possible problems. Those that are too hard may be fibrotic as a result of injury or disease. The prepuce or sheath that protects the penis may be a source of trouble. Ulcerative prosthitis or pizzle rot may prevent copulation. This disease causes ulcers of the skin and the mucosa. It is thought to be caused by bacteria and urea. Ulcerative dermatosis, caused by a virus, is found throughout the western United States. It affects the skin, face, feet, prepuce, and penis. This disease has no treatment to date.

7.10.7. Structural defects Defects of the locomotor system can be detrimental to ram performance. Severe structural faults, such as hocks too straight or postlegged, may prevent the ram from mounting a ewe. Lameness from foot rot may also be detrimental and should be avoided.

7.10.8. Behavioral defects A dominant, infertile ram may cause poor performance in small groups of ewes in confinement, even though a fertile ram is present. Under pasture conditions, the effect may be reduced. Some sheep breeders avoid the problem by using two groups of rams on alternate days.

Some rams that are apparently healthy and have high-quality semen are ineffective because of failure to mate. They may simply not mount the ewes, or they may mount without ejaculation. It is essential to observe the flock at mating time to be certain the rams are copulating. There are some indicators that homosexuality among young rams may have a carryover effect when they are with ewes, but the evidence is as yet not clear-cut.

7.11 MEASURES OF SEMEN QUALITY

If for any reason the sperm furnished by the ram is inadequate, fertilization cannot result. The only true test of ram fertility is to mate him with ewes, but that is impractical for obvious reasons. Examination of the semen may not prove that a ram is fertile, but it can demonstrate that it is improbable that he can function. No single test is adequate, but if several are negative, a breeder should be assured that the odds are that the ram is either sterile or has impaired fertility. The following characteristics are indicative of the quality of sperm:

1. *Volume:* This may vary with the means of collection, with more obtained from the electro ejaculator than the artificial vagina. Normal volume should be 0.7 to 2 ml per ejaculation.
2. *Motility:* Microscopic examination under low and high power should show the wave motion and provide an estimate of the sperm that are moving. High-quality semen should have marked wave motion and 90% of the stream moving. This is rated as the single most important characteristic of good semen.
3. *Concentration:* This is estimated by use of a hemocytometer or other counting device to calculate sperm per unit of volume. High-quality semen should contain 1.8 billion sperm/ml[3].
4. *Morphology:* Sperm may be stained to differentiate between those that are alive and dead. Dead sperm should be less than 10%. Abnormal sperm, as shown in Figure 7.3, should constitute no more than 10%, with little or no abnormal heads.

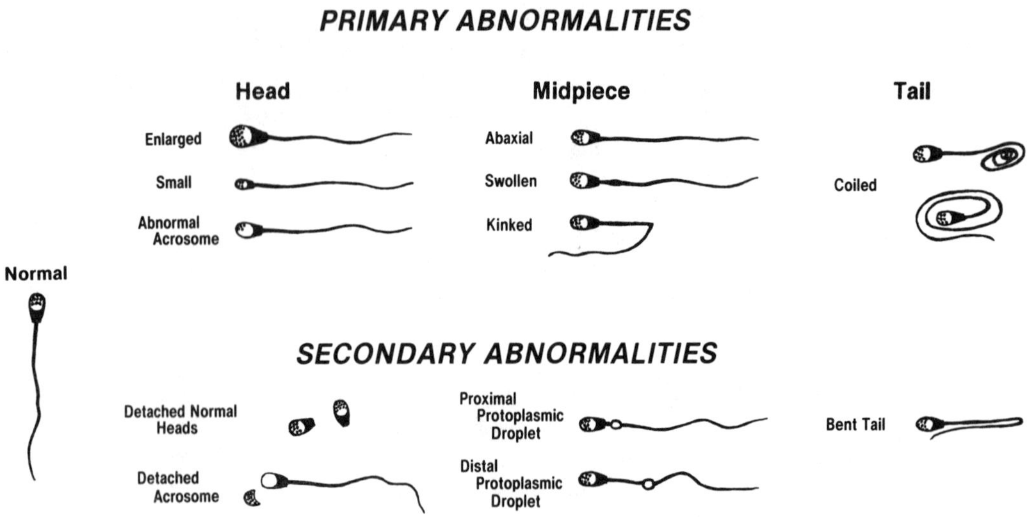

Figure 7.3 Normal and abnormal sperm.

7.12 MATING BEHAVIOR OF SHEEP

7.12.1. Indication of estrus in the ewe A ewe in estrus, or heat as it is commonly called, shows few physical signs. The vulva may be slightly enlarged and secrete mucus, but these are not reliable indications that she is physiologically as well as psychologically ready to accept the ram. If a ram is present, ewes in estrus will stay near him or will stand at a fence separating them. The best indication is the ewe's willingness to be mounted.

7.12.2. Courtship and mating Sheep have a courtship procedure unique to the species. The ram approaches the ewe from alongside and slightly behind the shoulder. He usually makes a low-pitched, growling sound accompanied by licking and biting her wool. He may attempt to paw her side with his front foot. The usual indication that the ewe is ready to receive the ram is that she stands still, drops her back slightly, elevates her rump and wiggles her tail. The ram then moves to the rear and attempts to mount. The initial attempt, if not successful, is usually repeated immediately. The ram may make a few tentative thrusts before his penis makes contact with the vulva. Once the contact is made, the ram hunches his pelvis forward in a deep thrust accompanied by throwing his head backward. When this occurs, the ram has ejaculated. The actual coitus only requires a few seconds. The period immediately following coitus is called intromission, while the ram recovers and is ready to mate again. There is some evidence that a ram may prefer to mate with a ewe that has not had contact with a ram if a number of ewes are in estrus at the same time. The ewes in estrus will tend to form a harem around the ram, and the more aggressive ones will be bred more often than others. Studies have shown that in small groups services per ewe may average just under four. It has also been shown that rams tend to mount three to four times for each successful service, and that conception rates are higher for ewes served three times versus once. Ewe lambs tend to be served fewer times than mature ewes, possibly because they have a shorter duration of estrus and are less aggressive than mature ewes. There appears to be individual differences in rams. Some may be capable of only a few services per day and others as many as 40 services. Although the quality does not appear to be affected, the number of sperm per ejaculate is reduced with frequent services. Sexual activity generally varies throughout the day, especially when the weather is warm. Mating is reduced during darkness and resumes with dawn. Often the height of activity occurs early in the morning, subsides during midday, and is resumed in the afternoon or evening.

7.12.3. Preference in mating Behavioral research has demonstrated that the most powerful rams tend to breed the most ewes if several are running with the flock. In experiments in which rams were tethered in a paddock with ewes in estrus, ewes perferred some rams over others. There also is evidence that rams and ewes of certain breeds prefer partners of the same breed to those of another.

7.13 HAND MATING

There may be occasions when it is important to extend the numbers of ewes serviced by a valuable ram. Then, if ewes in standing heat are located and brought to the ram one at a time or in small groups, the ewe may be removed when she has been serviced. In this procedure, young, vigorous rams equipped with marking harness and an apron or that have had vasectomies are let loose with the flock. These "traveling salesmen," as they are frequently termed in Australia, locate the ewes so that they can be sorted off. A vigorous, mature ram may service as many as 100 ewes in a 6-week breeding season by hand breeding.

7.14 GROUP MATING

Under practical conditions, ewes are bred in groups. A mature ram should be capable of settling 50 to 60 ewes in a 6-week breeding season.

7.14.1. Marking the ram There are distinct advantages in using a marking harness or painting the floor of the chest of rams. This procedure indicates the date ewes are in heat and that rams are active. It is useful in predicting the lambing date, and it may offer good evidence that conception is not occurring. If a marking harness is used, make certain that the chalk is not so hard that it fails to leave a mark. If the ram is painted, the paint should be renewed daily. A practice that works well is to change colors at 15-day intervals, starting with lighter colors and changing to darker ones, for example, yellow, red, blue, and black. A cheap and effective paint can be made of low-grade motor oil and pigment. A swab made of burlap tied to a short length of broomstick is satisfactory for administering the paint.

7.15 MECHANICS OF SPERM TRANSPORT, FERTILIZATION, AND IMPLANTATION

When the semen is deposited in the vagina of the ewe during copulation, the sperm moves rapidly by vaginal contractions into the cervix to form sperm reservoirs. They leave the cervix by means of their own motility or by cervical and intrauterine contractions. Sperm are transported through the uterus, largely by contraction, and into the oviduct. The oviduct carries the sperm in one direction and the ova in the other till they meet at the site of fertilization. Research has shown that 8 minutes is required for sperm to travel from the vagina to oviducts.

7.15.1. Passage of ova The ova, when released by the rupture of the follicle, pass into a funnel-shaped organ called the infundibulum, which leads into the fallopian tubes. The ova move by contractions of the fallopian tubes and the cilia inside the tubes. The ova remain in the oviduct for approximately 72 hours. Fluids inside the tract prepare the ova for fertilization and, later, implantation.

7.15.2. Fertilization The total number of sperm in the ejaculate of the ram is in the billions. Not all the sperm survive to reach the ampullary–isthmic junction of the oviduct where fertilization occurs. Sperm require changes called capacitation before they can penetrate the outer membrane of the ova. Enzymes produced by the sperm cause changes in the protective membranes of the ova, which allow the penetration of the sperm. One sperm penetrates and unites with the ova. The joining of the male and female gametes is called syngamy. It has been estimated that it requires 16 to 21 hours from the penetration of the sperm to the first cleavage division.

7.15.3. Cleavage of the embryo Cellular proliferation of the embryo begins in the oviduct and continues in the uterus. During the period before implantation, the embryo exists as a free-living body nourished by the yolk in the ectoplasm and fluids in the uterus. The embryo cells continue to divide, with no increase in size until implantation.

7.15.4. Implantation of the embryo Implantation occurs when the embryo becomes attached to the wall of the uterus. This has been estimated to require 10 to 22 days from coitus to implantation. After implantation, the embryo begins to receive sustenance from the ewe.

7.16 GESTATION

An embryo becomes a fetus after the body parts and organs develop. Even before implantation, there are changes in the uterus. The blood supply increases, and there is more secretory activity to provide nourishment for the embryo before the placenta is formed. Progesterone from the corpus luteum is essential for implantation and maintenance of pregnancy. The uterus produces hormones, prostaglandins, that may destroy the corpus luteum, but the fetus and placenta secrete hormonelike substances to prevent that from occurring.

7.16.1. Fetal nutrition The embryo, which is surrounded by membranes, has no direct connection with the bloodstream of the ewe, but has its own independent circulatory system. The transfer of nutrients from the ewe occurs through the caruncles, which are the attachments between the uterus and the placenta. Although there is no direct connection between the circulatory systems of the ewe and the fetus, the blood supply between the two is close enough together at the junction of the chorion for oxygen and nutrients to pass from one to the other and waste products from the fetus to be voided. Water and electrolytes pass directly through fetal membranes from one system to the other. Energy is transferred as fatty acids and protein as amino acids. Water-soluble vitamins pass the fetal barrier more readily than the lipid-soluble ones.

7.16.2. Growth and development of the fetus Genetic factors affect growth. Lambs from larger breeds grow faster and are larger at birth than those from small breeds. Size of dam, nutrition, litter size, sex of lambs, and maturity of the dam all affect growth of the fetus. Inadequate nutrition of the ewe, especially late in gestation, may result in weak and stunted lambs. Lambs from ewes fed diets deficient in iodine may be dead or deformed. High ambient temperature during pregnancy may cause dwarfing of lambs.

7.16.3. Changes in the ewe during gestation The cervix remains tightly sealed to protect the uterus during pregnancy. The uterus gradually expands to permit growth of the fetus. There is a gradual relaxation of the ligaments in the pelvic region. The ewe tends to increase in weight not only as a result of growth of the fetus, but there is a tendency to store fat. The mammary glands develop and milk secretion occurs just before parturition.

7.16.4. Duration of gestation Length of the gestation period varies from 140 to 159 days, with an average of 147 days. The length is influenced by genetic, maternal, and fetal factors. Length in Finnsheep is only slightly more than 140 days; Dorsets and Southdowns, 144 days; and Karakuls, 151 days. Medium-wooled breeds have been observed to require 147 days, while Rambouillets and Merinos were 149 days. Younger ewes have a shorter term, as a rule, than older ones. Singles are carried longer than twins or triplets and males longer than females. There are individual differences in ewes, but generally ewes of the same breed are similar.

7.17 PREGNANCY DETECTION

Accurate, early diagnosis of pregnancy may result in a saving of $5 to $20 per head on feed cost of each open ewe that is removed from the flock before the heavy feeding period of late gestation. There is also some reduction in labor, and the space can be used by the rest of the flock.

If ewe lambs have been bred, a pregnancy check can be used to screen out those that failed to settle. They can then be finished and marketed, rather than held for the next breeding season. This procedure allows the breeder to select the ewes with the greatest potential for production, and the others can be sold at favorable market prices.

Pregnancy checking may be a useful tool in an accelerated lambing system in which ewes are bred more frequently than yearly intervals. If an open ewe is discovered, she can be changed to another group and rebred much earlier.

7.17.1. Natural methods Some indication of the number of ewes that are not pregnant can be gained at the end of the breeding season by the use of a marking harness or painting a vasectomized or aproned "check" ram. If the color is different from those used during the breeding season, the ewes marked are probably not pregnant. Unfortunately, marking by rams is not precise. Some ewes may be marked but not actually bred, others may have been in estrus only once or twice and did not settle, and others may accept a ram and fail to maintain pregnancy. Thus, marking the rams only indicates nonpregnant ewes that are cycling.

Late in the gestation period, it is possible to determine pregnancy by inspection of the ewes, but there is much variation in ewes. Some may not be detected until a few weeks before parturition.

Physiologists use laparotomy, vaginal biopsy, bioassay for progesterone in blood, and x-rays with excellent accuracy. But all are too complicated, slow, and expensive for practical use.

Only within recent years have equipment and techniques been developed for accurate, rapid pregnancy detection. The three described next are all satisfactory and are being used commercially.

7.17.2. Ultrasonic scanning Commercial pregnancy checking equipment is similar to that used for determining back fat and loin eye area in swine and cattle. A transducer transmits ultrasonic sound waves through the body of the ewe and receives an echo based on the Doppler effect. If the ewe is pregnant, the presence of a fetus registers on the machine either by light or sound.

The ewe is either held by hand or in a stanchion or stand in a normal standing position. The operator places the transducer on the bare portion of the abdomen on the right side 2 to 3 in. (5.1 to 7.6 cm) in front of the right nipple. The skin must be bare, and the transducer must be in close contact with the skin. Nondetergent oil is applied to the skin and the transducer so that sound waves are transmitted. The transducer is aimed forward at an angle of 35° and across the body at 45° to a point just behind the last rib on the left side as the operator stands behind the ewe.

The equipment, being demonstrated in Figure 7.4, appears to have a high degree of accuracy at 60 days of gestation. Some operators claim 98% at 6 weeks. The equipment cannot differentiate between single or multiple pregnancy. Very little training is required to use the equipment, but it is desirable to check 10 to 20 ewes with an experienced operator.

7.17.3. Interrectal Doppler ultrasound probe Equipment designed to use fetal heartbeats to diagnose pregnancy is available. Lindahl (1966) reported excellent results with the equipment when used at the beginning of the second trimester of gestation. Ewes are put on dry feed for 2 to 3 days and on the day of the examination are given tranquilizers. The ewe is placed on her back in a cradle with hindlegs secured. A rectal probe is lubricated and inserted gently into the rectum on a horizontal plane below the uterus. As the probe is inserted, sounds are amplified on the machine. First should be maternal heartbeats (90 to 110 heartbeats per minute) as the probe passes the uterine artery. Next is the swishing or windlike sound of the blood flow in the placenta, and finally the very rapid (130 to 160 beats per minute) heartbeats of the fetus.

Figure 7.4 Pregnancy checking a ewe.

This equipment is fairly accurate and simple to use, but it requires some training to operate it. An inexperienced person should work with a qualified operator before attempting to use it. Accuracy after 60 days by experienced operators is equal to that of the ultrasonic sound system, but it requires somewhat longer to check each ewe. The machine does not indicate whether the ewe has a single or multiple pregnancy.

7.17.4. Rectal–abdominal palpation This method has been described by Hulet (1972) and involves the use of a plastic probe to palpate the abdomen of a ewe. Essentially, the procedure consists of restraining the ewe on her back in a cradle or similar device with her hindlegs secured firmly, but in a manner that allows the ewe to relax the muscles of the abdomen. Approximately one-half pint of a warm soap solution is injected into the rectum to remove feces and a plastic rod is used for palpation. The plastic rod used is ⅝ in. (1.6 cm) in diameter with a bullet-shaped tip and 21 in. (53 cm) long. The plastic rod is inserted gently into the rectum, beginning with an angle slightly toward the backbone, for 14 to 16 in. (36 to 41 cm), and then pressed gently but firmly upward to the region of the uterus. The rod is used to manipulate the uterus so that the fetus can be felt with the operator's fingers through the abdominal wall. This procedure requires training and should not be attempted without instruction from an expert. Trained operators can diagnose as many as 100 to 120 ewes per hour without injury.

7.18 PARTURITION

The delivery of the fetus from the ewe is the final sequence in reproduction. In the ewe, birth is the result of the interaction of a number of hormonal and mechanical factors.

7.18.1. Hormonal changes prepartum During the last days of pregnancy, there is a lowering of blood levels of progesterone and an increase in estrogens and prostaglandins. Apparently, the endocrine system of the lamb triggers the chain of events that occurs during parturition. The fetal hypothalamus may cause the release of hormones that stimulate the pituitary gland to release hormones that act on the adrenals. The adrenals, in turn, generate corticoid hormones, which transfer across the placenta and cause the release of prostaglandins and estrogens in the ewe. There, hormones are responsible for the contractions of the uterus and abdomen, the dilation of the cervix, stretching and distention of the vagina and vulva, and expulsion. Labor consists of dilation of the cervix, expulsion of the fetus, and expulsion of the placenta. Passage of the placenta should follow parturition by a period of 30 to 60 minutes. The lamb is moved through the birth canal by peristalsis.

7.18.2. Normal presentation The normal presentation of the lamb is with the front feet first and the nose of the lamb lying on the front legs. A number of abnormal presentations will be discussed in the chapter on management at lambing.

7.19 BEHAVIOR AT PARTURITION

Ewes that are about to give birth characteristically show signs that lambing is imminent. They may separate themselves from the group, are restless, and may stamp their feet, bleat, and look behind themselves. Physical signs are sinking in front of the hips, relaxation in the pelvic region, and milk in the udder. From the beginning of heavy contractions to delivery should usually require less than 1 hour.

7.20 MANAGEMENT TO ENSURE FERTILITY OF SHEEP

It should be remembered that the natural breeding season for most sheep occurs during the cool nights of the fall. Ewes have an estrus cycle of approximately 17 days. They remain in estrus for 30 to 35 hours and ovulate near the end of estrus. The gestation period is approximately 147 days. The ram ejaculates 0.7 to 2.0 cm^3 of sperm containing 1 to 2 billion sperm/ml. Fertility is affected by genetic, environmental, and other factors. Management can overcome many of the problems that reduce reproductive effectiveness. The following are practices that should be beneficial to sheep breeders.

Practices for Maintaining High Reproductive Performance in a Sheep Flock.

1. Maintain a breeding flock with the genetic potential for high reproductive efficiency. Consider crossbreeding.
2. Choose only sound, healthy animals for breeding. Pay particular attention to the testicle size and soundness of the ram.
3. Maintain a healthy flock of sheep.
4. Provide adequate nutrition to the flock.
5. Keep records on ewes, and eliminate loafers and poor breeders.
6. Keep rams and ewes separate until near the breeding season.

7. Maintain ewes and rams in medium, not fat, condition during the anestrus period before breeding.
8. Shear rams 6 to 8 weeks before the breeding season. Shear ewes before the breeding season if they have heavy fleeces.
9. Worm and trim feet, and remove manure tags from all ewes at least 2 weeks before breeding.
10. Flush ewes on good grass pasture or on grain in drylot from 2 weeks before and through two heat periods after the beginning of the breeding season.
11. Protect the ram from heat and stress prior to and during the breeding season.
12. If the ram has a heavy load of ewes, separate him from them from midmorning to late midafternoon, and feed him grain and good hay. Also, leave a ewe with him for company.
13. Protect ewes from heat and stress during the breeding season. A cool, shady place with a breeze is ideal. Take special precautions during the first week postbreeding.
14. Use marking harness or paint rams to see that they are working and to obtain breeding dates. Actual observation is important to see how the ram works.
15. Do not overload rams. A yearling can handle 40 to 50 ewes and a mature ram 50 to 60 ewes in a 6-week breeding season.
16. Breed young ewes separately from mature ewes.

REFERENCES

Anderson, G. B., G. E. Bradford, and P. T. Cupps, "Length of Gestation in Ewes Carrying Lambs of Two Different Breeds," *Theriogenology,* 16, 1981, p. 119.

Braun, W. F., J. M. Thompson, and C. V. Ross, "Ram Scrotal Circumference Measurements," *Theriogenology,* 13, 1980, p. 221.

Dickerson, G. E., and D. B. Laster, "Breed, Heterosis and Environment Influences on Growth and Puberty of Ewe Lambs," *J. Anim. Sci.,* 41, 1975, pp. 1–9.

Dutt, R. H., "Fertilization Rate and Early Embryo Survival in Sheared and Unsheared Ewes Following Exposure to Elevated Air Temperature," *J. Anim. Sci.,* 18, 1959, pp. 1308–18.

Hafez, E. Z. A., *Reproduction in Farm Animals.* Philadelphia: Lea and Febiger, 1980.

Hight, G. K., and K. E. Jury, "Hill Country Sheep Production. XIII. Relationship of Hogged and Two-Year-Old Ovulation Rate to Subsequent Fertility in Romney and Border Leicester × Romney Ewes," *New Zealand J. Agr. Res.,* 19, 1976, p. 281.

Hulet, C. V., "Management of Reproduction in Sheep," *Proc. Symp. Mgmt. of Reproduction in Sheep and Goats,* University of Wisconsin, Madison, 1972, pp. 119–33.

———, W. C. Foote, and R. L. Blackwell, "Relationship of Semen Quality and Fertility in the Ram to Fecundity in the Ewe," *J. Reprod. Fert.,* 9, 1965, p. 311.

———, E. L. Wiggins, and S. K. Ercanbrack, "Estrus in Range Lambs and Its Relationship to Lifetime Reproductive Performance," *J. Anim. Sci.,* 28, 1969, p. 246.

Levine, J. M., R. R. Phillips, and W. Hohenboken, "Ewe Lamb Conception as an Indicator of Future Production in Farm Flock Columbia and Targhee Ewes," *J. Anim. Sci.,* 46, 1978, p. 19.

Lindahl, I. L., "Detection of Pregnancy in Sheep by Means of Ultrasonic Sound," *Nature,* 212, 1966, p. 642.

Lundstra, D. D., and R. K. Christenson, "Synchronization of Ewes during Anestrus: Influence of Time of Year and Interval of Onset of Estrus on Conception Date," *J. Anim. Sci.,* 53, 1981, p. 448.

______, and ______, "Fertilization and Embryonic Survival in Ewes Synchronized with Exogenous Hormones during Anestrus and Estrus Seasons," *J. Anim. Sci.,* 53, 1981, p. 458.

Mickelson, W. D., I. Paisley, and J. J. Dahmen, "The Effect of Season on the Scrotal Circumference and Sperm Motility and Morphology in Rams," *Theriogenology,* 16, 1981, p. 45.

______, ______, and ______, "The Effect of Scrotal Circumference, Sperm, Motility and Morphology in the Ram on Conception Rates and Lambing Percentage in the Ewe," *Theriogenology,* 16, 1981, p. 53.

Ott, R. W., and M. A. Memon, "Sheep and Goat Manual," *Theriogenology,* 10, 1980, p. 1–50.

Owen, John B., *Sheep Production.* London: Cassell and Collier, Macmillan Publishers Ltd., 1976.

Schanbacher, B. D., "Increased Lamb Production with Rams Exposed to Short Day Length during the Nonbreeding Season," *J. Anim. Sci.,* 49, 1979, p. 927.

______, "Relationship of Day Length and Prolactin to Resumption of Reproductive Activity in Anestrus Ewes," *J. Anim. Sci.,* 50, 1980, p. 293.

Scott, G. D., *The Sheepman's Production Handbook,* Sheep Industry Development Program, Inc. Denver, Colo.: Abegg Printing Co., 1982.

Sorenson, A. M., Jr., *Animal Reproduction.* New York: McGraw-Hill Book Co., 1979.

Terrill, C. E., and J. A. Stohr, "Reproduction in Range Sheep," *Proc. Amer. Soc. Anim. Prod.,* 369, 1939, p. 1939.

Whiteman, J. V., "Relationship between Ovulation Rate and Lambing Ratio," *J. Anim. Sci.,* 25, 1966, p. 1029.

______, 1982. Observation of different ages in twin lambs. Personal communication.

Williams, S. M., and others, "Variations in Length of Estrus Cycles and Breeding Season in Ewes," *J. Anim. Sci.,* 15, 1956, pp. 978–89.

QUESTIONS FOR STUDY AND DISCUSSION

1. How does reproductive performance directly limit returns from a sheep operation?
2. Why should multiple births be more economical than singles?
3. Which measure of breeding efficiency is the most important to the producer?
4. Draw a diagram of the reproductive organs of the ewe.
5. The "egg" produced by the ewe begins development on what location of the ovary?
6. What hormones are secreted by the ovary and what are their function?
7. What organs are connected by the oviducts?
8. What are the functions of the oviducts?
9. Where does the ova and sperm unite in fertilization of the egg?
10. Describe the function of the cervix.
11. What sex hormones affect the testes?
12. What hormones are produced by the testes and what is their function?
13. What is the function of the filiform appendage in a ram?

14. What fluids constitute the seminal plasma and what is its function?
15. Why is reproduction in the ewe more complex than in the ram?
16. What hormone is responsible for the initiation of the estrus cycle?
17. What hormones are responsible for the following reproductive functions:
    a. Cause development of secondary sex characteristics?
    b. Cause the ewe to be in estrus?
    c. Trigger ovulation?
    d. Are responsible for maintenance of pregnancy?
18. What are progestagens and how are they used in controlling the reproduction of sheep?
19. Name five factors that affect puberty in sheep.
20. Why should ewe lambs be bred separately from mature ewes?
21. Why should ram lambs be bred to only a limited number of ewes?
22. What is the effect of high temperatures on newly bred ewes?
23. How may ewes be induced to cycle by the use of photoperiodism?
24. What nutritional deficiencies affect reproductive performance in ewes and rams?
25. What measures are used to estimate the semen quality of rams? How are these made?
26. What are the reasons for using marking harness or painting the breast and brisket of a ram?
27. What is the role of the fetus in the initiation of parturition?
28. What factors affect length of gestation?
29. Describe the various stages of parturition in the ewe.
30. What are the external signs that a ewe is nearing parturition?

8 Sheep Nutrition

The major problem in sheep nutrition is to provide the essential nutrients in adequate amounts to satisfy the requirements of the animal at an economical cost. Thus, the feed supply must occupy a high priority in the planning of a prospective sheep operation.

An understanding of the physiology of a sheep, the nutritional requirements, and the nutrient content of feeds is of great value to the producer. Knowing what can and cannot be done quite often is the margin between success and failure.

It has been estimated that as much as 90% of the problems of production are in some way related to the nutritional state of the animals. Well-fed sheep never have the range and complexity of problems encountered by those fed inadequate or insufficient rations.

8.1 DIGESTIVE SYSTEM OF SHEEP

Like goats, cattle, deer, camels, llamas, moose, buffalo, antelopes, and other ungulates, sheep are ruminants. Unlike the pig, chicken, and dog, which have comparatively simple monogastric mechanisms for digesting and assimilating highly concentrated feeds, the ruminant relies on its ability to utilize fibrous vegetative matter that is far more difficult to break down into the required nutrients. The digestive system of a mature sheep is illustrated in Figure 8.1.

8.1.1. Anatomy and physiology of the digestive tract The digestive system of a sheep consists of a mouth, gullet, the four compartments of the stomach (rumen, reticulum, omasum, and abomasum), small intestine, cecum, large intestine, and anus.

The mouth of a sheep is well adapted to discriminate grazing. The muzzle is narrow and the lips are split, thin, and mobile. The sheep can control right and left lips independently. The tongue is small, pointed, and very sensitive. Instead of upper incisors, there is a tough cartilagenous dental pad. The upper jaw is wider than the lower. Thus a sheep must chew on one side or the other, never both at the same

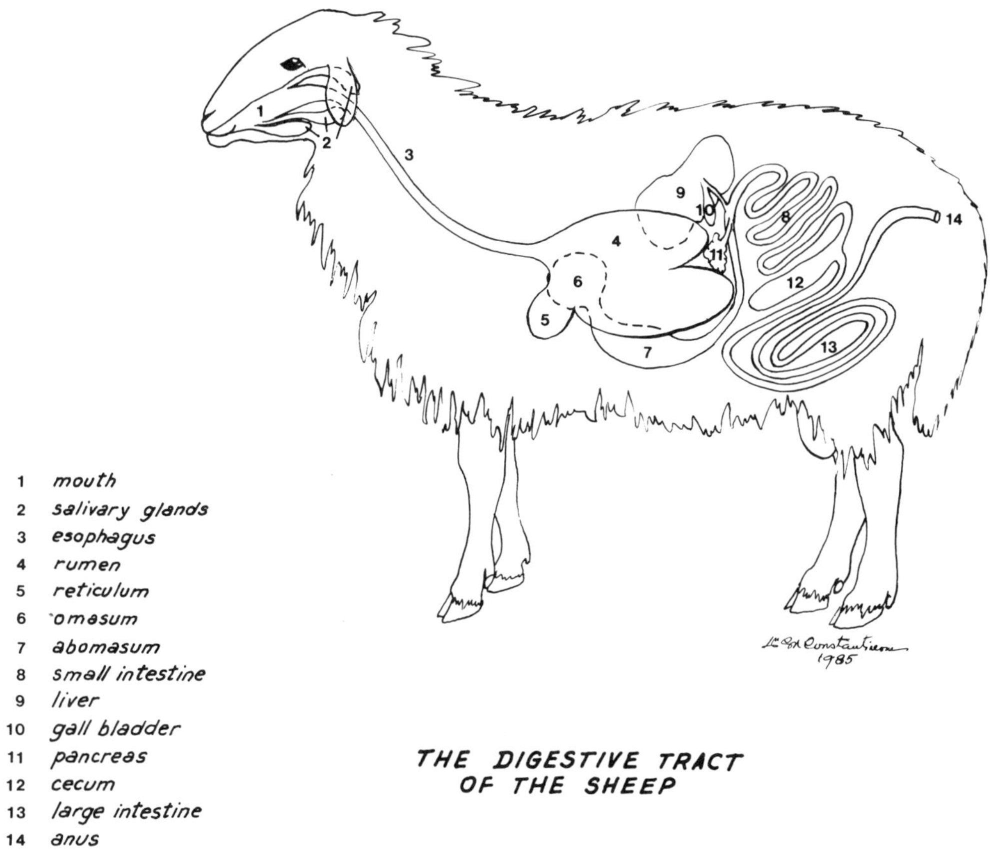

Figure 8.1 Diagram of the digestive system of a sheep. (Courtesy Dr. E. M. Constantinescu, Veterinary Biomedical Science Dept., University of Missouri, Columbia)

time. The salivary glands located beneath the tongue are quite active. The esophagus extends from the throat to the caudia valve in the rumen, which controls the passage of fibrous matter in and out of the rumen. The rumen, which is much the largest compartment of the stomach, is the storage vat for the newly consumed feed and water. It is inhabited by countless millions of bacteria and protozoa that enjoy a symbiotic relationship with the sheep. A number of muscles attached to the rumen wall allow it to contract and expand and move the contents within. Folds in the rumen allow the animal to retain or release particles. The reticulum is not truly separated from the rumen, but folds in the rumen serve to keep them apart to some extent. The reticulum moves ingested feed into the rumen or out through the omasum. It plays an important role in regurgitation of ingesta during rumination. The reticulorumen makes up 88% to 93% of the capacity of the four compartments.

The omasum is the smallest of the four compartments. The interior is partially filled with longitudinal folds or leaves, which may function to reduce particle size and absorb nutrients, but there is still some question about what actually happens inside the compartment.

The abomasum is the true stomach, which compares with those in monogastric animals. It connects directly to the omasum by an orifice (opening), which usually remains open. The region closest to the omasum is called the fundus. It has many folds with a large surface area. It is filled with glandular tissues that secrete the gastric juices. A constriction divides the fundus and pyloric regions. The pyloric region has few secretory cells.

The abomasum is connected to the small intestine by a valve called the pylorus. This valve allows the ingesta to move only in one direction.

The small intestine of a mature sheep is approximately 85 to 90 ft (26 to 27 m) long. It secretes digestive juices that break down organic compounds such as proteins, starches, sugars, and fats into simple compounds that can be absorbed into the body. Much of the absorption of nutrients occurs in the small intestine.

The cecum is connected to the intestinal tract, but has a minor function in digestion and absorption of nutrients.

The colon is larger in diameter than the small intestine and only a fourth as long. It is the site of some digestion and absorption of volatile fatty acids (VFA), but mostly it absorbs water.

Table 8.1 gives the estimated capacity of each part of the digestive system of a mature sheep.

TABLE 8.1 CAPACITIES OF THE PARTS OF THE DIGESTIVE TRACT

	English	Metric
Body weight	176 lb	80 kg
Reticulorumen	4.5 gal	17 liters
Omasum	1.06 qt	1 liter
Abomasum	2.12 qt	2 liters
Total stomach	5.3 gal	20 liters
Small intestine	6.4 qt	6 liters
Cecum	1.06 qt	1 liter
Large intestine	3.18 qt	3 liters
Total digestive tract	6.62 gal	25 liters

8.1.2. Development of the digestive tract Initially, the digestive system is similar to that of monogastric animals like the pig. It is adapted to the ingestion and digestion of milk. The undeveloped reticulorumen is largely bypassed by the esophageal groove, which connects the omasum and the esophagus. This structure is essentially a tube formed by a fold in the wall of the reticulum. As the lamb develops and consumes dry feed, the reticulo groove eventually becomes nonfunctional and feed ingested goes directly to the rumen.

The rumen develops rapidly in the young lamb and is functional at 5 to 8 weeks of age. The speed of development is dependent on the ration consumed. Dry concentrates stimulate development, but not as much as roughages. Continued feeding of a liquid diet tends to retard development.

8.2 DIGESTIVE PROCESS

The various functions of an animal in the ingestion, mastication, rumination, digestion, and absorption of nutrients are all part of the process of digestion. Essentially, in the digestion of feed, carbohydrates, proteins, and fats are degraded into simpler units that can be absorbed into the body of the host animal. The welfare of the animal is dependent on its ability to consume, digest, and use the feeds available to it.

8.2.1. Eating, grazing, and mastication In the natural state, sheep like cattle, deer, and other ruminants obtain most of their nutrition from plants. Sheep eat grasses, broad-leaved plants, leaves from trees, and shrubs. They will also eat hay, silage, grains, and other concentrates.

A sheep grazes differently than cattle with much less dependence on the tongue. It is possible for a sheep to graze quite discriminately because of the construction of its mouthparts. It can not only select plants, but even various parts of a plant. The same physical characteristics allow a sheep to pick out certain ingredients in a mixed feed if they can be distinguished. Perhaps as a carryover from the wild state, grazing sheep like to begin early in the morning and continue until they are full. They grasp the forage between the incisor teeth and the dental pad and, by moving their head, sever it from the rest of the plant. Grazing is accompanied by a copious flow of saliva. Only enough chewing is done to reduce particle size sufficiently to swallow it. Sheep eat hay in much the same way, although they appear to use their tongue somewhat more. Concentrates, such as shelled corn, are ground between the molars before they are swallowed. Ground feed is picked up by the lips and tongue, mixed with saliva, and swallowed.

8.2.2. Eructation In a ruminant, much pregastric preparation occurs in the rumen. Mastication not only includes reducing particles to a size that can be swallowed, but also mixing the ingested feed with saliva. When the newly eaten feed arrives at the rumen, it is mixed with the rest of the contents and is subject to fermentation by the various organisms that inhabit the organ. During the fermentation process, much gas is formed. The gases are released from the body by belching or eructation. The mechanism of eructation consists of relaxation of the valve at the lower end of the esophagus and contraction of the reticulum to force the gas out. Approximately two-thirds goes to the lungs and one-third passes out through the nostrils. Blockage of the valve at the esophagus causes excessive accumulation of gas, called bloat.

8.2.3. Rumination A very important step in the breakdown of fibrous feeds by a sheep is the process the sheep uses to rework feed previously eaten. Usually the animal seeks a comfortable place where it can relax and chew its cud. This process of regurgitation of previously eaten feed, chewing, remixing with saliva, and swallowing is called rumination. The bolus is forced into the esophagus by contractions of the reticulum and carried to the mouth by a wave of antiperistaltic action. Excess fluid is immediately swallowed.

Generally, rumination proceeds in a more leisurely manner than grazing or eating. From actual observations, it appears that a ruminating sheep may chew 74 times per minute compared with 100 times per minute when eating. The average time of the bolus in the mouth is approximately 50 seconds, followed by 7 to 12 seconds for swallowing, regurgitation, and swallowing the excess liquids. Sheep may have 15 to 18 rumination periods lasting from 30 seconds to 2 hours, with 7 to 9 hours spent daily in that activity. Rumination is influenced by the degree of fullness of the digestive system. Tactile stimulation (scratching effect) of roughages is stimulatory, as contrasted to concentrates or finely ground roughages, which tend to inhibit rumination. Some other factors that appear to affect rumination are frequency of feeding, level of roughage, physical form of the feed, availability of water, and stress.

8.2.4. Ruminal activity The rumen is essentially a fermentation vat maintained under conditions favorable for the function of the microorganisms that inhabit it. The rumen is a muscular organ that has the capability of circulating the contents in an orderly fashion into the various parts. The amount of time feed spends in the rumen depends on the fineness of particles, the amount of fiber or ease of digestion, and the specific gravity of the material. Generally, it appears to require 0.8 to 2.2 days for feed to clear the organ. Certain feeds pass through faster than others. Slow passage is accompanied by higher digestibility of fiber, but increased feed intake with a resultant faster passage provides more nutrients to the animal.

8.2.5. Digestion and absorption of carbohydrates and fats. Within the rumen, complex organic materials such as cellulose, hemicellulose, and lignin are subjected to attack by the microorganisms. Hemicellulose is apparently the most digestible, followed by cellulose and finally lignin, which is largely indigestible. The end products produced by the organisms are volatile fatty acids (VFA), carbon dioxide, and methane. The volatile fatty acids may be absorbed directly through the walls of the rumen or lower in the digestive tract. These are the primary source of energy to the animal. Of the volatile fatty acids, acetate is produced in the greatest quantities and has the lowest usefulness for production; most is used as a source of energy for the rumen epithelium and muscle. Butyrate and propionate are metabolized into ketone bodies as lactate and pyruvate, respectively. They are then transported by the blood stream to the liver, where they are converted to products that can be used by the body.

Although sheep consume very few fats after they are weaned, forages and concentrates do contain lipids. The rumen organisms hydrolyze the triglycerides and galactolipids, releasing free fatty acids, which may be absorbed lower in the digestive tract. The freed glycerol and galactose can then be fermented to volatile fatty acids. Unsaturated fatty acids are hydrogenated to a saturated form.

8.2.6. Digestion and absorption of protein and nitrogenous compounds Protein in feed ingested is subject to attack by organisms that break it down into peptides, amino acids, ammonia, and VFA. The microorganisms can use ammonia, in the presence of adequate energy, to grow and reproduce. In doing this, they synthesize amino acids, which may be useful to the host when the microorganisms are washed into the lower digestive tract and digested.

When protein not digested by microbes reaches the abomasum, it is subject to the hydrochloric acid of that organ. This activates the proteolytic enzymes, pepsin and rennin, which have the ability to digest protein. Further digestion occurs in the small intestine due to enzymes secreted by the pancreas. The protein is degraded in the small intestine into amino acids, which are absorbed and available to the animal. Protein not digested in the abomasum and small intestine passes into the cecum and large intestine, where it is continuously attacked by microorganisms. In those sites, it may yield volatile fatty acids which can be absorbed.

Nonprotein nitrogen, such as urea, is rapidly changed to ammonia in the rumen. The ammonia can be absorbed by the organism if it has a sufficient quantity of readily digestible carbohydrates available. It is interesting that the microbial protein formed from the feeding of urea is of very high quality. This explains why the amino acid requirements of a mature sheep are much less critical than those of a monogastric animal, such as a pig.

Rumen ammonia in excess of the amount that can be used by the microorganisms is absorbed into the blood stream, transported to the liver, and converted to urea, which can either be removed by the kidneys or recycled to the rumen.

The fact that some proteins are subject to immediate digestion in the rumen is actually detrimental to the utilization of some protein-rich feeds. The microorganisms degrade the protein into ammonia so rapidly that some of it escapes, because they cannot use the ammonia as fast as it is produced. For this reason, a protein capable of resisting degradation in the rumen could be more efficient if it were later digested in the abomasum and the small intestine. Proteins capable of resisting degradation in the rumen are called *escape* or *bypass* proteins.

Heat treatment or the use of formaldehyde, tannin, or encapsulation may also be used to protect protein in the rumen. Examples of differences in degradability of protein sources as shown by the work of Miller (1973), Burroughs, Nelson, and Mertens (1975), Chalupa (1975), and Mertens (1977) are included in Table 8.2.

From a practical standpoint, feeds high in escape (bypass) proteins should be more efficiently used than those containing highly degradable proteins if they have a similar amino acid content. Extent of degradability of the protein in feed should surely become more important as more is learned about ruminant nutrition.

8.2.7. Digestion and absorption of minerals Minerals, especially calcium and phosphorus, are released when feed is degraded by fermentation and later by gastric juices. Rumen organisms can also use inorganic minerals such as salts of sodium, calcium, phosphorus, potassium, magnesium, and sulfur. The body absorbs minerals mainly in the small intestine.

Cobalt, copper, iron, and molybdenum have been demonstrated to be stimulatory to cellulose digestion by microorganisms. Other micronutrients are known to be required by the animal and must pass through the rumen and be absorbed in the small intestine.

TABLE 8.2 COMPARISON OF ESCAPE PROTEIN IN FEEDS FOR RUMINANTS BY VARIOUS INVESTIGATORS

Source of protein	Protein not degrated (%)			
	Burroughs[a]	Chalupa[b]	Mertens[c]	Miller[d]
Urea	0	0	0	
Casein		10		
Dried milk	15		4	
Alfalfa (fresh)			21	
Linseed meal	25		24	
Rye (grain)	30		30	
Peanut meal	25	35	30	20
Wheat middlings	30		30	
Alfalfa hay	5	40	30	
Wheat bran	30		32	
Corn gluten feed	25		34	
Sunflower meal	25	25	35	25
Wheat grain	30		35	
Distillers dried grains	25		37	
Oat grain	30		37	
Corn silage		40		
Barley grain	30	20	41	10
Meat and bonemeal	25		42	
Soybean meal	45–51	40	44	45
Corn grain	38	60	44	
Corn gluten meal	25		47	
Cottonseed meal		30	47	
Milo grain		60	48	
Brewers grain			49	
Fish meal			70	70

[a]Burroughs, Nelson, and Mertens (1975).

[b]Chalupa (1975).

[c]Mertens (1977). Data calculated from solubility as follows: degregation = 1.0(solubility %) + 0.5(100 − solubility %).

[d]Miller (1973).

8.3 FACTORS AFFECTING INTAKE OF FEED

Whether or not a feedstuff can be used effectively is determined not only by the nutrient content but also by the acceptance and consumption by the sheep. Regardless of the chemical composition, if the animal will not eat the feed, nothing can be accomplished. Except under laboratory and other conditions that are exceptional, the feeder does not control the total intake by the sheep. It should be borne in mind that a number of factors affect feed intake as discussed next.

8.3.1. Physical There are limits to the amount of feed an animal can consume and digest. Low-quality feed with much indigestible matter must be processed through the digestive tract before there is space available in the rumen. It appears that rate of passage of digested material is the most limiting factor.

8.3.2. Physiological In addition to the physical factor of space, which may limit feed intake, other factors may be in control. There are physiological limits to the energy that can be successfully used by the animal. An animal properly brought up to full feed on a diet will involuntarily limit intake to the amount the system can handle through the action of the central nervous system. Overloading or too rapidly increasing concentrate causes the animal to become sick, go off feed, and scour.

8.3.3. Other factors Other factors affecting feed intake are:

1. The concentrate to roughage ratio
2. Quality of the roughage
3. Physical form of the diet (such as hay and whole grain, meal, or pelleted)
4. Nutrient content (particularly protein and minerals)
5. Previous treatment of sheep (whether deprived of nutrients, including water)
6. Age and development of the sheep (young lambs have less ability to adapt)
7. Physiological state of the animals (pregnant, lactating, or open)
8. Taste and odor factors, especially those associated with decay or degradation or contamination with body waste

While it is true that much is known about the factors affecting voluntary intake, a most useful step is to observe whether the sheep will eat the feed alone or in mixtures. It could also be that a period of adjustment is needed to accustom the sheep to a new and different feed.

8.4 NUTRIENT REQUIREMENTS FOR SHEEP

The National Research Council has published nutrient requirements for sheep as recommended by the subcommittee on Sheep Nutrition (1985). They are shown in Tables 1 through 7 in the appendix. These requirements are based on published nutrition experiments in the United States and abroad. They show the requirements for water, energy, crude protein, macrominerals, and vitamins A and D. These are discussed as an aid in using the requirements in practical production situations.

8.4.1. Water While some nutritionists are not convinced that water should be classed as a nutrient, it is indispensable to the health and well-being of the animal. A young lamb has a body water content of more than 70%. This decreases with age and fatness, but it is invariably 50% or more of the composition of any sheep.

Water is so important because it makes up the major part of essential body fluids. Saliva, blood, and milk are but three that are of paramount importance. Transport of nutrients and removal of wastes from the body both require water. Water is also essential for the regulation of body temperature and is part of many chemical compounds in the body.

Under most circumstances, if clean pure water is available, sheep will consume adequate quantities to meet their needs. This will vary with a number of environmental and other conditions. Factors affecting water consumption are:

1. *Environmental temperature*: As environmental temperatures increase above 70 °F (21 °C), consumption tends to increase especially if the water temperature is below that of the air. Low temperatures tend to decrease the voluntary intake especially below 40 °F (4 °C). Shading waterers during hot summer months tends to increase consumption.

2. *Water in feed:* Sheep on lush pasture tend to consume little water. Feeds high in moisture such as silages, wet beet pulp, wet brewers grains, and the like reduce water intake. Very dry feed, on the other hand, stimulates water consumption.

3. *Type of ration fed:* Sheep on high concentrate rations tend to consume less water than those on high roughage.

4. *Physiological status:* Pregnant ewes tend to consume more water than open ewes. Lactation greatly increases water consumption.

5. *Mineral content of feed and water:* A ration containing 1% salt is well tolerated by sheep, but increasing the salt level to 1.3% doubles water intake. It has been estimated that each gram of excess salt requires 40 ml of water to excrete it.

6. *Nitrogen consumption:* Excessive nitrogen consumption increases the need for water to excrete the products of metabolism.

7. *Dews, showers, snow:* Precipitation in any form on pasture forage tends to reduce water consumption. It has been demonstrated that ewes on winter range may derive enough water from snow cover to satisfy their needs. Sheep can adapt to varying conditions of aridity by reducing the amount of water excreted in feces, concentrating urine, and reducing the losses by respiration and insensible losses over the body from evaporation or sweating. It has been demonstrated that it requires 2 to 3.5 lb (0.9 to 1.6 kg) of water for each pound of dry matter consumed. On dry feed, the ratio may be 1 to 4. Sheep like clear, cool water and will consume somewhat more if it is dripping or running than if it is stagnant. Clay, silt, or animal waste in water reduces consumption. Particular care should be taken to guard the water supply from contamination with toxic substances, pathological disease organisms, or parasites. It has been estimated by practical sheep herders that at least 1 gal (3.8 liters) of water should be provided per ewe daily, even during periods of ideal weather.

8.4.2. Energy One of the most common deficiencies in the ration of sheep is energy. This occurs because the animal does not have access to an adequate quantity or quality of feed. Generally, rations deficient in energy are also lacking in other nutrients.

Sheep that are deficient in energy do not show distinctive symptoms other than slow growth or loss of weight. Reproductive failure and increased mortality in lambs may result from feeding inadequate energy to pregnant ewes. Starving sheep will usually have an unthrifty appearance, with thin, narrow bodies, prominent backbones, and light, characterless fleeces, and will appear listless. Resistance to disease is lowered, and death losses are increased. Nutritionists have made extensive studies of energy, and thus much is known about it. The following are definitions of terms used in discussing energy metabolism:

1. *Gross energy (GE):* Total energy in feed is determined by combustion in a bomb calorimeter.

2. *Fecal energy:* That amount voided in feces and thus not digested.
3. *Digestible energy (DE):* Gross energy (GE) minus fecal energy
4. *Metabolizable energy (ME):* DE minus energy lost from urine and gases.
5. *Heat increment (HI):* Energy lost from digestion and metabolism of feed.
6. *Net energy (NE):* ME minus HI.
7. *Basal metabolic rate (BMR):* Energy required for maintenance of body heat and functions.
8. *Productive energy (PE):* NE minus BMR.

Energy metabolism is illustrated in the following example: A lamb consumes 4,000 kcal of GE in feed. The partition of the energy is diagrammed below.

- GE 4,000 kcal
  - Fecal loss 1,284 kcal (32.1%)
  - DE 2,716 kcal (67.9%)
    - Urinary loss, 140 kcal (3.5%)
    - Methane loss, 292 kcal (7.3%)
    - ME 2,284 kcal (57.1%)
      - HI loss 559.6 kcal (23.1%)
      - NE 1,863 kcal (34%)
        - BMR loss 752 kcal (18.8%)
        - PE 680 kcal (15.2%)

Note: Only 15.2% (680) kcal) of the original 4,000 kcal is available for production functions.

Energy is needed for the following maintenance functions:

1. Heat production and temperature control
2. Digestion, assimilation, and passage of feed
3. Function of muscles, circulation of blood, and respiration
4. Elaboration of glandular secretions
5. Activity increment
6. Nervous impulses and responses
7. Elimination of body wastes

Production functions requiring energy are:

1. Body growth
2. Reproduction
3. Lactation
4. Growth of wool
5. Finishing or fattening

Since the primary nutrient requirement of sheep, in the sense of greatest amount needed, is for energy, it is important to satisfy that need first in the ration. No real production is possible unless energy requirements are met. Energy feeds contain compounds composed primarily of carbon, hydrogen, and oxygen in varying amounts and configurations. These are carbohydrates (cellulose, starches, and sugars), fats, and proteins. All are constituents of roughages and concentrates. Generally, grains and other concentrates have more available energy than roughages. This is usually in the form of starches. Comparable energy levels of concentrates, hays, and silage are shown in Table 8.3.

Roughages, such as silage, hay, and pasture, generally furnish energy at less cost than those from concentrates. Energy is most commonly expressed as digestible energy (DE) or total digestible nutrients (TDN). DE can be converted to TDN by dividing by 2 megacalories. (A megacalorie = 1,000 kilocalories.) Thus, 2,000 kilocalories or 2 megacalories of DE = 1 lb (454 g) of TDN.

TABLE 8.3 COMPARABLE ENERGY LEVELS PER POUND (454 g) OF FEED

Feed	DE (mcals)	TDN (lb)	TDN (g)
Concentrates	1.4–1.6	0.7–0.8	318–363
Hays	0.8–1.1	0.4–0.55	182–250
Corn silage	0.4–0.6	0.2–0.3	91–136

In figuring energy requirements for a sheep, maintenance should be considered first. This varies somewhat with ambient temperature. Usually HI is wasted, but during cold weather it is put to use warming the animal. HI can be a problem when the environmental temperature is high. Then the animal must expend extra energy to dissipate the heat, as in panting.

After the basic requirement of energy for the maintenance of body functions BMR have been met, the animal uses the remaining energy PE for body growth, wool growth, reproduction, lactation, exercise, and fattening. Roughages usually provide energy at lower levels than concentrates, and thus some concentrates may be needed to provide adequate energy levels to support gestation, lactation, and fattening. Fattening is essentially a process of storing energy in the form of fatty tissues.

The availability of the energy in feeds is also related to other nutrient requirements, such as protein and minerals. Thus, if there is not a balance of nutrients, the full value of the energy in feeds may not be realized. Even though there may be sufficient dry matter (DM) available to sheep, they may be energy deficient if they fail to consume enough, due to low palatability or energy concentration, to satisfy their needs. Then the palatability of the feed becomes important.

8.4.3. Protein Proteins are complex organic compounds that contain carbon, hydrogen, oxygen, and nitrogen; they may also contain sulfur, phosphorus, and iron. The basic building stones of proteins are amino acids. The character of the protein is determined by the kind and amount of amino acids it contains. The content of nitrogen in protein averages 16%. The usual analysis for crude protein consists of determining nitrogen content and multiplying that figure by a conversion factor of 6.25. This is called crude protein (CP).

Protein is essential in the nutrition of sheep because it is the main constituent of the protoplasm of body cells. Muscles are largely protein and so is hair, wool, horns, hooves, bone marrow, and almost all the body parts. Protein is essential for growth and repair of body tissues. It is a part of vital substances in the body, such as antibodies responsible for disease resistance, hormones that regulate reproduction and other functions, and enzymes that are vital to the health of the animal.

Lambs require all the essential amino acids needed for monogastric animals until the rumen is functional. No specific amino acids are required by mature sheep except sulfur-bearing ones needed for proper wool growth. The importance of the quality of protein is much less in ruminants than in monogastric animals because of the ability of microorganisms to synthesize most of the amino acids from ammonia and the by-products of the digestion of carbohydrates. They also can use inorganic sulfur to synthesize the sulfur-bearing amino acids. This is the reason that simple nonprotein nitrogen compounds such as urea, biuret, and some others can be included in the diet of sheep, except preruminating young lambs, to satisfy protein requirements. It should be noted, however, that urea must be limited to approximately one-third of the total protein requirements because it is converted almost immediately in the rumen to ammonia, which escapes unless it is used promptly by rumen microorganisms. Thus, some readily digestible carbohydrates such as starches and sugars should be fed at the same time as nonprotein nitrogen so that the organisms have a supply of energy available to enable them to use the compound.

If sheep producers are experiencing trouble with their sheep, it is usually a good procedure to find out if they are feeding enough to satisfy energy and protein requirements. In fact, extension workers claim that they always suspect protein as well as energy deficiency when ketosis or pregnancy disease is diagnosed.

The reasons why it is so important to furnish sufficient protein to ewes are:

1. Ewes very often produce twins or triplets.
2. The gestation length of a sheep is relatively short, and thus there is a heavy demand for protein by the fetus, which is largely composed of protein and water.
3. The ewe is growing wool as well as developing a fetus, and later she is subject to the drain of lactation.

4. It provides nitrogen required by bacteria to digest forage.
5. It stimulates feed intake.

Common symptoms of protein deficiency are:

1. Loss of appetite
2. Cessation of growth
3. Poor development of growing lambs
4. Mature ewes thin and unthrifty
5. Poor wool growth, tender spots on wool, loss of wool
6. Wool eating
7. Impaired reproduction, lambs small and weak at birth
8. Low milk flow
9. Pregnancy disease (also called ketosis, twin lamb disease, or lambing paralysis) is often observed in ewes fed inadequate protein.

While it is true that pregnancy disease is associated with energy metabolism, crude protein deficiency may cause enough reduction in total feed consumption to precipitate the disease. Experiments in the upper Peninsula of Michigan by Blakeslee and Loeffel (1941) demonstrated that feeding a 0.5 lb (23 kg) of oilmeal per ewe daily virtually eliminated pregnancy disease.

If insufficient protein is fed, sheep are usually restless and will try to find a source. A case in point was observed by the author in lamb-feeding experiments with high- and low-protein corn. The lambs were fed a mineral mixture composed of salt and bonemeal ad lib and low-quality hay. Almost immediately, the lambs on low-protein corn began consuming excessive quantities of mineral in an obvious effort to satisfy their requirements of protein from the bonemeal. Their mineral consumption was approximately equal when dicalcium phosphate was substituted for bonemeal.

In an experiment with pregnant ewes at the Missouri Experiment Station (Ross, Dyer, and Krieg, 1967), a group on inadequate protein produced wool with very weak fibers that was easily pulled loose; many ewes shed most of their fleeces. Some of the ewes developed the habit of eating the wool in an effort to compensate for the deficiency.

Protein deficiency is much more critical for growing lambs than mature sheep. Pregnant ewes tend to draw on body stores as long as possible to satisfy the needs of the fetus. It is true that protein is used most efficiently at minimal levels, but this usually results in reduced production. It is safe to conclude that stinting on protein is seldom an economical practice. A ration somewhat in excess of requirements is much to be preferred to one that is inadequate.

Much is being learned about protein in nutrition, especially the use of nonprotein nitrogen supplementation and the use of feeds rich in slowly degraded protein, which can occur naturally or by treatment with formalin, tannin, or heat. These developments foretell increased efficiency of protein supplementation in the future.

8.4.4. Minerals Sheep require minerals for normal growth and development and to maintain body functions. Minerals are responsible for rigidity and strength of bones, hardness of teeth, and regulation of pH of body fluids and are essential components of many enzyme systems. These functions are of vital importance for the maintenance of healthy and productive animals.

The following elements, called macrominerals, are needed in relatively large amounts in the body: sodium, chlorine, calcium, phosphorus, magnesium, potassium, and sulfur. Eight others are required in much smaller amounts and are known as micronutrients or trace elements. These are iodine, iron, copper, molybdenum, cobalt, manganese, zinc, and selenium. Other elements are found in the body of the sheep, but have not been proved to be essential. It seems reasonable to assume that other trace elements will be shown to be essential in the future.

Sodium and Chlorine (Salt). Salt as the principal source of sodium and chlorine is needed by ruminants and thus should be provided, except in unusual circumstances. Sodium and chlorine serve many regulatory functions in the body, such as maintenance of osmotic pressure in cells and transfer of nutrients and waste materials. Chlorine is required as part of the acid in the abomasum. Information is available on the requirements for sodium, but there is little, if any, on chlorine.

Symptoms of salt deficiency are loss of appetite and condition and depraved appetite (chewing wood or eating dirt and weeds not normally found attractive). The amount of salt consumed by sheep depends on the salt content in water and feed. In many range areas, the amount in the drinking water is sufficient to meet requirements. In those areas, too much salt may be a problem.

Range operators estimate ewes will consume 0.5 to 0.75 lb (0.23 to 0.34 kg) of salt per month. Sheep producers in farm states usually feed a mineral mixture composed of salt, plus calcium and phosphorus supplements. A simple mineral mixture may contain salt, bonemeal or dicalcium phosphate, and limestone. In complete mixed diets, 0.5% salt is added, and in concentrate supplements, 1%. These have been shown to be adequate for optimal production.

The form of salt fed sheep is important. Salt blocks on the range tend to dry out and become quite hard, and it may be difficult for the animal to lick enough to satisfy needs. Loose, granular salt is more satisfactory. Salt should always be fed in protected mineral feeders near a source of water.

Sometimes salt is used to limit the intake of very palatable feeds such as soybean oilmeal. The addition of 10% to 15% salt to a concentrate should reduce consumption to 0.5 to 1 lb (0.23 to 0.45 kg) of concentrate daily, but it should be remembered that a generous water supply must be available if sheep are consuming excess salt in feed.

Iodine. This element is found in minute amounts in the body, mostly in the thyroid gland. It is required for the formation of thyroxine, an essential hormone for the regulation of body functions. A deficiency in the diet of pregnant ewes may not be noticed until parturition. Then dead and weak lambs or lambs without wool cover or with goiters are proof of a deficiency. In mature animals, problems of sterility and poor wool growth are symptoms. Soil deficiency in iodine occurs in the Northwest, the Great Lakes region and areas in Pennsylvania, Nebraska, Colorado, Idaho, Utah, and Montana. This condition can be prevented simply by feeding iodized salt to pregnant ewes. Generally, if there is any question about the area, iodized salt is used since it is very little more expensive than the pure product. In cases where salt is used to limit feed intake, iodized or trace-mineralized salt should not be used.

Calcium and Phosphorus. Calcium and phosphorus are important in the formation of normal bones and teeth. It has been shown that 99% of these two elements is found in those two locations. It has also been demonstrated that more than half of the minerals in milk are calcium and phosphorus. Phosphorus is also an important constituent of body compounds, especially those related to reproduction.

Symptoms of calcium deficiency are:

1. Unthriftiness, reduced milk production or milk fever (hypoglycemia) in ewes, slow growth and unthriftiness in lambs.
2. Lowered efficiency, weak bones, lameness, rickets, abnormal muscle contractions, and increased incidence of calculi in lambs.

Symptoms of phosphorus deficiency are:

1. Lack of appetite and listlessness, depraved appetite (eating dirt, chewing bones, wood, or other objects), loss of flesh, and knock knees in growing lambs. Figure 8.2 shows a phosphorus deficient lamb.
2. Poor milk production and breeding failures occur in ewes.

Vitamin D is necessary for normal metabolism of calcium and phosphorus. Thus, if sheep do not have enough of this vitamin or are sheltered from sunshine, the result may be similar to a deficiency of the two elements. Sources of calcium are leafy legumes, limestone, oyster shell or similar materials, dicalcium phosphate, or bonemeal. Grasses grown on soils high in calcium have a higher level than those grown where the soil is deficient. Corn silage, grass hay, and silage are low in calcium, and so are most cereal grains.

Figure 8.2 A phosphorus deficient lamb. (Courtesy J. M. Asplund, Animal Science Department, University of Missouri, Columbia)

Sources of phosphorus are seeds, steamed bonemeal, dicalcium phosphate, monosodium phosphate, monoammonium phosphate, and defluorinated rock phosphate. Mature forage plants are notably deficient in phosphorous. Not only is it necessary to provide minimal quantities of calcium and phosphorus in the diet of sheep, but, under certain conditions, the ratio between the two is important. At the Missouri Experiment Station (Ross, Dyer, and Krieg, 1967), a satisfactory ratio of calcium to phosphorus for pregnant and lactating ewes ranged from 2 to 1 to 1 to 2. When the calcium source was leafy legume hay, the ratios were much wider but were quite satisfactory. Apparently, the ratio is not important if the source of calcium is legume hay. Proper calcium to phosphorus ratios for growing lambs are essential for maintenance of satisfactory growth rates and also to prevent urinary calculi formation in male lambs. A ratio of 2 to 1 or 2.5 to 1 has been shown to effectively reduce the incidence of the condition. A ratio narrower than 1.6 to 1 should be considered near the danger zone for growing finishing lambs. The addition of ammonium chloride or sulfate to feeding rations to prevent urinary calculi is now approved by the Food and Drug Administration for its nutrient content (i.e., nitrogen).

Magnesium. Magnesium requirements for sheep have been estimated at 0.04% to 0.08% of the diet dry matter. It is required in the enzyme systems and for the functioning of the nervous system. Magnesium is important for normal calcium and phosphorus metabolism.

Symptoms of a deficiency are hyperirritability, stiff hind legs, and retracted head. Although it appears to be uncommon in the United States, hypomagnesemia (low levels of magnesium in the blood called grass tetany) is common in the British Isles. It occurs during the peak of lactation, especially in ewes suckling twins. It is characterized by sudden incoordination of muscles, staggering, muscle twitching, and later by death in intense convulsions.

Since most roughages contain 0.1% magnesium and the requirements for the element is approximately 0.06% of the ration, it is seldom a problem except under conditions of stress during cool, wet weather in the spring. Magnesium oxide appears to be the most satisfactory mineral supplement, or magnesium sulfate can be injected using 100 ml of a 25% solution administered subcutaneously or intravenously. If the magnesium solution is to be injected, it is desirable to have the assistance of a veterinarian to avoid overdosing.

Copper. Copper is found in adequate amounts in the forage grown over most of the United States, with the exception of Florida and some of the coastal plains areas of the southwestern United States. A minimum level of 5 ppm is adequate for pregnant and lactating ewes. In young lambs, a deficiency is associated with the absorption of iron and, thus, may be accompanied by anemia. Young lambs from ewes on copper-deficient diets may be born too weak to nurse and thus fail to live. Muscular incoordination and partial paralysis may occur. Copper sulfate is a source, which may be added to salt at the rate of 0.5%. Copper is toxic at levels of 25 mg per head daily; thus it should not be provided unless positive evidence of a need has been shown.

Cobalt. Cobalt is essential for the formation of vitamin B_{12} in the rumen. Cobalt has been found in adequate amounts over much of the United States, except in Florida, Wisconsin, Michigan, Massachusetts, New Hampshire, New York, Pennsylvania, and Ontario in Canada. It seems possible that pockets of deficient soils may also be found elsewhere.

Symptoms of the deficiency are loss of appetite, unthriftiness, weakness, anemia, and decreased fertility. Deficiencies are most common during the spring or a period of lush pasture growth. Mineral sources are cobalt chloride or cobalt sulfate, which may be added to the diet to provide 0.007 ppm on a dry basis. An alternate way to prevent a deficiency is by adding cobalt to a salt mixture at the rate of 5.7 g/45 kg of salt fed free choice. Cobalt levels of 200 mg/100 lb of bodyweight have been shown to be toxic to sheep.

Sulfur. Sulfur content of most grasses and forages has been shown to be adequate to satisfy the needs of sheep. Sheep need the element to synthesize sulfur-bearing amino acids, which are essential for wool growth and lamb production. Sources of sulfur are sulfates or elemental sulfur, either in organic or inorganic form. Sometimes mature grasses or hays may be deficient in sulfur, especially where urea is depended on for a substantial part of the protein requirement.

Selenium. There appears to be a close relationship between selenium and vitamin E metabolism in the prevention of muscular dystrophy (white muscle disease) in lambs. Areas in the United States where selenium is deficient in the soil and thus in feed are Florida, the Pacific Northwest, Idaho, western Montana, Nevada, southern California, Utah, the Northeast (including states bordering on the Great Lakes), New England, and the southeastern seaboard. The U. S. Food and Drug Administration has recently approved the use of selenium supplements in the rations of sheep at levels of 0.1 ppm in total diet and 39 ppm in salt.

Other Essential Elements. Potassium, zinc, molybdenum, and iron have all been demonstrated to be essential to sheep. Natural supplies in feeds have, to date, provided adequate amounts to prevent problems. Some of the elements, such as iron, may be needed when lambs are reared in confinement. A trace mineral salt designed for sheep may be used to satisfy the needs. Care should be taken not to use trace mineral salt designed for swine rations, because the copper content is too high for sheep. Lambs born in confinement should be given iron injections at birth.

Other elements found essential for some species are silicon, tin, vanadium, nickel, and manganese, but none of these has been shown essential for sheep. Fluorine has not been considered an essential element for sheep, but it has been shown to be beneficial for the formation and maintenance of healthy teeth in some animals.

Excess Minerals May Be Harmful. High levels of fluorine, selenium copper, cobalt, molybdenum, and lead found in certain areas of the United States may cause problems with sheep production and management.

Fluorine is cumulative in the body and may not be observed for years. The animals exhibit anorexia and their bones become thickened and soft and break easily. Teeth become discolored and erosion occurs. Continued intake of water or feed with toxic levels of fluorine results in reduced feed consumption and degenerative changes in soft tissues and organs. Breeding ewes can tolerate 150 ppm and finishing lambs, 100 ppm.

Levels of 10 to 20 ppm of selenium in feed may cause toxicity in sheep if they remain on the feed for long periods. Ewes with the toxicity have loose, weak-fibered wool and may have malformed lambs. Although it is cumulative, the animal rapidly eliminates accumulated selenium when changed to rations low in the element. Areas where forage is high in selenium are parts of Montana, Wyoming, Colorado, New Mexico, Minnesota, Arizona, North and South Dakota, Nebraska, and Kansas. Some success has been achieved in treating selenium toxicity with arsenic compounds.

Only a few areas have the element molybdenum in excess. Toxicity has been observed in parts of California and Florida in the United States and in Manitoba in Canada. Symptoms of toxicity are similar to those of copper deficiency, including anemia, diarrhea, and emaciation. Addition of copper sulfate to the diet at levels of 0.25 g/100 lb of body weight should prevent scouring of animals on high levels of molybdenum in forage.

Lead is a poison that accumulates in the body. Lead-based paints, batteries, used motor oil, and linoleum are possible sources of the element. In areas where industry is working with lead, as in smelters, the amount in dust and air can be dangerous. Sheep appear to be able to tolerate levels below 200 ppm in feed.

8.4.5. Vitamins

Vitamin A. Sheep require a dietary source of vitamin A or its precursor, carotene, to maintain normal epithelial tissues. These are the tissues lining the respiratory, digestive, and reproductive tracts. The eyes and bones are also affected. Generally, the first symptoms of vitamin A deficiency are elevated spinal pressure and night blindness. Later the eyes weep excessively and become bloodshot. A deficiency may result in poor breeding performance or weak, dead, or malformed lambs at birth. It may also result in abortion in severe cases. A sheep suffering from vitamin A deficiency is often prone to respiratory or other infections. Carotene found in the green leaves of plants can be converted into vitamin A in the body. It is also stored for long periods in the liver and other tissues. Thus a sheep that has access to good pasture can very well store enough vitamin A for 6 months or longer. Problems usually occur after prolonged droughts or when hay or grain that has been stored for long periods is fed in drylot or confinement. Feeds with high vitamin A potential are leafy legume hay, dehydrated legumes or grasses, and yellow corn. Commercial sources are available both as a feed supplement or in an injectable form. The carotene to vitamin A equivalence is 1 mg of carotene = 400 IU of vitamin A for sheep.

Vitamin D. Vitamin D is essential for the proper metabolism of calcium or phosphorus to form bones and teeth. This vitamin is most important for young, growing animals. If they are exposed to sunlight, they will usually receive adequate amounts of the vitamin by irradiation. Rickets results from the lack of vitamin D, with characteristic bowed legs or misshapen bones. In mature animals, it appears to be a problem only in pregnant ewes, which under extreme deficiency may give birth to lambs with congenital malformations. Rapidly growing lambs during extended periods of cloudy weather or lambs reared in confinement may have the problem. Diagnosis may be made in postmortem by observing how easily the ribs can be broken. Lambs deficient in vitamin D or calcium or phosphorus tend to gain less than those on adequate levels. Vitamin D injections usually give immediate response. These are often sold as a vitamins ADE complex. Most feed stores also sell vitamin ADE supplements that can be added to the feed. Sun-cured legume hay is considered to be a satisfactory source of the vitamin.

Vitamin E. Vitamin E is required by sheep at a level of 15 to 20 IU Vitamin E/kg of diet daily, but is seldom a problem except in young suckling lambs. Then a deficiency in the diet of the ewe may result in muscular dystrophy (also called white muscle or stiff lamb disease), which may be fatal to lambs. It is characterized by muscular weakness in live lambs and degeneration of the skeletal muscles and sometimes the heart. It may be recognized in lambs by stiffness and incoordination in both hind legs. There is a close relationship between vitamin E deficiency and inadequate selenium in the diet of the ewe. In areas where the problem exists, it may be prevented by adding supplements rich in vitamin E and selenium to the ration of the pregnant ewe. Most feeds contain vitamin E, especially seeds, but the vitamin is destroyed rapidly by oxidation; thus feeds stored for long periods may be deficient in it. Lambs showing symptoms of the disease (see Figure 8.3) may be saved if treated immediately with oral doses of 500 mg of alpha tocopherol, followed on alternate days with one-fifth of that amount.

B Vitamins. B vitamins are synthesized in the rumen of sheep and except for B_{12}, which requires cobalt for synthesis, have not been shown to be essential in the diet. Young lambs before they have a functional rumen require thiamin, riboflavin, and folic acid, but these are plentiful in the milk of the ewe. B-complex vitamins should be provided in milk replacers.

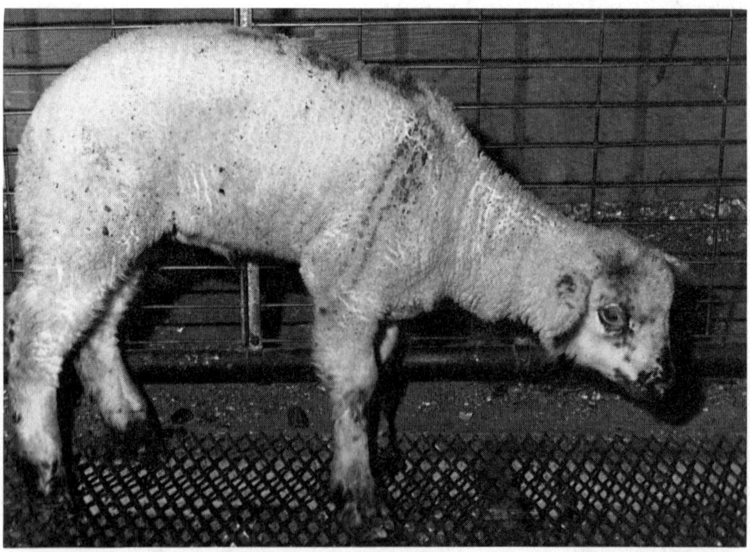

Figure 8.3 Lamb with stiff lamb disease due to vitamin E deficiency. (Courtesy Animal Science Department, Ohio Research and Development Center, The Ohio State University, Wooster)

TABLE 8.4 COMPARISON OF NUTRIENT REQUIREMENTS OF 110-POUND (50-KILOGRAMS) EWES WITH A FINISHING LAMB OF SIMILAR WEIGHT

	DM		TDN		DE	CP		Ca		P	
	(lb)	(kg)	(lb	(kg)	(mcal)	(lb)	(g)	(lb)	(g)	(lb)	(g)
Dry ewe (maintenance)	2.2	1.0	1.2	0.55	2.4	0.21	95	0.004	2.0	0.004	2.0
Pregnant ewe last 4 weeks of gestation	3.5	1.6	2.1	0.94	4.1	0.38	175	0.013	5.9	0.01	4.8
Ewe suckling twins first 6 to 8 weeks of lactation	5.3	2.4	3.4	1.56	6.9	0.86	389	0.023	10.5	0.016	7.3
Finishing lambs 4 to 7 months old	3.5	1.6	2.7	1.23	5.4	0.35	160	0.012	5.6	0.007	3.0

8.5 EWE NUTRITION

The requirements of a ewe for energy and protein change radically during a year, depending on her status. The energy requirements increase from the time she is dry and open until she is lactating. In fact, her peak demand is more than double that of the dry state. Requirements also change during gestation and lactation, and a ewe suckling twins requires much more nutrients than one suckling a single lamb. Energy, crude protein, calcium, and phosphorus all increase appreciably as the ewe progresses from breeding through gestation and lactation. This emphasizes the need for nutrients to produce a lamb plus a wool crop. It is also of interest to compare the energy and protein requirements of a ewe and a finishing lamb, as shown in Table 8.4. On the basis of similar weights, a lactating ewe requires more nutrients than a finishing lamb. Actually, a heavy milking ewe cannot eat enough to provide all the nutrients secreted in her milk, and thus she must draw on body stores to make up the difference. She will tend to lose weight and stored fat so that, at the end of the suckling period, she should be suckled down to thin condition. This is the reason why ewes should generally gain 15 to 30 lb and be fleshy when they lamb. The extra flesh is needed to provide energy for formation of the fetus and lactation. Ewes suckling twins require 20% to 25% more energy and 25% more protein than those with singles. Rations for ewes must include feeds that are palatable, with sufficient energy per unit of weight so that she can consume enough to meet her needs.

8.6 LAMB NUTRITION

Suckling lamb requirements are satisfied initially by milk, providing it is furnished in adequate amounts. Table 8.5 shows a comparison of the milk of a ewe and cow. Ewe milk is more concentrated and contains more protein, fat, and minerals than

TABLE 8.5 COMPOSITION OF MILK OF THE EWE AND COW

	Ewe, %	Cow, %
Total solids	18.4	12.1
Crude protein	5.6	3.25
Fat	7.5	3.5
Lactose	4.4	4.6
Ash	0.87	0.75
Calcium	0.19	0.12
Phosphorus	0.15	0.10
Sodium	0.07	0.05
Potassium	0.19	0.15

milk from a cow. Only in lactose is cow's milk somewhat superior. This is the reason why rearing lambs on cow's milk is often not satisfactory. Commercial milk replacers now on the market have been formulated to provide more protein, fat, and minerals than cow's milk and are therefore much more satisfactory. On a dry basis, the needs of a suckling lamb are quite high for protein and energy. Suckling lambs need high-quality protein and easily digested sources of energy, along with minerals and vitamins, much the same as animals with a simple stomach, such as the pig and human. Not until the rumen is developed and functioning can they use fibrous feeds and nonprotein nitrogen. On a percentage basis, early-weaned lambs have a higher protein requirement than older and more mature lambs. As the lamb matures, the protein requirement tends to decrease and the energy level of the diet increases. The same can be said for the calcium–phosphorus and energy content of the diet. The ratio of calcium to phosphorus remains wider than 1.6 : 1.

Growing lambs to be used for replacements require decreasing concentrations of energy and protein as they mature, but the level of calcium and phosphorus and the relationship between the two changes very little.

REFERENCES

Blakeslee, L. H., and W. J. Loeffel, "Cull Beans and Cottonseed Meal for Wintering Breeding Ewes," *Mich. Agr. Exp. Sta. Quarterly Bull.*, 24, 1941.

Burroughs, W., D. K. Nelson, and D. R. Mertens, "Evaluation of Protein Nutrition by Metabolizable Protein and Urea Fermentation Potential," *J. Dairy Sci.*, 58, 1975, p. 611.

Chalupa, W., "Rumen Bypass and Protection of Proteins and Amino Acids," *J. Dairy Sci.*, 58, 1975, p. 1198.

Church, D. C., *Digestive Physiology and Nutrition of Ruminants*, Vol. 1. Corvallis, Ore.: D. C. Church. 1969.

_______, *Digestive Physiology and Nutrition of Ruminants,* Vol. 2. Corvallis, Ore.: D. C. Church, 1972.

_______, *Livestock Feeds and Feeding*. Corvallis, Ore.: D. C. Church, 1977.

Coop, Ian, *Sheep and Goat Production*. New York: Elsevier Scientific Publishing Co., 1982.

Ensminger, M. E., *Sheep and Goat Science.* Danville, Ill.: Interstate Printers and Publishers, Inc., 1986.

Ensminger, M. E., and C. B. Olentine, Jr., *Feeds and Nutrition*, 1st ed. Clovis, Calif.: Ensminger Publishing Company, 1978.

Haresign, William (ed.), *Sheep Production.* London: Butterworths, 1983.

Lassiter, J. W., and H. M. Edwards, Jr. *Animal Nutrition*, Reston, Va.: Reston Publishing Co., 1982.

Maynard, L. H., and J. K. Loosh, *Animal Nutrition*, 7th ed. New York: McGraw-Hill Book Co., 1979.

McDonald, Peter, R. A. Edwards, and J. F. D. Greenhalgh, *Animal Nutrition.* New York: Hafner Publishing Co., 1981.

Mertens, D. R., *Importance of Measurement of Protein Insolubility in Ruminant Diets*, Proceedings of the Georgia Nutrition Conference, 1977, p. 30.

Miller, E. L., "Bypass Protein," *Proc. Nutrition Soc.*, 32, 1973, p. 79.

Morrison, F. B., *Feeds and Feeding,* 22nd ed. Ithaca, N.Y.: Morrison Publishing Co., 1956.

Orskov, E. R., *Protein Nutrition in Ruminants*, New York: Academic Press, 1982.

Ross, C. V., A. J. Dyer, and K. L. Krieg, "Rations for Wintering Ewes," *Mo. Agr. Exp. Sta. Bull. 861*, 1967.

Shalk, A. F., and R. S. Amadon, "Physiology of the Ruminant Stomach," *N. Dakota Agr. Exp. Sta. Bull, 216*, 1928.

Scott, George E., *The Sheepman's Production Handbook.* Denver, Colo.: Sheep Industry Development Program, Inc., 1970.

Subcommittee on Sheep Nutrition, *Nutrient Requirements of Sheep*, 6th rev. ed. Washington, D.C.: National Academy of Sciences, 1985.

Wilson, P. N., and T. D. A. Brigstoeke, *Improved Feeding of Cattle and Sheep.* London: Grenada Publishing Ltd., 1981.

QUESTIONS FOR STUDY AND DISCUSSION

1. What is an ungulate, and what are their identifying characteristics?
2. How do the digestion systems of a pig and sheep differ?
3. Draw a diagram of the digestive system of a sheep showing the activity that occurs at each location.
4. How are the ingesta moved through the digestive system of a ruminant?
5. What is the significance of rate of passage in ruminant nutrition?
6. What is the total capacity of the organs in the digestive system of a sheep?
7. What purpose is served by saliva secretion by the ruminant?
8. What happens to the gases given off by the rumen?
9. What microorganisms inhabit the rumen?
10. The abomasum in a ruminant corresponds to what organ in the human?
11. Volatile fatty acids are most important in the metabolism of which of the following: energy, protein, or minerals?
12. In what ways can a protein be treated so it will have slow degradation in the rumen? Why is this important?

13. List two feeds that have highly digestible proteins and two that have a high level of escape (bypass) proteins.
14. Discuss factors affecting feed intake of a sheep.
15. Support the statement "Water may be the most limiting nutrient for life."
16. What is the difference between net energy and productive energy?
17. Under what circumstances is HI useful in energy metabolism?
18. Why is quality of protein less important to ruminants than monogastric animals?
19. Iodine deficiency is associated with what syndrome in newborn lambs?
20. Under what conditions may lambs have vitamin D deficiency?
21. How do requirements for energy of a ewe change throughout a production cycle?
22. Relate selenium deficiency with a syndrome in baby lambs.
23. Relate pregnancy disease to a nutritional deficiency.
24. Urinary calculi are often related to the intake of two nutrients in an incorrect ratio. What nutrients are involved, and how can the syndrome be prevented?

9 Feeds and Feed Formulation for Sheep and Lambs

9.1 IMPORTANCE OF FEEDS

The largest item of expense in a sheep operation is usually feed. Sheep are adapted to the consumption of large amounts of bulky feeds, mostly of plant origin, but there are critical periods in the production cycle when concentrates are needed to satisfy nutritional requirements. A smart producer should be well acquainted with the various types of feed and know their costs and capabilities. Often, some options may be much more favorable than others. A constant study of feeds should pay good dividends.

9.2 CLASSIFICATION OF FEEDS

Feeds for sheep are classified as concentrates, roughages, or pasture. Feeds are classed according to the amount of nutrients they contain. Concentrates generally have more available energy and less crude fiber than roughages. The dividing line is based on fiber content. Feeds containing 18% or more of crude fiber (CF) are classed as roughage; those below are classified as concentrates. Pasture differs from roughage in the dry matter content, and it is not harvested mechanically. Most of the diet of a sheep is pasture and roughage. Concentrates are seldom needed except for growing and finishing lambs and pregnant and lactating ewes.

Concentrates are nearly always more expensive than roughages per ton but may be less expensive on the basis of energy content. Thus they should be fed judiciously. Some of the concentrates may have to be purchased off the farm, which may put a strain on the budget.

9.3 CLASSIFICATION OF CONCENTRATES

Concentrates can be classified as follows:

1. *Carbonaceous:* Grains, by-products, molasses, fats from plants and animals

2. *Nitrogenous:* Oilmeals, by-products from plants and animals, urea and other nonprotein nitrogen
3. *Minerals:* Salt, calcium, phosphorus, and trace-mineral supplements
4. *Vitamins:* A, D, and E supplements

9.3.1. Carbonaceous concentrates Most high-energy feeds for sheep are carbonaceous concentrates. These consist primarily of grains such as corn, barley, sorghum, oats, wheat, and rye; but they may also be by-products, such as corn gluten meal, wheat bran, hominy feed, brewers dried grains, and beet pulp. Each source is discussed in the following.

Corn (Zea mays). Of the 4 to 6 billion bushels of feed grains produced annually in the United States, corn is by far the most important. Corn is produced in greatest quantities in the north-central region of the country, but it is shipped to many parts of the United States and to foreign countries. Corn is very palatable to sheep, and they do well on it. It is high in energy but relatively low in crude protein (7% to 9%). The protein content may vary widely depending on the variety and the fertility of the soil where it is grown. It is low in calcium and medium in phosphorus content. It is the only grain that has vitamin A potency. Xanthophyll in corn will provide some vitamin A, providing it has not been stored for long periods.

Specifications for number 2 shelled yellow corn, the standard grade for futures and grain sales, are minimum test weight of 54 lb (24.6 kg) per bushel, moisture content of 15.5%, no more than 3% broken kernels or foreign material or 5% damaged kernels. Permissible heat damage is 0.5%. Corn showing weevil damage, mold, cob rot, or insect damage is graded lower and discounted on the market. All these defects can decrease the palatability and feed value of the grain. Corn is the grain used as the standard by which others are measured. Except for old ewes and young lambs or in mixed diets, there is no need to grind or otherwise process corn.

Grain Sorghum (Sorghum bicolor subglabrescens). Sorghum (milo) is grown and fed extensively in the southwest part of the United States. It accounts for 15% to 20% as much grain as corn in the United States. Milo has a higher protein content than corn, 9% to 14% versus 7% to 9%, but is lower in energy (3.66 versus 4.14 Mcal DE/kg). It is as low in calcium and about the same in phosphorus as corn. It is devoid of carotene. Milo is not as palatable as corn, but has 85% to 100% the feeding value. It is satisfactory as the sole grain. Unless it is for very old sheep with poor teeth and suckling lambs or where diets need to be thoroughly mixed, there is no advantage in grinding or other processing of the grain. Number 1 sorghum must have a test weight of 57 lb (25.9 kg) per bushel, 13% or less moisture content, less than 2% damaged kernels, and less than 4% broken kernels, foreign material, or other grain.

Barley Grain (Hordeum vulgare). Only 5% to 6% as much barley as corn is fed to livestock. Most of it is produced in the north-central, northwestern, and western states in irrigated valleys. It is higher in protein (12% to 14% versus 7% to 9% for corn), calcium, and phosphorus and lower in energy than corn, and devoid of carotene. It may be used as the only grain for sheep and except in unusual cases needs no processing. Specification for grade 1 are test weight per bushel of 47 lb (21.4 kg), minimum sound grain of 97%, maximum limit for foreign material of 1%, broken grains, 5%, heat-damaged kernels, 0.2%, total damaged kernels, 2.0%, and discolored grain, 0.5%. Barley grown in the Pacific Coast states is lower in protein than the grain from the rest of the United States.

Wheat (Triticum aestivum). Wheat grain is not often fed to sheep because it is in such demand for human food. It is produced over much of the United States. It is higher in protein, calcium, and phosphorus but slightly lower in energy than corn, and is devoid of carotene. Although it can be fed alone, better results can be obtained by combining with another grain, preferably as no more than 50% of the mixture. Grinding or processing wheat may allow easier mixing with other feeds, but it is questionable if it is worth the cost. Specifications for grade 1 wheat are test weight per bushel of 60 lb (27.3 kg), damaged kernels at no more than 2% (heat damaged, 0.2 percent), foreign material, 0.5%, and shrunken and broken kernels, 3%.

Rye (Secale cereale). Rye may be fed to sheep, and they eat it better if it is unground. It is similar in composition to barley. It is approximately equal in feed value to corn when it makes up 50% or less of the grain mix. Weight per bushel is 56 lb (25.45 kg) for grade 1 and 54 (24.5 kg) for grade 2. Availability is limited because it is seldom used for animal feed. It is not as palatable as corn, barley, and the sorghums. Rye may contain ergot, which can cause pregnant ewes to abort.

Oat Grain (Avena sativa). Oats are higher in protein and lower in DE than corn. They are higher in crude fiber and calcium and about the same in phosphorus as corn. Oats are the safest of all grains for sheep, and are the preferred grain for breeding flocks and growing lambs and for starting feeder lambs on feed. They need no processing before being fed to sheep. The standard weight for oats is 32 lb (14.55 kg) per bushel. Specifications for U. S. grade 1 are minimum sound grains of 97% and maximum limits of 2% foreign materials, 2% other grains, and 0.1% heat-damaged kernels. There is much variation in the weights of oats throughout the United States. Heavy oats are considerably more valuable than those under the standard weight for the grain. Oats are grown in a wide area of the United States, but supply only 10% to 12% as much grain as corn.

Other Carbonaceous Concentrates. Beet pulp is high in carbohydrates and a good feed for sheep. It is low in dry matter and protein when it is fed wet. The dried product is a very palatable, high-energy concentrate. Most dried beet pulp is considerably more expensive than grains and is seldom fed to sheep, except those being fitted for show and sale. There is also a product called beet pulp with molasses, which contains more energy but is slightly lower in protein. Because of the cost of the products, they are usually only fed in combination with less expensive grains.

Dried citrus pulp is sometimes available in areas where oranges and grapefruit are grown. It is similar in composition to beet pulp, but lower in protein. Sheep require time to become accustomed to citrus pulp, but then they eat it readily. It is advisable to use citrus pulp only as part of a mixture with grain and a protein supplement. It appears to be approximately equal to barley in finishing rations, containing 50% corn or barley.

Grain Milling Products. A number of grain milling by-products are used for livestock feed. Screenings are sometimes available and can be used as energy sources. These consist of broken, small, or shriveled grains or weed seeds. Screenings may vary in nutrient content and are inferior to wheat as an energy source for sheep. They are usually ground before use.

Most by-products of milling wheat are used for feeding monogastric animals and are therefore too expensive for feeding sheep. Wheat bran, however, is often used in the rations of pregnant ewes and suckling lambs because it is bulky and rather laxative. It contains approximately 16% crude protein and is rich in phosphorus and vitamin E.

Bakery waste may be cheap enough to be fed in combination with grain (usually no more than 50%) with satisfactory results. On a dry basis, it may be expected to be similar to grain in nutritive value. Unless it is kiln dried, it may spoil rather quickly. It handles better in mixtures with grain if it is coarsely ground. Dried apple pomace can be used in combination with grain in lamb finishing and ewe rations and has a feeding value of approximately 75% of dry beet pulp.

Molasses. Cane, beet [wood molasses (hydrol)], and citrus molasses are used in sheep rations to provide energy, increase palatability, and reduce dust in mixed diets. They are all of approximately equal feeding value, provided the dry matter is similar. Their feeding value is approximately 75% that of corn, provided they are limited to no more than 10% of the diet. Under practical feedlot conditions, more satisfactory results are obtained if the level is kept to 7% or below in lamb rations.

The level of sugar in cane, beet, and citrus molasses indicates the feed value. Brix values are expressed in degrees, with each degree equal to 1% sucrose (see Table 9.1).

TABLE 9.1 BRIX VALUES FOR MOLASSES

		Weight per gallon (3.8 liters)	
	Brix (degrees)[a]	lb	kg
Cane	79.5	11.75	(5.34)
Beet	79.5	11.75	(5.34)
Citrus	71.0	11.29	(5.34)

[a]The brix scale represents the percentage, by weight, of sugars in solution as measured by hydrometer at standard temperature.

Animal and Other Fat. Fats may be added to lamb rations to advantage if mixing facilities are available. They are much higher in energy (225%) than corn, but digestive disturbances and reduced feed intake are possible if the total level exceeds 5% to 7%. Fats in a mixed feedlot diet may appreciably increase the palatability by decreasing the dusty character of the feed.

9.3.2. Nitrogenous concentrates

Soybean Meal. Soybeans are produced in abundance in the United States. The raw beans are usually processed to remove the oil for human food, and the remaining residue is called soybean meal. Most soybeans are processed by the solvent process because it is more efficient in removing the oil than the expeller method. It is the most widely used protein supplement in the central part of the United States. Soybean meal is very palatable to sheep. It contains 48% to 52% protein and 3.53 Mcal DE/kg, depending on the efficiency in removing the oil.

Cottonseed Meal. Cottonseed meal is the result of dehulling and extracting the oil from cottonseeds. It may vary from 36% to 41% crude protein depending on the efficiency of removing the hull. Most of the cottonseed meal available contains 41% protein. It ranks second only to soybean meal as a protein supplement. The energy content is 3.0 Mcal DE/kg, which is somewhat less than for soybean meal. It is a safe feed and is widely used in the southwestern and western states for supplementing sheep rations.

Linseed Meal. Linseed meal is the ground product after the oil has been extracted from flax seed. It contains 35% protein and 3.44 Mcal DE/kg. It is produced in the Dakotas, Minnesota, and Texas. Many of the older, purebred sheep breeders preferred linseed meal to any other protein supplement for feeding show sheep. This was because the expeller process used in the early years was less efficient in removing the oil and thus much was left in the meal. The linseed oil caused sheep to have a bloom not observed with other supplements. Present linseed meal is produced by the solvent method. It is well liked by sheep and fully equal to soybean and cottonseed meal on a comparable protein basis.

Other Nitrogenous Concentrates of Plant Origin. A number of other sources of protein can be used in sheep diets, but they are limited in quantity and to certain geographical locations. There are brewers grains, distillers grains and solubles, peanut meal, safflower meal, sesame meal, and sunflower meal. If the energy situation should change appreciably, it is conceivable that much more distilling will be done to provide alcohol as a gasoline additive. Then the amount of residues would be much increased. Both brewers and distillers grains may be available in wet form, but the very high water content and the difficulty of storing the products dictate a location for feeding near the site of production. Sunflower seed meal could become more plentiful if the petroleum situation remains critical. It appears to be almost equal to soybean meal in value as a protein supplement.

Nitrogenous Concentrates of Animal Origin. Protein supplements of animal origin are primarily used for feeding poultry and swine and are seldom available in quantity or at a price that will allow them to be used in ruminant diets. Under unusual market conditions, lamb feeders have effectively used such diverse animal protein products as tankage, meat scraps, fish meal, feather meal, and dairy by-products in diets. These may not be immediately palatable; therefore, they can be incorporated most simply by using them in complete mixed diets. Since many animal proteins, such as fish meal, are slowly degradable, they may be used to a greater extent in ruminant rations in the future.

Nonprotein Nitrogen. Urea is manufactured using natural gas and air. It may contain either 42% or 45% nitrogen, which gives it a protein equivalent of 262% or 281%, respectively. It has no energy, vitamin, or mineral value and requires an adjustment period of 2 to 3 weeks before the animal can use it efficiently. It should never be fed to lambs before they are weaned. It should never be fed to sheep unless they are receiving a source of readily digestible carbohydrate, such as starch or grain. The general recommendation is that urea should account for no more than one-third of the protein in the ration. Urea tends to work best in feed mixtures for finishing lambs. It is ineffective in high-roughage diets unless some grain is added. Urea must always be thoroughly mixed with other feed. Clumps may cause death by ammonia toxicosis.

Biuret. Biuret is made by combining two molecules of urea into one compound. It is not nearly as soluble as urea and therefore may be used where sheep are on limited feed. It is considerably more expensive than urea.

9.3.3. Characteristics of mineral concentrates There are a few places in the world where no supplemental minerals are required, but these are rare. Almost all livestock require at least salt in their ration or available free choice. The kind and amount of supplementation needed are dependent on the function being performed by the animal and the mineral content of the ration. The various mineral sources may be furnished by the concentrates discussed next.

Salt. Salt is a source of sodium and chlorine, which are required by all animals. It is available in loose, granulated form or in blocks. Salt is an excellent carrier of other elements and thus is often fortified with trace minerals. Trace-mineral salt should never be used for sheep unless specifically designed for the species or unless the feeder is knowledgeable or skilled enough to understand and compensate for the elements that may be in excess. Salt may be fed free choice, alone or in mixtures with calcium and phosphorus concentrates. It may also be incorporated into mixed diets or supplements.

Calcium Concentrates. If high levels of leafy, legume hay are being fed, there is little danger of a calcium deficiency in a ruminant. However, carbonaceous roughages and high levels of cereal grains may provide insufficient calcium to sheep and lambs. Finely ground limestone is usually the cheapest and best source of calcium. High-quality limestone is essentially pure calcium carbonate that contains 38% or more of calcium; however, the average calcium content is 33% to 34%. Dolomitic limestone may also be used, but it is a calcium magnesium carbonate and has a calcium and magnesium content of 22.4% and 10%, respectively. Ground oyster shell, marl with a low clay content, precipitated calcium carbonate, and coral are all sources of calcium, but are less satisfactory than ground limestone.

Phosphorus. Sources of phosphorus are steamed bonemeal, dicalcium phosphate, diammonium phosphate, phosphoric acid (feed grade), and defluorinated rock phosphate. Nearly all phosphorus supplements provide some calcium in addition to phosphorus. Steamed bonemeal has historically been a favorite of livestock feeders, but it is usually very expensive and difficult to find. Dicalcium phosphate is somewhat less palatable than bonemeal but an excellent supplement. Defluorinated rock phosphate is a satisfactory source also, but is somewhat less palatable than bonemeal or dicalcium phosphate. Other sources are even less acceptable to sheep. Table

TABLE 9.2 CONTENT OF CALCIUM AND PHOSPHORUS IN MINERAL SUPPLEMENTS

Product	Calcium, %	Phosphorus, %
Steamed bonemeal	30.7	12.9
Dicalcium phosphate	24–27.1	18–19.3
Diammonium phosphate	0.59	20.6
Phosphoric acid (feed grade)	—	31.6
Defluorinated rock phosphate	32.0	16.25
Limestone	34.0–38	—
Dolomitic limestone	22.1	—
Oyster shell	38.0	—

9.2 shows levels of calcium and phosphorus in products used for mineral supplementation.

Iodine. Generally, iodized salt is available at very little expense above the pure product and should be used in states of the Northwest, the Great Lakes region, the Midwest and Mountain States.

Selenium. This element is important in the prevention of muscular dystrophy (white muscle or stiff lamb disease) in lambs. Since 1978, it has been approved as a supplement not to exceed 0.1 ppm of the complete feed or total ration. Commercial salt mineral mixtures are available for use in formulating rations for pregnant and lactating ewes.

Cobalt. Areas on the Gulf Coast, the Great Lakes, parts of New England, and Alberta, Canada, have soils deficient in cobalt. Isolated localities in other states are also deficient. Incorporating cobalt chloride or cobalt sulfate at a level to provide 5.7 g of cobalt per 100 lb of salt or 0.1 ppm has been recommended.

9.3.4. Characteristics of vitamin concentrates Since sheep and lambs rarely have a deficiency of vitamins, not many products designed especially for sheep are available. Vitamins and their sources are discussed next.

Vitamin A. Deficiency of vitamin A can be prevented or controlled by feeding a high level of carotene, such as dehydrated alfalfa meal, high-quality, green legume hay, or the vitamin mixed with a dry carrier. Intermuscular injections of the vitamin are also available. Ewes on good pasture will store enough of the provitamin to last them 6 months.

Vitamin D. Sources of the vitamin are sun-cured hay, fish liver oils, and calciferol, which is an irradiated plant product. A dry supplement is available commercially or the vitamin can be injected.

Vitamin E. Since vitamin E is fat soluble and the germs of cereal grains contain the vitamin, a ewe being fed grain will not often need supplementation. Pregnant ewes sometimes are deficient and may need some in their ration. There are commercial preparations that provide A, D, and E in concentrated form or intramuscular injections can be given.

9.4 CLASSIFICATION OF ROUGHAGES

Harvested roughages can be classified as follows:

1. *Carbonaceous:* Those derived from the various grasses and processed to produce hay, haylage, and silage.
2. *Nitrogenous:* Those made from the various legumes and processed to produce hay, haylage, and silage.
3. *Combination of carbonaceous and nitrogenous:* Mixed grass–legumes processed to produce hay, haylage, and silage.
4. Other carbonaceous roughages.

9.4.1. Characteristics of carbonaceous roughages Preserved products in this category are from the grasses, including cereal grains. They are generally higher in fiber and lower in protein than the nitrogenous roughages.

Grass Hay. Hay made from meadow grasses varies in quality depending on its stage of maturity when harvested and the way it is cured and stored. In areas of the country that have high rainfall in the spring, it is difficult to harvest and cure at the correct time; frequent showers may make it impossible to work in the fields, and high humidity prevents the grass from curing. Grasses are highest in nutrients when they are in the vegetative stage. Once they have formed seeds, their quality decreases rapidly. Therefore, grass hay cut during the bloom stage is much higher in crude protein than it is when it is more mature. Generally, grass hay needs to be supplemented with protein, energy, and minerals for sheep, except open ewes on maintenance rations. Bright green grass hay may be fed in any quantity without digestive disturbances. Stemmy grass hay tends to be unpalatable, low in protein and digestible nutrients, and high in fiber.

Moisture in hay must be reduced to approximately 22% before it can be safely baled without excessive fermentation, which causes heating and molding. Addition of proprionic acid to hay at the rate of 1 to 2 gal/ton will allow baling at higher moisture levels.

Grass Haylage. Haylage by definition is forage that has been allowed to cure enough to reduce the moisture to 50% or less before it is ensiled in an airtight structure. The grass may also be preserved by using carbohydrates (such as the sugar in molasses), starch (in grain), organic acids, or other preservatives. Haylage is less subject to losses of nutrients than hay and can be harvested with less regard to the weather. It is well liked by sheep and generally presents very few problems in feeding. Protein and mineral, particularly calcium, supplementation may be required.

Grass Silage. Grass silage may be stored in airtight silos or preserved by the use of fermentable carbohydrates or other preserving agents. The moisture content may be high, and the feed may require supplementing with protein, minerals, and even energy if the ewe is pregnant or lactating.

Corn Silage. Many sheep raisers effectively use corn silage for sheep. It is palatable and provides energy and other nutrients. It is low in protein and calcium and thus must be supplemented for pregnant and lactating rations; but there are few if any crops that can match corn silage in the amount of energy produced per acre. Some producers are reluctant to use corn silage because of fear of listerosis, a bacterial disease. The organisms grow profusely on poorly ensiled forage.

Corn silage is made by chopping the whole corn plant when it is at the early dent stage of maturity (32% to 40% moisture) and the leaves are still green. The chopped forage is put into a silo, where it is thoroughly tramped or compacted so that very little air remains in it. The sugars in the corn then ferment to provide acids that act as a preservative. The value of corn silage is closely related to the amount of grain in it. The protein and calcium content of the silage can be increased by adding urca or anothcr nonprotcin nitrogcn compound and limestone when it is ensiled. Corn silage is an excellent feed for pregnant and lactating ewes if it is properly supplemented. It is not a fattening feed, and thus it is seldom fed to finishing lambs.

Sorghum Silage. Milo and other sorghums are also ensiled for sheep. They are well liked, but they furnish less energy than corn silage.

Other Carbonaceous Roughages. Straw from oats or wheat is not as well used by sheep as cattle. Both are low in crude protein, calcium, phosphorus, and vitamins A and D and are not palatable. Straw can be incorporated into mixed diets if ground or chopped and properly fortified.

Cottonseed hulls are a source of energy, but are quite low in protein, phosphorus, carotene, and vitamin D and are not palatable to sheep when fed alone. They have been useful in ground mixed sheep and lamb diets when the other feeds contain the needed nutrients. It is especially important to include some legume hay or dehydrated alfalfa meal to compensate for lack of calcium and carotene in the cottonseed hulls.

Gin trash from cotton gins can be chopped or ground and used in the same way as cottonseed hulls, but it is a better cattle than sheep feed. Feeds such as oat and rice hulls have limited value as roughage components in lamb diets.

9.4.2. Characteristics of nitrogenous roughages Nitrogenous roughages are the product of harvesting legumes. If legumes are properly harvested and stored, they are usually an economical source of nutrients. There is not much difference in the energy value between grass and legume hays, but the protein content of high-quality legume roughages is appreciably higher. The calcium content is also greater in most legume forages.

Legume Hay. Alfalfa (*Medicago sativa*), clovers (*Trifoliums*), trefoil (*Lotus cornicultus*), lespedeza (*Lespedeza* sp.), soybeans (*glycine max*), cow peas (*Vigna* sp.) and other legumes may be harvested for hay. Of all the legumes, alfalfa is the most valuable crop. Alfalfa and some clovers may be cut more than once during the year for hay. The first cutting can be expected to have coarser stems and lower feeding value than hay harvested later on in the season.

The best sheep hay is leafy, well cured, green, fine stemmed, and free of mold. If hay is rained on in the field after it has been cut, it tends to brown and have a different smell than hay properly cured. It is lower in feeding value than green hay because there is some leaching of nutrients, and usually there is a loss of leaves and carotene. Often a more serious problem is mold.

Legume Haylage. Legume haylage is harvested in the same way described for grass haylage. The nutrient content is higher and is it relished by sheep. If facilities are available, haylage is an excellent way of harvesting the first cutting of alfalfa or clover. The weather conditions are not nearly as critical to the harvesting and storing of the crop. Of course, the energy and vitamin D content is not as great as hay on an as fed basis, but on a DM basis it is often higher in nutrients since few leaves are lost and the losses in storage are less.

Legume Silage. Legumes may be ensiled in the same way as the grasses. The protein, mineral, and vitamin content is usually higher in legumes than grasses. If properly ensiled, the nutrients are better preserved than in hay and the silage is very palatable. Spoilage may be high if the silo is shallow, or there may be seepage if the silage was wet when stored. Failure to use the proper preservatives or to thoroughly pack the silage may result in losses due to molding.

9.4.3. Characteristics of mixed carbonaceous–nitrogenous roughages Many livestock farmers use grass–legume mixtures in their pastures. During rapid-growth periods in the spring, they harvest excess growth as hay, haylage, or silage. This mixed forage, if cut at the proper time and preserved without spoilage, has a feeding value proportional to the grass–legume content. It is usually more suitable for ewes during early gestation and lambs early in the feeding period.

9.5 LOSSES IN STORAGE AND FEEDING ROUGHAGE

Regardless of the method of preserving forages, losses of nutrients occur. These should be minimized by doing a thorough job of harvesting and preserving the forage. Silos are usually expensive to construct, and it requires more labor and equipment to make and feed the silage than hay. On the other hand, hay is more difficult to make in rainy seasons, and it loses more nutrients in storage. Round bales have become popular for putting up grass hay. Sheep will eat round bales, but will waste as much as 50% of it if they have unrestricted access to the bales in the field. Losses can be controlled by limiting access to only a few bales until they are consumed. Very large, round bales may also be used, but losses from spoiling are great if the bales are subjected to the weather. Care must also be taken to cure legume hay sufficiently before it is made into large bales. Otherwise, unless an inhibitor is used, it may mold excessively.

Ensiling does not magically increase the nutritive value of a poor forage. If it has low quality when it is put in the silo, it will invariably be lower when it comes out. Ensiling does help to preserve forage not easily stored. For example, by-products such as the stalks of corn after high-moisture corn has been harvested may be chopped and ensiled. The Ohio Experiment Station has used the stalk silage, which they call "stalklage," very effectively for feeding ewes in confinement. It is possible to fortify such forages with calcium and other nutrients when they are being ensiled. It is probable that such practices may increase as more by-products of grain cropping are utilized.

9.6 FEED PROCESSING

9.6.1. Processing concentrates Grains may be fed whole or processed to make them more palatable, to increase their utilization, or to increase ease of feeding. Some of the processes used are cracking, grinding, cooking, heating, dry or steam rolling, flaking, popping, and pelleting. From the standpoint of increased digestibility and efficiency of utilization, it is doubtful if any of this processing can be shown to be economical except in the case of young lambs or sheep with poor teeth. There are some advantages to grinding, cracking, or rolling grain to be used in mixtures with other ingredients. Then it may be worthwhile, especially if there are unpalatable ingredients to be added to the diet.

Grains are more palatable if they are cracked, rolled, flaked, or the like, than ground very fine. Sometimes it is desirable to mix a number of ingredients together to provide a concentrate such as a range pellet that can be spread on the ground. If ingredients are to be pelleted, they must be finely ground first. There appears to be no nutritional advantage to pelleting high-quality feeds, but there may be management advantages such as ease of handling and storage and being sure that the desired ingredients are consumed by the sheep.

9.6.2. Processing roughages Forages are processed during harvesting. Corn and forages must be chopped to make silage or haylage. Alfalfa may be chopped, dehydrated, and then ground into meal. Hay may be dehydrated before storage.

Are there reasons why preserved forages need preparation before feeding? The answer depends on the age and condition of the sheep and the function it is performing. If complete ground mixed diets are being used, the forage must be ground or chopped to obtain a homogenous mixture. Coarse, stemmy hay may be consumed more readily if it is ground, and thus feed wastage is reduced.

Some other advantages of grinding, chopping, or shredding roughages are:

1. Processed hay can be handled more effectively by machinery.
2. Processed hay that is pelleted, cubed, or made into wafers requires less space for storage and transportation than the whole forage.
3. Ground forage included in pellets can be consumed in greater quantities than in its natural state.
4. Ground forage lends itself to automation in confinement operations.

Some disadvantages of processing forages are:

1. Processing may require special equipment and much energy and thus be expensive.
2. Pelleting requires feed to be ground fine with a loss of leaves. It is also very expensive.
3. If high-concentrate, low-roughage pellets are used in finishing lamb rations, parakeratosis may occur in the rumen. The condition results in sloughing off of the surface tissue of the rumen.
4. Grinding alone often produces dusty feed, which may be unpalatable.

9.7 ADDITIVES TO FEEDS

Methods have been devised to preserve feedstuff from the time of harvest until it is needed for feeding livestock. Forages and grains may be dried to safe limits and protected from insects and rodents, or they may be ensiled and protected from oxygen or preserved by acids or other compounds to prevent spoilage. Some of the additives used on feed are discussed next.

9.7.1. Preservatives Cornmeal or cane molasses is used effectively on grass or legume silage to provide the media for the formation of acids. Organic acids, propionic, acetic, formic and others, are used to preserve silage. The organic acids may also be used on high-moisture grains to inhibit mold and bacterial action.

Hay containing as much as 28% moisture may be preserved effectively by spraying it with propionic acid. Anhydrous ammonia has been used to preserve hay baled at 30% moisture. Propionic acid or ammonium isobutyrate is effective in preserving hay containing 30% moisture.

9.7.2. Fortifications Corn silage can be fortified with urea or biuret and limestone at the time of ensiling to add protein equivalent and calcium to the forage. It also may reduce danger of listerosis.

9.7.3. Antibiotics Research has shown the advantages to the inclusion of tetracyclines to the rations of suckling and growing finishing lambs and pregnant and lactating ewes. Inclusion of 40 to 50 mg of oxytetracycline (Aureomycin) or chlorotetracycline (Terramycin) per lamb daily in feed or 10 mg/lb of mixed feed for feeder lambs has been shown to increase daily gains and feed utilization and decrease death losses. Inclusion of the tetracyclines in rations for ewes reduces incidence of udder infections and abortion and results in more uniform lambs.

9.8 FORMULATING RATIONS AND DIETS FOR SHEEP AND LAMBS

The terms ration and diet are often used interchangeably, but there is actually a technical difference. A balanced ration is a feed or mixture of feeds that, when fed in amounts the animal will consume in one day, will supply the nutrients required for a given physiological function. A balanced diet is a feed or mixture of feeds that will supply required nutrients for a physiological function when the animal consumes it free choice (ad libitum).

A study of nutrient requirements and composition of feeds will show that there are some single feeds such as legume hay that will satisfy the needs of an open ewe, a ewe during early gestation, or growing weaned lambs. But when the requirements are more critical, for example for ewes in early lactation or finishing lambs, a mixture of concentrates and roughage will probably be required to provide enough nutrients within the capacity of the ewe or lamb to consume them. Not only must the lactating ewe and finishing lamb obtain the proper nutrients, but they are ruminants that require bulk in order for their digestive systems to function properly. A ration composed of corn silage alone would provide adequate bulk, but not enough energy and protein to meet requirements. Thus, if corn silage is to be used, feeds richer in energy and protein must be added. No hay ration would satisfy requirements for the lactating ewe or finishing lamb.

9.8.1. Balancing a ration For purposes of illustration, let us balance a ration for a 154-lb ewe suckling twins during the first 8 weeks of lactation. The steps in ration formulation are:

1. Choose basal feeds (usually these are grown on the farm); start with the roughage.
2. Determine how much of the required nutrients would be satisfied if the basal feeds supplied all the dry matter. This should indicate how much supplement will be required to balance the ration.
3. Balance the ration for energy. Note amount of protein needed.
4. Balance for protein.
5. Calculate mineral and vitamin content.
6. Balance for mineral and vitamin content.
7. Make final adjustments.

Step 1. Choose basal feeds. In this example, assume there is available from farm or ranch production sun-cured orchardgrass hay cut in early bloom and number 2 yellow corn. These will be the basal feeds. The basal feeds will be supplemented with soybean meal, dicalcium phosphate, (dical) and ground limestone in case the ration is deficient in protein, calcium, and phosphorous.

Step 2. Start with requirements. NRC requirements for a 154-lb (70-kg) ewe suckling twins in the first 8 weeks of lactation are:

Body weight	Daily gain or loss (lb) (kg)	Dry matter		TDN lb (kg)	CP lb (g)	Ca (g)	P (g)	Vitamin activity (IU/day)	
		lb	as % of body weight					A	E
154	−0.13 (0.06)	6.2 (2.8)	4	4 (1.82)	0.92 (420)	13.4	9.5	7000	40

Note that the ewe is losing weight. She has the capacity to consume 4% of her body weight in dry matter or 154 × 4% = 6.16 lb (rounded to 6.2), which must contain 4 lb of TDN and 0.92 lb of protein. Remember, the above is on a dry basis. The ewe will need to eat more on as as fed basis. The nutrient content of the basal feeds and supplements is as follows:

Feed Basis of composition	DM, % as fed	TDN, % dry	Crude protein, dry	Ca (g/lb)	P (g/lb)	Vitamin (IU/lb) A	 E
Basal feeds							
Orchardgrass hay (early bloom IFN 103286)	89	52	15	1.23	1.54	11,762	
Corn no. 2 grain, IFN 4-02-931	88	87	10.1	0.09	1.59	613	11.36
Supplemental feeds							
Soybean meal (solvent extracted IFN 5-04-604)	90	88.0	49.9	1.54	3.18		2.04
Limestone	100			154.0			

Would 6.2 lb of orchardgrass hay provide enough TDN and CP to satisfy requirements?

	Dry matter (lb)	TDN (lb)	CP (lb)
NRC requirements	6.2	4	0.92
Nutrients in 6.2 lb of orchardgrass hay	6.2	3.22	0.93
Difference		− 0.78	+ 0.01

A ration of orchardgrass hay is deficient in TDN and approximately correct in protein. It is also unlikely that a ewe could consume enough hay to provide 6.2 lb of dry matter (assuming the hay contains 90% DM); then 6.2/0.89 = 6.97 lb must be consumed to equal 6.2 lb DM.

Even though the ewe consumes 6.97 lb of orchardgrass hay, she would be deficient in TDN (she needs 4 lb, but 6.2 × 52.0% TDN = only 3.22 lb). A combination with corn, 87% TDN, is needed. This can be determined by trial and error or as demonstrated in the Pearson square method. First determine the percent of TDN needed in the ration as follows:

$$\frac{4.0 \text{ lb TDN}}{6.2 \text{ lb feed}} \times 100 = 64.5\% \text{ TDN required}$$

Set up the square as follows:

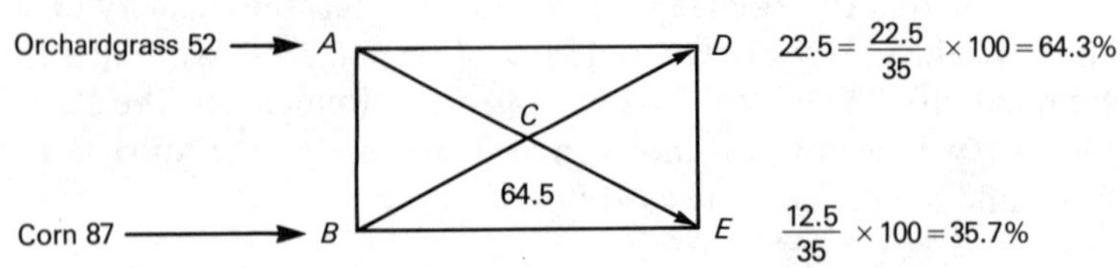

Let A = TDN of orchardgrass, B = TDN of corn, C = desired TDN, D = proportion of orchardgrass, and E = proportion of corn. Subtract diagonally the larger from the smaller (i.e., A from C and C from B). Proportions of each feed in the mixture are D = 22.5 for orchardgrass and E = 12.5 for corn. D and E added equal the total parts of 35. Divide the parts of orchardgrass and corn by 35 and multiply by 100 to obtain the percent of each in the mixture.

The weight of each ingredient in the ration can be obtained by multiplying the percentage by 6.2 as follows:

$$\begin{aligned} 6.2 \times 64.3\% &= 3.99 \text{ lb of orchardgrass hay} \\ 6.2 \times 35.7\% &= 2.21 \text{ lb of corn} \\ \text{Total fed} &= 6.2 \text{ lb} \end{aligned}$$

Next, compare the TDN and CP with the requirements:

	DM (lb)	TDN (lb)	CP (lb)
NRC requirements:	6.2	4	0.92
Feed ingredient			
Orchardgrass hay	3.99	2.07	0.60
Corn	2.21	1.92	0.22
Total	6.2	3.99	0.82
Difference	—	—	−0.10

The ration is adequate in TDN but 0.1 lb low in CP. Since soybean meal is available, the ration may be balanced by substituting some soybean meal for corn to bring up the protein. This works if TDN is similar for the two ingredients; in this case it is 87% for corn and 88% for soybean meal. We need 0.10 lb of CP.

1 lb of soybean meal contains	0.499 lb CP
1 lb of corn contains	0.101 lb CP
Difference	0.398, which can be rounded to 0.4

For every pound of corn replaced by soybean meal in the ration, 0.4 lb of CP will be added. To find the amount of the substitution needed, divide 0.1 by 0.4 as follows: 0.1/0.4 = 0.25 lb. Thus 0.25 lb of soybean meal should be added and 0.25 lb of corn subtracted from the ration. The ration now should contain:

Orchardgrass hay	3.99 lb
Corn	1.96 lb
Soybean meal	0.25 lb

This ration is balanced for TDN and CP, but must be checked for calcium, phosphorous, and vitamins.

	DM (lb)	TDN (lb)	CP (lb)	Ca (g)	P (g)	Vitamin activity IU/day A	E
NRC requirements:	6.2	4	0.92	13.4	9.5	7,000	40
Orchardgrass hay	3.99	2.07	0.60	4.91	6.14	46,930	
Corn	1.96	1.70	0.20	0.18	3.04	1,201	22.26
Soybean meal	0.25	0.22	0.12	0.38	0.80		
Total	6.2	3.99	0.92	5.47	10.05	48,131	22.26
Balance	Yes	Yes	Yes	−7.93	+0.55	+44,131	−17.74

The ration is deficient in calcium and somewhat above requirements in phosphorus. It is adequate in vitamin A but lacking 17.7 IU of vitamin E daily. Calcium can be furnished by adding limestone, which contains 34% calcium (154.4 g/lb). If the ewe needs 7.93 g, then 7.93/154.4 = 0.052 lb of limestone daily would satisfy her requirements. The vitamin E requirements can be met by refiguring the ration using alfalfa leaf meal or more easily by adding a vitamin supplement. These are usually ADE supplements and highly concentrated. One commonly available supplement contains 4 million IU of vitamin A, 800,000 IU of D_3, and 1200 IU of E per pound. Since the ration lacks 17.74 IU of vitamin E daily, it will require 0.0148 lb of the supplement (17.74 ÷ 1200 = 0.0148 lb) or 0.0148 × 454 = 6.7 g daily. Since this and the limestone are both small amounts, it is usually more satisfactory to premix them with soybean meal to obtain a uniform mixture and to allow them to be fed easier. The weight would be added to the 0.25 lb of soybean meal as follows:

	Pounds	Percent of mixture
Soybean meal	0.25	78.9
Limestone	0.0520	16.4
Vitamin ADE supplement	0.0148	4.7
Total	0.3168	100.0

The soybean meal–limestone–vitamin supplement must then be fed at the rate of 0.316 lb daily to provide the CP minerals and vitamins needed, but the proportions of the ration are somewhat changed and thus another ingredient must be reduced in order to add limestone and the vitamin supplement. The simplest solution is to reduce the orchardgrass hay by 0.066 (rounded to 0.07) lb. The ration can then be compared with the NRC requirements.

	DM (lb)	TDN (lb)	CP (lb)	Ca (g)	P (g)	Vitamin activity (IU/day) A	E
NRC requirements:	6.2	4	0.92	13.4	9.5	7,000	40
Orchardgrass hay	3.92	2.04	0.59	4.82	6.04	46,107	
Corn no. 2	1.96	1.70	0.20	0.18	3.04	1,201	22.26
Soybean meal	0.25	0.22	0.124	0.38	0.80		
Limestone	0.052			8.03			
ADE supplement	0.015					59,200	17.74
Total	6.197	3.96	0.914	13.41	9.88	106,508	40

The ration is complete for practical purposes when figures are rounded. Crude protein is 0.06 lb too low, but analyses have that much or more variation between two samples of the same ingredient. The ration apparently contains much more vitamin A than needed and generous amounts of vitamin D_3. The nutrients are often very low in grass hay and thus it is well to have an excess in the ration, especially in years of dry weather or if the ewes have not had extensive exposure to sunshine in the winter. The next step is to convert the ration to an as fed basis. This is done by dividing the weight of the dry feed ingredient by the dry matter content. For example, orchardgrass hay contains 89% dry matter. If 3.92 lb of dry hay is fed, then 3.92 ÷ 0.89 = 4.4 lb as fed.

Ingredient	Weight (dry)	DM content (%)	As fed weight (lb)	As fed (%)
Orchardgrass hay	3.92	89	4.4	63.4
Corn	1.96	89	2.2	31.7
Soybean meal	0.25	90	0.28	4.0
Limestone	0.05	100	0.05	0.7
Vitamin supplement	0.015	100	0.015	0.2
			6.94	100.00

9.8.2. Formulating diets NRC requirements are given on the basis of composition of nutrients fed free choice (ad lib) in a mixture of feed ingredients. Many lamb feedlots routinely feed mixed diets to their lambs for the sake of convenience, efficiency of handling feeds, the improved consumption of mixed feeds, and the ease of incorporating drugs and otherwise unpalatable feed ingredients into the mixture.

The steps to follow in balancing a diet are:

1. Choose basal feeds.
2. Decide what supplements to use.
3. Tabulate the composition of feeds to be used.
4. Tabulate the nutrient requirements.
5. Determine if the roughage will satisfy the requirements alone.
6. Work out the energy requirements.
7. Use substitution to satisfy the protein requirements.
8. Correct the mineral and vitamin deficiencies.

The following is a typical problem in formulating a diet for 66-lb (30-kg) finishing lambs. The basal feeds will be sun-cured alfalfa hay, cut at mid bloom, cottonseed hulls, corn, and cane molasses. The lamb feeders would like to feed a 1 to 2 mixture of alfalfa hay and cottonseed hulls. They also desire to feed 3.75% cane molasses in the completed diet on a dry basis. They expect to use solvent soybean meal as a protein supplement and limestone if there is a calcium deficiency. The composition of the feed ingredients is as follows:

Feed ingredient	DM (as fed)	TDN (%)	CP (%)	Ca (%)	P (%)	Vitamin activity (IU/lb)	
						A	E
Sun-cured alfalfa hay, IFN 1-00-063	90	56	17	1.41	0.24	10,215	
Cottonseed hulls, IFN 1-01-599	91	49	4.1	0.15	0.09		
Corn No. 2, IFN 4-02-931	88	87	10.1	0.02	0.35	619	17
Cane molasses IFN 4-04-696	75	79	5.8	1.0	0.11		4.8
Soybean meal (solvent), IFN 5-04-604	90	88	49.9	30	0.70		2.0
Limestone, IFN 6-02-632	98			34.0			

The diet will be prepared by grinding the hay and corn and mixing all ingredients so they can be fed free choice (ad lib) in self-feeders.

The nutrient requirements for finishing lambs 4 to 7 months are:

Body weight		Weight change per day		Diet proportions		TDN	CP	Ca	P	Vitamin activity (IU/lb)	
(lb)	(kg)	(lb)	(kg)	Concentrates (%)	Forage (%)	(%)	(%)	(%)	(%)	A	E
66	30	0.65	0.295	60	40	72	14.7	0.51	0.24	493.2	6.8

The diet will be formulated on the basis of 100 lb of feed on a dry basis for ease of calculating.

Since it has been stipulated that the forage will be included in the diet at a 1 to 2 ratio of alfalfa to cottonseed hulls, it simplifies figuring to calculate a conglomerate of the two feeds and treat it like one ingredient. This will appear as AFL–CSH forage.

	DM (%)	TDN (%)	CP (%)	Ca (%)	P (%)	Vitamin activity (IU/lb) A	E
Alfalfa (1 part)	90	56	17	1.41	0.24	10,215	—
Cottonseed hulls (2 parts)	182	98	8.2	0.30	0.18		
Total (3 parts)	272	154	25.2	1.71	0.43	10,215	—
Avg. composition ALF–CSH forage (total/3)	90.7	51.33	8.4	0.57	0.14	3,405	—

The initial step should be to determine the adequacy of the diet if it is composed of 60% concentrates and 40% forage, and corn and cane molasses are the concentrates. It was stipulated that cane molasses would be included at 3.75% of the mixture. Then the mixture will contain 40% of ALF–CSH forage, 56.25% corn, and 3.75% cane molasses. The nutrient content of the mixture is as follows:

Ingredient	Weight (lbs)	TDN (lb)	CP (lb)	Ca (lb)	P (lb)
ALF–CSH forage	40	20.53	3.36	0.23	0.06
Corn	56.25	48.94	5.68	0.01	0.20
Cane molasses	3.75	2.96	0.22	0.04	0.004
Total	100.00	72.43	9.26	0.28	0.264
NRC		72	14.7	0.51	0.24
Difference		+0.43	−5.44	−0.23	+0.024

The diet is slightly in excess of TDN, very low in CP, low in Ca, and slightly in excess in P. Since the diet is almost correct in TDN, substitution of soybean meal for corn will be used to raise the CP. This should have little effect on the TDN because the two ingredients are similar in energy. The substitution factor is obtained by subtracting the CP in the lower feed from the higher. Crude protein in soybean meal is 0.499% and corn is 0.101. Thus the factor is 0.399, rounded to 0.40. For each pound substituted, 0.4 lb of CP will be added to the diet. Next divide the number of pounds needed by the factor to obtain the weight of substitution needed: 5.44/0.40 = 13.6 lb of soybean meal required to meet the requirement for CP. After subtracting 13.6 lb from the corn and adding 13.6 lb of soybean meal, the composition of the diet is:

Ingredient	Weight (lb)	TDN (lb)	CP (lb)	Ca (lb)	P (lb)
ALF-CSH forage	40.	20.53	3.36	0.23	0.06
Corn	42.65	37.11	4.31	0.01	0.15
Cane molasses	3.75	2.96	0.22	0.04	0.004
Soybean meal	13.6	11.75	6.79	0.046	0.10
Total	100.0	72.57	14.68	0.33	0.31
NRC requirement		72	14.70	0.51	0.24
Difference		+0.57	−0.02	−0.18	+0.07

The diet has sufficient TDN, CP, and P, but is low in calcium. The Ca to P ratio is, however, 1.06 to 1 (0.33/0.31), which is too narrow and may cause problems with urinary calculi in wether lambs. To achieve a 2 to 1 ratio in accordance with NRC recommendations, the content of calcium must be 0.62 lb (2 × 0.31 lb P). This will require adding 0.29 lb of Ca to the 0.33 lb in the diet. The simplest way to increase the Ca is to add limestone, which contains 34% calcium. The limestone can be substituted for corn with a factor of 0.3398 or 0.34 lb substituted (0.34 − 0.0002 = 0.3398, rounded to 0.34). The amount to substitute is calculated by dividing the amount needed by the factor 0.29/0.34 = 0.85. Thus 0.85 lb of limestone will be added and a like amount subtracted from the corn. The composition of the diet corrected for calcium and including vitamins A and E is:

Ingredient	Weight (lb)	TDN (lb)	CP (lb)	Ca (lb)	P (lb)	Vitamin activity (IU)	
						A	E
ALF-CSH forage	40	20.53	3.36	0.23	0.06	408,600	
Corn	41.8	36.37	4.22	0.01	0.146	258.74	710.6
Cane molasses	3.75	2.96	0.22	0.04	0.004		18
Soybean meal	13.6	11.97	6.79	0.046	0.10		27.7
Limestone	.85			0.29			
Total	100	71.83	14.59	0.62	0.31	408,858	756.3
Vitamin content (IU/lb)						4088.6	7.5
NRC/100 lb of diet		72	14.7	0.62	0.31	493.2	6.8

If the figures are rounded, the diet is adequate for TDN, Ca, and P with a 2 to 1 Ca to P ratio. It also has plenty of vitamins A and E. The addition of 0.85 lb of limestone caused a reduction of 0.11 lb of CP. This can be easily corrected by substituting soybean meal for corn. Remember the substitution factor is 0.4 lb CP/lb. We need 0.11 lb CP. The weight of substitution = 0.11/0.4 = 0.28 lb. Subtract 0.28

lb from corn: 41.8 − 0.28 = 41.52. Add 0.28 lb to soybean meal: 13.6 + 0.28 = 13.88. The composition of the completed diet is:

Ingredient	Weight	TDN	CP	Ca	P
ALF-CSH forage	40	20.53	3.36	0.23	0.06
Corn	41.52	36.06	4.19	0.01	0.145
Cane molasses	3.75	2.96	0.22	0.04	0.004
Soybean meal	13.88	12.21	6.93	0.05	0.10
Limestone	0.85			0.29	
Total	100	71.75	14.7	0.62	0.31
NRC/100 lb of diet		72	14.7	0.62	0.31

The diet meets NRC requirements when values are rounded. No effort has been made to carry out figures beyond two decimal places. This is because analytical methods are not completely precise. Thus, if the values are within 1% of the requirement, it should be satisfactory. The next step is to convert the dry matter of the diet to an as fed basis. This is done by dividing parts by the dry matter of the feed. Then the proportions are added and divided by the total times 100 to obtain the percentage of each ingredient on an as fed basis.

The complete diet for a 66-lb (30-kg) finishing lamb 4 to 7 weeks old, as fed, and proportions per ton are as follows:

Ingredient	Wt/100 lb of diet (lb)	DM content (%)	Proportions of diet as fed		
			lb	%	lb/ton
Alfalfa	10	90	11.1	9.8	196
Cottonseed hulls	30	91	33.0	29.2	584
Corn	41.52	88	47.5	42.1	842
Cane molasses	3.75	75	5.0	4.4	88
Soybean meal	13.88	90	15.42	13.7	274
Limestone	0.85	98	0.87	0.8	16
Total	100		112.89	100	2000

9.9 EVALUATING FEEDS

The ability to evaluate feeds skillfully may make a great deal of difference in operations where many purchases are made. It pays to be able to judge quality and to know that fair value is being received for money spent on feeds. Not only is it important to estimate the value accurately among samples of a feed, but it may be just as valuable to compare alternative sources.

9.9.1. Comparison of grains Previously in this chapter, it was pointed out that corn is the standard grain over much of the United States. On a scale of 0 to 100, grains usually rate as follows for sheep:

Corn	100
Sorghum (milo)	85–100
Barley	85
Wheat	95–98
Oats	75–100
Rye	83–85

This rating is subject to change with the prices of energy and protein feeds. On the basis of nutrient content, the differences are not clear. The following is their nutrient content:

Basis of composition	DM (% as fed)	TDN (%, dry)	DE (Mcal/kg, dry)	CP (%, dry)	Ca (%, dry)	P (%, dry)
Corn (no. 2), IFN 4-02-931	88	87	3.84	10.1	0.02	0.35
Sorghum (milo), IFN 4-04-444	89	83	3.88	11.3	0.05	0.34
Barley, IFN 4-00-549	88	86	3.79	13.5	0.05	0.38
Wheat, IFN 4-05-211	89	87	3.84	16	0.04	0.42
Oats, IFN 4-03-309	89	77	3.4	13.3	0.07	0.38
Rye, IFN 4-04-047	88	85	3.75	13.8	0.07	0.37

To compare values of grains, it may be useful to figure their cost per pound (kilogram) and then calculate the cost per unit of TDN. If, for example, no. 2 corn is selling at \$3.00 per bushel and no. 1 barley at \$2.50 per bushel, which is the least expensive source of TDN?

weight per bushel of no. 2 corn = 54 lb (24.5 kg)

dry matter = 88%

Thus

54 lb (24.5 kg) × 88% = 47.52 lb (21.6 kg)dry corn/bushel

Then

47.52 lb (21.6 kg) × 87% TDN = 41.3416 (18.79 kg) TDN per bushel of no. 2 corn

Corn at \$3.00 per bushel ÷ 41.34 lb (18.79 kg) = \$0.0726/lb (0.16/kg) of TDN.

weight per bushel of no. 1 barley = 47 lb (21.36 kg)

dry matter in barley = 86%

Thus

47 lb (21.36 kg) × 86% = 40.42 lb (18.37 kg) of dry barley per bushel

Then

$$40.42 \text{ lb } (18.37 \text{ kg}) \times 86\% \text{ TDN} = 34.76 \text{ lb } (15.8 \text{ kg}) \text{ TDN per bushel of barley}$$

Barley at \$2.50 per bushel ÷ 34.76 lb (15.8 kg) TDN = \$0.0719/lb (0.158/kg) of TDN. On the basis of TDN purchased, barley is slightly less expensive than corn.

Some grains are less palatable than others. When it can be purchased economically, wheat can be fed to sheep but is not as acceptable as corn or barley. Rye is also not as palatable as corn. Both wheat and rye are more satisfactory for feeding sheep if they are mixed with corn. Oats are very palatable for sheep, but are lower in TDN and usually more expensive on the basis of TDN than corn, barley, or sorghum.

Quality of grains should be a major purchasing consideration. Damage caused by overheating in drying, mold, smut, insects, and rodents decreases the feeding value and should be noted when grain is inspected. Broken kernels and foreign material such as other grains, weed seeds, sand, stones, dirt, and plant material are indications of lower quality in grains.

Weight per bushel is an excellent indication of the feeding value of grains. Lightweight grain is usually discounted. Heavy-weight grain is generally higher in starch and lower in fiber than a lightweight sample.

Moisture in grain is a major factor in trading in grain. Not only does a high moisture content reduce the nutrient content, but it may cause damage when the grain is stored.

9.9.2. Comparison of roughage Quality in hay is based on factors related to feeding value and wholesomeness. A suggested scorecard for evaluating hay is as follows:

- Leafiness, 30%
- Color and aroma, 30%
- Foreign material, 20%
- Condition, 20%

Leafiness is important because the leaves contain 70% of the protein, 60% of the energy, and 90% of the vitamins and minerals. The best hay is dark green and has a clean, wholesome smell. Color and aroma are closely associated with the carotene content.

Foreign material, such as weeds and plant residues, is undesirable because it is usually unpalatable and low in nutrition and may be a source of weed seeds. The condition of hay relates to the cutting, curing, and storage. The presence of dust and mold is a sure indication that the hay was not properly cured or stored. Factors affecting the value of corn silage are usually given the following weights:

- Grain content, 50%
- Color, 25%
- Aroma, 25%

Corn silage is fed primarily for the energy it provides. Thus, the higher the grain content is the more energy available. An olive-green color is most desirable. Dark-brown or black silage results from excessive heating and indicates loss of nutrients. A yeasty odor is most desirable. Mustiness or a strong butyric acid or ammonia odor indicates improper preservation and loss of nutrients.

High-moisture grass and legume silage or haylage usually has a stronger odor than corn silage. It should not smell putrid, musty, or moldy, and the color should be green rather than black.

9.10 PURCHASING FEEDS

The ability to evaluate and purchase feeds worth the money may contribute a great deal to a sheep operation. Care should be taken to assure that the purchased feed is adequate for its intended purpose and meets the specifications under which it was bought. In feedlot operations where many tons of feed are purchased, errors in ascertaining quantity or quality delivered may be costly.

9.10.1. Purchasing grains Grains should be purchased on the basis of moisture content and quality factors used in determining grade. A grain probe is an important piece of equipment if much is to be purchased. Probing in each corner and the middle of the truck is recommended. Equipment for obtaining test weight and moisture should be available for use either on the farm or nearby. Most elevators have such equipment and will test grain for a fee or as an accommodation.

The time to purchase grain will depend on a number of factors, such as when it will be needed, how much is to be purchased, the storage available, and the size of the crop. Generally, grain is cheapest at harvest and increases throughout the year. The increase in price is related to storage costs and to interest rates. There are some advantages to owning storage facilities so that grain can be purchased and processed when the price is most favorable during harvest. The producer knows what the grain will cost and that it is available when needed. It is also possible to be somewhat more selective when the greatest volume is on the market. The disadvantage is the high cost of purchasing storage and equipment and the high carrying charge for holding grain.

9.10.2. Purchasing hay Hay should be inspected carefully before it is purchased. If any considerable amount is needed, the hay should be sampled and sent to a laboratory for analysis. It is recommended that at least 12 bales picked at random be sampled. A core sampler that fits in a hand drill is a very satisfactory way of obtaining samples. The best technique is to drill diagonally from the ends of the bales. If this is not possible, several grab samples can be taken from each of the randomly selected bales and sent to the laboratory. The samples should also be inspected for color, odor, and leafiness.

If hay is properly cut and cured, it is usually cheapest during the haying season. Whether or not to buy it then will depend on the crop, storage, and carrying charges. Hay goes through a "sweat" and loses about 5% in dry matter during storage due to enzymatic action. This must be considered, along with the advantage of having an assured supply on hand when needed.

9.10.3. Purchasing protein and mineral concentrates Protein and mineral concentrates should be bought on the basis of guaranteed analysis. Usually protein concentrates are less expensive in the fall than other times of the year. Purchasing in bulk usually results in considerable discount in price. (Sacks add $12 to $16/ton). If a protein concentrate contains nonprotein nitrogen, the DE (TDN) of the product may be considerably lower than an oil meal. In that case, consider both the cost per pound (kilogram) of protein and TDN (DE) when comparing the two.

Buy minerals based on the calcium, phosphorus, and other mineral content. Remember, salt is usually very inexpensive and so is limestone as a source of calcium. Many producers find it saves money to prepare their own mineral mixtures using salt, limestone, and dicalcium phosphate or some other source of phosphorus.

REFERENCES

Church, D. C., *Livestock Feeds and Feeding.* Corvallis, Ore.: D. C. Church, 1977.

Ensminger, M. E., *Sheep and Goat Science.* Danville, Ill.: Interstate Printers and Publishers, Inc., 1986.

______, and C. G. Olentine, Jr., *Feeds and Nutrition.* Clovis, Calif.: Ensminger Publishing Company, 1978.

Morrison, F. B., *Feeds and Feeding.* Ithaca, N.Y.: Morrison Publishing Comany, 1957.

National Academy of Sciences, *Nutrient Requirements of Sheep,* 6th rev. ed. Washington, D.C.: The Academy, 1985.

Scott, George E., *The Sheepman's Production Handbook.* Denver, Colo.: Sheep Industry Development Program, Inc., 1970.

United States-Canadian Tables of Feed Composition. Washington, D.C.: National Research Council, 1984.

QUESTIONS FOR STUDY AND DISCUSSION

1. What is the basis for classifying feeds as concentrates or roughages?
2. Why are concentrates usually more expensive than roughages?
3. What is the importance of other nutrients in a protein concentrate besides protein?
4. What are the advantages of buying corn on a grade and moisture basis?
5. Compare corn, barley, and sorghums as feeds for sheep.
6. What is the basis for grading grains?
7. What are reasons, other than nutritional, for using molasses in diets for sheep?
8. Explain why fats should be limited in rations or diets for sheep.
9. What processes are used in manufacturing oil meals?
10. Why aren't protein concentrates of animal origin used widely in sheep feeding?
11. Explain the limits to the use of nonprotein nitrogen products for sheep.
12. Why is biuret safer to feed than urea?
13. Why shouldn't trace-mineral salt formulated for swine be used for sheep?
14. How would you furnish cobalt to sheep if it is needed?
15. When is vitamin A deficiency likely to occur?
16. What is calciferol?

17. What are the differences among hay, haylage, and silage?
18. Describe the ideal roughage for a sheep.
19. Describe the principles involved in preserving silage.
20. Under what circumstances should grains be processed?
21. What are advantages of processing hay?
22. Why are antibiotics used in feeding sheep?
23. Why are rations balanced?
24. Assume you wished to feed equal parts of fescue and alfalfa hay plus corn to sheep to obtain a mixture containing 11% crude protein. Show how you would proceed using the Pearson square.
25. When is the most favorable time to purchase feed?
26. On what basis can the value of two grains be compared to determine which should be purchased?

10 Pastures and Pasture Management

10.1 IMPORTANCE OF PASTURES

Sheep are uniquely adapted to the utilization of growing herbage. Throughout the world, pastures and rangeland supply the majority of nutrients to sheep for the production of meat and wool.

All ruminants have the ability to convert fiberous plants into products useful to humans, but sheep have some exceptional advantages over cattle in that they have the ability to graze steep slopes, they eat more broad-leaved weeds (forbs), and they can derive much nutrition from the leaves and twigs of shrubs, trees, and woody vines (browse). Sheep also are less affected by arid conditions than cattle. Sheep are adaptable to the use of high-quality, high-yielding pastures on fertile land, and they are superior to all farm animals in their ability to produce marketable products from forage alone (Terrill, Lindahl, and Price, 1973).

Not only can pasture furnish most of the nutrients required by sheep, but it provides cover to the soil to reduce erosion and conserve water and to prevent loss of soil fertility. When pastures are properly managed, organic matter is renewed in the soil and tilth is improved. The nutrients from pasture are generally cheaper than those from other sources.

10.2 FACTORS AFFECTING PASTURES

Both grasses and legumes have greater nutritive value when they are in a rapidly growing, vegetative state. Part of the technique of maintaining good pastures is to keep them grazed or clipped to prevent heading out and forming seed. The species and productivity of plants used for pasture are determined by a number of factors. These are soil topography, environment, natural enemies, and grazing management.

10.2.1. Effect of soil and topography Soil as the source of plant nutrients is of great importance. The level and balance of minerals and nitrogen in the soil limit the growth, composition, and productivity of plants.

On land that can be tilled, it is possible to apply fertilizer to correct soil deficiencies before the forage is established. Deficiencies in nitrogen, calcium, phosphorus, potassium, and other elements can be determined by soil tests and the fertilizer applied at the recommended levels. If the young plants have adequate fertilizer, they have a much better chance of living.

Most soils used for pasture are deficient in nitrogen (N). Nitrogen is usually the most limiting element for grasses, especially the tall-growing species, but adequate levels of calcium (Ca), phosphorus (P), and potassium (K) must be available to achieve the most response from nitrogen. Nitrogen fertilization of grasses causes earlier growth in the spring, sustains growth later in the fall, and increases the overall carrying capacity. Nitrogen fertilization is not usually effective if the sward contains more than 50% legumes.

Grass–legume mixtures respond to calcium, phosphorus, and potasium fertilization by increasing the proportion of desirable species and the density of the sward. If levels of Ca, P, and K are adequate, little benefit has been shown from the application of N.

Nitrogen fertilizer is most effective in the spring and fall. Lime, P, and K can be applied any time of the year, but it is wise to follow recommendations of extension agronomists. On rough land where machinery cannot be used, brush and timber may be killed by herbicides and burned before forages are seeded in the ashes. Fertilizing by air may be necessary to correct soil deficiencies. New Zealand uses the technique quite effectively in hilly, rough areas.

According to Martin and Berry (1970), five conditions must exist before a range operator can profitably use fertilier:

1. The soil must be low in fertility.
2. Climatic conditions must be favorable for growth when needed.
3. Plant species must respond to increased available nutrients.
4. Management must be adjusted to forage production and maintenance.
5. Fertilization must be a more favorable option, economically, than other steps that might be taken.

Not only the chemical composition but the physical characteristics of the soil are important. Drainage is critical for many plants.

The topography of the land determines the forage that may be grown by limiting cultural practices. On steep, rough, or rocky land, it may not be possible to use machinery to prepare the soil in order to establish forages used for pasture.

Progress has been made in correcting soil deficiencies and drainage on accessible land, and seeding and fertilizing by air have broadened the selection of available species.

10.2.2. Environment The environment under which plants exist is largely climatic with emphasis on the following cardinal factors: light (intensity and duration), temperature, and moisture.

Light intensity affects plant growth, but quantity is more important. Sunlight is required for photosynthesis to occur. Flowering and the development of seed are related to day length.

Temperature influences all physiological processes. Plants differ in their ability to function when temperatures are extreme. There are heat-tolerant grasses and legumes, and others that do better when it is cool. These are classified as warm- and cool-season grasses and legumes. Generally, higher temperatures hasten maturity and decrease the nonstructural carbohydrate content and disgestibility of the herbage. On the other extreme, plant growth is easily destroyed by frost.

Water is essential for the roots of a plant to absorb and transfer plant nutrients from the soil. Cell division and other physiological functions are impossible without water. Either excessive or insufficient water may be harmful to plants. There are wide differences between plant species in their requirements for water. Total precipitation and its distribution is very important in selecting pasture crops.

10.2.3. Natural enemies Diseases may destroy or damage plants. Viruses, bacteria, and fungi may be the cause of damaging disease. Progress has been made in developing disease resistance in some forage plants, while chemical treatments are sometimes effective in others.

Parasites cause losses in pastures in some areas. Controls vary from biological to chemical. Generally, spraying is expensive. Fortunately, there is some difference in resistance to parasites, and this offers possibilities for choices of species.

Weeds are natural enemies of forage crops. When weeds appear, it may indicate incorrect selection of pasture crops, use of contaminated seed, poor cultural practices, lack of fertility, overstocking, or poor management of livestock. Weed control is essential to the maintenance of good pastures.

10.2.4. Use of pastures Grazing management has much to do with total productivity and difficulty of maintenance of the sward. Animals cannot graze without damaging the plants in a pasture. Grazing results in defoliation, which reduces the leaf area exposed to the sun for photosynthesis. Livestock trample the plants and foul the pasture with their dung. They develop paths that are subject to erosion. Overstocking prevents plants from recovering from grazing and results in destroying the most desirable species. Understocking may result in excessive growth and unpalatable vegetation or, if continued, spot grazing in which the same field has over- and undergrazed areas.

Time of grazing is important with some species of forage. Some grasses and legumes must be rested before a frost in order to survive over the winter.

The sheep producer is limited in controlling many of the factors affecting pasture production, but it is possible to select adapted species and varieties of crops to use. Judicious use of fertilizer may be practical. Grazing at the proper times and levels and the control of weeds are all within the reach of managers. Under special circumstances, some special efforts to control disease and parasites may be indicated.

10.3 MANAGEMENT OBJECTIVES

Pasture and forage production deserve a lot of thought and should be integrated with the overall farm or ranch plan, since these feed sources may be expected to constitute the major part of nutrition to the sheep operation. Two objectives should be:

1. To furnish inexpensive feed of adequate quality when it is needed for a long grazing season.
2. To protect the soil from loss of nutrients and damage by erosion.

10.4 CHARACTERISTICS OF GOOD PASTURE

Good pastures should have the following characteristics:

1. Be easily established and maintained.
2. Hold up well under grazing and recover rapidly.
3. Provide an abundance of nutritious, palatable feed over a long grazing season.
4. Be safe from poisonous plants and not cause bloat.
5. Contain no seeds, burrs, awns, or other vegetable matter that may contaminate wool or injure the sheep.
6. Protect the soil from leaching or erosion.

Few plant species meet all the criteria for a good pasture in the temperate zones of the world. The alternative is to use a mixture of species or to grow species adapted to seasons and to schedule grazing for those times when the plant species is most productive.

10.5 PLANT SPECIES USED FOR PASTURES

Among the thousands of grass and legume species grown in the world, only a few have the characteristics needed to be a good pasture crop. Grasses may be perennial, biennial, or annual. There is wide variation between species in their requirements for moisture, fertility, and tolerance of wide variations in temperature.

Grasses are valuable in pastures because they produce much forage, resist overgrazing, tend to hold soil and prevent erosion, make early growth in the spring, and are nutritious late in the fall. Grasses are safe feed and seldom cause digestive disturbances. They resist parasites and diseases in plants.

Legumes provide palatable and nutritious herbage that retains its feeding value during hot weather. They also build up nitrogen in the soil where they are grown.

Some legumes in solid stands can cause bloat in ruminants. Although they may produce a heavy volume of forage, they are subject to parasite and disease damage. Since plant species are so dependent on climatic conditions, it is important to understand those that exist in the United States. Agronomists have divided the country into regions according to climatic conditions.

10.6 ADAPTATION OF PLANT SPECIES AND VARIETIES

The ability of plants to thrive under the environmental conditions of any geographical location is a measure of their adaptability. It is important to choose species or combinations of species that are adapted to local conditions. Fortunately, the wide diversity within and between species allows a selection of species best suited to the situation encountered on the farm or ranch.

Some of the forage crops available in the United States and their adaptation are shown in Table 10.1.

TABLE 10.1 FORAGE CROPS IN THE UNITED STATES

Drought resistant

- Bromegrass (*Bromus arvensis*)
- Tall oatgrass (*Arrhenatherum elatius*)
- Switch grass (*Panicum virgatum*)
- Big bluestem (*Andropogon gerardi*)
- Wheatgrasses (*Agropyron desertorium*)
- Tall fescue (*Festuca arundinacea*)
- Lespedeza (*Lespedeza stipulaceo*)
- Burr clover (*Medicago*)
- Sweet clover (*Melilotus alba*)
- Alfalfa (*Medicago sativa*)
- Bird'sfoot trefoil (*Lotus corniculatus*)

Tolerant to poor soil

- Orchardgrass (*Dactylis glomerata*)
- Canada bluegrass (*Poa compressa*)
- Tall oatgrass (*Arrhenatherum elatius*)
- Redtop (*Agrostis alba*)
- Tall fescue (*Festuca arundinacea*)
- Big bluestem (*Andropogon furcatus*)
- Little bluestem (*Andropogon scoparius*)
- Switch grass (*Panicum virgatum*)
- Indian grass (*Sorghastrum nutans*)
- Bentgrass (*Agrostis*)
- Sheep fescue (*Festuca ovina*)
- Lespedeza (*Lespedeza stipulacea*)

Tolerant to shade

- Orchardgrass (*Dactylis glomerata*)
- Tall fescue (*Festuca arundinacea*)
- Red fescue (*Festuca rubra*)
- Red clover (*Trifolium protense*)

Cool weather required

- Kentucky bluegrass (*Poa pratensis*)
- Canada bluegrass (*Poa compressa*)
- Timothy (Phleum pratense)
- Bromegrass (*Bromus arvensis*)
- Wheatgrasses (*Agropyron desertorium*)
- Field peas (*Pisum sativum*)
- Winter vetch (*Vicia villosa*)
- Red clover (*Trifolium pratense*)
- Alsike clover (*Trifolium hybridum*)
- Crimson clover (*Trifolium incarnatum*)
- Hop clover (*Trifolium agrarium*)
- Yellow trefoil (*Medicago lupulina*)
- Orchardgrass (*Dactylis glomerata*)

Tolerant if soil has adequate calcium

- Sweet clover (*Melilotus alba*)
- Burr clover (*Medicago*)
- Mammouth clover (*Trifolium pratense*)
- Bird'sfoot trefoil (*Lotus corniculatus*)
- Strawberry clover (*Trifolium fragifarum*)
- Field peas (*Pisum sativum*)

Rich soil required

- Kentucky bluegrass (*Poa pratensis*)
- Timothy (*Phleum pratense* L.)
- Bromegrass (*Bromus arvensis*)
- Perennial ryegrass (*Lolium perenne*)
- Alfalfa (*Medicago sativa*)
- Red clover (*Trifolium pratense*)
- Alsike clover (*Trifolium hybridum*)
- Ladino clover (*Trifolium repens*)
- White clover (*Trifolium repens*)

Warm weather required

- Sorghums (*Sorghum bicolor*)
- Sudangrass (*Sorghum vulgare*)
- Bahiagrass (*Paspalum notatum*)
- Bermudagrass (*Cynodon dactylon*)
- Pangola digitgrass (*Digitaria decumbens*)
- Soybeans (*Glycine max* Merr.)
- Cowpeas (*Vigna sinensis*)
- Lespedeza (*Lespedeza stipulacea*)
- Hubam sweet clover (*Melilotus alba*)

Tolerant to wet soil

- Reed canarygrass (*Phalaris arundinacea*)
- Timothy (*Phleum pratense* L.)
- Tall fescue (*Festuca arundinacea*)
- Meadow foxtail (*Alopecurus pratensis*)
- Canada bluegrass (*Poa compressa*)
- Redtop (*Agrostis alba* L.)
- Alsike clover (*Trifolium hybridum*)
- White clover (*Trifolium repens*)
- Ladino clover (*Trifolium repens*)
- Big trefoil (*Lotus Uliginosus*)
- Orchardgrass (*Dactylis glomerata*)

Tolerant to sandy soil

- Brome grass (*Bromus arvensis*)
- Italian ryegrass (*Lolium multiflorum*)
- Tall oatgrass (*Arrhenatherum elatius*)
- Canada bluegrass (*Poa compressa*)
- Meadow foxtail (*Alopecurus pratensis*) (Tolerant if soil is wet)
- Reed canarygrass (*Phalaris arundinacea*)
- Redtop (*Agrostis alba* L.)
- Alfalfa (*Medicago sativa*)
- Winter vetch (*Vicia villosa*)
- Cowpeas (*Vigna sinensis*)

10.7 SELECTION OF PASTURE CROPS

Choice of pasture crops to be used should be made only after careful study and knowledge of local conditions. Not only must they be adapted to the environment, but they must fit the needs of the livestock program.

A valuable resource available in most states is the Cooperative Agricultural Extension Service. A visit with the county or area agronomist should be useful in determining the adaptation of varieties, mixtures that have been successful, cultural practices, and sources of seed.

10.7.1. Deciding whether or not to use a single crop or mixture of forages In some areas, a monoculture is most satisfactory, but if the environment is favorable, a mixture of grasses and legumes as shown in Figure 10.1 is more satisfactory. As previously stated, grasses and legumes each have their advantages. When they are properly combined into a pasture mixture, there can be a symbiotic relationship as follows:

1. Legumes in a mixture, when properly inoculated, fix nitrogen in the soil, which can be used by grasses.
2. Grasses tend to protect the legumes from heaving during freezing and thawing.
3. Grass holds the soil against erosion.
4. Grass in the mixture reduces the danger of bloating.
5. Legumes provide highly nutritious pasture during periods when most grasses are low in feeding value. Thus, the nutritive level for the whole pasture season may be maintained.

Figure 10.1 Mixed grass legume pastures are excellent for late lambs in summer. (Courtesy Animal Science Department, Ohio Agricultural Research and Development Center, The Ohio State University, Wooster)

TABLE 10.2 CLASSIFICATION OF PASTURES

Type	Estimated duration
Permanent pasture	Indefinitely
Supplementary pasture	Indefinitely
Cropland	
Short term	2–4 years
Long term	5–8 years
Temporary	3–6 months
Crop residues	2–4 months

10.8 CLASSIFICATION OF PASTURES

Pastures have been classified in a number of ways. For example, they may be designated as permanent and supplementary, with the bulk of grazing on permanent pasture supplemented by other crops that may have more than one use. It is also desirable to include the duration of the crop in the classification, as shown in Table 10.2.

10.9 PERMANENT PASTURES

Permanent pastures may be defined as fields or areas maintained indefinitely for grazing livestock. Characteristically, on farms or ranches the least fertile, most rolling, and roughest land is maintained in permanent pasture. The manager expects to use it extensively with as little upkeep as possible. Permanent pasture may be valuable in winter, as shown in Figure 10.2.

10.10 SUPPLEMENTARY PASTURES

In most pasture systems, the permanent pastures are expected to furnish the bulk of feed for grazing livestock. Under favorable seasons, it may be possible for livestock to remain on permanent pasture throughout the grazing season, but it may not be the most economical use of the forage or result in a high level of production. If other sources of pasture are available, it should benefit the sheep as well as the permanent

Figure 10.2 Winter pasture like this fescue reduces feed costs and provides adequate nutrition for ewes until late gestation.

pasture to use them. Supplementary pastures are those that are available for use when the permanent pasture is deficient in volume or quality of production or for special uses.

10.10.1. Cropland pastures In farming areas, it is desirable periodically to change tillable fields from row crops to pasture or forage production. This may be done in an effort to build up the field, or it may be necessary from the standpoint of the total feed supply, especially if there is a shortage of land available for permanent pasture. In many sheep operations, a number of fields may be used for pasture and some fields for producing grain. The fields are rotated so that pastures can be renovated periodically. This may call for a longer or shorter period in pasture, depending on the cropping and livestock programs. Sometimes rotation may be affected by government programs. Some extra pasture that has a dual use is good insurance for the sheep business.

10.10.2. Temporary pastures Temporary pastures are those in use for less than a year. Examples of temporary pastures are winter wheat and barley. These can be grazed in the fall if carefully managed without reducing the grain yield. Or, if the pasture is limited, they may be pastured out completely. Combinations of oats and rye are found useful by many sheep producers for summer grazing. Sudan, if it is allowed to become well established, resists drought very well (see Figure 10.3). Care must be taken to use only certified seed to prevent problems with hydrocyanic acid poisoning. In recent years, field turnips have been seeded in the summer and pastured in the fall to fatten lambs and maintain pregnant ewes.

Whether or not you use temporary pastures depends on the cost versus the benefits. It is usually expensive to plow and prepare a seedbed just for a crop that may be used for only a limited time. It is well to look at other alternatives before doing so.

Figure 10.3 Sudan is a very satisfactory temporary pasture for sheep in summer.

10.10.3. Crop residues and other sources of pasture Meadow aftermath, the residue after removing hay crops and grain stubble, has been used for open ewes during the summer when they are being maintained in medium to low condition. This is a good use of otherwise waste feed, and it prevents ewes from becoming too fat. Producers should be cautioned to ascertain that there is something to eat in the fields besides unpalatable or poisonous weeds. It is also necessary to furnish minerals, water, and some shade for the sheep.

Beet tops are relished by lambs during the sugarbeet harvest. It is desirable to provide mineral and water and a small amount of legume hay to them. If there is much of the beet left on the top, lambs may come off the fields in satisfactory condition for slaughter. Care must be taken to remove the lambs before the beet tops are gone.

Corn Fields. In years past, many Corn Belt farmers purchased lambs to run in their fields. They preferred 60-lb (27.3-kg) western lambs that had never been fed grain. The lambs would eat weeds and lower leaves of the corn and make satisfactory gains. The farmers prepared the lambs ahead of time by shearing and worming them, if needed, tagging and vaccinating them for enterotoxemia and soremouth a week before turning them in. Usually the lambs were turned into the fields when the corn silks were dry and the kernels dented, about September 15, and allowed to graze until they learned to ride down the stalks. Stocking rate was 2 to 4 lambs per acre depending on the corn crop and weeds and grass between the corn rows. Farmers provided water and a salt-protein supplement mixture (usually 1 to 5) to limit the lambs to ⅛ lb (57 g) of supplement daily.

Not many lambs are put in corn fields today because corn land is often not fenced and many fields are clear of weeds, but under proper conditions very cheap gains may be put on lambs.

Gleaning Corn Fields. After corn is picked, lambs can be turned to the fields to pick up the fallen corn. If there are no weeds and grass in the field, it is advisable to feed them 1 lb (0.45 kg) or more of legume hay daily, along with a mineral mixture and water. It is also prudent to vaccinate the lambs for enterotoxemia at least a week before they are turned into the fields. It is important to check frequently to see that corn is still left in the field and remove them when they have gleaned it.

Other Sources of Pasture. Sheep can be used to clean up weeds in fence rows and waterways, between buildings, or along lanes. A pure diet of weeds may be harmful to sheep. During dry weather, forcing sheep to eat weeds may cause them to eat poisonous ones. While it is true that sheep have a well-deserved reputation as scavengers, there are limits to their ability to handle unpalatable, fibrous plants.

10.11 GRAZING MANAGEMENT

Management of grazing is an important factor affecting the productivity and longevity of a pasture. Research results have shown that grazing can be beneficial to both the pasture and sheep depending on the methods used. Methods of grazing management are discussed next.

10.11.1. Continuous grazing The simplest and easiest method is to put the sheep on pasture and leave them there throughout the season. This method is called continuous grazing and is used most frequently. Once the watering facilities, mineral feeders, and shades are in place, very little labor is required to manage the sheep. There are, however, some unfavorable consequences of continuous grazing:

1. Sheep tend to spot graze. Some grasses grow more rapidly than others and, as they mature, they become less palatable. Sheep prefer shorter, more tender vegetation; so they tend to overgraze some spots and undergraze the tall forage as shown in Figure 10.4. The tall forage also tends to shade the surrounding plants and reduce their growth.
2. Fields grazed continuously may produce an overabundance of forage in the spring when conditions are most favorable for plants and not enough during the hot, dry months of summer. This results in feast and famine, with too much and too little feed available. The total production useful to livestock is reduced.
3. Certain plant species, especially the clovers, need a recovery period after grazing or they may not survive.
4. It is difficult to fertilize pastures while sheep are grazing them.
5. Internal parasites build up on continuously grazed pastures, especially in areas around watering facilities.

10.11.2. Rotation grazing Rotation grazing is designed to intensify the pressure on an area for a limited time and then allow it to recover before it is grazed again. For example, if a field is subdivided into three areas, area 1 can be grazed for 7 days and rested for 14 while areas 2 and 3 are being grazed. Then the rotation can be repeated. If four fields are used, the grazing time may be as short as 5 days. Some producers run cows in fields, after sheep are removed, to clean up tall growth and others clip fields for the same purpose. Many New Zealand sheep producers are using rotation grazing effectively in what they call "mob" grazing. They have many

Figure 10.4 Continuous grazing results in spot grazing and loss of desirable plants, especially legumes.

pastures and turn in a great many sheep for only a few days and then move them. By using extreme pressure for a short time, excess grass growth is grazed down.

One possible innovation to reduce the labor in rotation grazing is to locate a corral that is equipped with mineral feeders and waterers and is shaded in the center of a field. Fences are then radiated from the corral to the perimeter to form smaller triangular paddocks. Such an arrangement also has the advantage that the corral can be built so that it is predator proof. Then the sheep can be confined at night. The sheep can be rotated to the next field merely by opening the gate to it.

The advantages of rotation grazing, then, are:

1. It increases the carrying capacity of mixed forages.
2. It is beneficial to the maintenance of legumes in pastures.
3. It assists in reducing the buildup of internal parasites.

10.11.3. Mixed species grazing Investigators in widely different geographical locations have obtained improved performance from grazing sheep with other species. This refutes the belief by some pioneer cattle ranchers that sheep ruin the range for other livestock. Rather than ruining the range, a combination of sheep and cattle, sheep and goats, or all three species may be advantageous over a single species. Combination grazing has also resulted in improved production for both species of livestock where cattle and sheep are grazed together on improved pastures.

Feeding Preference Differs between Ruminant Species. Sheep, goats and cattle are all ruminants adapted to obtaining their livelihood from growing plants. All three species are capable of existing on grasses, legumes, and browse, but their grazing patterns differ and are remarkably compatible on diverse ranges and pastures.

Sheep like short, tender grass and wideleaf herbaceous plants, such as legumes and many weeds. Goats choose the leaves of shrubs or small trees and weeds. Cattle prefer taller grasses and legumes. On the range, sheep are mobile and prefer low-growing, broad-leaved herbaceous plants that can be grazed close to the ground at repeated intervals. They graze in a predictable circular pattern. Goats are more mobile than sheep and select their diet from a wider range of plant types and parts of plants. Cattle prefer to spend most of their time grazing less hilly, more fertile meadows or canyon bottoms where grasses grow more profusely.

Since no two species are competitive for feed, there is excellent complementarity in mixed grazing with sheep and cattle, as shown in Figure 10.5. It has been observed that fields were more uniformly grazed by a mixture than a single species of livestock.

10.11.4. Research on mixed species grazing Nolan and Connolly (1977) made an extensive review of mixed cattle and sheep grazing. Some general observations were:

1. Mixed grazing is most favorable on pastures or ranges in which there is a mixture of herbage.
2. Sheep eat many weeds (forbs) and browse not palatable to cattle. Some of the forbs such as tansy ragwort and threadleaf groundsel are poisonous to cattle but not to sheep.

Figure 10.5 Because they graze differently, cattle and sheep complement each other and both benefit.

3. Output per unit area is generally favorable to both species for mixed grazing.
4. In some cases where cattle performance was not improved, the presence of cattle caused enough better sheep production to make the combination worthwhile.
5. Stocking rate is very important to the success of grazing. If pastures or ranges are overstocked, sheep appear to suffer less than cattle.
6. Stocking ratio is important to the success of combined grazing. Ratios wider than one steer to eight ewes appear to be unfavorable and may result in the loss of legumes from the pasture. Ratios of 1 to 5 to 1 to 6 appear to be optimum.

10.12 MAINTAINING PASTURES

The amount and quality of feed produced by a pasture deserves much emphasis, but the persistence of the pasture is also very important. Renovation is invariably very expensive and should be avoided if possible.

10.12.1. Evaluating pastures A good operator constantly observes pastures to determine when they are ready to graze, when they are under- or overgrazed, and for their overall well-being. On a long-term basis, it is desirable to note the changes of species in the plant population. Unless weather conditions are responsible, loss of legumes, invasion of unpalatable weeds, and undesirable grasses may indicate overstocking, the need for fertilizer, or problems with disease and parasites.

10.12.2. Controlling weeds Weeds are usually a problem in permanent pastures or rangeland. They may become so serious as to force renovation. Know the harmful species in your area and how to control them. Control depends on the kind of weeds and the type of pastures grown. If weeds are a problem, it is desirable to consult with the extension agronomist for suggestions.

Perhaps the first consideration should be to exercise care when the pasture is being established. Following the proper procedures in preparation of the soil, fertilizing, and management is necessary for the plants to grow and develop into a suitable pasture. It is important to use certified seed to avoid introducing unwanted plants.

After the pasture has been established, weeds and brush may be controlled by chemical and mechanical means. Chemical control of weeds is possible with some species, but is highly selective and should be used with caution. Use only those herbicides that are recommended by the Cooperative Agricultural Extension Service. Follow directions to prevent harming other crops and livestock.

Mechanical control usually consists of mowing pastures. Mowing is very effective on some species if it is done just before the seeds are formed. In much of the United States, optimum dates to mow for weed control are May 15 and September 15. However, dates may vary with the type of weed and local conditions.

In some areas, burning pastures is quite effective. This method should be employed only after consulting with people who have used it. It is necessary to know the effect on the species and when to burn. State and federal laws must be complied with when a burn is done. Fire lanes, equipment, supplies, and personnel to control the fire must be on hand to ensure control. In addition, it is necessary to manage the stocking and time of grazing after burning to obtain the full benefits.

10.13 PLAN FOR PASTURES AND GRAZING

If pastures are to be used most efficiently by sheep, it is important to develop a workable plan so that forage is available when it is needed. The class, age, and stage of production of the sheep should be determined and put on a schedule; then the forage that will be used to satisfy the needs can be added. A simple form can be used and modified as required. An example is given next.

10.13.1. Pasture plan for sheep

- Class (includes ewes, lambs, rams, and other species)
- Number (number of head)
- State of production (open, flushing, early gestation, late gestation, suckling lambs, etc.)
- AUM (animal units/month; five ewes = 1 AU)
- Fields (fields to be used for pasture)
- Acres
- Dates

Very good pasture will carry 1 AUM, but this is exceptional except during the most favorable months. It is generally unrealistic to expect permanent pastures to handle more than 0.5 AUM for extended periods. If supplementary pasture is available, an acreage of one-third to one-half of the permanent pasture, if seeded to a grass–legume mixture, can be harvested for hay or silage when the season is good and permanent pasture is carrying the sheep. Then, if the weather is dry, the supplementary field can be used for pasture. If the drought is severe, the forage harvested in the spring can be used to tide the sheep over until pastures can be used. Some sheep producers harvest the excess spring growth as round bales and use them as emergency feed. Others routinely plant sudangrass to be used in emergencies.

Records should be kept on the carrying capacity of the fields or areas so that there is information for planning in future years. Some excess of feed is quite desirable compared with the stress of locating it when seasons are unfavorable.

10.14 SUPPLEMENTING PASTURES

There is evidence that it is beneficial to provide some additional protein to sheep during dry seasons when forages have ceased growing and have low feeding value. Weir and Torell (1967) observed improved performance in replacement ewe lambs fed cottonseed meal or soybean meal in combination with salt to limit the intake of protein supplement to 0.25 lb (114 g) per lamb daily. Alfalfa pellets were equally effective in preventing weight losses and increasing wool production. In work at the Missouri Experiment Station, Ross, Riley, and Shelton (1967) reported improved gains and resistance to internal parasite damage, as shown by higher erythrocyte* levels and lower worm egg counts per gram of feces than controls, by lambs which received 0.25 lb (114 g) of soybean meal per head daily in a self-fed mixture with salt. Generally, supplementing with grains or other carbonaceous concentrates has not been as effective as protein concentrates except for lambs grazed in the spring and early summer on lush pastures.

REFERENCES

Bennett, D., and others, "The Effect of Grazing Cattle and Sheep Together," *Aust. J. Exp. Agric. Anim. Husb.,* 10, 1970, pp. 694–709.

Brown, E. M., "How to Seed New Pastures," *Missouri Agric. Exp. Sta. Bull. 739,* 1959.

______, "Missouri Pastures," *Missouri Agric. Exp. Sta. Bull. 750,* 1960.

Curll, M. L., J. L. Davidson, and M. Freer, "Efficiency of Lamb Production in Relation to the Weight of the Ewe at Mating and during Pregnancy," *Aust. J. Agr. Res.,* 26, 1975, pp. 553–65.

Hamilton, D., and J. G. Bath, "Performance of Sheep and Cattle Grazed Separately and Together," *Aust. J. Exp. Agr. Anim. Husb.,* 10, 1970, pp. 19–26.

Heady, H. F., *Rangeland Management.* New York: McGraw-Hill Book Co., 1975.

Heath, M. E., R. F. Barnes, and D. S. Metcafe, *Forages, the Science of Grassland Agriculture,* 4th ed. Ames, Iowa: Iowa University Press, 1985.

Heinemann, W. W., "Dual Grazing of Irrigated Pastures by Cattle and Sheep," *Int. Grassland Cong., Surfer's Paradise, Qld.,* 1970, pp. 810–13.

Martin, W. E., and L. J. Berry, "Effect of Nitrogen Fertilizer on California Range as Measured by Weight Gains of Grazing Cattle," *Calif. Agric. Exp. Sta. Bull. 846,* 1970.

Moore, R. M., N. Barrie, and E. H. Kipps, "Continuous and Rotational Grazing by Merino Sheep," *Bull. Council Sci. Ind. Res., No. 201,* 1946, 69 p.

Morley, F. H. W., *Grazing Animals* (World Animal Science Series). New York: Elsevier Scientific Publishing Company, 1981.

*Erythrocyte levels, also called hematocrits, are determined by drawing a blood sample in a capillary tube, spinning in a centrifuge, and reading the percent of packed red cells on a scale designed for that purpose.

———, D. Bennett, and G. I. McKinney, "Effects of Intensity of Rotational Grazing with Breeding Ewes on Philaris subterraneous Clover Pastures," *Aust. J. Exp. Agric. Anim. Husb.,* 9, 1969, pp. 74–81.

Nolan, T., and J. Connolly, "Mixed Stocking by Sheep and Steers. A Review," *Herbage Abs.,* 47, 1977, pp. 367–74.

Ross, C. V., J. G. Riley, and G. C. Shelton, "Practical Tests with Anthelmintics for Grazing Lambs," *Missouri Agric. Res. Bull. 927,* 1967.

Stoddart, L. A., A. D. Smith, and T. W. Box, *Range Management.* New York: McGraw-Hill Book Co., 1975.

Suckling, F. T. A., "Pasture Management Trials on Unploughable Hill Country" (Te Awa, N.Z.), *J. Exp. Agric.,* 3, 1975, pp. 351–436.

Terrill, C. E., I. L. Lindahl, and D. A. Price, *Forages, the Science of Grassland Agriculture,* M. E. Heath, D. S. Metcafe, and R. F. Barnes, eds. Ames, Iowa: Iowa State University Press, 1973, pp. 703–15.

Vallentine, J. F., *Range Development and Improvement.* Provo, Utah: Brigham Young University Press, 1971.

VanKeuren, R. W., and C. F. Parker, "Better Pasture Utilization—Grazing Cattle and Sheep Together," *Ohio Exp. Sta. Rept. 57,* 1967, pp. 12–13.

Weir, W. C., and D. T. Torell, Supplemental Feeding of Sheep Grazing on Dry Range," *Calif. Agric. Exp. Sta. Bull. 832,* 1967.

QUESTIONS FOR STUDY AND DISCUSSION

1. Explain why pasture is so important to a sheep enterprise.
2. What are the advantages of sheep over cattle as grazing animals?
3. In what ways does soil fertility limit the type of forages grown?
4. What is meant by adaptability of plants, and why is it important?
5. What factors affect the adaptability of forages? Which are the most limiting?
6. Discuss the characteristics of good pasture.
7. The United States may be divided into regions of forage crop production. What is the basis for the division?
8. What are sources of information on adapted species and cultivars of pasture forage?
9. Discuss the symbiosis that may occur with the use of grass–legume mixtures for pastures.
10. What is the reason for using both permanent and supplementary pastures on a farm or ranch?
11. What are some advantages of using cropland pasture in addition to permanent pasture?
12. Explain why temporary pastures are of value in a production system.
13. Explain why rotation grazing is superior to continuous grazing.
14. Why is mixed species grazing superior to individual species grazing?
15. What are the criteria for evaluating a pasture? When should it be done?
16. What is the purpose of developing a pasture plan?
17. What are the advantages of dual use of fields for hay or silage and pasture?
18. When should permanent pastures be mowed to be most effective in controlling weeds?
19. Speculate on why supplementing summer pastures with protein supplement should result in superior gains and resistance to internal parasites.
20. Why should tall grass be clipped or grazed down by cows after sheep have grazed a field?

11 Care and Management of the Breeding Flock

In any sheep operation, the degree of success achieved depends to a large degree on how well the animals react to the environment on the farm or ranch. It is the primary duty of the sheep producer to provide conditions conducive to their health and well-being. Regardless of the hereditary components of sheep, they cannot properly express their potential unless the feed, care, and management are favorable. It should be remembered that environment is responsible for 50% to 90% of the expression of any trait.

Sheep have been produced successfully for centuries using fundamentals developed mostly by trial and error and passed on to each succeeding generation. It is only recently, in terms of history, that science has made any considerable contribution to the technology of the industry.

To be successful in the competition of our present agriculture, it is necessary to use principles passed along from sheepraisers of the past, plus the scientific information being generated by our research organizations throughout the world. No scientific discovery will make a real contribution to the sheep industry until it has been put to practical use.

In the final analysis, the objective of sheep production is to convert plant products to protein-rich food and fiber for clothing. The care and management of sheep have always been the main contributing factor to the success of an operation. The degree to which a sheepraiser can equal or exceed the average of the operators in any phase of the operation can usually be directly related to the profits of the enterprise. Knowing how and when to take action is the reason for the study of sheep production and management.

11.1 IDENTIFICATION

If you expect to understand your sheep business and to make real progress, it is important to know the identity of each animal in the flock. This is much less important in commercial operations than in purebreds, but there are distinct advantages in having an identification for all breeding stock, especially if replacements are to be produced. Purebred sheep are commonly identified by metal eartags showing the individual number and usually the year of birth. These may also be used on commercial ewes with success. There are some satisfactory plastic tags that can be used, but these may be lost during the lifetime of the sheep. Some plastic tags may include numbers that can be easily read when the sheep are in groups. Usually it is worthwhile to have a permanent identification, such as a metal tag, in addition to the plastic one (Fig. 11.1). Sheep also may be tattooed in the ear using dark ink for white-faced breeds and green for black faces (Fig. 11.2). An ear-notching system for individual numbers, in the same manner that swine are marked, is shown in Figure 11.3. Temporary identification is possible with paint brands, although caution must be taken to use only paint made for that purpose. Sheep-branding paint is available and is not expensive.

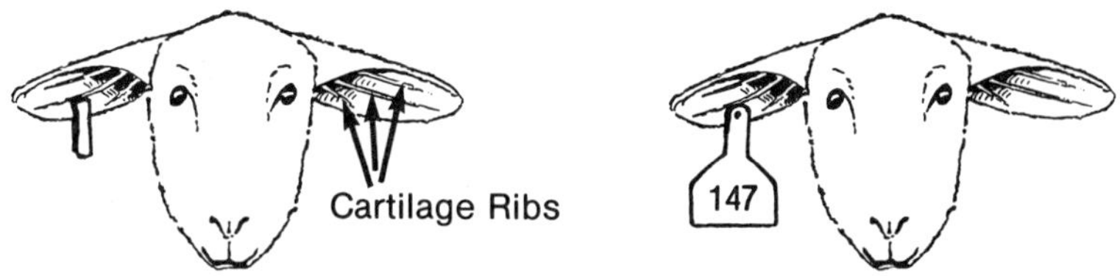

Figure 11.1 Metal ear tags (left) and plastic tags (right) should be inserted between the middle and lower cartilage ribs.

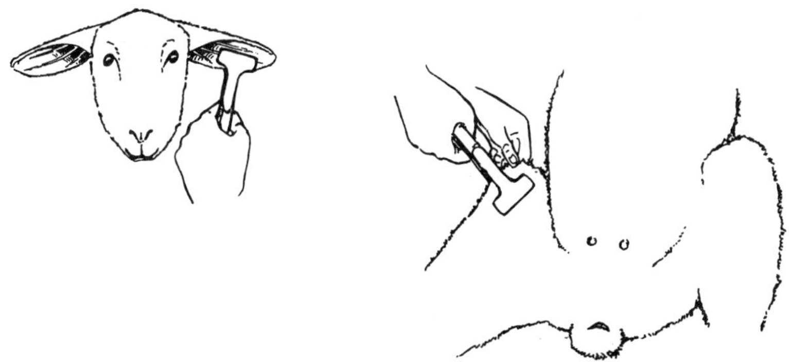

Figure 11.2 The dies should pierce the ear between cartilage ribs and the bare skin on the inside of the hind flank.

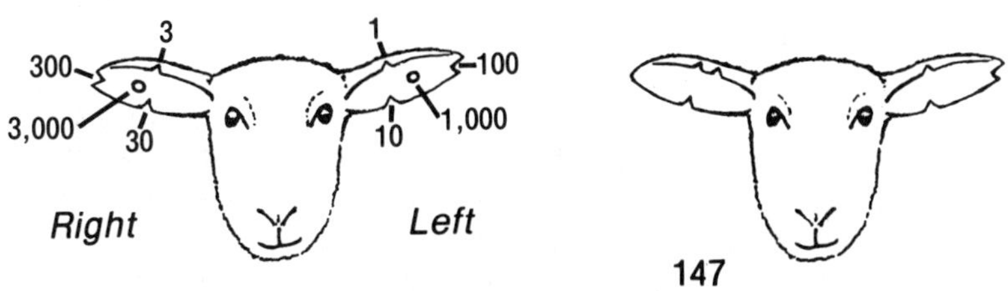

Figure 11.3 A large number of sheep can be identified by ear notching. Note each number can be doubled with a second notch in the same location.

11.2 RECORD-KEEPING SYSTEM

Records are not easily kept, but they are essential to a well-run operation. If they are too complicated or laborious, they not only will not be kept, but they may be inaccurate and incomplete and in many cases worse than none. A practical system should be set up that can be followed routinely with a minimum of effort. Furthermore, if records are kept, the sheepraiser must resolve to analyze them and use them as a basis for making management decisions.

11.2.1. Production records Reasons for keeping production records are:

1. To measure flock productivity
2. To identify high producers so that their progeny can be kept for replacements
3. To find the loafers so that they can be culled
4. To evaluate ram performance
5. To obtain factual data to supplement selection by sight
6. To provide permanent records so progress can be measured

The detail of records kept must be decided by the operator based on how much they are needed and the cost in time to keep them. Obviously, the purebred breeder must have individual records in order to register breeding stock. The more progressive breeder of the future should have more to offer than the mere fact that the animals are purebred.

In the past, much emphasis has been put on showring records, but that era appears to be ending. In the future, a purchaser should expect to be shown some evidence that the stock he or she buys will make a substantial contribution to the improvement of the flock. Records showing superior production should be the key to success for the purebred breeder. Some of these are:

1. Lambs born and weaned per ewe
2. Weaning and 180-day weights
3. Wool quality and production per ewe
4. Lambing dates
5. Postweaning gains and feed efficiency on rams
6. Carcass characteristics on rams (if they can be estimated by scanogram, whole body counter, probe, or carcass data on half-sib rams)
7. Health records

The commercial producer should be interested in knowing the number of lambs born, weaned, and marketed and the dates of sales and prices received for lambs. The fleece weights and prices received for wool should be kept. Death losses and causes of mortality should be recorded. All these records are needed if such efficiency factors as lambing and weaning percent and lamb production per ewe per year can be figured. The wool weight is necessary if ewe indexes are to be calculated.

11.2.2. Other records Any good farm record system includes a budget for estimating production and returns. Almost every extension service has a farm management program that is useful in evaluating an enterprise. At least in the beginning, it should be useful to enter such a program to learn how to estimate and record feed, labor, and other costs and how to determine the cash flow and profitability of the enterprise. In the end, a comparison should be made between the estimates and the actual performance and an analysis made of reasons for the results obtained. These records should be kept permanently so that the progress made can be measured.

A top-notch producer will be able to give such information as lambing percentage, death losses, yearly production per ewe, and costs to produce a lamb, for instance. It is the mark of a professional to know and use data on the efficiency factors of a sheep operation.

11.3 GENTLING THE FLOCK

11.3.1. Handling groups of sheep It has been wisely said that a sheep flock reflects the personality of the owner. If you believe that sheep are completely stupid and are incapable of being influenced by your actions, you very much underrate them. Sheep have the ability to observe, and they can remember very well. Most are willing to obey the master, providing they can understand what is required. Just be sure when you handle sheep that you know what is to be done. Most troubles in handling sheep occur because the operation has not been planned, equipment is not adequate, or the sheep have been mistreated. It is a joy to see sheep well handled, either with or without a dog. It is a horrible experience to see sheep mishandled by people, dog, or horse. When you take control of a flock of sheep, resolve that you will not injure them in any way or allow anyone else to mistreat them. Hitting, kicking, beating, chasing, scaring, or raising your voice are all sure to cause sheep to distrust and avoid you. Then it becomes more difficult for you to approach them close enough to make an inspection without extra work on your part. Then, too, when you have to run them through a chute, they always lose weight, which may very well mean loss of money. Start out by making it a practice to walk quietly among them to let them know that you are to be trusted. Be quiet and avoid sudden moves.

Establish routines and follow them. Even though it may take a little longer, handle sheep gently, but firmly. Think ahead and plan and construct equipment so that it is easy to move them. Locate gates in the corners of a field so that the fences will help you control them. Use panels to reduce the chances of escape. Provide swing gates for putting sheep into a chute. Give them a chance to see where they are going rather than forcing them.

11.3.2. Handling individual sheep Individual animals must be handled frequently for such things as close inspection, treatment for parasites and disease, foot trimming, and shearing. How single sheep are handled affects their disposition. Expect the sheep to remember a bad experience and be harder to handle in the future. First, learn how to catch and hold a sheep properly. Remember it is nearly always easier to catch one in a group than an isolated individual. A lamb alone in a pen is usually very nervous and may become berserk and injure itself by running into the fences. It is best to have the individual in a group in a small pen or chute. Let the group mill around until the one you want is between other sheep and facing away from you. Then move quickly and deftly behind it and grasp it by the loose skin in the right hind flank if you are right handed, and vice versa if you are left handed. Hold the skin firmly, lifting upward to control the sheep. If the sheep are wedged in close and facing you catch the head with one hand under the throat and the other behind the neck. The hand at the throat should grasp the loose skin and wool. That is, incidently, the only place on a sheep where it is permissible to grab the wool. Once the sheep is caught, hold its head up or it may escape. If the sheep is very wild, it may be better to hold one hand on the flank and the other at the throat.

More control can be exercised by setting the sheep up. With small ones it can be done by holding one hand under the neck and the other at the hind flank with the side of the sheep pressing against your leg. Lift the sheep up until its feet clear the ground; then turn the head up and the rump down with the sheep's back to you. keep the sheep tilted backward so that the hind feet cannot make contact with the ground.

Larger animals are too heavy to lift and could be injured; thus they should be handled in another way. Hold the dock with one hand, the right if you are right handed. Stand with your legs against the shoulder and side of the sheep. Grasp the muzzle of the sheep with your left hand and twist its nose back sharply toward the shoulder. As the neck twist is applied, step back slightly and push down on the top of the rump. If this is done properly, the sheep will sit down. When it does, pull it backward with its back toward you and leave it tilted back enough so the hind feet cannot touch the ground.

To move a sheep, hold it by the skin under the throat and place the other under the dock. Lift up on the dock and the sheep will move. Large sheep or very wild ones can be controlled more completely by the use of a rope halter.

A shepherd's crook will assist in catching sheep. It will hold the sheep until you can get your hands on it, if it is facing away from you. A leg hook works very well, but it is dangerous unless the one using it is directly behind the sheep. It is easy to displace the stifle joint if the user is too far to the side. The hock joint may also be injured if the sheep is moving fast when hooked.

It is often necessary to lift a sheep into a pickup truck or over a fence. An excellent method with smaller sheep is to straddle it facing forward, place both arms around the sheep just behind the front legs, grip the sheep tightly to your chest, and stand up so that the hind feet clear the ground. Boost the sheep with one knee to clear the fence or endgate of the truck. Larger sheep can be lifted by two people facing each other with the sheep in between them. The two people pass the arm nearest the front of the sheep under the breast bone just in front of the forelegs and grip wrists. They pass their other arms under the sheep at the hind flank and lock wrists. It is best for each person to squat down then and lift with the legs instead of the back. Large sheep should be loaded with a chute instead of lifting them.

Moving a ewe and her lamb into a lambing pen can usually be accomplished best by carrying the lamb and letting her follow. Be patient and she will finally understand what is wanted. Hold the lamb with one hand grasping both front legs at the shanks. Carry it just off the ground so the ewe can see it.

11.4 PREPARATION FOR BREEDING

To a large extent, the successful breeding of the ewe flock may be the result of preparation of the ewes and rams ahead of time. Rams should be sheared and protected from heat stress 4 to 6 weeks ahead of the breeding season. Their feet should be trimmed, and they should be wormed. They should be in medium flesh and healthy. Be sure their testicles are normal and that they have no abnormalities of the penis.

The ewes should be wormed and their feet trimmed. If they have heavy fleeces, they should be shorn. If ewes have manure tags, they should be removed. During the anestrous period, ewes should have been pastured or fed so that they approach the breeding season in no more than medium condition. Then they may be turned to a better pasture for 2 to 3 weeks to flush them before rams are turned in. Flushing will cause them to produce 8% to 10% more twins, and they will be more likely to settle. The same effect can be obtained by feeding them 0.5 lb (.23kg) to 0.75 lb (.34 kg) of concentrates per head daily. Legume pastures may not be satisfactory during years of heavy rainfall due to the plant estrogens they contain. In these kinds of years, grass pastures are more satisfactory. Consider using vasectomized teaser rams 10 days before turning in fertile rams. This may result in bunching the lamb crop.

11.5 CARE DURING THE BREEDING SEASON

If ewes are to be bred on pasture, they may require very little attention, but they should be inspected frequently to see that the rams are working and that they are being settled. The use of a marking harness or painting the rams should provide that information.

11.5.1. Managing rams Rams work better in the cool of the mornings and afternoons. Separating rams from ewes during the heat of the day and keeping them in a cool, shady place will prevent much heat stress and allow them to be fed some grain and legume hay to prevent them from losing weight. One pound (0.45 kg) of concentrate mixture of corn and oats should be sufficient for the purpose. If a single ram is being used, he will fret less if he is given a ewe for company.

11.5.2. Care of ewes If ewes are flushed, they should be continued on the same level of nutrition through the breeding season, especially if they were thin before they were flushed. Cut back on grain after the breeding season if that was the method used, but it should be gradual and good pasture should be available. Ewes should be protected against heat stress by allowing them to graze early and late in the day and providing a shady place where a breeze can flow during the heat of the day. Excessive exercise and exposure to direct sunlight should be avoided.

11.6 FEED AND CARE OF PREGNANT EWES

Probably the most precarious time for the fetus is during the first 5 days after conception. Then, according to the work of Dutt, Ellington, and Carlton (1959), high temperatures or stress of any kind that results in an elevation of body temperature may destroy the embryo. Thus, for at least a week after breeding, avoid excessive exercise, long exposure to direct sun, or overcrowding. Allowing the ewes to graze in the early morning and late afternoon, with a shady, cool place to rest during the middle of the day, is ideal.

11.6.1. Ewes should gain in weight and condition If ewes are bred in the late summer or fall when pastures are growing, management during early gestation is no problem. Ewes should generally gain 15 to 30 lb (7 to 14 kg) from breeding to parturition depending on a number of factors, such as age and development, breeding condition, size of breed, and whether or not they produce multiple births.

Young ewes are expected to continue growing when they are pregnant. Thus, they must be fed more than mature ones. On the other hand, old ewes may be exceptionally thin and need building up in condition before they lamb.

From a practical viewpoint, few producers will weigh their ewes individually, so it is more sensible to discuss the feeding of pregnant ewes on the basis of body condition (fat) of the ewe when she lambs. Ideally, the ewe should be in medium to thin condition prior to breeding. Then, if she is flushed, she should respond. She should then continue to gain weight gradually until she lambs. When she is handled early in pregnancy, it may be possible to feel each vertebrae in her back and each of her ribs; but if her early condition exists near parturition, there is cause for alarm because she is entirely too thin. If the shepherd has made it a policy frequently to walk quietly among the flock, it should not cause them any stress to bunch them up rather close and feel along their backbones and back ribs. It is important to note the progress they are making and to remember that they should be fleshy, although not fat, when they lamb. Scoring the ewes for condition may be useful. A score of 1 is thin and 5 is overly fat, with 2, 3, and 4 between the two.

The very young ones, ewe lambs and yearlings, should never be fed with vigorous, mature ewes because they will invariably be pushed aside and not receive their fair share. A possible solution may be to group the aged and very young ones together, but they, too, should be watched carefully to make sure they are all eating their rations.

11.6.2. Pasture and roughage for early gestation For the first 4 months of gestation, pasture is usually the best feed for ewes. If it is of high quality, it should contain adequate nutrients, be easily digested, provide exercise for the ewes, and require very little labor or out of pocket expense. Dry weather may cause a number of problems for pregnant ewes. If the rainfall has been limited for a long period, the forage may be low in palatability and nutrients. At the same time, sheep will have grazed down the most succulent vegetation, leaving coarse, fibrous growth and unpalatable plants that may be injurious to sheep. A good example is the nightshade, which is unpalatable, except under extreme conditions and poisonous to sheep. Management on good pasture consists of providing water and minerals and protection from other farm animals, predators, and parasites.

11.6.3. Water Rapidly growing grass usually contains a high level of water, and thus ewes will drink very little; but it should be available for them if they want it. Automatic water fountains are excellent, but tubs or tanks that are kept clean and refilled frequently are satisfactory. If ponds are used, they should be fenced and the water piped to a tank or waterer. Generally, ewes require 3.5 lb (1.6 kg) of water for each 1 lb (0.45 kg) of dry matter consumed. Inadequate water intake either in quantity or quality will reduce feed consumption.

11.6.4. Minerals Minerals can usually be self-fed on pasture. Except in those areas that have special deficiencies, such as of copper, cobalt, or other trace minerals, iodized salt mixed with a source of calcium and phosphorus can be provided ad lib. A simple mineral mixture of equal parts of dicalcium phosphate, iodized salt, and ground limestone will usually provide for the needs of ewes. Bonemeal may be substituted for dicalcium phosphate if it is not too expensive. Wherever the ranch or farm is located, a practical, simple mineral mixture adapted to the locality can usually be obtained from agricultural extension personnel.

Some precautions in providing minerals are (1) locate the feeders near the water source, (2) the box in the feeder should be under cover to protect the mineral mixture from the weather (once a mixture gets wet, it usually hardens so that it is very difficult for ewes to get enough to satisfy their needs), and (3) mineral feeders or boxes should be checked and refilled several times a week to be sure the sheep are getting what they need. When sheep are eating dirt, it is a good sign they are not receiving sufficient minerals.

Commercial mineral mixtures may also be provided for the ewes. They are designed to be quite palatable, and sheep may eat more than they need. The salt content of the mixture can be used to figure the consumption per day. If consumption of salt exceeds 0.4 oz (11.4 g) per ewe daily, it indicates that the ewes are eating more than they need. Excess consumption may not be harmful, but it does represent an unnecessary expense.

11.6.5. Protection from other livestock Farm animals may injure pregnant ewes and cause their death or force an abortion. It is desirable to separate rams from the flock after breeding. Horses, on cool mornings, are frequently in high spirits and may want to run the ewes and play with them. Sheep should never be grazed with horses or hogs. Cattle and sheep are usually quite compatible, but young bulls and some cows may be aggressive toward sheep. Careful observation should satisfy the operator whether or not cattle and sheep can be grazed together.

11.6.6. Protection from predators Elsewhere there is a discussion on predator control. Pregnant ewes are easy prey to coyotes, dogs, and other large carnivora. Even though the ewes may not be killed, they may be forced to injure themselves if overly excited by being chased by playful dogs. Many dogs start out merely playing with sheep, but it soon develops into attacks. Well-built perimeter fences prevent most attacks. If, in addition, it is possible to construct a two-wire electric fence outside the perimeter fence, it usually assures protection, or the ewes may be confined at night, when the attacks usually occur.

11.6.7. Ewes should be handled with care A good shepherd will work with the ewes quietly and gain their confidence, but will restrict handling through chutes increasingly as gestation progresses. Some precautions should be taken to build gates and doors wide enough (at least 8 ft) to prevent wedging ewes in them. Also, keep ewes away from machinery storage, or they may get lodged between or under the equipment and may also contaminate their fleeces with grease and oil.

11.6.8. Feeding during late gestation One of the three most critical times in the life of a ewe is during the last 4 to 6 weeks of gestation. Initially, the fetus grows rather slowly. But during the last 20 to 40 days, fetal growth increases 60% to 80% until parturition. The nutrient requirements for pregnant ewes should be followed, but care must be taken to formulate rations the ewes will consume. Feeding a ewe heavy with lambs on coarse, low-quality roughage may very well look satisfactory on paper, but it could be beyond her capacity to eat enough to fulfill her needs. Refer to the appendix for suggested rations for ewes. Virtually all ewes need some concentrates during the last 4 to 6 weeks of gestation because their capacity is being reduced as the fetus grows.

Inadequate nutrition during late gestation may result in the following problems:

1. Pregnancy disease in ewes that are too fat or thin
2. Low birth weights
3. Weak, unthrifty lambs with high death losses
4. Slow gains by lambs
5. Reduced milk yields

Bright, leafy, fine-stemmed legume hay plus grain may be enough for the ewe during gestation. Grass hay may be high quality, but too often it is coarse and low in nutrients. If grass hay is to be used, balance the ration to make certain that adequate protein is fed. Also, it is quite common for rations containing grass hay to be too low in calcium. A calcium to phosphorus ratio that varies from 1 to 2 to 2 to 1 is satisfactory for pregnant ewes, but ewes on good legume hay may do very well on rations containing Ca to P ratios of as much as 6 to 1. Under those circumstances, excessive calcium does not appear to be harmful; but it may be if calcium comes from inorganic sources such as limestone. Even though excellent mineral mixtures are placed before the ewes, they may still suffer from mineral deficiencies. A good rule to follow is that some extra protein or mineral supplements should be added unless half of the ration is composed of legume hay. Another safeguard is to feed 0.5 lb (0.23 kg) of oil meal per ewe daily if there is any doubt about furnishing enough protein. Many of the best sheepraisers insist on feeding at least 1 lb (0.45 kg) of good legume per ewe daily. Salt and clean water should be furnished free choice at all times. Pregnant ewes may be self-fed complete ground mixed rations if they are properly formulated and fed. This method of feeding should result in a saving of labor and good use of low-quality roughage, but it also may increase costs. Using such materials as ground cobs, cottonseed hulls, or straw to limit feed intake will be useful in preventing ewes from consuming so much feed that they become too fat. Another system is to limit the time ewes have access to feeders. This system of feeding works well in confinement, but some judgment must be exercised. It is important to know how much feed is being consumed and adjust the roughage portion or the time exposed to feeders to the situation. Research has shown that ewes need not be fed every day, provided they are given the same weekly allowances on a regular basis. This method of feeding seems most practical during early gestation or after weaning, but should be followed with caution during late gestation. Usually the biggest problem in feeding pregnant ewes is that of too little feed. For some reason, many inexperienced operators fail to provide adequate rations and do not handle ewes for condition until they start having problems. Then it may be too late to avoid losses.

11.6.9. Pregnancy disease (ketosis) In 20 years of research with wintering ewes at the Missouri Experiment Station (Ross, Dyer, and Krieg, 1967), it was observed that ewes on good winter pasture never developed pregnancy disease. Whether or not pregnancy disease is strictly caused by inadequate fat metabolism is still somewhat questionable. But, in the majority of cases, thin ewes that have been fed too little, especially protein, are the most common victims. E. S. Matteson, long-time livestock extension specialist in Missouri, has often said, "Show me a man who is having pregnancy disease in his ewes, and I'll show you someone who hasn't fed enough protein." Matteson claimed that 0.5 lb (0.23 kg) of soybean oil meal per day during late gestation was a good insurance policy against the disease. Other investigators (Outhouse, 1956) showed that inadequate energy, especially glucose, may be a major contributing factor to pregnancy disease.

11.6.10. Other problems with pregnant ewes In addition to pregnancy disease, other problems may occur, especially with immature yearling ewes or ewe lambs. Feeding inadequate protein and energy to young, developing ewes resulted in small, weak lambs, watery, abnormal colostrum, and poor milking of ewes after they lambed (Ross, Dyer, and Krieg, 1967).

11.6.11. Exercise for pregnant ewes Most pregnant females benefit from exercise in moderation, and ewes are no exception. If ewes are on pasture, it is no problem. But when snow covers the ground, sheep may not voluntarily go outside. The old-fashioned shepherd would often feed hay on the ground away from the barn to encourage ewes to exercise. It appeared to work for them but, in general, it is a questionable practice to feed anything on the ground where it can be contaminated with feces. Inexpensive hay racks could very well serve the same purpose.

It is not a good idea to have watering facilities located very far from shelter. If ewes must brave bad weather to find drinking water, they may not drink enough to satisfy their needs. Then, too, feed consumption is limited by water intake. Thus they may not eat enough either.

11.7 HOUSING FOR PREGNANT EWES

If ewes are properly fed, their requirements for housing are easily met. An open shed facing to the south that is dry and clean should be all they need. If the lots are dry, many ewes prefer to bed down outside rather than in an enclosed building. When it is raining, sleeting, or snowing, ewes need shelter. A good roof and a windbreak are sufficient. Closed barns, unless they have forced ventilation, are usually not healthy because of high humidity and may result in respiratory problems.

Space requirements for ewes should include 10 to 12 ft² (0.93 to 1.1 m²) under shelter and 25 to 40 ft² (2.3 to 3.7 m²) lot space per ewe depending on the drainage of the lot. If the lot is poorly drained, they will need more space. Another alternative is to use earth mounds. Feeder space will depend on the management system. Unless ewes are self-fed, allow adequate space for all ewes to eat at the same time. Allow 16 to 20 in. (40 to 50 cm) per ewe, but observe them closely and provide more if needed. Ewes in the fleece require more than sheared ewes. Allow one automatic waterer per 40 to 50 ewes or 15 to 25 ewes per ft (30.5 cm) of water trough. If ewes are in confinement on solid floors, allow 12 to 16 ft² (1.1 to 1.5 m²) per ewe or if on slotted or expanded metal floors, 8 to 10 ft² (0.75 to 0.93 m²) per ewe.

11.8 PREPARING EWES FOR LAMBING

Four to six weeks ahead of lambing, ewes should be inspected carefully and prepared for parturition.

11.8.1. Shearing, crutching and facing, and foot trimming Ewes should be sheared or crutched and faced and their feet inspected and trimmed if needed. Whether or not to shear will depend on facilities for protecting ewes if weather conditions are severe. Many producers in the northern and western states swear by the practice, but they warn that a good barn is needed if a blizzard should occur within a week of shearing. The main consideration is to have a good roof overhead and a windbreak so that ewes will not become wet and chilled. The advantages of shearing are:

1. It eliminates the need of crutching and facing.
2. Lambs can nurse easier, and it is more sanitary.
3. Fleeces are cleaner.
4. Less space is required for ewes.
5. It is easier to judge imminence of lambing and to observe those with lambing problems.
6. Ewes are more responsive to weather and may become better mothers.

The disadvantages are:

1. Rough handling may cause premature births and injure udders.
2. Protection is needed if weather is cold. Otherwise use Wyoming combs, which leave ¼ in. (0.63 cm) of stubble for protection.

If ewes are to be sheared, insist on a careful shearer. Be there when the job is done and insist that the ewes be handled gently and that teats and udders not be injured. Provide a good holding pen and a shearing floor so that the shearer can handle ewes rapidly without strain and injury.

If it is impractical to shear ewes before lambing, they should be crutched and faced. Ewes are crutched by shearing around the udder, between the legs, and around the dock, as shown in Figures 11.4 through 11.6. If the ewe has excessive wool covering on the face, shear the wool around the ewe's head as shown in Figures 11.7 and

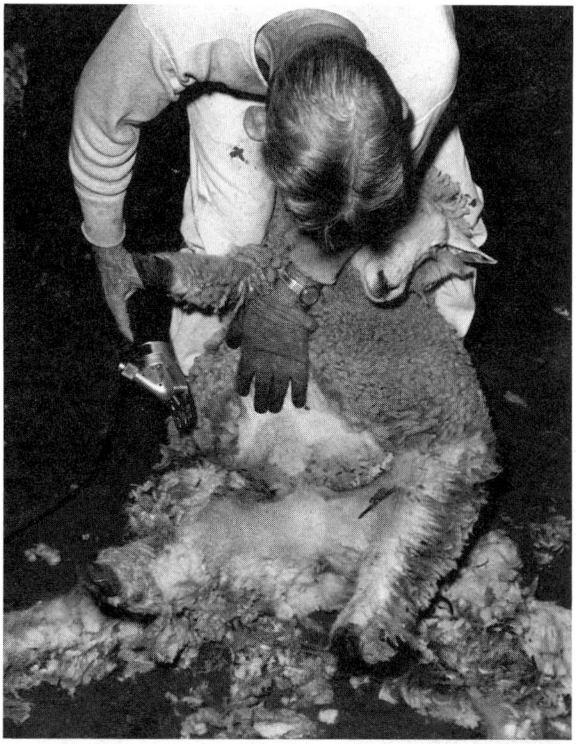

Figure 11.4 Ewes are crutched by removing the wool from around the vulva, teats, and udder.

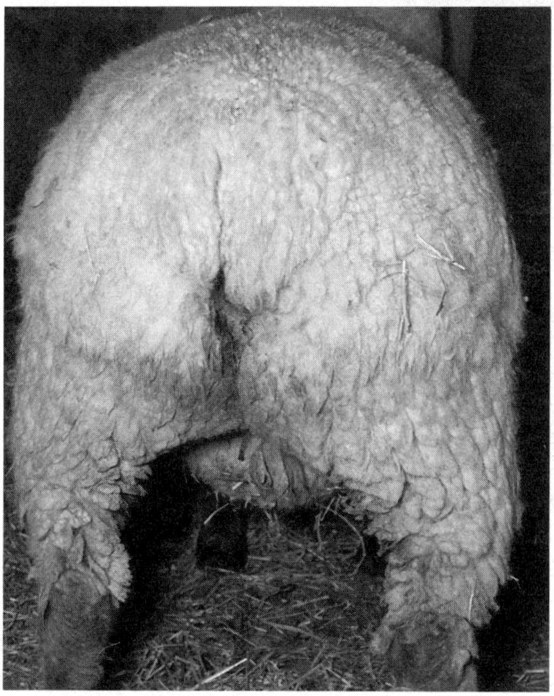

Figure 11.5 A ewe before crutching.

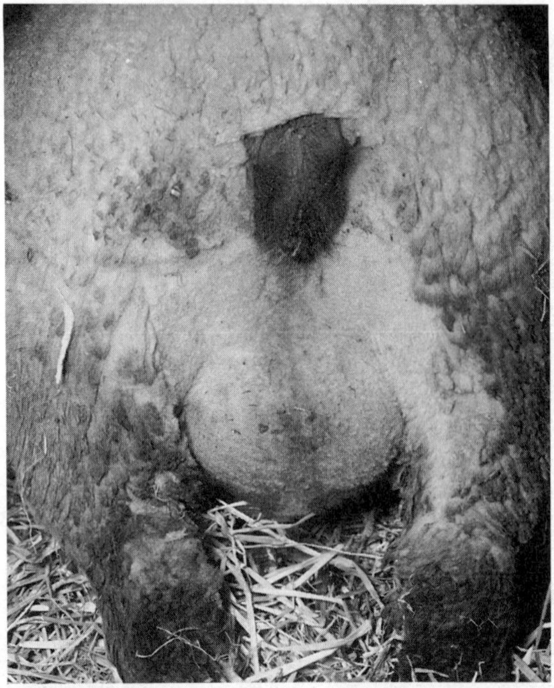

Figure 11.6 The same ewe after crutching.

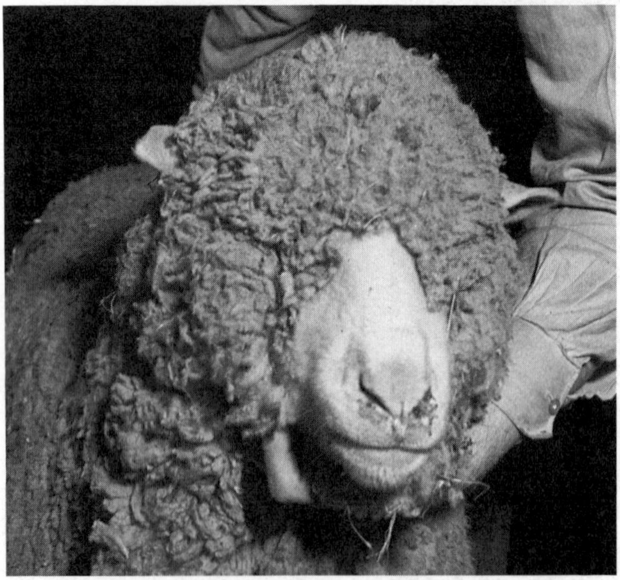

Figure 11.7 A wool blind ewe.

11.8. Check feet and trim those that need it, as shown in Figures 11.9 and 11.10. Figure 11.11 shows the same feet as those in Fig. 11.10 after they were trimmed.

11.8.2. Dividing into groups Ewes should be sorted into groups based on breeding records as well as visual inspection. Ewes with large udders and taut teats

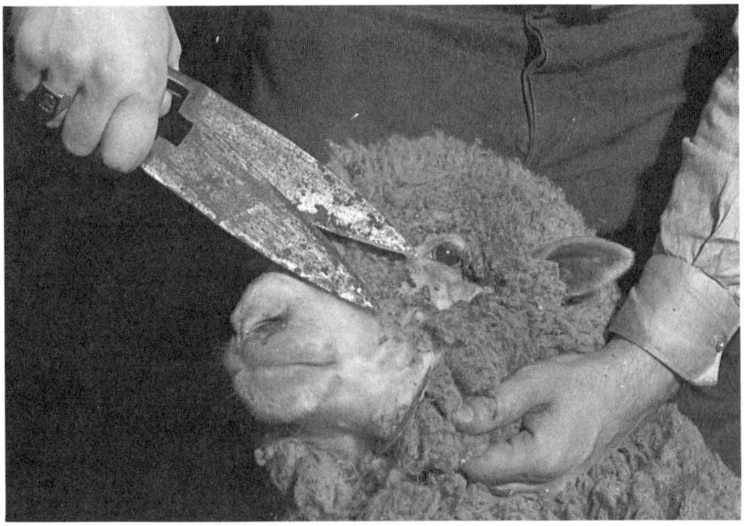

Figure 11.8 Facing a wool blind ewe.

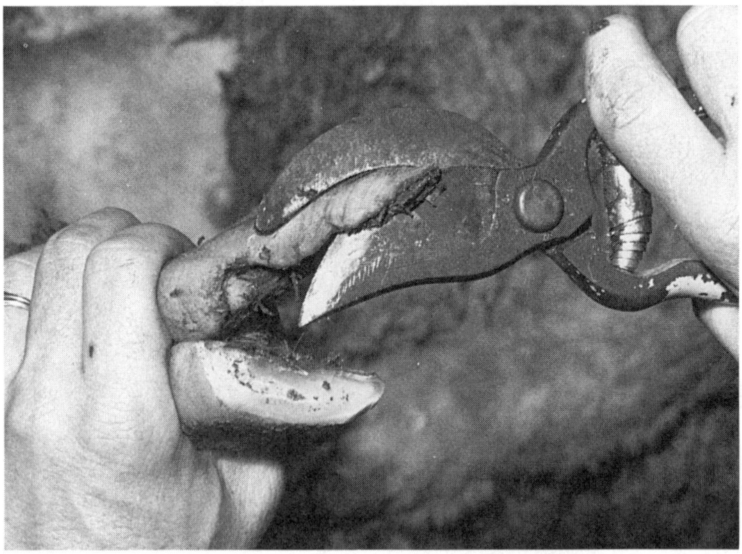

Figure 11.9 Trimming feet with a pruning shears.

that are heavy with lamb should be in one group, and those that are less developed in another. The ewes that are near parturition should be penned near the lambing quarters so that they can be moved in easily.

11.8.3. Vaccination for soremouth (contagious ecthyma) and enterotoxemia It is a good practice to vaccinate replacement ewes for soremouth and enterotoxemia when they are first brought to the farm. They should never have to

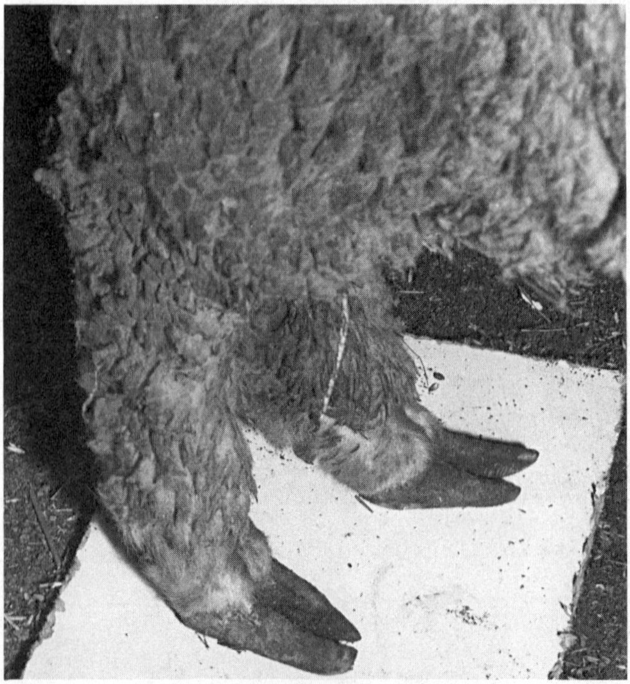

Figure 11.10 Feet that need to be trimmed.

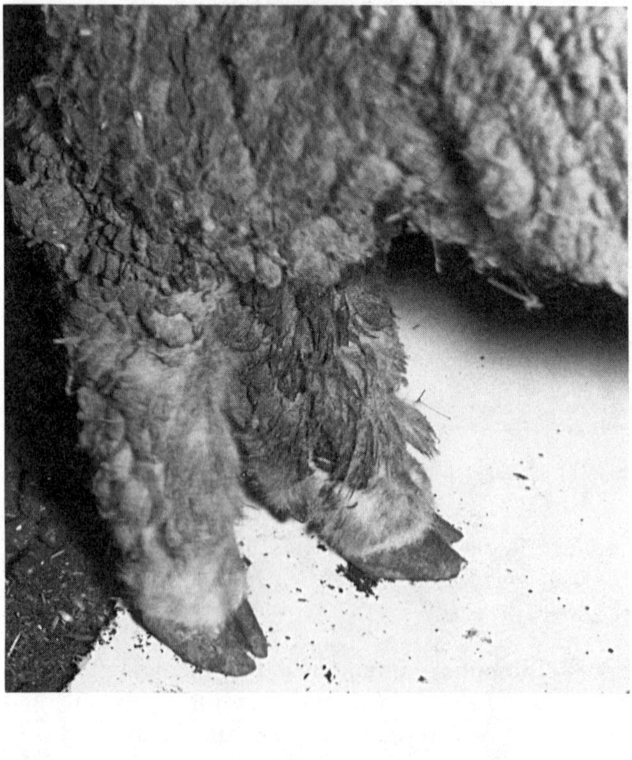

Figure 11.11 The same feet after they were trimmed.

be vaccinated again for soremouth, but it may be necessary to give a booster shot for enterotoxemia if lambs under 3 to 4 weeks are having difficulty with the disease. Unvaccinated ewes may also die of the disease when they are put on high concentrate levels during the last week of gestation. Local conditions may warrant routine revaccination.

In some areas, tetanus is a problem with ewes and lambs. Here, a program of vaccination for both the ewe and lambs may be needed. This is a problem that should be worked out with the local veterinarian. Since the spores of tetanus have an unbelievable lifetime, it is another reason not to use a horse barn for sheep or to build a barn where a horse barn once stood.

11.9 PREPARATION OF LAMBING QUARTERS

Where winters are severe, lambing quarters should be enclosed and warmer than the outside temperature. If a number of ewes are held inside the lambing quarters, the temperature will usually be above freezing, but some ventilation may be needed to keep the humidity down.

The room should be clean and dry. Weeks before lambing, the room, including the walls and ceiling, should be swept down and sprayed with disinfectant. A concrete floor allows scrubbing down thoroughly and disinfecting. Then, the room should be allowed to dry thoroughly. Panels used for lambing pens (also called jugs, jails, claiming pen, or maternity pens) should be repaired, cleaned, and put in place. It is important that enough panels be available. Usually, one lambing pen per ten ewes is sufficient.

11.9.1 The shepherd's kit. The shepherd's kit should be stocked with drugs and equipment well ahead of the birth of the first lamb. The contents of the shepherd's kit may vary depending on the area of the country, but certain things should be common to all of them. A list of articles that may be found in a shepherd's kit and their purposes is given in Table 11.1. It should be emphasized that not everyone will want or need all the articles listed. A visit to a successful producer should provide information on what is needed in the locality.

11.10 CARE OF EWES AND LAMBS AT PARTURITION

Lambing time has been rightly called the shepherd's harvest. Of course, some may argue that the true harvest comes when the lambs are sold and the money is in the bank. But there is little evidence that any factor is more important to the financial success of lamb production than the percentage of lambs saved at birth. The best shepherds save many more lambs than the poor ones. The difference is greater when weather and other conditions are worst. Talent is important, but almost anyone who is willing to learn, is perceptive, and enjoys working with sheep can become competent lambing out ewes. Even the most expert shepherds say that they continue to learn after years of experience.

If the ewes have been properly fed and managed, problems at lambing are usually not serious. If, in addition, the lambing quarters have been prepared ahead of time

TABLE 11.1 SHEPHERD'S KIT

Name	Use
Automatic drenching gun	For drenching sheep for internal parasites
Balling gun	For administering boluses, capsules, or pills
Drenching syringe	4-oz (118 ml) size used for drenching with liquids
Glass jar	2 in. (5 cm) in diameter and 2-in. (5 cm) deep for applying iodine to navels of newborn lambs.
Hypodermic syringe and needles	For injecting drugs either subcutaneously or intramuscularly
Needles and nylon thread or sutures	For sewing open turned-under eyelids
Nipple tubes (Lam Bar and Nursette)	For feeding orphan lambs
Rectal thermometer	For determining body temperature of sheep
Stomach tube	For relief of bloat or administering medications
All-in-one docking and castrating tool	For docking and castrating lambs
Burdizzo	For castrating large lambs and for docking
Castrating knife	For castrating lambs
Pruning shears	For trimming feet
Elastrator pliers and rubberbands	For docking and castrating lambs
Electric docking irons	For docking and for cauterizing excessive bleeding
Emasculator	For docking lambs
Hand shears	For trimming wool, tagging, facing, etc.
Shearmaster	Shearing, facing, and crutching
Bearing retainers	For preventing vaginal prolapse in ewes
Lamb puller	For pulling difficult to deliver lambs
Nylon cords	For assisting in pulling lambs
Branding paint	Special paint that will scour out of wool
Ear tags (metal or plastic)	For permanent individual identification of sheep
Ear tag punch	For applying eartags
Marking chalk	For identifying sheep when sorting
Branding irons	For temporary individual identification
Clean burlap bags	For drying and cleaning the newborn lamb
Paper towels	For drying and cleaning newborn lambs
Rope halters	For restraining ewes
Shepherd's crook	For handling sheep
Turkish toweling	For drying and cleaning newborn lambs and for cleaning wounds
Heat lamps	To prevent chilling of lambs
Marking harness for rams and extra chalk	For marking ewes as they are bred
Nylon rope, ⅜ in. (1 cm)	Used in pulling lambs
Anthelmintics	For worming sheep and lambs (Loxon, Levamisol, Ivermectin, Phenothiazine, Thiabendazole)

TABLE 11.1 (continued)

Name	Use
Argyrol (30% solution)	For sore eyes
Cresol or Lysol disinfectant	For disinfecting premises
$CuSO_4$ crystals	Used in water solution for treating foot rot or as an anthelmintic
Enterotoxemia vaccine	To prevent enterotoxemia
Epsom salts	Treatment for constipation
Fly repellent	To prevent fly strike
Formalin (commercial 40% solution)	Used in footbath for foot rot
Insecticides (Toxaphene or others approved by FDA)	For control of external parasites of sheep
Iodine tincture (7% solution)	For disinfecting navel cords and cuts
Isopropyl alcohol	For disinfecting wounds or instruments
Liquid antibiotics, including combiotic, penicillin, streptomycin, sulfa methazine, sulfathalidine	For treating infections (should be refrigerated)
Milk of magnesia	For constipation in lambs
Mineral oil	For lamb delivery and constipation
Nasalgen (P13)	Vaccination to prevent pneumonia
Powdered alum	For clotting blood
Propylene glycol	For treating ketosis
Talc	For clotting blood
Petroleum jelly	For lubricating hands when assisting a ewe at lambing
Zinc sulfate	Used in solution with water in footbath or as a topical treatment for foot rot
ADE injectible vitamins (optional)	Treatment for vitamin deficiency
Ammonium chloride	Feed additive to prevent urinary calculi
Antibiotic supplements (Aureomycin or Terramycin)	Feed additives to stimulate growth and reduce infections
BO-SE	For treating selenium-deficient sheep
Ewe milk replacer	For rearing orhpan lambs
Frozen colostrum (ovine or bovine)	For feeding weak or orphan lambs (must be stored in freezer)
Liquid detergent	For cleaning utensils, etc.
Liquid soap	Used as a lubricant when helping a ewe lamb

and are adequate and all the supplies and equipment are on hand, lambing should be a pleasure. It should be approached with anticipation, because the lambs are the result of your creative efforts. It should be a challenge to save every one you possibly can. Schedule your time so that you or some other competent person is present and free to work with the ewes and lambs. Attention, when it is needed, may only take a few minutes but will pay big dividends.

11.10.1. Behavior of ewes at parturition Studies have been made of the time of day ewes lamb. There appears to be a difference, with 70% to 75% of lambs born during the hours from 6:00 A.M. to 6:00 P.M. Time of feeding of the ewes has some effect on when they lamb. It has been observed that the number of daylight births can be increased by feeding in the afternoon rather than the morning.

Symptoms of imminent lambing are sagging in front of the hips, relaxation in the area of the vulva, milk in the udder, and nervousness as shown by stamping her feet, continually looking behind herself, and separating herself from the rest of the flock.

From the time contractions begin, it should be only a matter of 45 to 60 minutes until the ewe lambs. The placenta (afterbirth) is normally expelled within another 2½ hours. If the ewe is lying on her side when she lambs, she should be on her feet and seeking the lamb promptly. The ewe should normally lick the lamb, unless she is giving birth to twins. Then she may ignore the first lamb during contractions and intermittently clean it. Normal lambs will seek the teat as soon as they can walk. The ewe will attempt to position herself so that the lamb can nurse.

11.10.2. Assisting the ewe at birth If the ewe can lamb without assistance, she should be allowed to do so. It may be easier in the long run to gently move the ewe into a lambing pen when signs indicate she is about to lamb. Then problems of claiming lambs, being sure the lambs nurse promptly, and disposal of the afterbirth are simplified. However, many ewes lamb soon after labor begins or they may escape observation until the lamb is on the ground. If the lambing quarters are well bedded, the ewe and lamb may be in good condition, but they should be moved to a lambing pen to avoid injury to the lamb or problems of claiming, especially with twins (Figure 11.12).

Figure 11.12 Lambing pens with heat lamps.

If ewes that have trouble during lambing are continually eliminated from the flock, problems of that sort gradually will be eliminated. The scientific name for difficult lambing is dystocia. This may occur in a number of forms, such as the lamb is too large, the head or shoulders of the lamb are too large, the cervix fails to dilate, or the lamb is not normally presented.

The normal birth presentation of a lamb is with the head between and slightly above the front feet or with the hindfeet first with tail between the legs. Commonly observed malpresentation are the head turned back or to the side, one or both legs turned back, the lamb comes hocks first, or the legs of twins are entangled. If the ewe has been in labor for as much as 40 minutes without delivering the lamb, she should be examined. Cleanliness is important to avoid infecting the ewe. Also, nails should be trimmed and hands and arms scrubbed with soap and a mild disinfectant. The area around the ewe's vulva should be washed with soap and a mild disinfectant to ensure that dirt and manure are removed. If help is available, it may be beneficial to have two assistants lock hands under the belly in front of the hind legs and elevate the hindquarters slightly. Lubricate the hand with liquid soap and insert the hand, with fingers held together, into the vulva and the vagina. Only insert the hand when the ewe is not straining. Feel for the feet and the head of the lamb. If it is in a normal presentation and is too large, place a noose of a lamb puller or snare over the ankle of each foot and another around the lower jaw of the lamb and exert a steady pull outward and downward when the ewe strains. Liquid soap applied between the lamb and vaginal wall will assist in the passage of the lamb.

If, upon examination, it is found that the lamb is not in a normal position, it must be pushed back when the ewe is not straining and straightened. If the noose of the lamb puller is placed over the feet, it will be helpful in pulling the lamb after it is in position. Caution should be exercised not to pull too hard or the ewe will be injured. Pressure on the cord attached to the lower jaw should be only enough to keep the head straight. Beginners should assist an experienced shepherd before attempting to lamb out ewes, or read the excellent discussion on the subject by Don Bailey in the March 1983 issue of *Sheep Magazine.*

11.10.3. Care of newborn lambs If the ewe gives birth without difficulty, the lamb should immediately attempt to get free of the membranes. Most lambs can do this without help, but some lambs, especially those born with difficulty, may have their nostrils clogged with mucus. If that is the situation, wipe the mucus from the nostrils with a piece of toweling and see that the lamb is breathing. Apply artificial respiration, blowing in the mouth of the lamb. Or, if an oxygen bottle is available, use a tube and release some oxygen into its mouth. Some shepherds have found that tickling the inside of the nostril of a lamb with a straw will initiate breathing. Immerse the navel cord in iodine as soon as possible. If white muscle disease is a problem, give vitamin E–selenium shots at birth. Check the teats and be sure colostrum is available. Some experts of the lambing shed make it a practice to strip each teat to be sure the wax plug is removed. It is important for the ewe to dry the lamb off and allow it to nurse as soon as possible.

It should be emphasized that colostrum is important for a newborn lamb because it contains antibodies, is a laxative, and contains the nutrients, particularly vitamin

A, needed by the lamb. Colostrum is secreted by the ewe for the first 3 days after lambing. In the event the ewe does not have colostrum, the lamb should be allowed to nurse a ewe that does. If a ewe with a newborn lamb is not available, many shepherds keep a supply of colostrum in the deep freeze for that purpose. Frozen colostrum should be thawed and warmed to approximately 100 °F (38 °C) and fed with a bottle and nipple. In an emergency cow colostrum can be used for lambs, but it is not as satisfactory as that from a ewe because the antibodies in cow colostrum are not specific for the lamb.

11.10.4. Care of chilled lambs Once a lamb is dry and full of warm colostrum, cold weather is not usually a serious problem, although the newborn lamb is still susceptible to cold for the first few days of life. Thus heat lamps suspended over the lambing pen are very useful.

If ewes are properly managed, very few lambs should be chilled. But sometimes it does occur in spite of precautions. It is well to remember that a newborn lamb can be revived even though it may seem hopeless at first glance.

First, see that the nostrils are clear of mucus and the lamb is breathing. Then take the lamb inside and immerse in warm water, making certain the nostrils are not submerged. As soon as the lamb is revived, remove from the water and dry it by rubbing vigorously with a rough towel or clean burlap. This should not only remove the water, but stimulate circulation. The next step is to take the lamb to the ewe and allow it to nurse. Weak lambs may be fed by stomach tube. Once the lamb is full of colostrum and under a heat lamp in a lambing pen, there should be little further difficulty.

11.10.5. Preventing unclaimed lambs The mothering instinct is naturally greater in some breeds and individuals than others. Ewes with first lambs may be expected to have difficulties that disappear with experience. One of the best precautions in this regard is to move the ewe into a lambing pen when she first shows signs of labor or immediately after she gives birth. Problems occur when a ewe has twins and the first born gets separated from her. The lamb may be adopted by another ewe, called a "grannie," or the mother may fail to claim it. Behavior studies have shown that the ewe identifies her lambs by smell to a large extent. If the ewe fails to claim one of her lambs, some of her milk rubbed on her nose and the rump of the lamb will often be sufficient. Some shepherds use household deodorants in the same way or some other strong odor such as kerosene, or a commercial product manufactured for that purpose may be used. Other devices used are to tie the ewe with a halter until she allows the lamb to nurse. A strange dog brought to the vicinity may also arouse her mother instinct. A more recent innovation is to maintain a fostering pen for that purpose. This method of getting the ewe to claim one of her lambs or even one from another ewe has been used with much success in Europe and in flocks in the United States. The main thing is to remember that gentleness and patience are required.

11.10.6. Fostering lambs What to do with the extra triplet or the lambs from ewes that for any reason cannot suckle them has always been a problem. While it is true they may be reared artificially, it is a laborious process and is more expen-

Figure 11.13 Fostering pens.

sive than allowing a ewe to nurse the orphan. If a ewe with a single lamb has plenty of milk, she should be able to rear two lambs satisfactorily. Often a ewe will accept another lamb if it is put in the lambing pen immediately. This can be tried first. Then if she objects, a fostering pen can be used. A fostering pen is constructed the same as a lambing pen but with a stanchion in one end. The side of the pen with the stanchion should be made of plywood or other material so the ewe cannot see behind her (Figure 11.13). She will be unable to see which lamb is nursing and thus, within 3 to 4 days, will usually claim both lambs. A feed box and water should be available so that the ewe can be restrained until the adoption is complete. Many sheep raisers attest that the fostering pen should be standard equipment in a well-run operation.

11.11 REARING LAMBS ARTIFICIALLY

It should first be remembered that fostering, if it can be done, is preferable to rearing on milk replacer. But lambs can be reared very well artificially if it is feasible from a financial and time standpoint.

Perhaps with new breeds and breeding procedures, the prolificacy of modern and future ewes may dictate such an operation. The following is an abbreviated description of the fundamentals of the operation:

1. Allow lambs to consume colostrum for 12 to 24 hours. Bovine colostrum may be used if necessary.
2. Use a milk replacer, of known quality, formulated for lambs. Many calf milk replacers are unsatisfactory because they are too low in fat and too high in lactose for lambs.
3. Make certain that lambs are in warm, dry quarters. Heat lamps may be necessary for the first week if the room is not heated.

4. Starting lambs on milk replacer is crucial to the success of rearing lambs artificially. Many problems of disease, vices, and death by exposure are possible if lambs are neglected. After extensive research, Fredrickson, Jordan, and Terrill (1980) of the U.S. Sheep Experiment Station of DuBois, Idaho, developed a procedure for rearing lambs artificially. They recommend using cold formula immediately after the lamb is removed from the ewe. For purposes of identification, this is called the DuBoise system. At the Missouri station, it has been more stisfactory to start lambs on warm formula and switch them to cold formula later. Both methods appear to be satisfactory.

   The Du Bois system of starting lambs follows:

1. Make sure the lamb receives colostrum.
2. Feed lambs three times for the first 12 to 16 hours. Feed lambs in a training pen separate from all the more mature lambs.
3. Use a feeder that is equipped with self-priming nipples. A piece of sheep pelt tacked to a board mounted above the nipples will help the lambs get used to nursing the feeder.
4. The lamb may be taught to suck by placing the palm of your hand over the lamb's head with your fingers on top of the nose. Open the lamb's mouth with the third finger and insert the nipple. Restrain the lamb with the other hand behind the rump.
5. Lambs may be taught to nurse the feeder quicker if one or two experienced lambs are put with the group.
6. When the lambs nurse without assistance (4 to 7 days), place them in a pen where they will remain for the rest of the time they are on formula.

   The Missouri system of starting lambs on warm formula follows:

1. Start lambs on warm formula (95° to 100°F; 35° to 37° C) fed at 6-hour intervals. Even-flow baby bottles are very satisfactory in the beginning. Vices, such as sucking each other, may be prevented by keeping lambs in small, individual compartments for the first 4 days. Continue feeding warm formula to the third day. Change to cold formula, 32° to 40°F (0° to 5°C) on the fourth day and use a nipple with a valve in it that requires very little suction to get the fluid started (Figure 11.14).
2. At the end of 4 to 7 days, lambs may be fed in groups. At least by the end of the first 7 days, provide nipple tube feeders that the lamb can nurse by suction. These are much less expensive and are just as satisfactory for older lambs as the nipples with a valve. Provide a nipple for each five lambs.
3. Place dry feed in the pen by the end of week 1. It should be a very palatable creep mixture. Also, provide clean water at all times.
4. At the end of 2 weeks, provide leafy, green alfalfa or other hay in a rack.
5. Discontinue formula feeding after 30 days. Later, after you have experience, you may wish to discontinue formula feeding at 21 days to reduce expenses.

Figure 11.14 Nipple tubes for rearing lambs artificially.

6. Follow these precautions:
   a. Keep everything clean, especially the formula feeders and utensils to prevent bacterial buildup.
   b. Don't feed sour formula. Consider adding 1 ml of a 37% formaldehyde solution per gallon of formula if the weather is warm enough to cause souring.
   c. Keep pens clean. If expanded metal floors are used, wash and disinfect weekly. If dirt floors are used, keep bedded with straw changed frequently.
   d. Vaccinate for enterotoxemia with *Clostridium perfringens* type C-D toxoid. On days 1, 10, and 30, give injections for white muscle disease. The lambs should also be vaccinated for soremouth (contagious ecthyma) by the time they are weaned.

      Expect a setback when the lambs are withdrawn from formula, but be patient. If they have been consuming some dry feed, they will recover within a week and eat very well.
7. Feed lambs a palatable, balanced growing finishing ration.

11.12 CARE OF THE EWE AFTER LAMBING

When the ewe casts the placenta (afterbirth) after she lambs, it should be removed from the pen. Ewes that fail to pass the placenta may smell bad and become infected. An old shepherd's treatment is to flush her with a warm salt solution containing 1 teaspoon of salt per quart of water that has been boiled and cooled. Use a funnel and hose to flush the uterus. Raising and lowering the funnel will assist in getting the fluid to go down. Flushing several times will usually cause her to clean herself, but if she continues to emit a bad odor, veterinary assistance may be required. Ewes that have had difficulty lambing should be given antibiotics either by injection or a bolus, such as sulfa urea, may be inserted into the vagina.

11.12.1. Feeding lactating ewes Feeds high in concentrates stimulate milk flow and may cause feverishness. It is usually preferable to feed lightly on the day of lambing, but allow plenty of clean, cool water in a bucket in the lambing pen. If the ewe seems hungry, a small amount (0.5 lb) (0.23 kg) mixture of bran and oats may be given along with leafy hay. The concentrate level should be increased gradually so that at the end of 3 days she is receiving 1½ lb (0.68 kg) of the lactation ration daily.

The amount of feed should be then increased gradually to meet the requirements for a lactating ewe. Attention should be given to requirements and the ewe fed according to the number of lambs she is suckling. A ewe suckling twins will produce as much as 50% more milk than one with a single, and she should be fed accordingly. Remember that a heavy milking ewe will invariably lose weight during lactation. A heavy ewe may lose as much as 30 lb (13.6 kg). Her requirements are so high she must be fed a high-energy ration. It is unreasonable to expect her to milk well on a high-forage ration. Digestive disturbances may occur if too much corn is fed. Therefore, it is advisable to add 30% oats or wheat bran to the concentrate mixture to provide some bulk and laxative effect. The same results may be obtained with ground hay mixed with the grain. If concentrates and hay are fed separately, make certain that adequate feeder space is available so that all ewes can get to the trough.

Many shepherds save their finest-stemmed, leafiest hay for feeding lactating ewes and suckling lambs. This is a good practice, since it is higher in nutrients and very palatable. The ewes should be fed all the hay or high-quality grain silage they will clean up without waste from the time they lamb until the lambs are weaned (Figure 11.15). Antibiotics may be included in the ewe's ration to reduce infections and to increase growth of lambs. Either Aureomycin or Terramycin will be equally effective. The level to feed should be 60 mg per ewe daily.

11.12.2. Udder problems Emphasis has been placed in previous chapters on the importance of sound udders in ewes. Lumps in udders, large teats, pendulous udders, and other defects are accentuated when the ewe begins to lactate. If any defects are noted, the ewe should be watched carefully to be sure she is producing milk and that the udder is not feverish and infected. If the ewe does have mastitis, she and the lamb should be separated from the group until the condition is cleared up. The ewe should be marked for culling when the disease is discovered. Udders may be bruised on doors with high sills or injured in chutes or on metal slotted floors in confinement. Handle ewes with care and remove obstructions that may cause injury. Also, if ewes are on concrete, make certain that bedding is sufficient in amount and changed often enough for it to be clean and dry. Watch the ewes and lambs to be sure lambs are nursing. If sore teats develop, treat them before they are so painful the lambs cannot nurse. Treatment of udder problems is discussed later.

Figure 11.15 Lactating ewes need a generous allowance of high-quality silage or hay.

11.13 LAMB MORTALITY

Statistics have shown that 50% to 70% of lamb death losses occur in the first 3 to 5 days after birth, and 80% to 90% occur during the first month. Good management in the lambing quarters and in the first few weeks of lactation will be well rewarded in extra income from those extra lambs saved. The principal causes of lamb mortality are weak lambs, starvation, stillbirths, and pneumonia. Most causes of death can be prevented by selecting ewes for mothering ability and proper feeding and management. It has been shown that crossbreeding improves vigor and livability of lambs, and inbreeding has the opposite effect. Other factors are that triplets may have greater losses than singles, and that ram lambs, even though larger than ewes, seem to have less will to live. Lambs from young ewes are less vigorous than those from mature mothers. Extremely large lambs, because of problems in delivery, often are lost, but so are very light (4 to 5.5 lb; 1.8 to 2.5 kg) lambs that may be quite weak. Nutrition of the ewe is a primary factor in producing strong, live lambs and plenty of milk. Make certain that ewes receive their nutrient requirements, and problems with lamb losses will be minimized. One last word of caution. If problems with lambs occur, find out what is the cause and do something about it. If a lamb dies of some unknown cause, have it posted to determine cause of death. If an examination pinpoints a disease and steps are taken promptly to prevent further infection, that lamb may be the most valuable one you produced. Work with your veterinarian, who may not have had experience with sheep, but if well-trained can become extremely valuable.

11.14 CARE OF LAMBS UNTIL WEANING

11.14.1. Docking lambs Lambs should be docked when they are 3 to 7 days old. Docked lambs are cleaner and less susceptible to disease and parasites. They present a much neater, more attractive appearance on the market. Buyers suspect poor management when they inspect lambs with long tails, and they tend to discount them. Long tails are a definite detriment to breeding ewes.

Docking, the cutting off of the tails, can be done with a knife, emasculators, pruning shears, a special docking–castrating tool, a hot chisel or hot docking irons; all appear to be equally effective, as does the rubberband (elastrator) (Figures 11.16 through 11.19), but if tetanus is a problem, elastrators should not be used. With all of them, it requires 17 to 26 days for the wound to heal.

If tails are to be docked, make sure the skin on the inside of the tail is pushed toward the body before cutting so the stub will be covered by a skin flap, which will assist in clotting (Figure 11.20). Tails should be cut no closer than 1½ in. (3.75 cm) from the body. Cutting at the juncture of the caudal fold on the underside of the tail seems to be a satisfactory place. Cutting tails of ewe lambs too close to the body may contribute to prolapse. Dock early on a bright, clear morning. Handle lambs easily so they are not excited any more than necessary. Place them in a well-bedded pen after they are docked and hold them there for several hours until you are certain they are not bleeding excessively. While it is true one person can dock a lamb, it is much easier with a helper to hold it.

The assistant catches the lamb and holds it with its back to him or her, head up and right front and hind legs held by the right hand and the left hand holding the left legs in the same way. The lamb will be somewhat more immobile if it is grasped

Figure 11.16 Docking with an all-in-one tool.

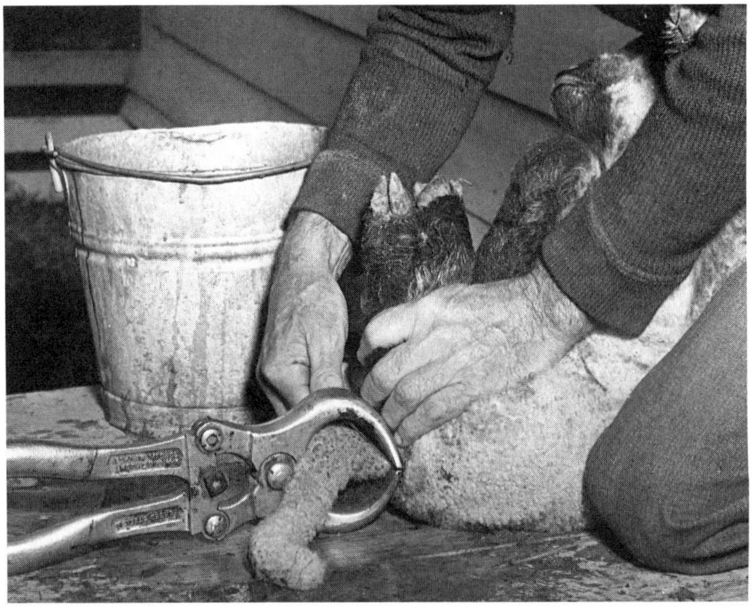

Figure 11.17 Docking with a Burdizzo and knife.

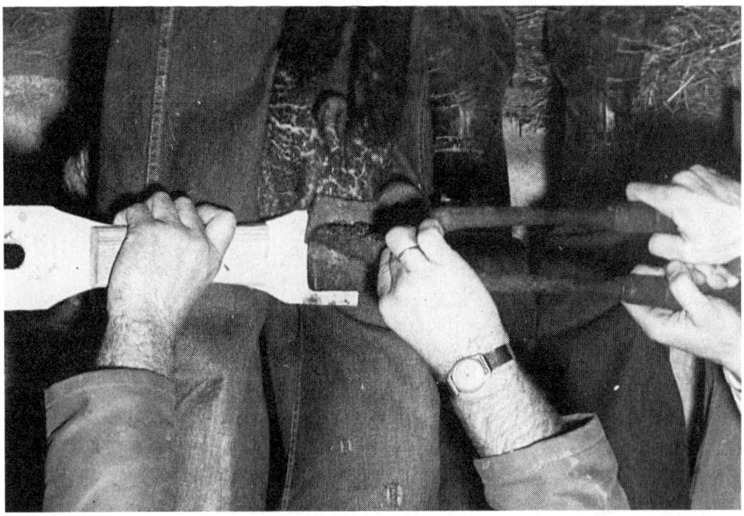

Figure 11.18 Docking with a hot iron.

above the hock of the hindleg and the knee of the front. The next step is to disinfect the tail and docking instrument.

After the cut is made, hold the skin flap over the stub until the blood clots. Then gently release the lamb into a well-bedded pen. Check the lambs to see that none are bleeding excessively. If that happens, catch the lamb and dust the cut with

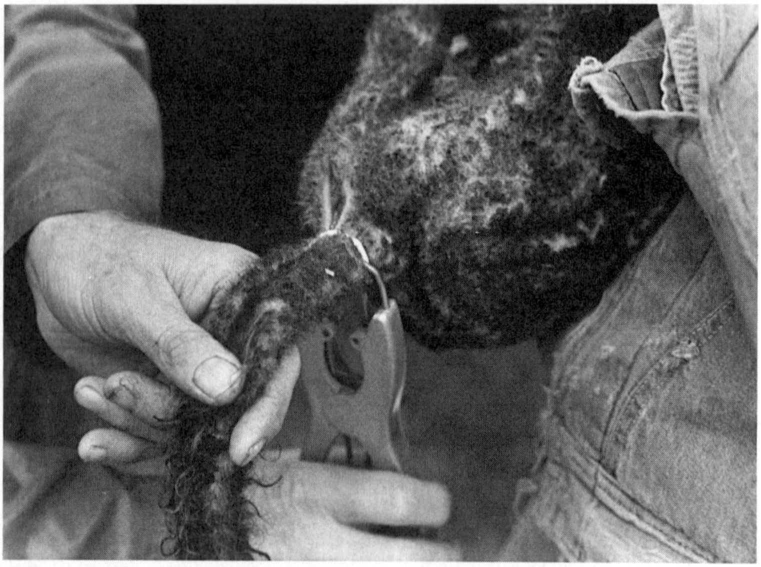

Figure 11.19 Docking with an elastrator (rubber band).

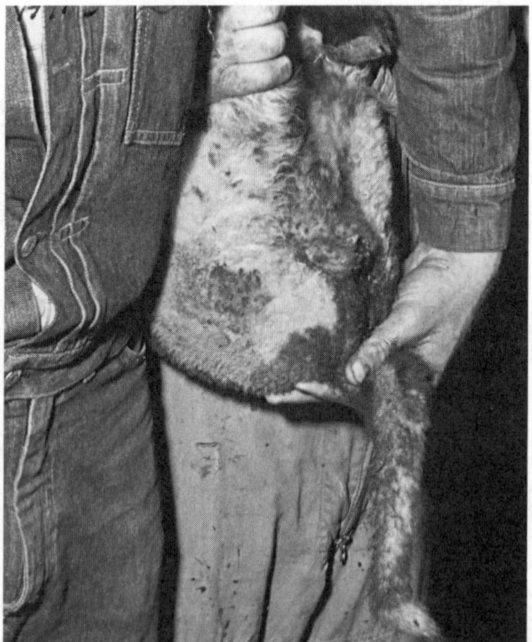

Figure 11.20 The skin of the tail should be pushed back toward the rump before docking.

talc or alum and again hold the skin together until the blood clots. If docking is done later in the spring, the wound should be treated with fly repellent. Heated chisels or pincers will sear the flesh and result in little blood loss. If they are used, the lamb's tail should be inserted through a hole in a board to prevent burning the rump. Comparisons of docking methods indicate that it requires longer for wounds to heal if heated instruments are used. Rubberbands, called elastrators, are now widely used and effective, especially with very young lambs. Observations at the University of Missouri (1982) indicate it requires approximately 14 days for the tail to drop off and somewhat more than another week for the wound to completely heal. Some shepherds prefer to cut the dead tail off after the elastrator has been on for a day, leaving the band on the dock.

11.14.2. Castrating lambs Whether or not to castrate lambs has been studied for years and is sometimes a bone of contention between lamb buyers and sheepraisers. Buyers claim that ram lambs (uncastrated) are not as fat and that they may show excessive development in the neck and shoulders.

Research has shown that ram lambs that are fed for maximum growth and are marketed when they are less than 5 months old are satisfactory in the carcass. They are somewhat leaner and have higher-yielding carcasses, and they gain 8% to 10% faster and 8% to 10% more efficiently than wethers; but wethers grade higher and have higher dressing percentages. If ram lambs are marketed by the middle of June and have sufficient finish to grade choice, there is usually no discrimination. But if they are to be held for later growth and finishing, they should always be castrated. Buyers of feeder lambs are much opposed to ram lambs in a group. Purebred breeders in ram testing programs need 60- to 90-day weights to assess growth rates as a basis for selection and may not want to castrate rams until they are older.

Castrating should usually be done at the same time lambs are docked. It is easier and safer to do both jobs when the lambs are young. There are a number of satisfactory methods for castrating, including the following: knife, the all-in-one docking and castrating tool, emasculators, and elastrators. If a knife is used, it is best to work with an assistant who should hold the lamb as described in docking. First, wet the scrotum thoroughly with disinfectant. Then with a clean, disinfected, sharp knife cut off the lower-third of the scrotum. This should expose the ends of the testicles (Fig. 11.21). Push down on the skin around the cord to better expose the testicles; then grasp them with your fingers, a pair of pliers, or the all-in-one instrument and pull them out, breaking the spermatic cord inside the body to prevent infection.

This is not the method taught to me by Luke McGlothan, a Scotchman, who was the shepherd at the University of Missouri for many years. Luke said, "Any sheepman worth his salt will expose the testicles, then grasp them with his teeth and pull them far enough out so he can hook them with his fingers." I have taught hundreds of students to castrate lambs using the old shepherd's method and others, and find that either is satisfactory if properly done. The final step in castrating is to disinfect the wound and apply a repellent if flies are in season.

Elastrators are effective but, on occasion, there have been cases of large-scale tetanus infection from their use. Lambs several months old with large testicles that are to be castrated in fly season should be clamped to prevent open wounds and excessive bleeding. The problem with clamps is that it requires some skill to be certain the spermatic cords have been crushed.

After castrating, keep the lambs under close observation to be sure they have not become infected and are not infested with maggots. One word of caution: If horses

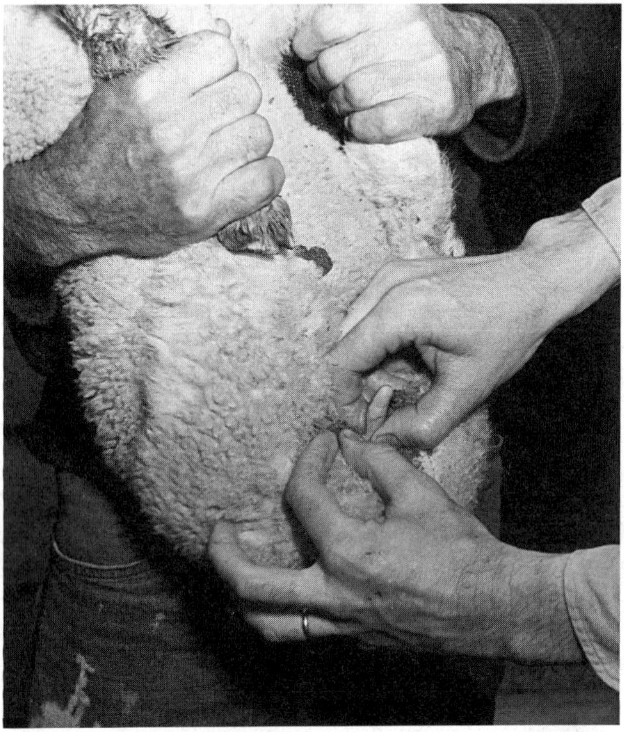

Figure 11.21 Exposure of the testicles in castrating a lamb.

are present on the farm or if they have been there previously, there may be enough danger from tetanus to warrant vaccination.

11.14.3. Problems with young lambs Problems may develop with young lambs. These are fairly simple to take care of, if observed early, but may lead to much trouble if they are neglected.

Pinning is a condition that occurs with very young lambs when their feces are quite sticky. The feces cause the tail to stick to the lamb's rear end. Treatment consists of washing the rear end of the lamb with warm water and detergent.

Inverted eyelids (entropion) occur in lambs mostly in breeds with wooly faces. The eyelid, usually the lower one, is rolled inward so that the eyelashes are pressed against the eyeball. If not corrected, it may cause permanent blindness. The condition can usually be alleviated by use of a needle and thread to sew the eye open, as shown in Figs. 11.22 through 11.24. Other treatments and procedures are equally as satisfactory. Make certain to use an eye ointment or powder to eliminate infection. Since the condition is inherited, the sire and dam should be eliminated from the flock. Sore eyes in newborn lambs can usually be treated with a few drops of a saturated solution of boric acid or a 10% solution of argyrol.

Stiff joints (navel ill) is often caused by infections resulting from the entry of organisms through the navel cord. Prevent this by disinfecting the navel cord at birth with tincture of iodine.

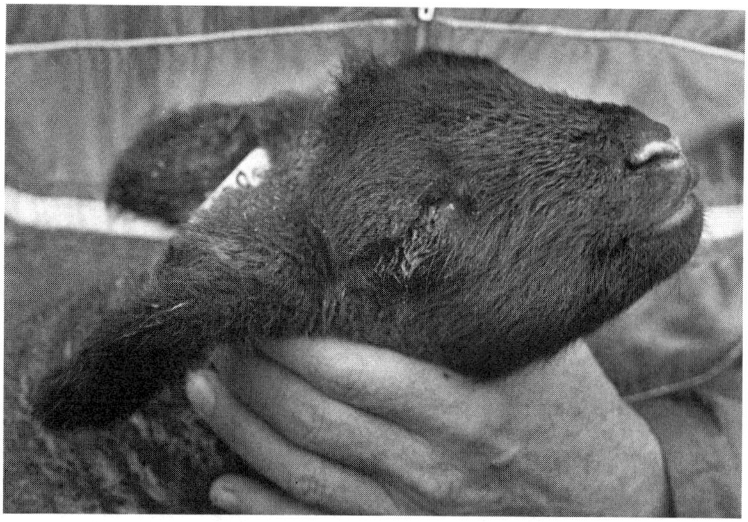

Figure 11.22 Turned under eyelids (entropion) causes blindness in lambs.

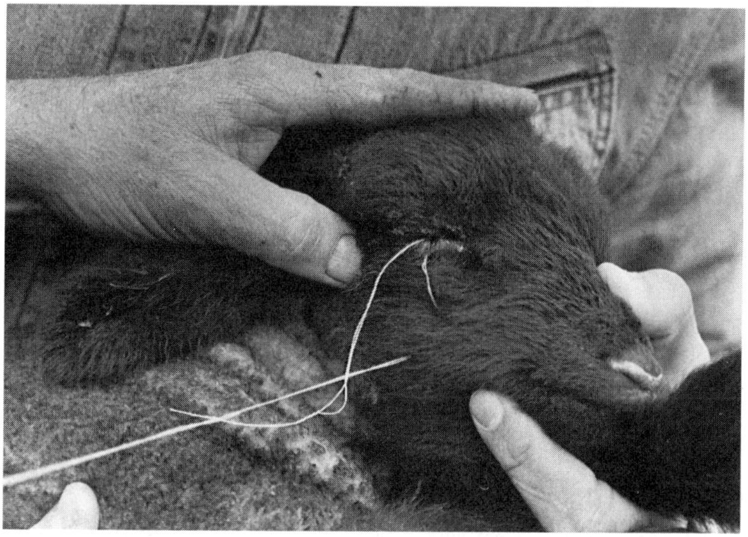

Figure 11.23 A thread is run through the eyelid and the skin on the face below the eye.

Enterotoxemia may be a problem if ewes are on exceptionally rich pasture or on high-concentrate rations. Vaccination of ewes (type B) and lambs (C and D) will prevent the disease. Also, inclusion of antibiotics in the lamb rations will reduce the incidence.

Stiff lamb disease and rickets may occur in fast-growing lambs on milk diets. These can be cured by injections of vitamin E–selenium and vitamin D at 1 week of age.

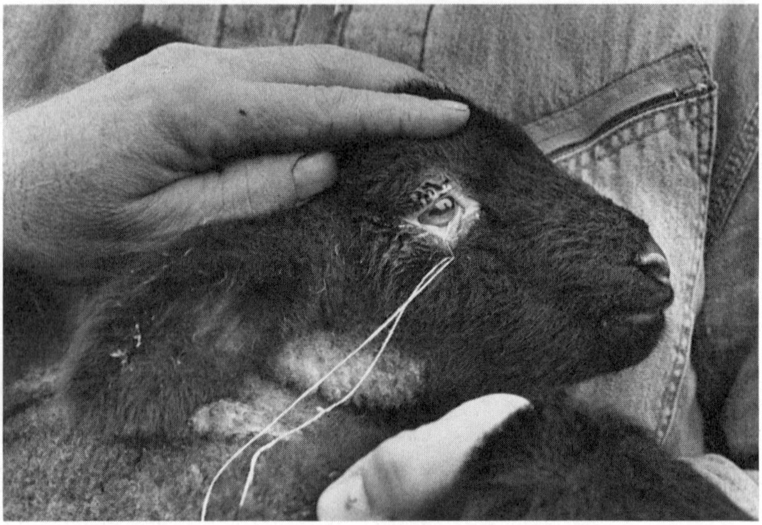

Figure 11.24 The thread is tightened and tied to hold the eye open. After a week the thread can be removed.

11.15 CREEP FEEDING LAMBS

The ideal feed for a lamb is mother's milk. Unfortunately, most ewes do not give enough milk to allow lambs to grow to their potential. It is the purpose of creep feeding to get more concentrates into the lamb. Young lambs are exceptionally efficient converters of feed into gain. Thus, it is of great importance for lambs to be provided supplemental feed as soon as they will eat it. Generally, twin lambs benefit the most from creep feeding, but it is an exceptional lamb that will not benefit from access to a creep (see Figure 11.25). The creep should be located in a comfortable, dry, clean area near the place where ewes loaf. The creep should have enough space for all lambs to fit into it comfortably. Openings to the creep must be small enough to keep ewes out. Openings 1 × 1 ft (30 × 30 cm) will prevent most ewes from entering. If the weather is cold, a heat lamp suspended over the feeder will serve as an inducement to lambs to enter the creep. Care should be taken that the heat lamps are plugged into an overhead outlet with short cords so that if they are knocked down they will become detached and not cause a fire.

Other equipment useful in a creep are a hay rack and water fountains or a tank if the water supply is far from the creep. Lambs should have feed available in the creep by the end of the first week. Research at the Missouri Experiment Station (Ross, Karr, and Pavey, 1961) and others have shown that lambs have a preference for certain feeds and that palatability of creep rations is important in obtaining maximum results. Soybean meal was found to be exceptionally palatable to lambs. Corn, wheat bran, and barley were all well-accepted by lambs. Generally, lambs do not start out well on oats or oat products, although later they eat them readily.

Formulate creep rations for palatability as well as cost. Don't let feed get stale in the creep. Many good shepherds clean out the creep mixture from troughs each day and feed it to their ewes. On the other hand, once started, don't let feeders run

Figure 11.25 Creep fed lambs make very efficient gains.

dry. Give the lambs your best leafy legume hay in the creep. Include antibiotics (Aureomycin or Terramycin) at 30 mg/lb (0.45 kg) of feed in the creep mixture. Creep mixtures are included in the appendix. Pelleting creep mixtures will generally increase palatability. It is not advisable to include cane molasses in creep feeds until the lambs are 2 or more weeks old since it does not agree with very young lambs. When lambs are older, sweet feeds appeal to them. Molasses should not constitute more than 7% of the mixture if it is to be used efficiently.

11.16 SHOULD LAMBS BE PASTURED?

Whether to allow lambs to run with their dams on pasture may depend on a number of factors such as season of the year, age and size of the lambs, amount of creep feed the lambs are consuming, number of twins, and the management system being followed. If the lambs were born in January or February and you plan to get them to market by the middle of June, it is doubtful if they should go to pasture.

Experimental results at the University of Missouri (Ross, Karr, and Pavey, 1961) have shown that suckling lambs weighing 40 lb (18.2 kg) or more and eating creep ration will decrease feed consumption radically when put on pasture and never equal that of similar lambs kept in drylot. Putting such lambs on pasture resulted in reduced rates of gain and lower grades and yields in each of several experiments.

One possible explanation for the poor performance of the lambs that were pastured with their dams was that they became parasitized. Ewes immediately after lambing have an enormous buildup of parasite eggs in their feces due to the exceptional activity of the parasites at that time. Thus, even though very few eggs or larvae can survive winters in the United States, the lactating ewe quickly seeds the pastures with parasite eggs. Another important factor on damage done to suckling lambs by internal parasites is that the young lambs have not had time to develop immunity to parasites before they are subjected to the trauma of infection. Finally, in the early spring the weather may be warm, with frequent showers or heavy dews, and the vegetation grows rank so that the larva are protected from the sun. This condition is optimum for the short life cycles of stomach worms. Even though lambs may appear to be healthy, many will have low levels of hemoglobin in the blood, indicating severe internal bleeding caused by the parasites. Sometimes lambs in good flesh and thrifty appearance die suddenly due to the infection.

Some alternatives may appeal to a producer who wants to make use of pasture. In the "in and out" system, ewes and lambs are kept in drylot together at night and grazed separately during the day, or the ewes may be turned out to pasture during the day and the lambs left in drylot with access to a creep.

Ewes may also be grazed at night and left in drylot with lambs during the day. When the "in and out" system is used, there is usually some apprehension relative to the time required to separate the two groups. Lambs and ewes soon learn what is expected, and the division requires less than 5 minutes for groups of as many as 200 ewes.

Late lambs (born after April 1) should probably not be creep fed. If they are to be grown on pasture during the summer and finished in the fall and winter, they may actually lose weight for a while after they are weaned and creep feeding discontinued. Expect that lambs will be parasitized if they graze with their dams. Consider grazing ewes in one field and lambs in one that has not been contaminated by ewes since the end of winter. The groups may run together at night. If ewes and lambs must graze together, rotate pastures frequently. Excessive concentration of sheep to control pasture growth is almost always accompanied by severe losses from parasites. Control excessive growth by using cattle to graze after sheep or by clipping. Check the lambs frequently, and use anthelmintics as needed. Also, be on the alert to spot injuries or evidence of external parasites.

11.17 MILK PRODUCTION BY EWES

During the first month of life, it has been estimated that milk consumption accounts for three-fourths of the variation in lamb growth (Owen, 1976). From then on, the supplemental feed consumption of the lamb and the growth potential become increasingly important. A number of factors affect milk production:

1. Heredity
2. Age of the ewe
3. Size of the ewe
4. Number, size, and vigor of lambs suckled
5. Nutrition during gestation and lactation
6. Health of the ewe
7. Stage of lactation

While it is true that milk production in domestic sheep has not been studied to any great extent, there is much variation in gains of suckling lambs by individuals

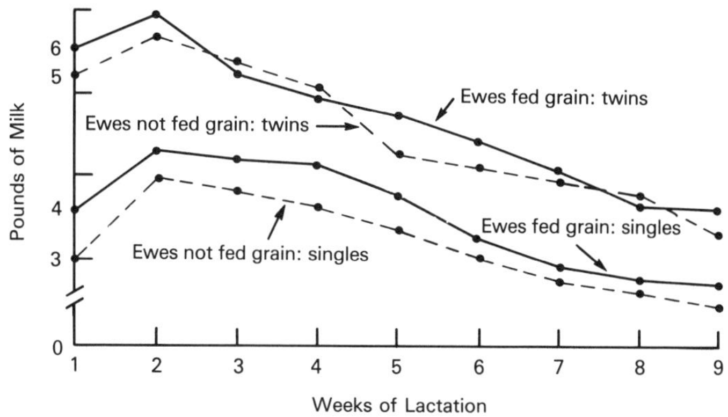

Figure 11.26 Average daily milk production (lbs) by ewes suckling singles and twins and fed grain versus no grain for 60 days prior to lambing.

and breeds. Gains up until weaning reflect the milk production of the ewe. Thus, selection for heavy weaning weights may also result in improving the milking ability of the ewe flock. Young ewes tend to produce less milk than their mature counterparts. Ewes will increase milk production until they are 5 to 6 years old. Size appears to be related to lactation. Larger ewes, in the sense of skeletal size rather than body weight, which may be an expression of fatness, tend to produce more milk than ewes with smaller frames.

Ewe lactation is affected by stimulation of the teats and udder. Ewes suckling twins or triplets give 40% to 50% more milk than similar ewes with singles. Ewes with ram lambs give more milk than ewes with ewe lambs. Crossbred lambs have been shown to receive more milk than purebreds suckling similar ewes.

Ewes that received inadequate nutrition, especially those that were immature when bred, gave less milk than their more adequately fed counterparts in Missouri research (Ross, Dyer, and Krieg, 1967).

Nutritional requirements for lactation are at least 20% higher than for gestation and 60% to 70% greater than for maintenance. They are also higher for ewes suckling twins and triplets versus singles.

Health of the ewe affects milk production, especially any infection of the udder. Other infections may also cause a reduction in milk flow. Nitrate poisoning will virtually stop milk secretion. Stage of lactation affects milk production. Figure 11.26 shows curves of daily milk yields by weeks. Note that maximum levels were reached between the second and third week of lactation and then gradually decreased. Only 25% of milk production occurs after the eighth week. These data indicate the reason for creep feeding and show that weaning after the eighth week may have little effect on lamb gains.

11.18 WEANING LAMBS

If lambs are eating 1.5 to 2 lb (0.7 to 0.9 kg) of the creep mixture per head daily at 8 weeks of age, it is usually more economical to wean them rather than continue feeding an expensive lactation ration to the ewes. Lambs can be weaned as early as

4 weeks if they are accustomed to eating creep feed, but it is the general consensus that 6 to 8 weeks is more satisfactory.

The following is a suggested weaning procedure:

1. Discontinue feeding concentrates to ewes 4 to 5 days before weaning. Only feed roughage.
2. Confine ewes and lambs to drylot with no feed or water for ewes 2 days before weaning. Lambs should have access to the ewes, the creep, and a water supply.
3. Move the ewes to an area where they cannot see or hear their lambs on weaning day.
4. Feed the ewes grass hay or put them on a low-quality grass pasture. Make certain they have water and a mineral mixture.
5. Check udders on ewes, and if they are uncomfortably tight and the teats red, relieve the pressure by partially milking them; but remember that some pressure in the udder will suppress milk secretion.
6. Be sure the lambs have plenty of the same ration they have been on plus good-quality hay, water, and minerals.

11.19 MANAGING LAMBS FROM WEANING TO MARKET

If lambs have been fed a palatable creep mixture, they should be eating from 1.5 to 2.0 lb (0.7 to 0.9 kg) of feed per lamb daily when they are weaned. After they are weaned, they should have no difficulty maintaining that level of consumption, provided they are not thrown off feed by radical changes in rations or their surroundings. It is best, if possible, to leave them in familiar quarters until they have been separated from their mothers for a week. Take special care to see that they have plenty of clean water and a mineral mixture. Hay, if fed, should be fine stemmed and leafy.

It is advisable to make no quick changes in the ration fed to weaned lambs. If it is desirable to alter the ration, do it gradually by blending it with the creep mixture. If the objective is to finish the lambs rapidly, increasing the amount of roughage in the mixture will result in decreasing the energy level and probably reducing the gains. However, reducing the roughage level below 10% may also be risky unless you have had experience in feeding high-concentrate diets. Follow the NRC recommendations for the weight of lambs. If they are exceptionally fast growers, it may be desirable to increase the protein content above recommendations, but the economics should also be considered.

The method of feeding lambs will depend on how they were fed in the creep, the equipment on hand to feed lambs, the feed available, and the feed-processing equipment. Suckling lambs will do very well on a high-concentrate mixture in a creep and thus will be on full feed when they are weaned. If, during the last part of the suckling period, you have anticipated what ration you will feed from weaning until they are ready for market, the transition may be started the last week to 10 days before weaning. For example, some sheepraisers provide a feed mixture in the creep, but also have a self-feeder filled with shelled corn. If it is planned to finish the lambs on shelled corn and alfalfa pellets, the creep mixture could be gradually blended with the alfalfa pellets over a 10-day period, and the lambs would be on self-feeders by the time they are weaned. This is usually a very inexpensive way of feeding lambs, but it does prevent adding antibiotics and other additives to the ration. Another strategy is to feed a mixture of pelleted protein supplement and shelled corn. Then additives can be included. In each case, it is advisable to provide some hay in racks for the lambs in addition to the feed in the self-feeders. Another alternative would be to provide a complete ground mixed diet with chopped hay, cracked grain, and protein supplement plus additives. Lambs do very well on complete ground mixed diets, but sometimes the feed tends to bridge over in the self-feeder and must be punched down, periodically.

Molasses is often added to a ground mixed diet to increase the palatability and cut down on dust, but it should be limited to less than 7% to be most efficient. The ground mixed complete diet may also be pelleted, provided the hay is chopped fine enough so that it will pellet. Pellets are usually very palatable, easily handled, and work well in a self-feeder or automated facility. But all too often pellets are too expensive for the extra gains or savings in labor to justify.

If the lambs were hand fed in the creep, the practice may be continued, but with changes in the formulation to bring them within recommended allowances. Since finishing out weanling lambs is similar to finishing feeder lambs after the transition from weaning, it is well to review that part of this book also, particularly the part regarding feedlot problems. These cardinal principles in finishing lambs should be emphasized:

1. Provide only wholesome feed. Damaged grain or by-products usually do not work with lambs.
2. If self-feeding, see that there is always feed in the feeders.
3. Protect feed from rain or snow, and do not let wet feed stay in feeders.
4. If grain is hand fed, always sweep out troughs before feeding.
5. Provide plenty of clean water.
6. Provide a mineral mixture.
7. Feed lambs of different ages and sizes separately.
8. Establish a routine and stick to it.
9. Don't handle lambs that are doing well unless it is necessary. Keep stress to a minimum.
10. Sell lambs when they reach market weight and grade.

11.20 MANAGING DEVELOPING LAMBS

Not all lambs should be pushed hard for an early market. Some producers like to lamb ewes in April when the weather is comfortable and graze them through the summer and fall and then finish them in drylot when the new corn is available in the fall. This is also desirable for the producer who is raising replacement ewes. If lambs are to be carried over the summer, it is rarely economical to creep feed them. If they can be kept growing on pasture without internal parasite infection, external parasite infestation, or damage from predators, they should ordinarily be produced inexpensively. If the lambs can be grazed on uncontaminated pastures, they may make

TABLE 11.2. EFFECTS OF SHEARING ON GAINS AND FEED CONSUMPTION OF LAMBS

Treatments	Unsheared	Sheared
Number of lambs	21	20
Average initial weight (lb)	75.5	72.0
Average final weight (lb)	94.2	94.0
Average daily gain (lb)	0.33	0.48
Average daily feed intake (lb)	2.7	3.2
Feed per pound of gain (lb)	10.2	7.5

moderate gains except when it is hot and dry. If the season is favorable and the pastures are good, some of them should be fat enough to slaughter directly off pasture.

Lambs to be kept over the summer should be sheared at least by July 1. They will eat more and gain faster and be much less susceptible to external parasites than those in full fleece. An experiment conducted at the Missouri Experiment Station (1972) indicates the effect of shearing on feeder lambs in drylot in the summer (Table 11.2).

If pastures are dry and low in protein, it may be beneficial to feed a small amount of protein supplement such as soybean meal daily. At the Missouri Experiment Station (Ross, Riley, and Shelton, 1967), it was found in a number of trials that lambs on orchardgrass pasture and given 0.25 lb (114 g) of soybean meal per head daily made greater gains, but also had higher hematocrits,* than controls on pasture alone. This indicated that the protein supplement not only resulted in superior gains, but also assisted them in resisting stomach worm (*Haemonchus contortus*) infection.

If replacement ewe lambs are being produced, they should have access to the best summer pasture available during the hot summer months. This should usually be a grass–legume mixture, since most cool-season grasses are low in nutritive value when temperatures are high. Make certain that the lambs are inspected frequently for evidence of internal and external parasites. Treat lambs as needed. It may also be necessary to tag the lambs if they have an accumulation of manure in the wool or their rear ends. Be on the lookout for injuries and treat promptly if they occur. Also, use a repellent on wounds to prevent fly strike. Protect lambs from predators and provide shade and plenty of clean, wholesome water and minerals.

11.21 MANAGEMENT OF DRY EWES

Once ewes have ceased lactating, they may be used as scavengers until it is time to prepare them for breeding. Weaning time is one of the best times to mark ewes for culling. Size of the udder, providing it is sound, is an excellent criterion of how well the ewe has milked. If she was a very good mother, she may not be pretty because she should be suckled down to a thin condition. Don't be fooled by the fleshy, good-looking ewe. She probably did not work very hard. Note lumpy udders, pendulous udders, very large teats, or infected areas. All these are excellent reasons for culling a ewe. Note, also, any deformities such as a sagging abdomen or foot problems and

*Hematocrit is a measure of red cells in the blood.

consider eliminating affected ewes from the flock. If you have kept a record of such things as prolapse, difficult lambing, inadequate milk production, failure to claim her lamb, or inherited defects, these are all reasons for culling.

If ewes are given access to high-quality pastures throughout the summer, they very likely will become fat and be difficult to breed. It is best to carry them in a low to medium condition until it is time to flush them.

Ewes may be used very effectively to graze stubble fields, fence rows, and other less desirable forage. Although sheep will eat 90% of all weeds, they should not be starved to the point that they are eating poisonous plants such as nightshade. Inspect them frequently to observe their condition and to make certain they have not been injured or are having difficulty with external or internal parasites. Follow a drenching program as outlined in the section on internal parasites. Ewes appreciate a cool, shady place where there is a good breeze. Provide plenty of cool, clean water where they can reach it and a mineral mixture. If temperatures are excessively high, consider opening up your barn so that the ewes can lie in it. Beware of such hazards as open culverts that ewes may become wedged in and suffocate. Look at your ewes everyday to be sure they are not having problems. An early morning inspection when they are just coming off their beds is the best time to spot trouble. This is also the best time to work sheep if it is needed.

REFERENCES

Bailey, Don E., "A Top Notch Veterinarian Looks at Lambing," *Sheep Magazine,* 4, no. 2, March 1983.

Collins, S. B., *Profitable Sheep.* New York: Macmillan, Inc., 1956.

Dutt, R. H., E. F. Ellington, and W. W. Carlton, "Fertilization Rate and Early Embryo Survival in Sheared and Unsheared Ewes Following Exposure to Elevated Air Temperature," *J. Anim. Sci.,* 18, 1959, pp. 1308–1318.

Ensminger, M. E., *Sheep and Goat Science.* Danville, Ill.: Interstate Publishers and Printers, Inc., 1986.

Fredrickson, K. R., R. M. Jordan, and C. E. Terrill, "Rearing Lambs on Milk Replacer Diets," USDA Farmers Bulletin 2270, 1980, 22 p.

Hall, K., J. Rhoades, E. Morgan, and C. V. Ross, "Effects of Method of Docking and Castrating Lambs on Gains and Healing Time," *Missouri Sheep Day Report 1982,* 1982.

Outhouse, J. B., Ketosis in Ewes, Ph.D. thesis, Purdue University, 1956.

Owen, John B., *Sheep Production.* London: Cassell and Collier-McMillan Publishers Ltd., 1976.

Parker, Ron, *The Sheep Book.* New York: Charles Scribner's Sons, 1983.

Ricketts, G. E., R. D. Scoggins, and D. L. Thomas, "Recommendations for a Sheep Management Program," *Ill. Agr. Ext. Circ. 1221,* 1983.

Ross, C. V., A. J. Dyer, and K. L. Krieg, "Rations for Wintering Ewes," *Missouri Agr. Exp. Sta. Bull. 861,* 1967.

______, M. L. Karr, and R. L. Pavey, "Creep Feeding Studies with Lambs," *Missouri Agr. Exp. Sta. Bull. 772,* 1961.

______, J. G. Riley, and G. C. Shelton, "Practical Tests with Anthelmintics for Lambs," *Missouri Agr. Exp. Sta. Res. Bull. 927,* 1967.

Scott, George E., *The Sheepman's Production Handbook*. Denver, Colo.: Sheep Industry Development Program, Inc., 1982.

Simmons, Paula, *Raising Sheep the Modern Way*. Charlotte, Vt.: Gardenway Publishing, 1976.

Townsend, W. M., C. V. Ross, and J. D. Rhoades, "Effect of Shearing on Feedlot Performance of Lambs," *Missouri Sheep Day Report*, 1972.

QUESTIONS FOR STUDY AND DISCUSSION

1. Why should individual sheep be identified?
2. What are reasons for keeping productivity records?
3. Why should records be more complete for a purebred than a commercial operation?
4. Discuss the value of special efforts to gentle a flock of sheep.
5. What is the reason for eliminating nervous and wild individuals from a sheep flock?
6. When should ewes and rams be prepared for the breeding season?
7. How can heat stress be minimized during the breeding season?
8. How can the producer determine that pregnant ewes are gaining in condition during the gestation period?
9. Discuss the value of good pasture for pregnant ewes.
10. In what ways can a shepherd determine whether or not the flock is consuming excessive quantities of minerals?
11. What dangers may be encountered in managing pregnant ewes, and how can these be minimized?
12. Why is nutrition so important during late gestation?
13. What are the consequences of inadequate nutrition during late gestation?
14. Describe pregnancy disease and how it can be prevented.
15. Why should young and old ewes be fed separately from the rest of the flock?
16. In what ways do requirements for shelter of pregnant ewes differ from ewes during lambing?
17. When should ewes be prepared for lambing?
18. Discuss the preparation of ewes for lambing.
19. What are the advantages and disadvantages of shearing ewes before lambing?
20. What vaccination program is advisable for ewes before lambing?
21. Describe suitable lambing quarters.
22. Describe the symptoms of imminent lambing by a ewe.
23. When should a ewe be helped during parturition?
24. Define the term *dystocia* and explain what may cause it.
25. Discuss the treatment needed for a chilled lamb.
26. Explain why lambing pens or jugs are valuable in lambing quarters.
27. What is a fostering pen and how is it used?
28. What precautions should be taken in rearing lambs artificially?
29. When is the period of highest mortality in lambs?
30. What are the principal causes of lamb mortality?
31. List some means of preventing lamb losses.
32. Why should lambs be docked?

33. Discuss the advantages of docking by the use of shears, emasculators, hot irons, and elastrators.
34. Under what circumstances may it be more profitable not to castrate lambs?
35. How should large lambs be castrated during the fly season?
36. Describe the treatment of young lambs for entropion, pinning, sore eyes, scours, and constipation.
37. Describe a satisfactory creep for lambs.
38. Why should winter lambs not run with their dams on pasture?
39. What factors affect milk production by the ewe? Which of these can be controlled?
40. What are the benefits of feeding a low level of protein supplement to lambs on summer pasture?

12 Finishing Lambs

12.1 GENERAL CONSIDERATIONS

12.1.1. Reasons for finishing lambs To satisfy consumer demands, slaughter lambs must weigh between 105 and 115 lb (48 and 52 kg) and have at least 0.05 in. (1.3 mm) of fat over the loin. If they have been creep fed and are suckling a good ewe, they should reach the weight and finish by the time they are 5 months old. When lambs come off the range in the fall, 40% to 60% of them may go directly to slaughter. Those remaining must be grown, developed, and finished before they can be marketed. These lambs are called feeders, because most of them will be put in the feedlot or on excellent pasture before they are marketed and subsequently slaughtered. Finishing, then, can be defined as the growth and fattening required to prepare lambs for market. Lambs require some fat on the carcass and within the tissues in order for the meat to have the flavor and eating qualities required by consumers. Also, if live lambs or carcasses are to be shipped, they must have some covering of fat in order for them to yield a satisfactory carcass.

12.1.2. History of lamb finishing Before 1890, 90% of all lambs slaughtered were at least 1 year old. The meat from these lambs was not as tender and juicy and carcasses were much more mature than present market demands. As transportation and the meat packing industry have improved, the size, age, and finish of lambs have changed, and now 97% of all lambs slaughtered are under 1 year of age. Lamb finishing on a large scale began in the milling centers of St. Paul, Minnesota, and Chicago, Illinois, where they had large quantities of a waste product called screenings. These are largely the small grains of wheat that are screened out in the process of flour milling. At that time, the screenings were available in large quantities and were very inexpensive. During the same period, there were many lambs on the range that were too small and thin for slaughter. Ranchers did not wish to carry over wethers for another year, and they sold them at low prices. Lamb feeding operations were carried on at or near the milling centers, and they were very profitable. This situation did not continue for many years, because it was discovered that screenings were satisfactory feed for dairy and beef cattle and the prices increased with demand. In the meantime, farmers found that they could profitably feed their surplus grain and hay to lambs and sell them on a good market. Farmer feeding was done over a wide area of the United States. Lambs were usually fed in small groups in sheds in New York, Pennsylvania, Michigan, Minnesota, and Wisconsin. Open-lot feeding was practiced in Colorado, Kansas, and Nebraska, and lot feeding with some shelter was the rule in the Corn Belt states. Through the years after World War II, the competition for feeder lambs grew, and feeding tended to shift to larger lots in Nebraska, Colorado, Kansas, Texas, California, South Dakota, and Minnesota. In 1982, the leading states in lamb feeding were Colorado, California, Texas, Wyoming, Iowa, Nebraska, and Arizona. Many of the modern feedlots were mechanized so that complete ground mixed rations were fed mechanically in fence-line bunks or self-feeders.

There has been a general trend toward less lamb feeding, as sheep numbers have fallen consistently since 1956, with a consequent reduction in the number of feeder lambs available to feed. Lamb finishing on pasture has also decreased with the reduction of available lambs. Pasture finishing was never as popular as feedlots.

12.1.3. Location of lamb finishing Areas of lamb finishing are related to the grain-producing regions of the United States. Barley is the standard in the northwestern and western states, milo in the southwest, and corn in most other states where lambs are fed. Corn is also fed in some of the western states where irrigation is practiced. The most popular hay used in feedlots is alfalfa, but it is by no means universal. Other legumes are also used effectively. Most large feedlots feed ground mixed complete rations with cracked or processed grain, chopped hay, a protein supplement, and a small amount of molasses. Small feed yards may simply feed whole grain and long hay. A few feed yards use pelleted complete rations. Some pasture finishing operations in central California use irrigated clover and others in the Imperial Valley of California and in Arizona use alfalfa during the winter season. In favorable seasons, winter wheat may be grazed in Kansas and the surrounding states.

12.2 CHARACTERISTICS OF LAMB FINISHING

It is difficult to generalize on lamb finishing, because it is so variable from year to year. Much depends on weather conditions during the grazing season. If summer pasture conditions are favorable in the mountain states where range production is practiced, the lambs may come off pasture somewhat later in higher condition. Then a high percent are sent directly to slaughter. This reduces the number of feeder lambs available, and quite often those for sale are so large they are penalized for excessive weight when finished.

Rainfall in the southwest United States also affects the size, condition, and the time of sale of lambs. In wet years, lambs may be heavier and fatter, with fewer lambs available for feedlots. Feed prices may also vary with the years due to the weather conditions. The following are some other features of lamb feeding that are unique to the business.

1. Lamb finishing is seasonal, with most of the feeders put in feedlots in September, October, and November.
2. Lambs are fed for a comparatively short time. Time in the feedlot varies from 50 to 70 days.
3. Lambs usually gain faster relative to body weights and more efficiently than cattle.
4. There is usually a narrow spread between price per hundred as feeders and finished lambs.
5. Lambs are sensitive feeders, and skill is required to finish them.
6. Death losses are often higher than those in cattle feedlots.
7. The market is very discriminating against underfinished or overweight carcasses. At times, ram lambs and undocked lambs are penalized.
8. Since the principal consumption areas are on the east and west coasts, either live lambs, carcasses, or cuts must be transported great distances from feedlot to retail markets.
9. Market outlets may be few and not as competitive as other species.
10. Lamb finishing has been traditionally a highly speculative business. Thus, financing may be difficult to arrange.

The characteristics of lamb finishing would appear to be mostly negative, especially during an era of constantly reduced ewe numbers. Under these circumstances, the feeder lamb supply has reached unprecedented lows. But there are lamb feeders who are able to stay in the business and make money. Lamb feeding is needed not only to change the thin, light lambs to animals suitable for slaughter, but also to even out the supply of lambs for slaughter. Within recent years, sheep numbers show some indication of increasing, and new technology may result in increases in efficiency sufficient to provide a profit for the intelligent operator.

12.3 FACTORS TO CONSIDER BEFORE GOING INTO LAMB FINISHING

Whether or not you should be in the business of lamb finishing will depend on a number of factors that determine whether or not the endeavor will be beneficial to your whole operation. No one should become a feeder without considering each factor and then evaluating the whole proposition. If it is not possible to estimate expenses and returns realistically, it may prove to be too much of a gamble to justify the risks. Careful consideration should be given to the factors that can be controlled by you and those that cannot. Then you should estimate the returns not only under the most favorable but also under less than optimum conditions. If you cannot show, on paper, how you can return a profit, it may be well to turn to some other venture. Some factors to consider are discussed next.

12.3.1. Facilities required by the operation If buildings, lots, and equipment are available, can they be adapted to a lamb feeding operation? Would they be adequate or would their cost be prohibitive? Remember the feeding operation must pay for the use of facilities or, if new ones are constructed, they must be depreciated systematically. Bear in mind that tying up large sums of money in physical facilities may decrease your operating funds and increase your vulnerability and dependence.

Feedlot location should be carefully considered. Under present laws, an operation causing concentration of livestock must comply with federal and state laws for the protection of the environment. Runoff from feedlots must be controlled to avoid polluting streams. Before building a feedlot, know what is and is not permissible. Consider neighbors or potential neighbors. Many feedlots have been located near a city or town and, over a period of time, the surrounding land developed into a residential area. Then the neighbors complained that the feedlot was a source of flies and pollution, with the result that the feedlot was forced to move at a considerable financial sacrifice. In most states, laws control land use and are regulated by county planning and zoning boards. Thoroughly investigate the laws and ordinances before building. The location of the feedlot should be protected as much as possible by natural barriers, such as a stream, a highway, or an area that can be controlled by the operator.

If possible, locate the feedlot on a south slope on well-drained soil. Sandy soil works best, but rocky soil is satisfactory. Locating on higher ground for good air drainage is better than a valley where air stagnates. Valleys are also subject to flooding and may endanger the lambs and the facilities. Locate near the residence but downwind, generally to the northeast, of the farmstead, accessible to a good road and the feed supply. Provide room for possible expansion. Fences around the perimeter should be dog proof if possible. It is recommended that they be 5.5 ft (1.68 m) high with barbed wire on the top and bottom. Consider using an electric fence around the outside of the lot to repel predators. The shelter should face either the south or east. Windbreaks are beneficial on the west side of the lot. Sheds should be at least 20 ft (6 m) deep. If longer than 70 ft (21 m), they should have a partition in the middle. If possible, have drop, overhead, or sliding doors in the north side of the shed so they can be opened for ventilation. Heights to the ceiling of the shed should be 8 to 10 ft (2.4 to 3.0 m) so that cleaning machinery can be used. Use of clear-span construction will also assist in manure removal. Recommended shed space per lamb for dirt floors is 10 ft^2 (0.93 m^2); if paved, 8 to 10 ft^2 (0.75 to 0.93 m^2); or 4 to 5 ft^2 (0.4 to 0.47 m^2) on slotted floors.

Hand Feeding

- Linear feeder space/lamb if hand fed grain: 12 in. (30 cm)
- Linear hayrack space/lamb if hand fed hay: 12 in. (30 cm)
- Width of grain trough if fed from one side: 14 to 16 in. (36 to 41 cm)
- Width of grain trough if fed from both sides: 18 to 22 in. (46 to 56 cm)
- Width of hay rack if fed from one side: 12 to 16 in. (30 to 41 cm)
- Width of hay rack if fed from both sides: 18 to 24 in. (46 to 61 cm)

Self-feeders

- Linear space/lamb for concentrates: 4 in. (10 cm)
- Linear space/lamb for ground mixed diets: 4 to 5 in. (10 to 13 cm)

- Linear space/lamb for pelleted diets: 4 in. (10 cm)
- Linear space/lamb for water trough: 1 in. (2.54 cm)
- Lambs/automatic waterer: 15

All specifications are subject to modification depending on conditions. If soil is heavy and drainage is poor, more lot space should be allowed. Under those circumstances, it is probable that mounds in the lots would be practical. Some feedlot operators cover the mound with coarsely ground limestone with a good effect. Others bed the mounds with straw in winter. Amount of space per lamb on slotted floors may also be varied depending on whether or not lambs were sheared.

Every feedlot should have a sorting chute, holding pens, and a good chute for loading and unloading lambs. It is not necessary to go to great expense to build handling equipment, but it should be substantial and practical. Feed handling and storage are important considerations because they may affect the efficiency of the operation both from the standpoint of labor and the ability to take advantage of price differentials that occur seasonally and annually. Some careful figuring before construction may save a lot of money in the long run. It is best to look at facilities at a number of successful operations, especially in cattle feedlots, before deciding what should be built. Plans are available from the agricultural experiment station or extension service that provide the principles for laying out feedlots.

12.3.2. Feed supply and availability Since feed represents one of the major costs of feeding, it is most important to know that a reliable supply is available at a reasonable price. Some feeders produce most of the grain and hay used in their feedlot. In many cases, it is possible to realize good returns on otherwise unmarketable farm products by feeding them to livestock. The same can also be said about hay and grain that could be sold but at less than the value received by putting them through livestock. It is also important to know that other sources are available without undue sacrifice.

12.3.3. Expenses other than feed Although feed represents the largest expense of feeding lambs, available labor, cost of operating machinery and equipment, insurance, interest, taxes, and depreciation on physical facilities and other expenses must be considered. Some farmers or business people have full employment for their skilled labor during the spring, summer, and fall, but they need to use them productively during the winter months in order to keep them on the payroll. Then they may be figured at a more favorable rate so that expenses can be spread over more numbers and reduce the cost per head. It may also be possible to mechanize the lot so that very few personnel are used. On the other hand, it is imperative that elaborate equipment be put to extensive use or it will not be feasible.

12.3.4. Costs of feeder lambs One of the most frequent pitfalls of the lamb feeder is to purchase the feeders too high. Then, when the market goes down, no profit is possible. It is usually not a safe proposition to buy high-priced feeders with the expectation that the market price will increase enough to ensure a profit. Ideally, the price of feeders would be less than that of finished lambs so that the value of the original weight purchased would increase during their time in the lot. This difference is usually based on the price per 100 lb (cwt).

Thus, if the purchase price was \$60.00 per 100 lb for 70-lb (32-kg) lambs, and they sold for \$65.00 per 100 lb when they weighed 110 lb (50 kg), the difference would be \$5.00 per 100 lb. This difference is called *spread* or *margin*. (In this case, it is positive, but it could also be negative.) Then the original 70 lb increased \$3.50 in value. If it cost \$60.00 per cwt to put on the 40 lb of gain, the profit would be \$5.50 per lamb, as shown:

Feeder cost = 70 lb × \$60.00/cwt	=	\$42.00
Cost of gain = 40 lb × \$60.00/cwt	=	24.00
Total cost of fat lamb	=	\$66.00
Sale of fat lamb = 110 lb × \$65.00/cwt	=	71.50
Profit per lamb = \$71.50 − 66.00	=	\$ 5.50

In this example, the spread was positive and actually accounted for \$3.50 of the \$5.50 profit per lamb. Lamb feeders generally would be pleased to have a \$5.00 spread, because it is seldom so favorable.

In addition to the \$3.50 per lamb that came from spread, there was a \$2.00 profit on gains (i.e., 40 lb at 60 cents per lb = \$24.00 for the gains and it sold for \$26.00; thus 40 × \$0.65 = \$26.00 − \$24.00 = \$2.00). The feeder, then, made a profit on both the spread and the gains. The next case is not so favorable. If the feeder lamb cost the same as before and sold for \$59.00/cwt with a cost of gain of \$55.00/cwt, the figures would be as follows:

Feeder cost = 70 lb × \$60.00/cwt	=	\$42.00
Cost of gain = 40 lb × \$55.00/cwt	=	22.00
Total cost of fat lamb	=	\$64.00
Sale of fat lamb = 110 lb × \$59.00/cwt	=	64.90
Profit per lamb = \$64.90 − 64.00	=	\$ 0.90

In this example, the spread is minus \$1.00 (\$60.00 − 59.00). Thus, the original 70 lb, which cost \$42.00, sold for only \$41.30, which was a loss of \$0.70 per lamb. The cost of 40 lb (18 kg) of gain was only \$22.00, and it sold for \$23.60 for a profit of \$1.60. Subtracting the 70-cent loss from the \$1.60 profit on gain gives a net profit of 90 cents per lamb. From these examples, it can be concluded that the profit in lamb feeding may come from spread or from economical gains.

The spread at the time lambs are put on feed is used as an indicator of the prospects for the venture. If the fat price is actually below the feeder price (a negative spread), the cost of gains must be extremely favorable to make the proposition attractive.

Some lamb feeders take some of the gamble out of lamb fattening by using pasture or some other cheap feed to put some inexpensive gains on the lambs. Assume the lambs weighed 60 lb (27 kg) and were purchased for 60 cents/lb (0.45 kg). If the purchaser could put 10 lb (4.5 kg) on them at 25 cents/lb (0.45 kg), then they would cost \$36.00 + \$2.50 or \$38.50/70 = 55 cents/lb (0.45 kg) or \$55.00/100 lb (45 kg). This is called "cheapening up" the lambs. Examples of the process are corn-

Figure 12.1 Inexpensive gains on pasture or corn fields before going into the feedlot is called "cheapening up lambs."

field grazing (as shown in Figure 12.1), cornfield gleaning, grazing fall wheat or other pasture, or beet tops or other crop residues.

Another possibility relative to purchasing feeders at an advantage is to become skillful in judging the "doing" ability of lambs by inspecting them. There are numerous auction sales, particularly in the Midwest, where mixed loads of lambs are brought to market and sold. Many times the finished lambs are sold with the option of sorting out those that are too thin to slaughter. These lambs may be bought at prices well worth the money if the lamb feeder has the time and patience to sit through sales. If the feeder has some reason for attending the auction anyway, such as a position of lamb buyer for a meat packer, the buying advantage may be great. Some of the best and worst buys of feeder lambs are made on locally grown lambs that can be taken home, wormed, doctored, and worked with to prepare them for the feedlot. Some lamb feeders become quite skillful at this practice, and they frequently are able to make spectacular profits. Unfortunately, the total volume of lambs may be limited and the risks of death too great for the inexperienced buyer to take. The point is that a person who is determined and a good judge of livestock may develop the skill to compete in the business.

There are marked differences in the ability of lambs to gain in the feedlot. Some of these are related to conformation and how the lamb has been handled previously. Lambs from certain regions are bred differently than others, and their performance may be affected. This should not be ignored. A skilled lamb feeder should know a lot about the types of lambs and their potentialities. A lamb feeder who studies his or her lessons and learns should discover how to be competitive in the field.

One last possibility is to contract feeders ahead. Some feeders know producers whose lambs do exceptionally well in the feedlot. If they have light shrinks and make fast, economical gains, it may pay to ensure your feeder supply. This sort of deal should be studied carefully before entering into it. It could offer stability to both the buyer and seller. It must be fair to both parties or it will not endure.

12.3.5. Markets and marketing Later in this book there is a section on marketing lambs. It is not proposed to preempt that discussion, but merely to say that markets and marketing are important to the success of the operation. Study the outlets and know the strengths and weaknesses of each before embarking on a lamb feeding operation. What are the costs of selling lambs? Are they reasonable? Is there any way you could have a marketing advantage? Too often, marketing is the Achilles heel of an operation. Regardless of the quality of the product you have for sale, the bottom line is closely related to getting it sold readily at a reasonable price. Know the situation locally, regionally, and nationally and how your position can be assured. Many feedlots have stayed in business simply because they had a firm contract with a packer for a certain number of lambs each year. While it was true that the feeder also fed other lambs, the fact that there were an assured number going through the lot each year offered the stability needed to ensure the continued existence of the lot.

12.3.6. Death losses Death losses are included in this discussion because they are a definite part of the whole proposition. They are usually greater in lamb finishing than they are with cattle. Their magnitude will depend on a number of things, such as the age, size, and previous treatment of the lambs. If the lambs are fresh from the range and have been well cared for in transit and at the feedlot, death losses should be low. But if the lambs are small, weak, and mistreated, they should be expected to be high. A good feeder will know what to expect with any type of lamb. Not many operators can do well with lambs that are unthrifty, but some feeders do feed culls and do a good job. They say that the spread is so favorable they can sustain a rather heavy death loss and still make money. This is probably too risky for most lamb feeders.

12.3.7. Skill and experience of the operator Not all the knowledge needed to run a successsful feedlot can be gained by reading the literature. Many daily operational decisions require some knowledge of lambs and how to handle them. There is some advantage to working in a feedlot for a while to gain some practical experience, but if that is not possible, it is advisable to start modestly and increase the scope of operations gradually so that a mistake is not disastrous. Be willing to make some mistakes, but first know the fundamentals of the operation. Don't be afraid to ask people to help you if you are having trouble. Some of them will turn you down, but a surprising number will help you to get started. Never feel that you know it all, because no one does. Be willing to learn and set up your operation so that you can benefit from experience.

In general, it can be said that a number of factors have a very strong influence on the success of a feeding operation. Any one of them, if bad enough, could cause the failure of a feedlot. But, at the same time, any one of the factors can be so favorable as to overcome some drawbacks in other factors. This is known as "the law of the outstanding advantage." Not only is it a fact that one exceptionally favorable factor could assure the success of an operation, but several outstanding advantages may be even better. Look for some unique advantages that you may have and exploit them. Remember, these are called trade secrets in business. If you share them, be sure it does not remove your advantages.

12.4 BUYING FEEDER LAMBS

A good job of buying feeder lambs may not assure success in the feeding operations, but a poor job can easily guarantee failure. If you have been in the feeding business for a long time, you should have many contacts as sources of information relative to availability and characteristics of feeder lambs. If you are a beginner, start making them as soon as possible.

12.4.1. Sources of feeder lambs Sources may be ranchers, order buyers in an area or on a market, local dealers who may ship in feeders and sell them at private treaty, dealers who service a wide area, community auctions, area auctions, and direct purchases from farmers. There are still a few stockyards where lambs can be bought through a commission firm.

If the buyer is inexperienced, it will usually pay to use a commission representative who knows the area and the ranchers. It is often true that a rancher will not sell except through a representative. If the commission representative is good, he or she can arrange for sorting the lambs, transportation, method of payment, and other details of the transaction. The buyer often pays the commission and should expect the services to include avoidance of problems such as disease or needlegrass, which may be prevalent in the area. The representative should be able to obtain favorable weighing conditions, so that shrinkage is not excessive, and a sort to avoid excessively large or small lambs. Regardless of where lambs are purchased, it behooves the feeder to fully understand how the system operates and the conditions of the sale. A record should be kept on every group of lambs bought so that the information can be used in the future.

12.4.2. Factors affecting costs of feeder lambs The following are items that, to a large extent, determine the cost of feeder lambs in the feedlot:

1. FOB cost of the lambs based on price per 100 lb (cwt) (45.4 kg) multiplied by the pay weight.
2. Buying commission, usually $0.50 to $1.00 per cwt (45.4 kg), depending on services performed.
3. Transportation, figured on the basis of so much a mile from the truck or on the basis of 100 lb (45.4 kg) liveweight.
4. Shrinkage (if transported as far as 1000 miles, it may be approximately 10%).
5. Intransit death loss, if not covered by insurance.
6. Feed cost for lambs in transit.

Weighing conditions are the terms under which the purchase weight of the lambs is figured. This is fully as important as the agreed price, but is sometimes overlooked by the neophyte. Weighing practices vary throughout the country. If the buyer does not understand what a term or condition means, it is most important to have it clarified. Shrink is the loss of weight from the time livestock are sold until they reach their destination. It is usually expressed as a percentage. For example, if 70-lb (31.8-kg) feeders lose 7 lb (3.18 kg) in transit, the shrinkage is 10%. Shrinkage is discussed in detail in Section 13.19.10.

Ranchers and farmers who have been in the business for a while understand the importance of weighing conditions very well. If the feeder does not understand how this is figured, it is a good plan to have the commission person do the negotiations initially until some skill is acquired.

If lambs are healthy, properly loaded, and cared for, death losses should be low. It must be emphasized that any lambs that die will raise the cost of the rest of them. Studies have shown that there are differences in the shrink and death loss in trucks and between drivers. For this reason, caution should be exercised in selecting the trucker to haul the lambs.

In most cases, truckers will attempt to drive through with a minimum of delay so that there is seldom a feed and yardage bill in transporting lambs. There is a law that requires lambs to be fed, watered, or rested for 5 hours after 28 hours in transit. The requirements can be waived by the shipper to 36 hours, but no more. Most truckers know the location of markets where lambs can be cared for if they are delayed for any reason, such as mechanical trouble.

12.4.3. Characteristics of feeder lambs

1. *Conformation and condition.* Good conformation in a feeder lamb is needed in order for it to yield a suitable carcass. Muscling in the back, loin, rump, and legs should be evident. Some length and scale is more desirable than short, pony-type lambs. Lambs that are exceptionally narrow, weak in the back, and short in the loin and rump are unlikely prospects to grade high when they are finished. Lambs that are too thin may never be finished satisfactorily. Feeder lambs should have some flesh on them when purchased. Lambs may have white, smuts, or black faces. There is a general preference for black faces because it indicates more mutton breeding. But good white-faced feeders do well in the feedlot and sell only slightly below black-faced lambs.

2. *Fleece condition.* Estimate the amount of wool on them. Know the value of the wool and how it will affect pelt credits when the lamb is sold.

3. *Health.* Thrifty, higher-priced lambs are usually cheaper in the end than apparent bargains that may have low gains and sustain high death losses. Look for lambs that have bright eyes and are alert. Look for life and luster in the wool and clean noses with no coughing; avoid weak, lifeless, undersized, gaunt lambs that are scouring or anemic. Know the origin of the lambs, and stay away from those that may have needlegrass or other troublesome weed seeds in the fleeces.

4. *Genuines or old crops.* Feeder lambs may be born throughout the year. Spring lambs may be available in the fall and winter and even into the next spring. Sometime after May, there may also be some feeder lambs from the January lambs of the present year. The lambs that were born last spring are then called *old crops,* and those from the January crop are called genuine spring lambs or *genuines.* Genuine feeders may be available from Texas and the southwest and from California.

5. *Previous treatment.* Some lambs are shipped from the range directly to the feedlot, but others may go to a pasture operation where they are fattened or may be grown and developed. Lambs that have been out to pasture may be known as wheatfield, alfalfa, beet top, or cornfield lambs. Usually the fats have been sorted

off and sold. The remaining lambs are called *comeback lambs,* because they have been marketed before and are again on the market. They usually are easy to feed and fleshy, and most of the weak and slow ones have been lost on the pasture operation. Thus, death losses may be low in them. Other kinds of lambs that are sometimes available are those that someone has started on grain but, for some reason, has decided to sell. These are called *warmed-up lambs.* Many feeders are very reluctant to buy such lambs for fear that they may be unusually unthrifty or may have some disease condition that makes them poor risks.

6. *Weights.* Lambs vary in weight from lightweights of under 60 lb (27.3 kg) to heavies, which may exceed 90 lb (40.9 kg). Light lambs have the advantage that it may be possible to graze them or to grow them on inexpensive feed and get some cheap gains, but too often those under 60 lb (27.3 kg) have something wrong with them and are difficult to handle. Buying whole loads of light lambs may result in securing the runts and sick lambs from a large band. It is advisable to get some experience handling light lambs before going much under 65 lb (29.5 kg) in purchase weights. The most serious problem with buying heavy lambs is the danger that they will be too large when they are finished. There is a general rule that if you buy 80-lb (36-kg) feeders, you will receive quite a few that weigh 70 lb (32 kg) and the same for 90 lb (41 kg). The 90-lb (41-kg) feeders will weigh about 120 lb (55 kg) when they are finished. That is often considered an undesirable weight, and lambs may be discounted as much as $10.00/cwt. Some feeders do very well with heavy lambs, but they assure themselves that the lambs are carrying a lot of flesh and will not have to get too big before they are finished. *Comeback lambs* are frequently heavy, but often fleshy enough to be finished at satisfactory weights if they are pushed in the feedlot.

12.4.4. Types of feeder lambs There is a terminology used in lamb feeding that allows the people in the business to communicate. To some extent the terms may be colloquial, but there are some general ones that apply throughout the United States.

1. *Natives:* This is generally a midwestern term for lambs from farm flocks. The lambs may be expected to have black faces, medium wool, and mutton conformation. Strong points for these lambs are:

1. They are available over much of the year.
2. May have good conformation.
3. Could have a light shrink.
4. Often sell at very low prices.

Weak points are:

1. Many of the lambs have not been docked and castrated.
2. Usually parasitized both internally and externally.
3. May be diseased. Foot rot and pneumonia are common in Natives.
4. They may lack uniformity in size, conformation, and age.

Figure 12.2 Native feeder lambs from Missouri.

5. It is not possible to buy large numbers of them in most locations. High-quality Native feeders are shown in Figure 12.2.

2. *Westerns:* These are from the range states in the Northwest. The lambs are usually called by the state of origin or even locally, within a state. For example, "Colorado west slope lambs," "Belle Fourche lambs," from the area of Bell Fourche, South Dakota. Western lambs may be either white or black faced and sired by Rambouillet, Columbia, and Targhee white-faced rams or Suffolk or Hampshire black-faced rams. The ewes of the region are predominantly of Rambouillet, Columbia, or Targhee breeding. The lambs may be expected to have length of body and good frames, medium conformation, and mostly smooth skin. In Figure 12.3

Figure 12.3 Blackfaced Western feeder lambs from Colorado.

are typical black-faced Werstern lambs. Generally, the strong points for Western lambs are:

1. Lambs are quite thrifty, free of parasites and disease, and make rapid gains.
2. Have good conformation, although not as desirable as the best Native feeders.
3. In season, they are available in large numbers.
4. Usually uniform in weight and condition.
5. All will be docked and castrated.
6. Most Western lambs are docile and easily handled.

Weak points are:

1. They are available only during the fall and early winter.
2. Sometimes they are too heavy.
3. Prices are highest for these feeders.

3. *Southwests:* These lambs are mostly from Texas, but they also may be from Arizona and New Mexico. They usually are identified by the location of their origin within the region, which may be the state, city, or valley. The lambs are also known as fall, winter, or spring lambs. If they have spent some time on alfalfa pastures in the Yuma, Arizona, area this information may be given. The lambs may be white or black faced. They are out of Rambouillet, Merino, Columbia, or Targhee ewes and sired by a Rambouillet, Columbia, or Targhee ram if they have white faces or a Suffolk or sometimes Hampshire ram if they have black or mottled faces. Most Southwest lambs have a predominance of fine-wool breeding, although it is not as easy to generalize as it once was. The lambs are smaller and not as stretchy as Western lambs. Their conformation is less desirable than Western lambs. They tend to be narrower and less well muscled. Most of the lambs will have smooth skin, but some will be heavy pelted and wrinkled. There seems to be more variation in type within Southwest than Western lambs, with a range from the small, fine-wool type to the more stretchy, medium-wool lambs similar to those in the Northwest. Strong points for Southwest lambs are:

1. They are available in larger numbers during the summer, fall, and winter since there is a wide range of lambing dates in the Southwest.
2. It is possible to obtain uniform groups at desirable weights.
3. Most of the lambs are healthy and have been dipped for external parasites.
4. Most will be docked and castrated.
5. They are usually lower in price than Northwest lambs.

Weak points are:

1. They may lack conformation.
2. They usually do not gain as fast or efficiently as Western lambs.
3. Some of the lambs may have internal parasites depending on rainfall and pasture conditions during the previous summer.

Figure 12.4 White faced Southwest feeder lambs from Texas.

4. They may be more excitable and wilder than Northwest or Native lambs. White faced Southwest feeders from Texas are shown in Figure 12.4.

4. *Californias:* These may be similar to Western lambs, but can be either old crops or genuines, depending on the season.

5. *Dakotas:* These lambs may be similar to Western lambs but, in certain areas, have more mutton breeding. These are seasonal and limited in numbers. But the best Dakotas are as good as any feeders available.

6. *Southerns:* These are not as numerous as they once were, and are more difficult to characterize since they are quite variable in quality and numbers. Generally, the lambs may be expected to be similar to Natives from the central United States. Most are sired by mutton-type rams and out of Southern or Southwest ewes. Strong points are:

1. Usually sell for low prices.
2. May be available through the summer and winter.

Weak points are:

1. May be parasitized both externally and internally.
2. May be diseased, especially with foot rot or pneumonia.
3. Usually have inferior conformation.
4. May not be docked or castrated.
5. May be expected to make the lowest gains and be least efficient of all feeder lambs.

12.5 CARE AND MANAGEMENT OF "GREEN" FEEDER LAMBS

Feeder lambs must be handled carefully in order for them to adjust to feedlot conditions. Most feeder lambs have been on pasture, and many have just been weaned. The lambs have often been subjected to stresses during marketing and transportation and possibly to other lambs that are not healthy. How they are treated enroute to the feedlot and upon arrival will usually have a lot of influence on their performance. Care during this most critical period will nearly always pay big dividends.

12.5.1. Management on arrival at the feedlot When lambs arrive at the feedlot, they may be expected to be nervous, tired, and thirsty. Most of them will have had very little to eat or drink for the last 48 hours. This is the time for tender, loving care to assist them in making the transition between two widely different environments. Make certain that facilities are prepared for them ahead of time so that they can be unloaded promptly without injury. Water troughs or fountains should be filled and ready. Hay racks should be filled with good-quality grass hay, if possible.

For some reason, lambs have less difficulty if started on grass or grass hay rather than legume hay. If the weather is warm, see that they have shade where they can rest. Be present when they are unloaded so that counts can be made and to see that lambs are handled with a minimum of stress.

A separate pen should be provided to handle lambs that are crippled, slow moving, gaunt, running at the nose, coughing, or showing any other condition that could indicate sickness or lack of thrift. These lambs should receive special treatment to overcome problems before they are returned to the group.

The lambs should be given limited access to water at first. Ideally, they would receive approximately 1 pint to 1 quart (0.47 to 0.95 liter) of water; then move them away and 2 hours later allow another quart (0.95 liter), followed by 2 hours, and then allow it free choice.

If hay racks are near, many lambs will eat some between waterings. Some lambs will camp by the water and drink too much if water is not limited. Under practical conditions, the lambs can be observed closely and moved away from the water after they have had time to drink for a few minutes. The condition of overfilling with water may make the lambs sick and is called water founder.

A mineral mixture should be provided free choice to the lambs. One composed of equal parts dicalcium phosphate and salt seems to do very well. Some commercial mixtures may be consumed excessively at first. If that occurs, it is advisable to feed salt and the mineral mixture in separate feeders or different compartments of the same feeder. If they continue to eat the mineral excessively, it is usually a good procedure to limit the amount daily, but leave the salt available.

12.5.2. Conditioning feeder lambs If the lambs are filling up well on grass hay and water, they may be shifted to legume hay over a period of 3 to 4 days. This can be done by blending the two hays in the racks and gradually decreasing the grass and increasing the legume portions. Any lambs that are still gaunt after the first 2 days should be removed to the sick pen and treated.

Treatment for symptoms is remarkably effective if done promptly. One word of caution learned the hard way; don't feed hay on the ground or force lambs to clean up hay they have pulled out of the racks. Sooner or later if they eat hay contaminated with droppings, they will contract coccidiosis. One experience with the black and bloody scours, the emaciation and death losses, and continued unthriftiness of lambs will leave an indelible impression.

Vaccinate lambs for enterotoxemia with *Clostridium perfrigens* type D toxoid on the second or third day, depending on how well they are eating, and again 10 days later if past history indicates a need. It is wise to seek veterinary help in deciding whether to use a C and D vaccine or a D alone. There is some disagreement about vaccinating for soremouth (contagious ecthyma). If feeder lambs always come from the same source where they were vaccinated as lambs, it is not necessary. But under most conditions where various groups are bought and possibly mixed, it is probably best to vaccinate when enterotoxemia injections are given. Examine the lambs carefully for anemia by looking at the insides of the eyelids and the lips and gums. If they are white instead of pink, it indicates that the lambs have internal parasites and should be drenched, usually at the end of the first week.

Most feedlot operators shear lambs if they carry as much as 5 lb (2.3 kg) of wool, if the weather is warm, or if lambs are to be fed in confinement in an enclosed building. Whether to shear depends to some extent on the cost of the shearer, value of the wool and the pelt, and how long the lamb is to be fed. If lambs are to be sheared, it is advisable not to do it within 10 days of vaccination. This is because the incidence of abscesses from vaccination is greater if the two operations are done close together. Lambs should be sheared with a high comb and protected against rain, sleet, snow, and cold winds for at least 1 week in the winter.

Most lambs are excitable, particularly those from the Southwest, during the first days in the feedlot. Sudden noises or attacks by predators may cause them to stampede, and many may be lost by piling up in corners. A nightlight will usually prevent the problem.

12.6 STARTING LAMBS ON FEED

Many lamb feeders are ready to start their lambs on concentrates at the end of 9 days after their arrival at the lot. Others are able to do it sooner, but it is best to be cautious and only change after some experience in feeding lambs.

Lambs must be adjusted over a period of time to concentrates. Once lambs are allowed to eat concentrates, they may be expected to gorge on them if they have the opportunity. This usually results in scours, impaction, founder, sickness with vomiting, and total lack of interest in feed. A sour smell in the pen with small piles of grain is a reliable indication that the lambs are getting too much concentrates before their digestive systems are ready to receive them. There are a number of methods of putting lambs on concentrates. The one chosen will depend on such things as the method of feeding to be used, the equipment in the feedlot, the kind of feed to be used, and the experience of the person doing the feeding. Basically, the procedure consists of feeding all the hay the lamb desires and gradually increasing the concentrates. A lamb is on full feed when it is eating 1.75 to 2.2 lb (0.8 to 1.0 kg) of grain or other concentrates daily, or, if lambs are on high-concentrate diets, when they can be allowed free access to self-feeders.

The way lambs are handled when they arrive and during the period they are started on feed is the most critical time of feeding lambs. Lambs that have been pushed too fast and have gone off feed are said to be "scorched" and may never do well thereafter. The best lamb feeder at the feedlot should be in charge of handling the green feeders. Close supervision will be rewarded by superior performance in the feedlot.

12.7 METHODS OF FEEDING LAMBS

There are a number of distinctly different methods of finishing lambs. These all have the same purpose, a lamb that will yield a desirable carcass. The procedures vary according to such factors as skill of the operator, feed supply, buildings and equipment, financing, and the scope of the operation. The methods of finishing will be discussed, beginning with the simplest and proceeding to the more complex operations.

12.7.1. Hand feeding in pens Requirements include hay racks or feeders, grain troughs, watering facilities, pens with shelter, feed storage, and a means of conveying feed to the troughs.

Expected Results. Feed consumption per lamb per day, 3.5% of average body weight; average daily gain (ADG) per lamb, 0.33 to 0.38 lb (0.15 to 0.18 kg); feed per pound of gain, 7 to 7.5 lb (3.2 to 3.4 kg); and death loss, 1.0 to 1.5%.

Procedure

1. Start lambs only after they are eating legume hay readily.
2. Sweep out troughs and feed lambs twice daily as far apart as possible, preferably early and late.
3. Always feed at the same times of day.
4. Always watch the lambs eat.
5. If any lamb fails to eat or the group leaves feed in the trough, hold the level of concentrates steady until all are on feed.
6. Make certain all lambs have room at the grain troughs, but very little extra space.
7. Feed the amount of loose hay daily that the lambs will clean up without waste. Turn the hay over at noon so the bottom leaves are available.
8. Start with 0.25 lb (114 g) of concentrates per head per day split into two feedings of 0.125 lb each (57 g). A mixture of 60% shelled corn, 30% oats, and 10% oil meal is very satisfactory.
9. Increase the concentrates 0.1 lb (45 g) per lamb per day as long as all lambs eat and clean up all concentrates within 20 minutes.
10. After two weeks, gradually phase out the oats. Estimate total feed consumption and regulate the ratio of protein supplement to grain so that the total ration (concentrates plus hay) contains 11% to 12% crude protein.
11. Provide a mineral mixture free choice and plenty of clean water.
12. Work the levels up to 1.5 lb (0.7 kg) of concentrates per head per day and hold steady for 2 to 3 days.

13. Raise concentrates slowly to 1.7 lb (0.77 kg) per head per day. When you have experience, go higher.
14. Under no circumstance, excite or handle lambs unnecessarily or change the routine.

The advantages of hand feeding are:

1. Equipment requirements are very simple and inexpensive. Out of pocket expenses can be low.
2. Very little feed preparation is needed.
3. Easy to initiate and terminate the operation.
4. Anyone can learn to observe lambs by hand feeding.
5. The method can be used to feed a very few to several hundred lambs.
6. It is adapted to the use of home-grown feeds.

The disadvantages are:

1. It requires much skill to do it well.
2. The method requires much hand labor.
3. Lambs make comparatively slow gains.
4. Feed efficiency is not high.
5. It is not adapted to large operations.

12.7.2. Hand feeding using a dining room This is similar to hand feeding in pens except that a pen, called a dining room, is set aside with grain troughs for feeding concentrates. The lambs from each pen are brought to the dining room twice daily and fed. The requirements are pens with hay racks, mineral feeders and watering facilities, a dining room with adequate grain troughs, feed storage and feed handling facilities, and conveyance for feed to the dining room. Performance is similar to hand feeding in pens. Procedures are the same as those described for hand feeding in pens. Advantages are similar, plus it is possible to feed larger numbers of lambs. The only additional disadvantages are that moving the lambs to the dining room may expose them to stress, and a sheltered dining room is needed. On the other hand, using a dining room reduces the number of grain troughs.

12.7.3. Self-feeding concentrates plus hay Requirements include hay racks, grain troughs, self-feeders for concentrates, watering facilities, mineral feeders, feed storage, and handling facilities.

Expected Performance. Average feed consumption, 4% of average body weight daily; average daily gain per lamb, 0.4 to 0.45 lb (0.18 to 0.2 kg); feed per pound of gain, 6 to 6.5 lb (2.7 to 3.0 kg); and death loss, 1.5 to 2.0%.

Procedure. Lambs must be started on feed by hand feeding either in pens or in a dining room. There is some advantage to starting lambs in the pens where they will be self-fed, because there is less stress. Self-feeders should be in the pens, although they should not be opened until later.

Suggested steps are as follows:

1. Hand feed lambs until they are consuming 2 lb (0.9 kg) of concentrates daily.

2. Hold at 2 lb (0.9 kg) for 2 to 3 days.

3. Raise amount daily by 0.1 lb (45 g) until the lambs go off feed (called *sticking*). Many refuse to eat.

4. Self-feeders should be filled with a concentrate mixture. This may be shelled corn, 90%, and pelleted oil meal, 10%; or shelled corn, 60%, and alfalfa pellets, 40%; or a mixture some feeders prefer containing 77.5% shelled corn, 10% shelled oats, and 12.5% pelleted oil meal. Sometimes there is an advantage to using additives in a pelleted protein supplement, particularly if some shipping fever is noted.

5. Open the self-feeders, but stay close by to observe what happens. Know how much feed was in the feeders and take some action if they begin eating excessively. The lambs may have severe digestive disturbances if they eat more than 2.5 lb (1.1 kg) of concentrates the first day. If they have been properly "stuck," they should eat only moderate amounts; but behavior is sometimes unpredictable. If they do begin to gorge themselves, one of the most effective ways of slowing them down is to put litter in the trough of the self-feeder. It is also possible to do the same thing by adding salt to the mixture, or if pelleted supplement is used, have some higher-fiber pellets available and mix them with the concentrate mixture.

6. Feed some legume hay daily, but limit it to the amount they will clean up, leaving only the coarsest stems.

7. Check feed consumption and determine the protein content and the calcium to phosphorus ratio of the feed they are consuming. Adjust the supplement to grain ratio to provide at least 11% crude protein and at least a 1.6 to 1 Ca to P ratio. Including Aureomycin or Terramycin in the ration should result in lower death losses and more uniform lambs when marketed, plus 5 to 10% faster gains.

8. Pay close attention to self-feeders, and if the lambs are wasting grain, adjust the hopper opening to prevent it.

9. Remember these ironclad rules and obey them strictly:

1. Never allow the feeders to be empty.
2. Keep feeders clean and the feed fresh. Remove rejected and moldy or wet feed at least twice a week.
3. Don't make changes or move lambs unless necessary or in any way stress them.
4. Observe lambs at least twice daily, preferably when they come off their beds in the morning and in the late afternoon. Take care of problems promptly.

The advantages are:

1. Lambs make comparatively fast gains and feed efficiency is good.
2. Requires little labor after lambs are eating from self-feeders.
3. Not much equipment is needed.
4. Feed preparation is usually simple and inexpensive.

5. Out of pocket expenses are modest.
6. The system is adapted to feeding several thousand lambs in a lot.
7. Can use mostly farm-grown feeds.

The disadvantages are:

1. Much skill is required to prepare lambs for self-feeding.
2. Death losses are comparatively high.
3. Sanitation must be practiced strictly.
4. Not adapted to extensive operations.

Lambs being self-fed shelled corn and alfalfa pellets in Colorado are shown in Figure 12.5.

12.7.4. Self-feeding complete ground mixed diets The requirements are pens, self-feeders or fence line bunks, mineral feeders, and watering facilities.

Expected Performance. Feed consumption, 5% of average body weight daily; average daily gain, 0.44 to 0.5 lb (0.2 to 0.23 kg); feed efficiency, 7 to 7.5 lb (3.2 to 3.4 kg) per pound (0.45 kg) of gain; and death loss, 1%.

This system uses the principle of feeding lambs ad libitum (free choice) from the beginning, but gradually changes the proportions of concentrates to roughage. The roughage part of the diet is usually hay chopped or ground not less than ½ in. (1.27 cm) in length. Other fibrous feeds such as cottonseed hulls, ground cobs, dried beet pulp, straw, almond hulls, gin trash, citrus pulp, and bagasse can be used in varying amounts. Molasses is often added to control dustiness and increase palatability, but it should be limited to no more than 7% of the diet. Fats can be included,

Figure 12.5 Self-feeding shelled corn and alfalfa pellets.

but they should be limited to no more than 5% of the mixture. Grain is usually coarse cracked, rolled, or otherwise treated to improve mixing. Protein supplement may include those of vegetable or animal origin. Nonprotein sources are satisfactory, but are limited to no more than one-third of the protein equivalent of the mixture, with no lumps of the substance. Antibiotics are usually included at 10 mg/lb (0.45 kg) of feed. Precautions to follow with this method of feeding are:

1. Mix feed often and thoroughly.
2. Keep feeders clean and free of wet and moldy feed.
3. Inspect feeders frequently for bridging. Bridging prevents downward flow of feed and must be corrected by punching down or other means.
4. Adjust self-feeders to prevent waste.
5. Make certain feed is available in the feeders at all times.
6. Protect feed from rain, snow, birds, and rodents.

The procedure for putting lambs on feed is:

1. The initial mixture fed the first week should contain 35% concentrates and 65% roughage.
2. The second week the mixture should contain 45% concentrates and 55% roughage.
3. The third week the mixture should be 55% concentrates and 45% roughage.
4. The fourth week the mixture should be 60% to 65% concentrates and 35% to 40% roughage.
5. The fifth week the proportions can be changed to 70% concentrates and 30% roughage if all lambs are eating well and are thrifty.
6. From the fifth week, the concentrate portion may be increased cautiously to as much as 85%, but the lambs must be observed carefully to be sure that they are all eating readily and there are no signs that any are off feed.

Figure 12.6 shows self-feeding complete ground mixed diets.
The advantages are:

1. The system requires very little skill to be used effectively.
2. Lambs eat very well and make rapid and efficient gains.
3. The method is well adapted to large operations with mechanization and large-volume feed handling.
4. Otherwise unpalatable feeds can be incorporated into mixed diets very effectively.
5. Feed additives are easily incorporated.

The disadvantages are:

1. Grinding and mixing diets usually requires expensive equipment and a high investment.

Figure 12.6 Self-feeding complete ground mixed diets.

2. Feed preparation may be expensive.
3. Sometimes there is extensive feed wastage.
4. This system requires a large volume of feeding to be economical. If conditions are unfavorable to feed many lambs, it may be quite expensive.
5. Sometimes, particularly with Western lambs, the gains will be exceptionally rapid and they will be finished rapidly. But when the lambs are slaughtered, many carcasses will have a bronze cast and be soft and oily and not firm up properly in the cooler. Then the packer will discount the lambs. The condition can be overcome by holding the lambs in the feedlot 2 to 3 weeks longer after the first lambs show it. The fatty acid content of the fat appears to change to more saturated ones with longer feeding.

12.7.5. Self-feeding complete pelleted diets The requirements are pens, self-feeders, mineral feeders, hay racks, and watering facilities.

Expected Performance. Feed consumption, 5% plus of average body weight per lamb daily; average daily gain, 0.5 to 0.55 lb (0.23 to 0.25 kg) per lamb; feed efficiency, 6 to 6.5 lb (2.7 to 3.0 kg) of feed per pound of gain; and death loss less than 1%. Finishing lambs on pellets, like finishing on ground mixed diets, uses the principle of starting lambs on a low concentrate diet and gradually changing the concentrate to roughage proportions by blending the mixtures to increase the energy level and bring the lambs on full feed rapidly. Most of the statements about the ground mixed complete feeds will apply to pellets. But grinding of roughage may have to be finer in order for it to pellet satisfactorily. It is necessary to produce intact pellets with a small percentage of fines to obtain the best performance with the method of feeding.

The procedure for putting lambs on pellets is:

1. Feed hay to the lambs in the morning; then give them 0.5 lb (0.23 kg) of 35% concentrate pellets.
2. On the second day, provide all the hay they will eat without waste and 0.75 lb (0.34 kg) of pellets.
3. Increase pellets 0.25 lb (0.11 kg) per lamb per day for 1 week. Then feed pellets free choice.
4. By blending for 2 days, change to 45% concentrate pellets.
5. Change to 55% concentrate pellets the third week.
6. Change to 60% concentrate pellets the fourth week. Higher concentrates in pellets may be fed, but it is doubtful if they are needed. Research at the Missouri Experiment Station by Rea, Ross, and Pfander (1963) and Moose, Ross, and Pfander (1970) demonstrated that temperature affected the feed utilization of pelleted rations. They demonstrated that in cold weather high-roughage mixtures were more satisfactory than they were in a warm environment. Under practical conditions, if high-concentrate diets are desirable, it is best to use them in warm temperatures.

The advantages of pellet feeding are:

1. Feed consumption is usually higher than with any other system.
2. Faster, more efficient gains than other methods.
3. It is possible to incorporate feed additives, unpalatable feeds, or low-quality roughage.
4. Pellets are easily handled and will store well.
5. Selectivity of feed by lambs is not a problem.
6. Level of skill to finish lambs on pellets is relatively low.
7. Feeding pellets may be highly mechanized with large units.

The disadvantages are:

1. Pelleting equipment is very expensive.
2. Pelleting is the most expensive method of preparing feed for lambs.
3. Much care in adjusting feeders is required to prevent feed wastage.

12.7.6. Feeding concentrates on pasture Lambs can be finished on concentrates on pasture. It works best if the pasture is a high-quality grass–legume mixture and weather conditions are satisfactory. The lambs are usually self-fed on concentrates, but they can be hand fed if it is desirable. The requirements are self-feeders or grain troughs, watering facilities, mineral feeders, shades or shelters, and a nutritious, palatable, safe pasture. Lambs must be vaccinated.

Expected Performance. Consumption of concentrates, 1.67% of average body weight daily per lamb; average daily gain per lamb, 0.36 to 0.40 lb (0.16 to 0.18 kg); feed efficiency, 4.6 lb (2.1 kg) of concentrates per pound of gain; and death loss, 1.5%.

The procedure is as follows:

1. Start lambs in drylot by hand feeding concentrates and hay as though they were to be self-fed concentrates in drylot. When lambs are consuming 2 to 2.25 lb (0.9 to 1.0 kg) of concentrates per head daily and are full, move to pasture.
2. Observe lambs carefully and adjust feeders if they begin wasting feed. Sometimes the lambs will root out the grain to get to the protein supplement. It may be necessary to feed the supplement each day in a grain trough.
3. Gradually decrease the hay for a week or more and discontinue.
4. Make sure that it is not possible for concentrates to get wet.
5. Remove wet, moldy, or picked over feed from the feeders.
6. Protect the lambs from predators.
7. Inspect the lambs frequently for internal and external parasites.
8. Never let feeders run dry.

The advantages of feeding grain on pasture are:

1. Hay requirements are very low.
2. Once lambs are on feed, labor requirements are low.
3. Lambs make efficient gains on grain with very little protein supplement.
4. Out of pocket expenses may be very low.

The disadvantages of feeding grain on pasture are:

1. Getting lambs to eat concentrates requires time and skill.
2. Gains are medium to low.
3. Predators may be a problem.
4. If rain is excessive, lambs do not eat well and parasites may be a problem.

12.8 LAMB FEEDING BUDGET

It is difficult to imagine how a lamb feeder can run a successful operation without a very intimate knowledge of the expenses as well as the possible returns that may be obtained. Very few bankers or other money lenders will seriously consider risking capital unless the applicant can produce a budget showing the prospective costs and returns. If budgets are made and records are kept and analyzed, it is possible to learn much about the business in a relatively short time. The budget should contain the information shown in Figure 12.7.

One of the best ways to achieve an understanding of lamb feeding is to study the results of past years and to estimate what will happen during a new feeding year. If each item in the budget is estimated and later the actual figures are obtained to compare with estimates, most lamb feeders soon improve in their predictions. Once a lamb finishing operation is underway, performance records are invaluable in building the budget and summarizing the results of the venture.

Investment	Value
Buildings	
Fences, corrals, working facilities	
Equipment (feeders, etc.)	
Land	
Machinery and trucks	
Total	

Direct costs	Amount	Cost/Unit	Estimate of Cost	Actual Cost
Feed				
Grain				
Hay				
Supplements				
Feeder lambs				
Fuel and oil				
Utilities				
Miscellaneous supplies				
Labor				
Veterinary and drugs				
Shearing				
Marketing (including transportation)				

	Estimated Expenses	Actual Expenses
Operating capital interest		
Supplies including fuel oil		
Personal property taxes		
Total direct costs		

Figure 12.7 Lamb feeding budget.

12.9 MEASURING PERFORMANCE IN LAMB FINISHING

Lamb finishing is competitive, and in every aspect it requires skill and good judgment for continued success. The following are efficiency factors commonly used in evaluating feeding operations:

1. Total gain (lb or kg)
2. Average days on feed
3. Average daily gain (lb or kg)
4. Average daily feed (lb or kg)
5. Feed per pound of gain (lb or kg)

Fixed costs	Estimated Expenses	Actual Expenses
Depreciation		
Buildings (25 years)		
Equipment (15 years)		
Machinery (10 years)		
Interest		
Buildings		
Equipment		
Machinery		
Land		
Taxes		
Buildings		
Equipment		
Machinery		
Land		
Repairs		
Buildings		
Equipment		
Machinery		
Insurance		
Buildings		
Equipment		
Machinery		
Total fixed costs		
Total costs (direct and fixed)		

*Sales	Estimated Income	Actual Income
Finished lambs		
Wool		
Wool incentive		
Total sales		

*Sales minus costs = profit, or it may be expressed as returns for labor and management of the operator.

Figure 12.7 Continued

6. Cost of feed daily ($)
7. Other expenses daily ($)
8. Cost per pound (kg) of gain ($)
9. Profit per head ($)

To obtain this information, it is necessary to keep records and summarize the information. The information should be reviewed and used in making future decisions.

TABLE 12.1 SUMMARY OF LAMB FEEDING OPERATIONS FOR 19___

	Group 1		Group 2	
	Estimate	Actual	Estimate	Actual
Description of lambs				
Number of lambs[a]				
Purchase price/cwt ($)				
Average purchase weight (lb[b] or kg)				
Death loss (%)[c]				
Average sale weight (lb or kg)				
Average sale price/cwt ($)				
Average gain/lamb (lb or kg)				
Average days on feed				
Average daily gain (lb or kg)				
Average feed/lamb/day (lb or kg)				
Feed/lb gain (lb or kg)				
Cost of feed/lamb/day ($)				
Other costs/lamb/day ($)				
Cost/lb (kg) gain ($)				
Average cost/lamb ($)				
Average sales/lamb ($)				
Average profit or loss/lamb ($)				

[a]Number of lambs paid for. Some lamb feeders reduce the number by the death loss but increase the price to cover the loss.

[b]This is the actual pay weight.

[c]The group must be charged for lambs lost either by increasing the purchase price to account for them or by using the number purchased to figure the final averages.

It is understood that it is unrealistic to expect a feedlot operator to keep elaborate records, but certain basic records are vital to running an operation. Otherwise, there is little reason to expect any improvement. If it is at all possible, it is valuable to feed types of lambs in groups. For example, Western lambs should be fed in a group and Natives in another. Lambs fed in the fall, winter, and spring should be recorded separately.

Table 12.1 shows the information needed to obtain efficiency factors for evaluating the feedlot.

More can be learned about the different kinds of lambs, the season lambs are fed, or other variables if two columns are provided for each group. Then the manager can estimate the performance of the lambs and fill in the actual figures when the group has been sold.

12.10 GOOD PRACTICES IN LAMB FEEDING OR HOW TO HAVE GOOD LUCK IN LAMB FEEDING

Many successful people will tell you that there is no such thing as good luck if you sit on your hands and hope. They believe that good luck happens most often to the person who does more things right than the rest of the pack. If you violate principles, you may expect to have bad luck.

The following are some practices that, if followed, should be helpful in building an efficient and profitable lamb feeding operation.

1. Learn how to do a good job of purchasing feeder lambs. Understand weighing conditions, shrink, transportation, and purchasing costs. Know what can be done with the various types of lambs and how they could fit your conditions.
2. Handle lambs properly on arrival and during the transition period.
3. Establish and maintain a sick pen and, most important of all, use it to advantage.
4. Vaccinate lambs as needed.
5. Shear lambs if they have 5 lb (2.2 kg) or more fleece or if conditions indicate it should be done.
6. Sort lambs into uniform groups (called shaping up) for feeding. This means lambs of the same size and condition and with a similar background.
7. Discover ways to cheapen up feeder lambs such as pasture, silage, or other inexpensive feed.
8. Feed balanced rations.
9. After the lambs have been on feed for 3 to 4 weeks, go through and sort off those that are doing poorly (called tailing out) and dispose of them. If they are not doing the job by then, they probably will not.
10. Run a low-key, comfortable, sanitary, efficient feedlot. Establish standard procedures and follow them. Be regular and dependable and teach the help to be the same way.
11. Watch the lambs carefully and sort off the fats (called topping out) when they are ready for market.
12. If you produce high-quality lambs, see that they get to market in good shape and that you sell them for what they are worth. Try to find out how your lambs do at slaughter.
13. Keep records on every group of lambs you feed with a description, what they cost, how they gained, cost of gains, selling price, and results of the venture. You will learn more if you estimate how they will do and write it down with a column for estimates and one for actual performance.
14. Evaluate your operation each year and ask yourself if you have some unique and outstanding advantage over your competitors. Try to discover one or more. If you can, your feedlot will prosper.
15. Keep up with new developments by joining a feedlot association. Work with your agriculture extension personnel.

REFERENCES

Botkin, M. P., R. A. Field, and C. LeRoy Johnson: *Sheep and Wool:* Englewod Cliffs, N.J.: Prentice Hall, Inc., 1988.

Collins, S. B., *Profitable Sheep.* New York: Macmillan, Inc., Company, 1956.

Ensminger, M. E., *Sheep and Goat Science,* 5th ed. Danville, Ill.: Interstate Printers and Publishers, 1986.

Kammlade, W. G., Sr., and W. G. Kammlade, Jr., *Sheep Science.* Philadelphia: J. Lippincott Co., 1947.

Kortan, L. J., *Lamb Feeding Guide,* S. Dakota S. U. Coop Ext. Civ. E. C. 634.

Moose, M. G., C. V. Ross, and W. H. Pfander, "Nutritional and Environmental Relationships with Lambs," *J. Anim. Sci.* 29, No. 4, 1970, pp. 617–627.

Rea, J. C., C. V. Ross, and W. H. Pfander. "Studies of Factors Affecting Performance of Growing Finishing Lambs," *University of Missouri Research Bulletin 827,* 1963.

Ross, C. V., and R. L. Pavey, "Concentrate Roughage Ratios, Pelleted Feed; Hexestrol for Fattening Lambs," *Missouri Agr. Exp. Sta. Bull. 773,* 1959.

Ross, James E., *Lamb Feeding,* University of Missouri, Columbia, Sci. and Tech. Guide Sheet 2625, 1968.

Scott, George, *The Sheepman's Production Handbook.* Denver, Colo.: Sheep Industry Development Program, Inc., 1982.

QUESTIONS FOR STUDY AND DISCUSSION

1. Why are lambs finished in a feedlot?
2. What appears to control the location of lamb feedlots in the United States?
3. In what ways does lamb finishing differ from that of cattle?
4. What are legal considerations before establishing a lamb feedlot?
5. Why is the type of construction important in the maintenance of lamb shelters in the feedlot?
6. While it is true feed accounts for the greatest percentage of the expenses of lamb feeding, what others are important?
7. What is meant by margin and spread? How is it used by lamb feeders?
8. What is meant by "cheapening up" feeder lambs, and how is it done?
9. What is meant by the "law of the outstanding advantage"?
10. Why is shrinkage important to the purchase of feeder lambs?
11. What are the dangers of purchasing light feeder lambs (under 65 lb or 29.5 kg)?
12. What is meant by pencil shrink?
13. What are the advantages of Northwest feeder lambs over Natives?
14. Outline a program for handling "green" feeder lambs.
15. From a physiological standpoint, what occurs in a lamb when it is put on feed?
16. What are the ironclad rules in self-feeding concentrates to lambs?
17. Which of the lamb feeding methods would fit the following?
    a. Requires the least equipment.
    b. Requires the most hand labor.
    c. Is the most efficient in feed per pound gain.
    d. Is best suited to large operations.
    e. Requires the least skill.
    f. May be too expensive because of feed preparation.
18. What is meant by the following feedlot terms?
    a. Shaping up
    b. Topping out
    c. Tailing out

    d. Scorching
    e. Shrink
    f. Fill

19. What are the differences in direct costs and fixed costs?
20. Why is a budget needed for lamb feeding?
21. What are the efficiency factors in lamb finishing?
22. Why estimate how lambs will perform when they are put in the feedlot?

13 Lamb and Sheep Marketing and Slaughter

13.1 INTRODUCTION

For a sheep operation to be successful, it must be profitable. Not only must salable products be produced in volume and with efficiency at reasonable costs, but they must be sold at a fair price. Marketing is thus the last step in most operations, but it may be the most important of all. It is at the marketplace that the producer realizes the rewards for the efforts made.

How well a sheepraiser is able to cope with the constraints of marketing may very well be the difference between success and failure. Thus, a knowledge of market alternatives and the mechanisms of marketing should be of great value to the producer. A continuous study of the subject is needed, because the industry has made great changes in the recent past and may be expected to continue to do so in the future.

Radical changes in any industry may be disastrous for people who only live in the present and are not adaptable to different conditions. On the other hand, the same situation may provide the golden opportunity for the innovator who understands the principles involved and uses them to advantage. Time spent studying markets and marketing will usually pay big dividends.

13.2 SHEEP POPULATION IN THE UNITED STATES

Sheep are unevenly distributed throughout the United States. Most are found in the western and northern parts of the country. The leading states in numbers in 1987 are shown in Table 13.1. Texas was the leading state with 18% of the total. California, Wyoming, Colorado, South Dakota, Montana, New Mexico, Utah, and Oregon each had more than 400,000 stock sheep on farms on January 1, 1987. These are all western states with extensive ranges. The 17 western states in recent years have accounted for 76% of the lamb production in the United States. Farm states such as Iowa, Minnesota, and Ohio also are important sheep states. Many other states have the potential for greater production. Production and income from sheep by states

TABLE 13.1 STOCK SHEEP AND LAMBS ON FARMS IN THE UNITED STATES ON JANUARY 1, 1987, BY STATES

State	Stock sheep and lambs	Ewes (one year or over)	Lambs on feed	All sheep
	1000 head			
Texas	1,780	1,300	150	1,930
California	795	685	185	980
Wyoming	690	530	86	775
Colorado	380	300	310	690
South Dakota	550	434	55	605
Montana	500	370	23	523
New Mexico	450	342	30	480
Utah	440	375	24	464
Oregon	320	255	95	415
Iowa	310	255	65	375
Idaho	299	245	15	314
Ohio	245	200	55	300
Arizona	219	144	64	283
Minnesota	165	130	72	237
Kansas	166	145	70	236
North Dakota	160	120	25	185
Nebraska	120	94	53	173
Virginia	120	98	48	168
Illinois	109	79	10	119
Missouri	102	80	8	110
Michigan	85	68	21	106
Oklahoma	80	60	25	105
Pennsylvania	104	73		104
Indiana	82	63	9	91
West Virginia	90	68		90
Nevada	86	69		86
Wisconsin	76	53	7	83
Total (all states)	8,823	6,840	1,504	10,327

Source: Livestock and Meat Statistics. USDA ERS Statistical Reporting Service, Agriculture Marketing Service. Statistical Bulletins 522, 715, and Lv Gn (2/87).

is shown in Table 13.2. Production and income are closely related to the number of sheep on farms.

13.3 CHARACTERISTICS OF THE SHEEP INDUSTRY IN THE UNITED STATES

13.3.1. Changes in numbers on farms Perhaps the outstanding characteristic in the history of sheep production in the United States has been the great fluctuations in numbers on farms and ranches since 1867, as shown in Figure 13.1. The most striking is the change that has occurred since World War II. Since the high point in 1942 of approximately 56 million head, there has been a steady decline to 11.5 million in 1982. There appeared to be a recovery in 1983, but there were record low

TABLE 13.2 PRODUCTION AND INCOME FROM SHEEP AND LAMBS BY LEADING STATES, 1985

State	Production[a] lb (000)	Production[a] kg (000)	Marketings[b] lb (000)	Marketings[b] kg (000)	Cash receipts[c]	Value of home consumption	Gross income
					1000 dollars		
Texas	103,150	46,885	135,670	61,668	78,164	313	78,477
California	73,145	33,247	82,708	37,594	51,771	2,405	54,176
Colorado	49,439	22,472	82,662	37,576	49,539	166	49,705
South Dakota	51,499	23,408	68,751	31,250	33,658	286	41,038
Wyoming	50,432	22,924	58,347	26,521	31,962	331	32,293
Utah	37,956	17,253	41,949	19,068	24,551	388	24,939
Iowa	30,487	13,858	37,731	17,150	24,303	65	24,368
Montana	31,661	14,391	46,895	21,316	23,129	220	23,349
Idaho	32,500	14,772	32,953	14,979	20,315	412	20,727
Oregon	27,240	12,382	28,778	13,080	17,221	379	17,600
Minnesota	19,730	8,968	28,180	12,809	16,387	241	16,628
New Mexico	21,577	9,807	25,090	11,404	13,341	1,396	14,737
Kansas	16,703	7,592	23,654	10,752	14,199	189	14,388
Nebraska	11,070	5,032	18,964	8,620	11,325	119	11,444
North Dakota	14,284	6,493	19,975	9,080	10,835	144	10,979
Ohio	17,205	7,820	16,255	7,389	10,203	416	10,619
Illinois	9,995	4,543	15,065	6,848	7,400	885	8,285
Arizona	10,910	4,959	18,795	8,543	6,885	931	7,816
Nevada	7,085	3,220	12,780	5,809	6,461	123	6,584
Missouri	8,035	3,652	10,746	4,884	5,781	147	5,928
Michigan	8,674	3,934	8,799	4,000	4,920	221	5,141
Indiana	7,775	3,534	8,488	3,858	4,430	354	4,784
Wisconsin	6,251	2,841	6,697	4,885	4,383	146	4,529
Virginia	8,785	3,993	8,625	3,920	4,256	127	4,383
Oklahoma	5,476	2,489	6,076	2,762	3,563	250	3,813
US Total	691,230	314,195	872,602	396,637	502,807	12,768	515,575

Source: Livestock and Meat Statistics. USDA ERS Statistical Reporting Service, Agriculture Marketing Service. Statistical Bulletins 522, 715, and Lv Gn (2/87).

[a]Adjustments made for changes in inventory and for in shipments.

[b]Excludes custom slaughter for use on farms where produced and interfarm sales within the states.

[c]Receipts from marketings and sale of farm slaughter.

numbers in each of 1984, 1985, and 1986. After reaching 9.9 million head, numbers recovered to approximately 10.4 million in 1986.

13.3.2. Value of production of lambs and wool Despite the drastic reduction in numbers on farms, the income from lambs and wool produced in 1983 was $429 million and $61.5 million, respectively. During the period from 1950 to 1984, value of wool production to the sheep operation declined from 40%, to 12.5% not including wool incentive payments.

13.3.3. Per capita consumption of lamb and wool Consumption of lamb and mutton has decreased from 6.7 lb (3 kg) during the period from 1931 to 1940

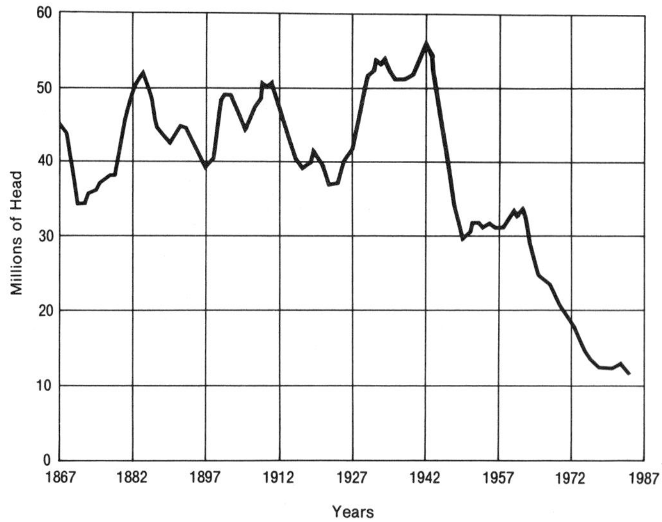

Figure 13.1 Stock sheep on farms 1867–1987. [Source: *Livestock and Meat Statistics.* USDA ERS Statistical Reporting Service, Agriculture Marketing Service. Statistical Bulletins 522, 715, and Lv Gn (2/87)]

to 1.4 lb (0.64 kg) in the 1980s. Americans eat all but 4 million lb (1786 metric tonnes) of the lamb and mutton produced and import 36 million lb (16,071 metric tonnes). Lamb and mutton are consumed in greatest amounts in urban areas of more than 1 million population. The northeast area of the United States consumed 43%, the West Coast, 21%, the Southwest, 19%, and the Central states, 17% in 1980.

Data on wool production, imports, and consumption indicate that most of the wool produced in the United States is consumed domestically along with imports of wool and wool products. The United States imports approximately the same amount of raw wool as it produces (56 million lb; 25,000 metric tonnes), plus the equivalent of 103 million lb (45,982 metric tonnes) as manufactured wool products such as woven fabrics, blankets, wearing apparel, carpets, and rugs. Approximately 14 million lb (6250 metric tonnes) are exported. In total, consumption in the United States is approximately 1 lb (0.45 kg) per capita. Wool represents only 2% of the fiber consumption, compared with 27% for cotton and 71% for synthetics.

13.3.4. Lamb feeding The peak in lamb feeding in the United States occurred in 1945, with 6,911,000 in feedlots on January 1 of that year. The trend in numbers of lambs fed may be observed in Figure 13.2. There has been a general decrease from 1955 until 1982, and then a slight increase during 1983 and 1984. Only California and Oregon had an increase in lambs on feed from 1970 to 1987. Most states fed less lambs as the available supply continued to decrease. At least 87% of lambs on feed are located in states west of the Mississippi River. Figure 13.3 shows that there has also been a change in the location of lamb feeding. Colorado was the leading lamb feeding state in 1970. It still ranks at the top of all states, but volume has decreased by 30%.

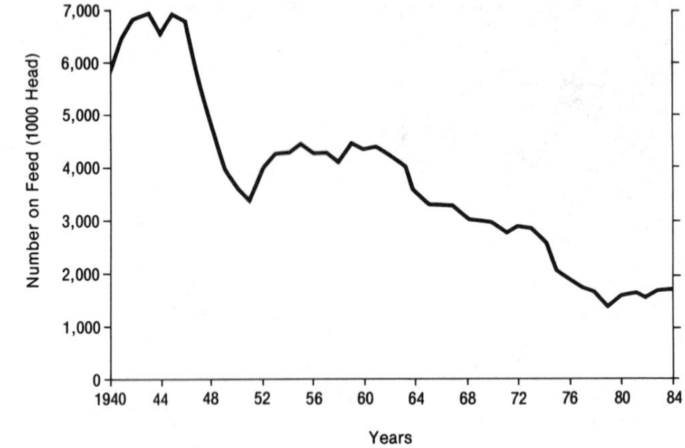

Figure 13.2 Numbers of lambs on feed January 1 in the United States from 1940–1984.

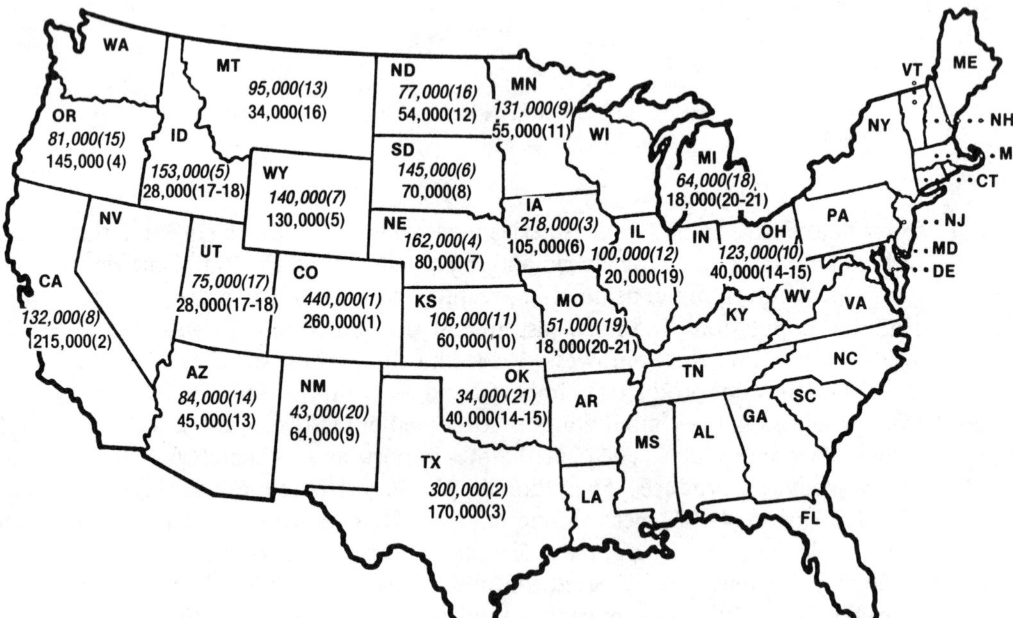

Figure 13.3 Sheep and lambs on feed January 1 and rank in the United States. Top numbers show number and rank in 1970; lower numbers show rank and number in 1987.

13.3.5. Packer feeding Many packers who are attempting to stay in business have turned to feeding lambs so that they have some control over the supply available for slaughter. The extent of packer feeding and trend in numbers is indicated in Table 13.3. The data show that the proportion fed varied from a high of 18.3% in 1973 to a low of 1.8% during 1982. Table 13.4A and B indicates the numbers of lambs fed in leading lamb feeding states and the extent of packer feeding. Colorado led all other states in packer feeding of lambs in 1985. They accounted for 47.9% of

TABLE 13.3 NUMBER OF FEDERALLY INSPECTED LAMB AND YEARLINGS SLAUGHTERED AND NUMBER OF LAMBS FED BY REPORTING SLAUGHTER PACKERS, 1955–1985 REPORTING YEARS

Year	Federally inspected lamb and yearling slaughter[a] (1000 head)	Packer feeding[b]		
		Feeding packers (no.)	Number (1000 head)	Percent of slaughter
1955	13,519	26	533	3.9
1956	13,314	16	724	5.4
1957	12,491	19	667	5.3
1958	11,678	23	713	6.1
1959	12,808	24	702	5.5
1960	13,159	24	826	6.3
1961	14,169	27	606	4.3
1962	13,762	26	551	4.0
1963	13,171	29	642	4.9
1964	12,230	29	867	7.1
1965	11,064	20	826	7.5
1966	10,795	17	1,296	12.0
1967	10,735	12	926	8.6
1968	10,198	16	952	9.3
1969	9,340	15	1,272	13.6
1970	9,298	13	1,234	13.3
1971	9,437	12	919	9.7
1972	9,227	9	1,440	15.2
1973	8,426	8	1,491	18.3
1974	7,987	9	1,334	16.6
1975	6,993	7	1,009	14.5
1976	6,058	7	1,064	17.3
1977	5,643	8	701	12.2
1978	4,810	8	755	15.5
1979	4,499	10	677	14.6
1980	4,970	8	504	10.1
1981	5,388	5	228	4.2
1982	5,820	5	103	1.8
1983	5,933	4	334	5.6
1984	6,035	8	300	5.0
1985	5,976	7	493	8.2

[a]U.S. Department of Agriculture, *Livestock and Meat Statistics Annual Supplement,* Economic Research Service, Statistical Bulletin 715.

[b]Includes lambs and sheep fed by or for meat packers and transferred from the feedlot for slaughter during the reporting year. Separate feeding activities by owners, officers, employees of meat packers, or nonreporting subsidiaries or affiliates are not included.

the lambs fed by packers. The numbers fluctuate radically from year to year. In 1982, Colorado only fed 16.5%, compared with 27.2% for Wyoming and 22.3% for California. A somewhat different picture is provided in Table 13.5A and B. On the basis of total lambs marketed in Colorado, packers fed 25.1% in 1983, while Wyoming fed 12%, Utah 6.9%, and Texas 4.1%. In 1984, they were respectively 25%, 0%,

TABLE 13.4A NUMBER OF LAMBS FED BY REPORTING SLAUGHTER PACKERS, SELECTED STATES, 1981–83 REPORTING YEARS[a]

	1981		1982		1983	
Feedlot locations[b]	1000 head	Pct. of total	1000 head	Pct. of total	1000 head	Pct. of total
Colorado	42	18.4	17	16.5	198	59.3
Wyoming	58	25.4	4	3.9	53	15.9
Texas	58	25.4	28	27.2	36	10.8
Utah	21	9.2	2	1.9	24	7.2
Idaho	0	0.0	15	14.6	10	3.0
California	0	0	23	22.3	7	2.1
Subtotal	179	78.5	89	86.4	328	98.3
Others	49	21.5	14	13.6	6	1.7
Total	228	100.0	103	100.0	334	100.0

[a]Packer feeding includes sheep and lambs fed by or for meat packers and transferred from the feedlot for slaughter during the reporting year. Separate feeding activities by owners, officers, employees of meat packers, or nonreporting subsidiaries or affiliates are not included.

[b]Location of feedlot but not necessarily slaughtered in that state. Ranked according to 1983 feeding volume.

TABLE 13.4B NUMBER OF LAMBS FED BY REPORTING SLAUGHTER PACKERS, SELECTED STATES, 1983–85 REPORTING YEARS[a]

	1983		1984		1985	
Feedlot locations[b]	1000 head	Pct. of total	1000 head	Pct. of total	1000 head	Pct. of total
Colorado	198	59.3	141	47.0	236	47.9
Texas	36	10.8	72	24.0	155	31.4
California	7	2.1	25	8.3	38	7.7
Wyoming	53	15.9	0	0.0	29	5.9
Minnesota	0	0.0	0	0.0	16	3.3
Utah	24	7.2	19	6.3	15	3.0
Subtotal	318	95.3	257	85.6	489	99.2
Others	16	4.7	43	14.4	4	0.8
Total	334	100.0	300	100.0	493	100.0

[a]Packer feeding includes sheep and lambs fed by or for meat packers and transferred from the feedlot for slaughter during the reporting year. Separate feeding activities by owners, officers, employees of meat packers, or nonreporting subsidiaries or affiliates are not included.

[b]Location of feedlot but not necessarily slaughtered in that state. Ranked according to 1985 feeding volume.

5.7%, and 7.3% for the same four states. But in 1985 packers fed 70% more lambs and Colorado had fed 41% of the lambs marketed in the state, Texas increased to 17.9%, Wyoming had 5.9%, and Utah 4.6%. The only other states where packers fed more than 1% of marketings were California with 6% and Nebraska with 13%.

TABLE 13.5A TOTAL LAMB MARKETINGS AND NUMBER OF LAMBS FED BY REPORTING SLAUGHTER PACKERS IN MAJOR STATES, 1981–83 REPORTING YEARS

Feedlot location[a]	1981			1982			1983		
	Total lamb marketings by farmers and feeders[b] (1000 head)	Packer feeding[c]		Total lamb marketings by farmers and feeders[b] (1000 head)	Packer feeding[c]		Total lamb marketings by farmers and feeders[b] (1000 head)	Packer feeding[c]	
		Number (1000 head)	Percent of marketings		Number (1000 head)	Percent of marketings		Number (1000 head	Percent of marketings
South Dakota	538	4	0.7	492	7	1.4	480	0	0.0
Iowa	244	0	0.0	291	0	0.0	311	0	0.0
Nebraska	176	5	2.8	171	0	0.0	194	4	2.1
Kansas	172	2	1.2	177	0	0.0	170	0	0.0
Texas	879	58	6.6	955	28	2.9	889	36	4.1
Wyoming	440	58	13.2	457	4	0.9	443	53	12.0
Colorado	744	42	5.6	725	17	2.3	788	198	25.1
New Mexico	308	9	2.9	223	4	1.8	224	1	0.5
Idaho	419	0	0.0	427	15	3.5	382	10	2.6
Utah	359	21	5.8	340	2	0.6	346	24	6.9
Nevada	199	28	14.1	221	0	0.0	241	0	0.0
California	685	0	0.0	820	23	2.8	634	7	1.1
Subtotal	5,163	227	4.4	5,299	100	1.9	5,102	333	6.5
Others	1,938	1	0.1	2,083	3	0.1	1,882	1	0.0
Total	7,101	228	3.2	7,382	103	1.4	6,984	334	4.8

[a]Location of feedlot, but not necessarily slaughtered in that state.

[b]Marketings of lambs include interstate sales but exclude custom slaughter for use on farms where produced and interfarm sales within the same state. U.S. Department of Agriculture, *Livestock and Meat Statistics Annual Supplement,* Economic Research Service, Statistical Bulletin 715.

[c]Includes lambs (and sheep) fed for or by meat packers and transferred from the feedlot for slaughter during the reporting year. Separate feeding activities by owners, officers, employees of meat packers, or nonreporting subsidiaries are not included.

TABLE 13.5B TOTAL LAMB MARKETINGS AND NUMBER OF LAMBS FED BY REPORTING SLAUGHTER PACKERS IN MAJOR STATES, 1983–85 REPORTING YEARS

Feedlot location[a]	1983 Total lamb marketings by farmers and feeders[b] (1000 head)	1983 Packer feeding[c] Number (1000 head)	1983 Packer feeding[c] Percent of marketings	1984 Total lamb marketings by farmers and feeders[b] (1000 head)	1984 Packer feeding[c] Number (1000 head)	1984 Packer feeding[c] Percent of marketings	1985 Total lamb marketings by farmers and feeders[b] (1000 head)	1985 Packer feeding[c] Number (1000 head	1985 Packer feeding[c] Percent of marketings
South Dakota	480	0	0.0	502	1	0.2	498	1	0.2
Iowa	311	0	0.0	320	0	0.0	322	0	0.0
Nebraska	194	4	2.1	188	9	4.8	154	2	1.3
Kansas	170	0	0.0	155	2	1.3	185	0	0.0
Texas	889	36	4.1	990	72	7.3	865	155	17.9
Wyoming	443	53	12.0	427	0	0.0	490	29	5.9
Colorado	788	198	25.1	563	141	25.0	575	236	41.0
New Mexico	224	1	0.5	264	0	0.0	172	0	0.0
Idaho	382	10	2.6	326	0	0.0	275	0	0.0
Utah	346	24	6.9	336	19	5.7	324	15	4.6
Nevada	241	0	0.0	215	11	5.1	92	0	0.0
California	634	7	1.1	737	25	3.4	629	38	6.0
Subtotal	5,102	333	6.5	5,023	280	5.6	4,581	476	10.4
Others	1,882	1	0.0	1,984	20	1.0	1,875	17	0.9
Total	6,984	334	4.8	7,007	300	4.3	6,456	493	7.6

[a]Location of feedlot, but not necessarily slaughtered in that state.

[b]Marketings of lambs include interstate sales but exclude custom slaughter for use on farms where produced and interfarm sales within the same state. U.S. Department of Agriculture, *Livestock and Meat Statistics Annual Supplement,* Economic Research Service, Statistical Bulletin 715.

[c]Includes lambs (and sheep) fed for or by meat packers and transferred from the feedlot for slaughter during the reporting year. Separate feeding activities by owners, officers, employees of meat packers, or nonreporting subsidiaries are not included.

There has been much discussion on the appropriateness of packers feeding not only lambs, but also other species of livestock. Some sheepraisers claim it gives the packer an unfair advantage, and the packers usually maintain that they need a cushion in order to keep their slaughtering crew busy. Until there is a much greater volume of lambs available, this question is not likely to be settled.

13.4 LAMB PRODUCTION IN THE UNITED STATES

Climatic conditions have much to do with the dates ewes are bred and lamb and with the methods of growing and finishing lambs. In the farm states, most lamb production is in small units. Many are one-ram flocks of 25 to 30 ewes, and the enterprise may be a minor one on the farm. In contrast, the flocks generally are larger in the western part of the country and may be the primary enterprise. In many of the more successful farm flocks, the ewes are bred to lamb early in the year, and the lambs are creep fed and pushed to reach the May and June markets. The western range operator must consider weather conditions and may lamb somewhat later. Producers in California, Arizona, and Texas have more options and may lamb out ewes in the fall or early winter or can even choose to produce spring lambs. While it is true that the greater proportion of lambs in the United States are born in the spring, there is some variation in the dates of birth and readiness for market of lambs due to climatic conditions or to the availability of buildings to shelter the newborn lambs.

Lambs that are grazed on the range during the summer may or may not have sufficient finish for slaughter by the fall. Generally, there are fat lambs and some that are too thin and small for slaughter that are sold as feeders. The feeder lambs may be grazed or fed roughage to grow and develop them, and then finished in the feedlot on concentrate and roughage.

Figure 13.4 shows a lambing and marketing schedule that demonstrates the difference in lambing and marketing dates. It can be seen that lambs from California, Arizona, and Texas are marketed earliest, followed by the farm states and then the mountain states. Then, fed lambs play a role in distributing the lamb supply over a year. This is most important to the slaughtering plants so that they may keep their crews occupied, and it allows the retailer to have a continuous supply of meat. But even with the natural flow being distributed throughout the year, there are serious voids that must somehow be filled.

13.5 SHEEP AND LAMB MARKETS

13.5.1 Public markets There has been a continuous decrease in the number of public markets for sheep and lambs. The largest markets now in operation are shown in Table 13.6. The 10 markets listed in the table accounted for 78.9% of all sheep and lambs sold in public markets in the United States in 1985. The marketing included stock sheep and feeders, as well as slaughter lambs. San Angelo, Texas, is the largest public market in the United States. It is much larger than South St. Paul, Minnesota, which is second. Billings, Montana, is third, and Sioux Falls, South Dakota, is fourth.

The number of terminal markets has declined steadily since 1937. In 1984, there were only 24 selling livestock in the United States. Few of them have facilities for

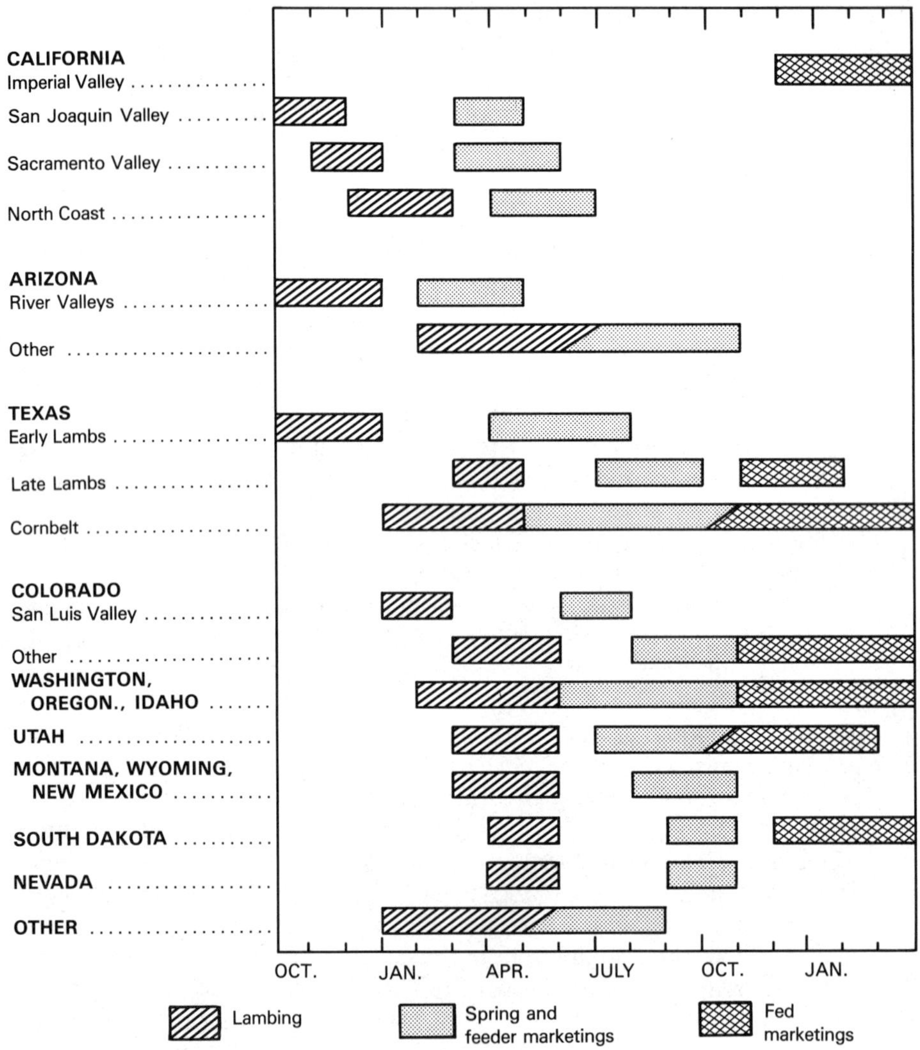

Figure 13.4 Lambing and marketing schedule in the United States (18-month cycle).

selling lambs. Many terminals that handle lambs are only open once or twice a week. On public markets in 1985, only 515,000 lambs were sold through terminals, compared with 643,000 through auctions. Historically, terminal markets for sheep and lambs have been far less popular than they have for hogs and cattle. Ten of the 24 terminals still in existence are located in the west north central region of the United States.

13.5.2. Direct marketing As shown in Table 13.7, in 1985 the Packers and Stockyards Administration reported that 80.4% of the slaughter sheep and lambs were purchased direct or through country dealers. Only 10.9% were bought through auctions and 8.7% on terminal markets. Since the number of packers has decreased

TABLE 13.6 SHEEP AND LAMBS SOLD AT THE 10 LEADING PUBLIC MARKETS IN 1980 AND 1986

Market	1980	Rank	1986	Rank
San Angelo, Tex.	642,686	1	429,353	1
Sioux Falls, S.D.	158,980	2	56,809	4
So. St. Paul, Minn.	146,247	3	87,287	2
Billings, Mont.	70,848	4	58,756	3
Norfolk, Neb.	—	14	46,502	5
West Fargo, N.D.	63,863	5	25,217	7
Lancaster, Pa.	7,773	12	30,984	6
Mexico, Mo.	24,280	7	19,046	8
Omaha, Neb.	52,168	6	13,938	10
National Stock Yards, Ill.	16,081	13	15,190	9

Source: Livestock and Meat Statistics Statistical Bulletin 522—Supplement for 1980. ERS Statistical Service, USDA and *Livestock and Meat Statistics 1983*, Economic Research Service Statistical Bulletin 715.

so radically, there are relatively few bidders for sheep and lambs in many areas of the country. This is a difficult situation, and in many locations there is very little competition for lambs.

13.5.3. Cooperative marketing In 1975, cooperatives handled about 32% of sheep and lambs sold at terminal markets and 11% at auctions. Some areas have pooled slaughter lambs and sold them on a dealer basis, but this amounted to only a small part of the total. Some cooperatives in the Corn Belt have acted as order buyers to obtain feeder lambs and breeding ewes, but numbers have been small.

A more recent method of selling cooperatively is through a system originally called tele-auction, now called electronic marketing. It was first instituted by the Eastern Lamb Producers Cooperative in the Virginia–West Virginia area in 1971. Another was organized by a cooperative in Oregon–Idaho in 1974. The Rocky Mountain Sheep Marketing Company was established in 1975 to serve the marketing needs

TABLE 13.7 SOURCES OF PACKER PURCHASES OF SHEEP AND LAMBS, 1976, 1978, and 1985

Method of Purchase	1976 No. (1000 head)	1976 %	1978 No. (1000 head)	1978 %	1980 No. (1000 head)	1980 %	1985 No. (1000 head)	1985 %
Direct purchases (including country dealers, etc.)	4925	75.2	4015	78.6	4335	79.6	4750	80.4
Terminal markets	642	9.8	516	10.1	441	7.5	515	8.7
Auction markets	980	15.0	578	11.3	703	12.9	643	10.9
Total	6548	100.0	5109	100.0	5449	100.0	5909	100.0

Source: Packers and Stockyards Statistical 1985 Reporting Year. P8 SA Statistical Report No. 85-1.

of large range producers. Others have been established in the Midwest. In a speech at the Purdue University Farm Science Days in 1982, David Holder, who is Program Leader, Livestock and Meat Marketing, Extension Services, U. S. Department of Agriculture, stated: "Electronic markets (tele-auctions and computerized sales) have proven to be successful in the East, western Corn Belt, Southwest, and Northwest where they are currently in operation. In 1981, they sold about 150,000 lambs. Each one has increased local competition and improved operating efficiency in the marketing of lambs and cull ewes."

13.6 MARKETING MECHANISMS

A market is defined as a place where buyers and sellers meet to transfer ownership. Lamb and sheep marketing may be very simple or an expensive and technical process that includes such diverse activities as assembling, transporting, sorting, buying and selling live animals, and slaughtering, processing, distributing, and merchandizing the meat and other products. When it is considered that sheep and lambs come to market from 140,000 farms and ranches and must be transported longer distances than any other species to market and slaughter, it is remarkable that the system is effective at all.

13.7 METHODS OF SELLING LAMBS

Like other farm livestock, lambs are sold in a number of ways throughout the country. Some areas prefer to use a particular method, while others would not be satisfied with that one at all. There have been great changes in marketing lambs, and there will probably be others as the sheep industry adjusts to reduced numbers being marketed and to slaughtering and merchandizing problems. In the following, methods of marketing lambs are discussed.

13.7.1. Spot sales This simply means that the price is negotiated on the spot between the buyer and seller. It is the most common method of transferring ownership between the two. It may be on the basis of live or carcass weight, by formula or by contractual arrangement. The advantages of spot sales are:

1. The price is determined quickly.
2. Ownership is transferred immediately.
3. The conditions of sale are known by both buyer and seller.
4. Marketing costs are known and paid at or soon after the transaction.
5. The seller is paid at the time of sale or soon thereafter.

13.7.2. Contract sales Sales may be made for future delivery and the agreement bound by contract. Many feeder lambs are contracted for in the summer for fall delivery. Slaughter lambs may also be sold for future delivery. The agreement usually contains date and time of delivery, number of lambs or sheep, weighing conditions, and the price or basis of pricing.

13.8 PRICE NEGOTIATIONS IN SELLING SHEEP AND LAMBS

The procedure for arriving at a satisfactory price for sheep and lambs is usually one of the following:

1. Private treaty
2. Auction
3. Terminal selling
4. Formula pricing
5. Selling on bids
6. Electronic sales

13.8.1. Private Treaty Direct negotiations between the buyer and the owner or the owner's representative have always been used extensively. It is still the basis for trading in terminal markets and in the country. A number of variations are in use. For example, on a terminal market, the producer is generally represented by a commission firm. This representative is expected to be knowledgeable about the market situation and adept in dealing with the buyers. If the producer has not had considerable experience and is not well informed on the current market, the buyer may have a decided advantage. Thus, an agent of the commission firm may be of considerable value. Also, at terminal markets, the commission firms usually handle the paperwork and pay bills such as trucking and yardage. The commission firm agents are paid by the producer on a per head or weight basis.

A variation of the function of commission agents is found in some western states where they are employed by the buyers to locate suitable lambs and sheep. Then the commission is paid by the buyer.

Many feeder lambs and breeding ewes are sold directly to the buyer on the ranch or farm. This may be on the basis of so much per 100 lb weighed or by the head.

13.8.2. Auction Auctions have been used for selling livestock in Great Britain for centuries. The method of selling was probably inherited from our European ancestors and has been used for generations in the United States. Farm, community, breed association, and co-op sales may be through auctions. Purebred breed associations have used auctions for many years, but it has only been since the 1960s that they have become an important factor in selling slaughter sheep and lambs. Many auctions, such as the one at San Angelo, Texas, are very well organized and run efficiently. They sell thousands of sheep and lambs each week. An auction, if attended by enough buyers and with sufficient numbers of sheep and lambs to justify the interest, is a very competitive method of establishing a selling price. If either of the necessary ingredients is missing, auctions may be unsatisfactory. Some of the most successful sales have been cooperatives in which lambs from many producers have been pooled for the sale. These are commonly called lamb pools. Sales expenses at auctions vary with the organization. Some charge yardage for services performed, plus a commission or per head charge for selling the sheep and lambs. Others merely levy a commission or selling fee that covers all the services performed by the market. The commission may be either as a percentage of the selling price or it may be per head.

13.8.3. Terminal market The physical facilities of a terminal market are owned by a stockyard company. Most stockyard companies provide such services as feed, water, scales, and transportation. For this, they are paid a fee, usually on a per head basis, which is collected when the animals are sold. At the terminal market, also called central markets, public stockyards, and public markets, are a group of commission firms whose function is to represent the owner in the sale of sheep and lambs. The commission firm usually consummates the sale, collects from the buyer, pays the yardage, and remits the remainder to the seller. For these services, the commission companies charge a service fee called commission. This is usually based on a percentage of the value sold. The commission rates are controlled by an organization of commission firms called the Livestock Exchange. The Exchange governs the trading practices and regulates the business hours. When receipts were large and packers were competing for lambs, terminal markets were, in most instances, a very fair and efficient method of marketing. Even though there was more competition, many packers preferred to buy at terminal markets. They claimed that the cost of purchasing was less than buying direct, because expenses were less. Now that the number of lambs has decreased, very few terminal markets can claim truthfully that they have the degree of competition necessary to arrive at fair market prices.

13.8.4. On the basis of formula This system of pricing is based on carcasses rather than the live animal. The basis of the price may be the carcass grade and weight. For example, the buyer may agree to pay $1.25/lb for all carcasses that grade choice and weigh between 50 and 55 lb (23 and 25 kg). The lambs are delivered to the plant and slaughtered. The chilled carcasses are graded and weighed, and the seller is paid on the agreed basis. There is usually some penalty assessed for overweight and low-grading carcasses. Other formulas may be on a live basis translated to the grade and yield of the lambs. In this method, lambs must grade according to terms of the agreement, and the carcass weight is used to determine the live weight. If the agreement is that the lambs will grade choice and yield 52%, the carcass weight is divided by 52% to obtain the pay weight. If the lamb carcasses weighed 56 lb (25.4 kg), the producer would be paid 56/52%, or for 107.6 lb (48.9 kg) at the current live prices, or an agreed-on figure. An alternate method is to base the price on the wholesale market quotations in the yellow sheet of the National Provisioner. There are regulations on selling by carcasses that are designed to protect the seller. These are contained in Title 9, Chapter 2, Part 201.99 of the Packers and Stockyards Act, and may be obtained from the Packers and Stockyards Administration, USDA, Washington, D.C., 20250.

13.8.5. By sealed bids Some sheep and lambs are sold by bids. The description of the animals is publicized or sent by direct mail to prospective buyers. The buyers are asked to submit bids, which may be sealed or open. At a designated closing time, the bids are assessed and the high bidder awarded the sale. An alternate method is for the seller to offer the animals at a set price and award the sale to the first one who accepts it.

13.8.6. By computer network The process begins about 10 days before a sale. The producers notify the cooperative of the number of lambs for sale. Qualified graders visit the farms and mark the lambs that meet the standards, or qualified producers may grade their own lambs and save $0.45 per head. The seller signs a consignment form committing the lambs to the sale. The cooperative summarizes the consignments and organizes the lambs into truckloads of uniform grade and weight and assigns them a lot number. On sale day, buyers are connected to the sale by a computer network and given the locations and descriptions of each truckload of lambs that will be sold. The computer asks for bids on the first load and sells to the high bidder. Each buyer has an assigned ID code for bidding. The buyer is usually allowed to choose a day within the next week for delivery. When the buyer designates a date, the sellers are informed of the place, date, and time to deliver the lambs. The buyer pays the cooperative, and they deduct the sale expenses and remit the rest to the producer.

Expenses for selling electronically vary with the cooperative. Cornbelt, which is a division of Equity Cooperative Livestock Sales, a large marketing co-op doing business in Wisconsin, Iowa, and adjacent states, has worked with producer groups in selling lambs. The following were expenses incurred in 1984 by members of the Cornbelt Electronic Market based in Baraboo, Wisconsin. The total charge was $1.95 per head broken down as follows: Cornbelt, $1.05; the loading point owner, $0.25; the coordinator, $0.20; and the grader, $0.45.

Cooperatives using the tele-auction (electronic marketing) claim the following advantages for the system:

1. Producers have a marketing system they can control and establish marketing procedures.
2. Lambs from different locations can be sold at one time, thus increasing efficiency for buyer and seller.
3. Buyers have reduced travel and expense and can compete for lambs.
4. Packers can purchase a week's supply and schedule delivery for efficient killing.
5. Sheep and lambs move quickly from the farm to slaughter with a minimum of handling and waiting in stockyards, thus decreasing shrink, death losses, and bruises.
6. Generally, there is more competitive bidding on lambs.

The success of tele-auction is very much dependent on the competence of graders and managers. The system has been highly successful in various parts of the country, and offers an attractive alternative when buying competition is lacking.

13.9 LAMB PRICE FLUCTUATIONS

Lamb prices, like those of other meat animals, are subject to seasonal and long-term changes.

13.9.1. Seasonal price changes Figure 13.5 shows the average monthly prices of slaughter lambs and sheep at Omaha, Nebraska, and feeder lambs at South St. Paul, Minnesota, from 1974 to 1983. Prices varied according to the supply available. The months of October and November were the lowest for slaughter lambs,

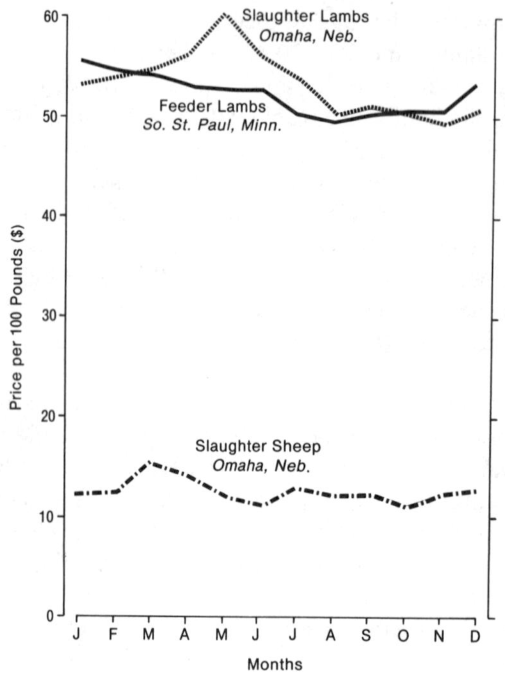

Figure 13.5 Average monthly prices of feeder lambs at South St. Paul, Minnesota, and slaughter lambs and sheep at Omaha, Nebraska, from 1974–1983. (Source: USDA, ERS Statistical Bulletin 715, *Livestock and Meat Statistics*)

and the highest prices were paid in May and June. The spread per 100 lb was $12.10 between the highest and lowest prices. Feeder lambs were the highest in January, February, and March and lowest in July, August, and September, with a spread of $6.54. Slaughter sheep prices varied less than the slaughter and feeder lambs. Highest prices were in January and lowest in June. The spread between high and low prices was $5.63/100 lb (45.4 kg). It should be said that markets for slaughter sheep are not large and information not easily obtained.

13.9.2. Long-term changes Figure 13.6 shows prices of live lambs and carcasses and value of wholesale carcasses and retail cuts from 1954 to 1981. Both live and carcass prices reached the lowest level in 1961 and increased until the high point for live prices in 1979 and for carcasses in 1980. Note that the trends in prices of live lambs and carcasses were similar during the whole period. Value of retail cuts had a similar trend to that of live and carcass prices until 1974; then the increase was greater. In 1954, the spread between carcass price and retail value was approximately $14/100 lb (45.4 kg); in 1964 it was $15; in 1974 it increased to $37; and in 1980 it was $70. This illustrates the extraordinary increase in processing costs.

The rise in live and carcass prices was spectacular from 1965 to 1979. Live prices decreased in 1980 as numbers of slaughter lambs increased. Whether or not the peak in lamb prices has been reached remains to be seen, although there are indications that this may have occurred.

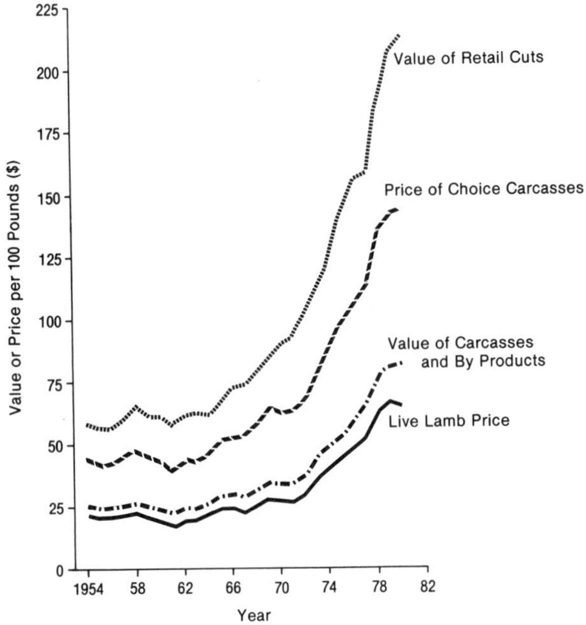

Figure 13.6 Prices of choice grade 100-lb (45.4-kg) live lambs and carcasses and value of wholesale carcasses and by-products and retail cuts from 1954 to 1980. (Source: USDA, ERS Statistical Bulletins 522 and 715, *Livestock and Meat Statistics Supplement for 1980*)

13.10 FACTORS AFFECTING LAMB PRICES

It has been wisely said that in the marketplace the price is mostly dependent on supply and demand. There is also a truism that the consumer controls the market for meat. Both statements are true to some extent, but actually the workings of the lamb market cannot be so simplified. The following are factors that affect lamb prices:

1. Supply of lamb available
2. Supply of other meats and protein-rich food
3. Demand for lamb

13.10.1. Effect of domestic production According to Armstrong, Courtenay, and Judge (1970), demand for lambs tends to comply with the traditional theory in which greater supplies of a commodity are associated negatively with price. Thus, when supplies are high, prices tend to be lower, and vice versa. Figure 13.7 illustrates the point. The supply of lamb and mutton has had a drastic reduction, with a corresponding increase in prices during the period from 1954 to 1980. The peak in lamb supply was in 1962 with 808 million lb (360,714 metric tonnes) produced domestically, plus 143 million lb (63,840 metric tonnes) imported, less 3 million lb (1340 metric tonnes) exported, for a net of 948 million lb (423,214 metric tonnes) available. The low point in production was 293 million lb (130,804 metric tonnes) produced domestically in 1979, plus 44 million lb (19,643 metric tonnes) imported, less 3 million lb (1430 metric tonnes) exported, for a net of 334 million lb (149,107 metric tonnes) available for consumption. This represents a reduction of 614 million

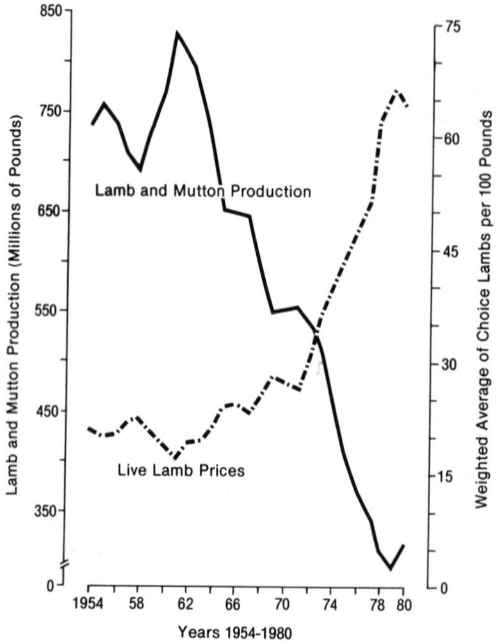

Figure 13.7 Lamb and mutton produced in the United States and live prices of choice lambs at leading markets. [Source: *Agriculture Statistics* (various years), USDA, ERS Statistical Bulletins 522 and 715]

lb (274,107 metric tonnes) or 65% from 1962 to 1980. The low point in prices was in 1961 at $16.78/cwt, and the high point was in 1979 at $66.68, or an increase of $49.9/cwt. This represented an increase of 397% between 1961 and 1979.

13.10.2. Lamb and mutton imports and exports The supply of lamb and mutton consumed includes imports and is reduced slightly by some exporting. Figure 13.8 shows the amount of lamb and mutton imported and exported from 1960 to 1983. Volume of meat imported is subject to wide fluctuations and was greatest from 1968 to 1972. The highest volume was in 1969 with 153 million lb (68,304 metric tonnes). This accounted for 27.8% of the amount produced domestically. In 1983, imports accounted for 4.8% of U.S. lamb production. The bulk of the lamb and mutton is imported from New Zealand and Australia since 1972. Lamb and mutton exports have not exceeded 8 million lb from 1960 to 1983. Most of the meat goes to Canada, Mexico, the Bahamas, and Venezuela. It rarely exceeds 0.3% of the domestic production.

Mutton production is relatively small compared with lambs. For the period 1970 to 1980, 92% to 94% of the kill under federal inspection was lambs. Mutton production is rather stable at 35 to 50 million pounds. Mutton imports have varied widely and appear to bear little relationship to slaughter ewe prices.

Whether or not imports of lamb affect lamb prices is questionable. In a study made in 1970 by Armstrong and others at Purdue University for the Sheep Industry Development Program, no definite conclusions were reached relative to lamb and mutton imports, but they stated that lamb was undoubtedly of greater importance in terms of its price impact on the U.S. sheep industry than mutton.

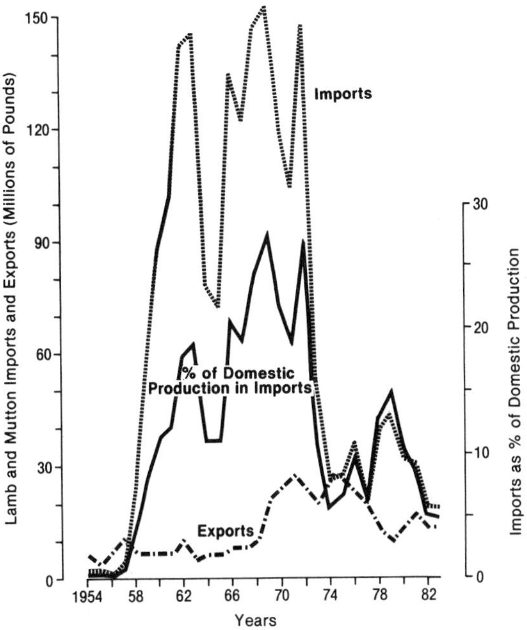

Figure 13.8 Volume of U.S. lamb and mutton imports/exports and production and imports as a percent of domestic production. [Source: *Agriculture Statistics* (various years), USDA, ERS Statistical Bulletins 522 and 715]

13.10.3. Effect of supply of other meats and protein sources on lamb prices Although they are not in direct competition with lamb, the supply of beef, pork, and poultry affects the prices of lamb. There is a closer relationship between beef and lamb than pork or poultry and lamb. If lamb is scarce and high, consumers may purchase beef instead of lamb. There have been shifts to other sources of protein from red meats since World War II. Per capita consumption of meats has changed, as shown in Table 13.8.

In addition, there has been marked increase in the consumption of seafoods and cheeses. These all exert some influence on the marketing of lamb since they represent options at the grocery.

TABLE 13.8 AVERAGE PER CAPITA CONSUMPTION OF RED MEATS AND POULTRY, 1950–1951 AND 1986–1987

Product	1950–1951		1986–1987	
	lb	kg	lb	kg
Beef	61.2	27.8	74.1	33.7
Veal	9.6	4.4	1.8	0.8
Lamb, mutton, and goat meat	6.0	2.7	1.4	0.64
Pork	70.7	32.1	56.8	25.8
All red meat	147.7	67.1	133.7	60.8
Poultry	21.8	9.9	78.7	35.8
Total red meat and poultry	169.2	76.9	212.4	96.5

Source: Agriculture Statistics (various years) USDA, ERS Statistical Bulletins 522, 715, and LPS-1.

13.11 DEMAND FOR LAMB

13.11.1. Type of demand If acceptance of a product is directly related to cost (i.e., if the price is lowered and it is purchased in increased amounts), it is said to be an elastic demand. If, on the other hand, a change in price has little effect on consumption, then demand is inelastic. Lamb does have some elasticity, but it now appears to be less than in the past. This may be because of the drastic reduction in the volume available for consumption.

13.11.2. The consumer determines the demand Previously it was stated that the consumer has many options in a modern supermarket. Whether lamb is purchased may depend on a number of factors, such as:

1. Background and beliefs of the buyer
2. Quality of the product
3. Reliability of supply
4. Price

Eating habits are to some extent regional. Lamb is consumed in greatest quantities in larger cities in the Northeast and the West Coast. Certain ethnic groups are very fond of lamb and will consume it readily if it is available. One drawback to lamb consumption, nationwise, was the experience of many members of the Armed Services during World War II. Personnel stationed in the Western Pacific and the Orient were fed mutton that was overfat and poorly prepared. This left them with an aversion for any meat from a sheep. Although it has been demonstrated countless times that people who have never tasted lamb like it when it is properly prepared, it is difficult to persuade consumers with no experience with lamb to try it on their own.

For consumers who like lamb, the quality of the product affects acceptance. The only criteria they have for judging quality is appearance and grades. Most packers either have lambs graded by the USDA or have their own grades, which are similar.

13.12 USDA GRADES OF LAMB

The USDA has formulated specifications for grades of lambs and mutton that are used in the trade. The grades of slaughter lambs as described in Service and Regulatory Announcement 168 from the Agriculture Marketing Service of the USDA are prime, choice, good, utility, and cull. The same grades are described for slaughter yearlings and sheep, except that prime grade is not used for mutton carcasses. In practice, only prime, choice, and good grades are used for lamb. But prices are not usually quoted for the good grade. There seldom is much spread between prime and choice prices, usually about \$1/100 lb, but much between choice and good. Most lambs that grade good are sold as feeders. The standards for the lamb grades follow.

13.12.1. Quality grades

Specifications for Official U.S. Standards for Grades of Lamb carcasses (quality)

1. Prime

a. Lamb carcasses having minimum conformation qualifications for the prime grade tend to be thickly muscled throughout, are moderately wide and thick in relation to their length and have moderately plump and full legs, moderately wide and thick backs, and moderately thick and full shoulders.

b. Minimum requirements for fat streakings within and upon the inside flank muscles are specified for two maturity groups that cover the entire range of maturity within the lamb class. Typical carcasses from the younger group have moderately narrow, slightly flat rib bones, moderately red, moist and porous break joints, and a slightly dark pink color of inside flank muscles. The minimum degree of flank fat streakings required for such carcasses increases with advancing maturity throughout this group from minimum small to maximum small.

c. Typical carcasses from the more mature group have slightly wide, moderately flat rib bones, slightly red but slightly dry and hard break joints, and a light-red color of inside flank muscles. The minimum degree of flank fat streakings required increases with advancing maturity throughout this group from minimum modest to maximum modest.

d. Regardless of the extent to which the conformation of a carcass may exceed the minimum requirements for prime, a carcass must have minimum prime quality to be eligible for the prime grade. However, a development of quality that is superior to that specified as minimum for the prime grade may compensate, on an equal basis, for a development of conformation that is inferior to that specified as minimum for prime, as indicated in the following example: A carcass that has evidence of quality equivalent to the midpoint of the prime grade (one degree of flank fat streakings more than required as the minimum for prime) may have conformation equivalent to the midpoint of the choice grade and remain eligible for prime. However, in no instance may a carcass be graded prime that has a development of conformation inferior to that specified as minimum for the choice grade. In addition, to be eligible for prime, the lean flesh and external fat of lamb carcasses must be not less than tends to be moderately firm. Also, to be eligible for prime, a carcass must have at least a very thin covering of external fat over the top of the shoulders and the outside of the center parts of the legs, and the back must have at least a thin covering of fat; that is, the muscles of the back may be no more than plainly visible through the fat.

2. Choice

a. Lamb carcasses having minimum conformation qualifications for the choice grade are slightly thick muscled throughout. They tend to be slightly wide and thick in relation to their length and to have slightly plump and full legs, slightly wide and thick backs, and slightly thick and full shoulders.

b. Minimum requirements for fat streakings within and upon the inside flank muscles are specified for two maturity groups that cover the entire range of maturity within the lamb class. Typical carcasses from the younger group have moderately narrow, slightly flat rib bones, moderately red, moist and porous break joints, and a moderately dark pink color of inside flank muscles. The minimum degree of flank fat streakings required for such carcasses increases with advancing maturity throughout this group from minimum traces.

c. Typical carcasses from the more mature group have slightly wide, moderately flat rib bones, slightly red but slightly dry and hard break joints, and a moderately light red color of inside flank muscles. The minimum degree of flank fat streakings required for such carcasses increases with advancing maturity throughout this group from minimum slight to maximum slight.

d. A development of quality that is superior to that specified as minimum for the choice grade may compensate, on an equal basis, for a development of conformation that is inferior to that specified as minimum for choice, as indicated in the following example: A carcass that has evidence of quality equivalent to the midpoint of the choice grade may have conformation equivalent to the midpoint of the good grade and remain eligible for choice. However, in no instance may a carcass be graded choice that has a development of conformation inferior to that specified as minimum for the good grade. Also, a carcass that has conformation at least one-third grade superior to that specified as minimum for the choice grade may qualify for choice with a development of quality equivalent to the lower limit of the upper third of the good grade. Compensation of superior conformation for inferior quality is limited to one-third grade of deficient quality. In addition, to be eligible for choice, the lean flesh and external fat of lamb carcasses must be not less than tends to be slightly firm. Also, to be eligible for choice, a carcass must have at least a very thin covering of external fat over the top of the shoulders and the outside of the center parts of the legs, and the back must have at least a thin covering of fat; that is, the muscles of the back may be no more than plainly visible through the fat.

3. Good

a. Lamb carcasses having minimum conformation qualifications for the good grade are slightly thin muscled throughout, are moderately narrow in relation to their length, have slightly thin, tapering legs, and slightly narrow and thin backs and shoulders.

b. Minimum requirements for fat streakings within and upon the inside flank muscles are specified for two maturity groups that cover the entire range of maturity within the lamb class. Typical carcasses from the younger group have moderately narrow, slightly flat rib bones, moderately red, moist and porous break joints, and a dark-pink color of inside flank muscles. The minimum degree of flank fat streakings required for such carcasses increases with advancing maturity throughout this group from minimum practically devoid to maximum practically devoid.

c. Typical carcasses from the more mature group have slightly wide, moderately flat rib bones, slightly red but slightly dry and hard break joints, and a slightly dark red color of inside flank muscles. The minimum degree of flank streakings required for such carcasses increases with advancing maturity throughout this group from minimum traces to maximum traces.

d. A development of quality that is superior to that specified as minimum for the good grade may compensate, on an equal basis, for the development of conformation that is inferior to that specified as minimum for good, as indicated in the following example: A carcass that has evidences of quality at least one-third grade superior to that specified as minimum for the good grade may have conformation equivalent to the minimum for the upper one-third of the utility grade and remain eligible for good. However, in no instance may a carcass be graded good that has a development of conformation inferior to the minimum for the upper one-third of the utility grade. Also, a carcass that has conformation at least one-third grade superior to that specified as minimum for the good grade may qualify for good with a development of quality equivalent to the lower limit of the upper third of the utility grade. Compensation of superior conformation for inferior quality is limited to one-third grade of deficient quality. In addition, to be eligible for good, the lean flesh and external fat of lamb carcasses must be not less than slightly soft.

4. Utility

The utility grade includes those lamb carcasses whose characteristics are inferior to those specified as minimum for the good grade.

5. Cull

The cull grade includes those lamb carcasses whose characteristics are inferior to those specified as minimum for the utility grade.

USDA carcass and live grades are an estimate of the eating qualities of lamb. In the live lamb, the amount of finish indicates the eating quality. Generally, the fatter lambs are said to have more quality, but this may be exaggerated since excessive finish contributes little to the carcass. Conformation is also a factor contributing to quality. The build or structure of the body varies between lambs. Some are narrow and shallow and have light musculature; others are heavily muscled in the regions of the most desirable cuts and thus yield more lean meat. There are, however, compensations where a lamb may have only a good grade conformation but is well finished and thus grades choice. Young milk-fed lambs are more tender than older, feedlot lambs. Carcass quality standards permit these lambs to grade choice or prime with less finish than the older lambs. In 1980, 96% of all lambs slaughtered were subjected to federal inspection and 72% of them were officially graded for quality. The data indicate that 21.3% graded prime, 78.6% choice, and 0.1% good.

13.12.2. Yield grades The quality grades of lambs are useful in the trade, but all too often carcasses of lambs of the same weight and grade when broken into retail cuts provide very different amounts of edible meat. After extensive research, the USDA, in 1969, published standards for yield grades for lamb and mutton carcasses.

Specifications for Official U.S. Standards for Grades of Carcass Lamb, Yearling Mutton, and Mutton (Yield). The following descriptions provide a guide to the characteristics of carcasses in each yield grade to aid in determining yield grades subjectively.

Yield Grade 1

A carcass in grade 1 usually has only a thin layer of external fat over the back and loin and slight deposits of fat in the flanks and cod or udder. There is usually a very thin layer of fat over the top of the shoulders and the outside of the legs. Muscles are usually plainly visible on most areas of the carcass.

A carcass of this yield grade that is near the borderline of grade 1 and grade 2, might have 0.1 in. (2.54 mm) of fat over the ribeye, 1.5% of its weight in kidney and pelvic fat, and an average prime leg conformation grade. Such a carcass with normal fat distribution would also have a body wall thickness of 0.5 in. (12.7 mm).

Yield Grade 2

A carcass in grade 2 usually has a slightly thin layer of fat over the back and loin, and the muscles of the back are not visible. The top of the shoulders and the outside of the legs have a thin covering of fat, and the muscles are slightly visible. There are usually small deposits of fat in the flanks and cod or udder.

A carcass of this yield grade that is near the borderline of grade 2 and grade 3, might have 0.2 in. (5.08 mm) of fat over the ribeye, 2.5% of its weight in kidney and pelvic fat, and a low prime leg conformation grade. Such a carcass with normal fat distribution would also have a body wall thickness of 0.7 in. (17.8 mm).

Yield Grade 3

A carcass in grade 3 usually has a slightly thick covering of fat over the back. The top of the shoulders are completely covered with fat, although the muscles are still barely visible. The legs are nearly completely covered, although the muscles on the outside of the lower legs are visible. There usually are slightly large deposits of fat in the flanks and cod or udder.

Yield Grade 4

A carcass in grade 4 usually is completely covered with fat. There usually is a moderately thick covering of fat over the back and a slightly thick covering over the shoulders and legs. There usually are large deposits of fat in the flanks and cod or udder.

A carcass in the grade that is near the borderline of grade 4 and grade 5 might have 0.4 in. (10.16 mm) of fat over the ribeye, 4.5% of its weight in kidney and pelvic fat, and an average choice leg conformation grade. Such a carcass with normal fat distribution would also have a body wall thickness of 1.1 in. (27.9 mm).

Yield Grade 5

A carcass in grade 5 usually has more external and kidney and pelvic fat and a lower conformation grade of leg than a carcass in grade 4.

Yield grades are based on the following:

1. Amount of kidney and pelvic fat
2. Amount of external fat
3. Conformation grade of the legs.

The yield grade of an ovine carcass or side is determined on the basis of the following equation: yield grade = 1.66 − (0.05 × leg conformation grade code) + (0.25 × percent kidney and pelvic fat) + (6.66 × adjusted fat thickness over the ribeye in inches). Using these factors, it is possible to estimate the percent of trimmed retail cuts in a carcass. This is called *cutability*. Lamb yield grades and the yields in trimmed cuts are shown in Table 13.9.

In an interesting article published in the July 1982 *Sheepbreeder and Sheepman* magazine, Henneman stated that there was a difference of $9.00/cwt of carcass in the retail value between yield grades. Figure 13.9 was used by Henneman to illustrate the relationship between external fat and yield grade of lambs.

Although there are definite advantages from the use of yield grades in the lamb

TABLE 13.9 CUTABILITY OF LAMB CARCASSES OF THE VARIOUS YIELD GRADES

Yield grade no.	Percent of boneless, trimmed retail cuts (leg, loin, rack, and shoulder)
1	47.3–49
2	45.5–47.2
3	43.7–45.4
4	42.0–43.6
5	Less than 41.9

Source: USDA Yield Grades of Lamb, Marketing Bulletin 52, USDA Consumer and Marketing Service.

trade, the system has not been accepted by the industry. Butchers claim a yield grade of 1 would not be satisfactory for shipping because more fat cover is necessary to protect the carcass. Furthermore, lamb buyers have not been trained to estimate the yield grade on live lambs. Most producers also have had no training in that skill. There seems to be little doubt that an emphasis on cutability of lambs would be an enormous incentive for breeders to improve their sheep and for feeders to make certain they do not overfinish lambs. Figure 13.10 shows the relationship between degree of finish, rate of gain, and feed efficiency of lambs. If sheep were bred to have more muscling and less fat, it should be possible to produce very acceptable heavier carcasses not requiring heavy trimming before they are ready for the consumer. It has been advocated that a yield grade of 2.7 (with 0.15 in. or 3.8 mm of external fat)

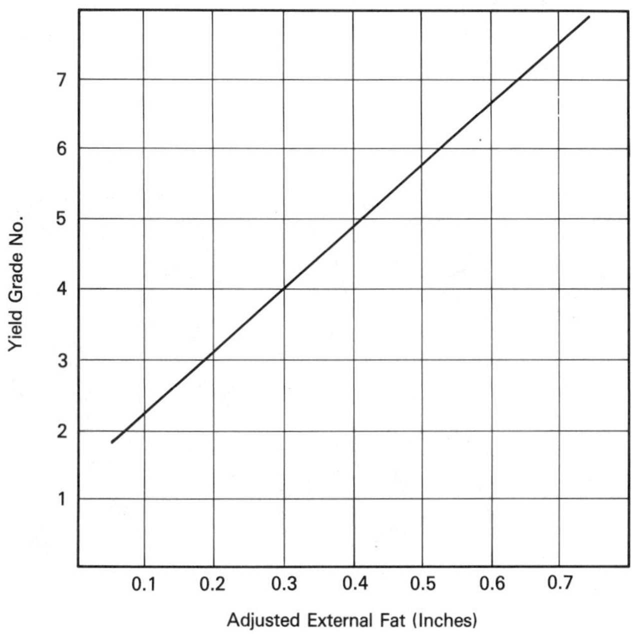

Figure 13.9 Relationship between external fat and yield grade of lamb carcasses. (Courtesy *Sheep Breeder and Sheepman*)

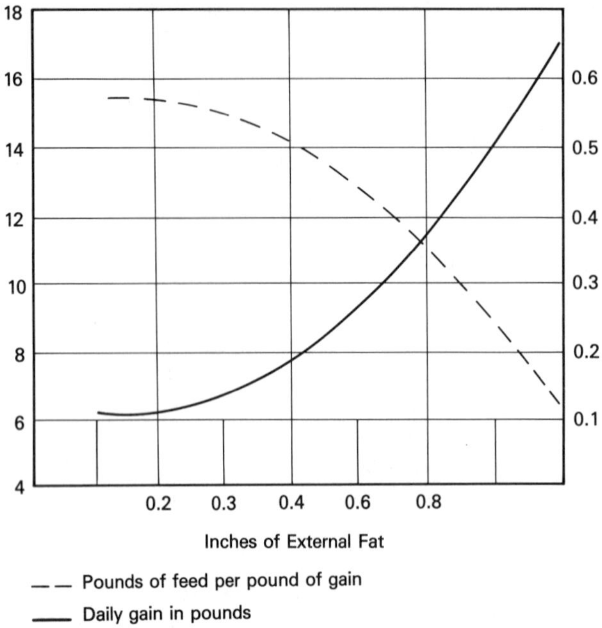

Figure 13.10 Relationship between degree of finish, rate of gain, and feed efficiency of lambs. (Courtesy *Sheep Breeder and Sheepman*)

should be very economical for sheepraisers to produce and for packers and retailers to process.

13.12.3. Effect of age on quality and acceptability of lamb Maturity affects the value of a carcass. The break joint, which is the cartilage joining the shank with the ankle joint, is used to determine whether or not the carcass is to be classified as a lamb or yearling. In lambs, the cartilage is still not calcified and can be snapped at the end of the shank bone. In older animals, it will not break and the feet must be severed at the ankle joint. This is called "breaking a spool" and automatically bars the carcass from being classified as a lamb (see Figure 13.11). Less finish is re-

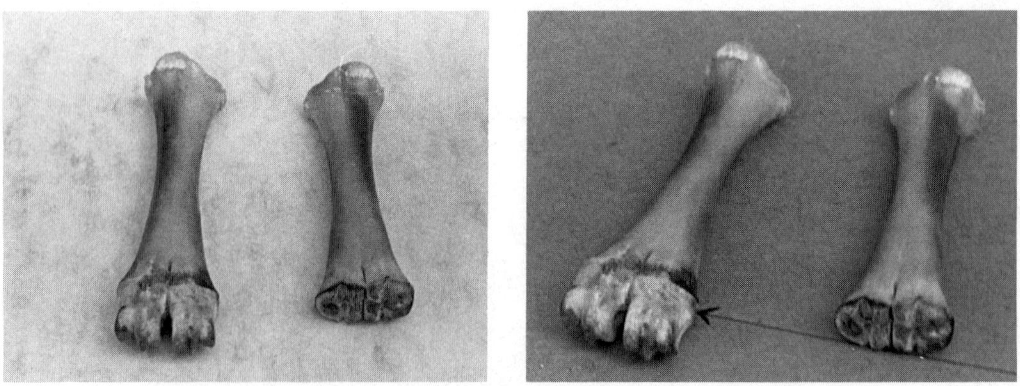

Figure 13.11 An intact shank bone with a spool end compared with one with a break joint is shown in the left photo. The location of the cartilage which snaps to form the break joint is indicated by the arrow in the right photo.

quired on young lambs in order to grade choice and prime than older ones. Not much mutton is marketed. Generally, the price is much lower for it than for lamb.

13.12.4. Effect of sex of lamb on acceptability Wether and ewe carcasses are similar and are equally acceptable in the lamb trade. Ram lamb carcasses are acceptable in the spring of the year and up until about the first of July. Then the price per hundred is reduced for them. Research has shown that ram lambs gain faster and more efficiently than wethers or ewes, and their meat is high in quality and palatability if they are slaughtered when they are young and weigh within the 105 to 110 lb (48 to 50 kg) range. Older, larger rams have excessive development of the neck and shoulders, which are objectionable to retailers.

13.12.5. Effect of carcass weight on acceptability of lamb Weight of the carcass affects the acceptability differently at certain times of the year. Lambs weighing 105 to 120 lb (48 to 54.5 kg) alive (50 to 60 lb or 23 to 27.3 kg dressed) are considered ideal. During the months of April through October when numbers of lambs available for slaughter are limited, there is usually no discrimination for carcasses weighing more than 55 lb (25 kg). During the period November through March, when many more lambs are available, heavy lamb carcasses sell from 10 to 19 cents/lb (0.45 kg) less than those within the most desirable weight range. January and February are usually the months when large numbers of fed lambs are marketed, and weight discrimination is most severe. Very light lambs are also not well accepted, except in the very early spring just before Easter and Greek Easter when they sell at a premium.

13.13 PRICE OF LAMB AND FACTORS DETERMINING IT

Regardless of how well the American family likes lamb, the price of it at the retail market may determine whether or not it is purchased. Among some ethnic groups, it is traditional to purchase lamb for certain holidays and occasions. Then the price may not be a major consideration. This type of consumption is important, but not nearly as much as a more steady demand by consumers who must stay within a budget. Price to them is important. It has been shown previously that there is a relationship between live and dressed carcass prices. These also relate to retail prices in the store. The chain from the producer to the consumer can then be shown as follows:

producer ⟶ slaughterer ⟶ wholesaler ⟶ retailer

Each link must be profitable to maintain the system. When lamb is purchased in the supermarkets, meat market, or other retail outlet, the price of the cuts includes the cost from each unit that handles it. Thus, the live price is only a part of the total cost of the meat sold across the counter.

Buyers purchase live lambs based on the cost of the dressed carcass in the cooler. Before a bid is made, the buyer must estimate the grade, carcass yield, the value of the pelt and other by-products, cost of slaughter, and cooler shrinkage in order to calculate the cost of the dressed lambs.

13.13.1. Estimating carcass yield Carcass yield is also called dressing percentage, since it represents the percent of live weight that the lamb yields in carcass.

The most important factors affecting carcass yield are the amount of fill (feed and water in the digestive tract), the amount of finish, and the weight of organs and the pelt removed at slaughter. Lambs with excessive fill will yield lower than those that have been fed less before shipping or after they arrive at the market. Most buyers have been well-trained to estimate yields and, thus, there is little advantage to excessive feeding either before shipping or on the market. On the contrary, heavy feeding before loading to go to market may result in digestive upsets, scouring, and heavy shrinkage.

The degree of finish or fat the lambs carry affects carcass yield. Fat lambs will yield more than thin ones of the same weight. Some fat is necessary, usually 0.1 in. (2.5 mm) to 0.15 in. (3.8 mm) for satisfactory eating quality, but excessive finish is a waste both to the feeder, because the rate of gain and feed efficiency tend to decrease as the lamb puts on fat, and the retailer, who must trim off the excess at a loss.

How much should lambs yield? The following are average carcass yields for wooled and shorn lambs by grade:

U.S. grade	Fleece condition wooled	shorn
Prime	52%	54%
Choice	50%	53%
Good	47%	50%

Wooled lambs grading utility dress 45% and culls 42% when they are slaughtered. This means that a choice 100-lb (45.4-kg) wooled lamb would produce 50 lb (22.7 kg) of carcass. If only 50 lb (22.7 kg) of meat is obtained from the lamb, what happens to the rest of the live weight purchased? Table 13.10 shows the products obtained from the original 100-lb (45.5-kg) choice lamb.

13.13.2. Value of lamb pelts In calculating the effect of pelt value on carcass costs, a buyer must pay attention to the amount of wool the lambs are carrying. Not

TABLE 13.10 PRODUCTS FROM SLAUGHTER OF A 100-LB. (45.4-KILOGRAM) LAMB

Products	%
Retail cuts	35.6
Pelt	15.6
Edible fats	8.9
Variety meats (liver, heart, etc.)	2.2
Blood	4.4
Inedible items (fats, bonemeal, and meat scraps)	22.2
Unsalable items (stomach contents, shrink, etc.)	11.1
	100.0

Source: H. B. Hedrick (1984), Food Science and Nutrition Department, University of Missouri, Columbia, Missouri.

TABLE 13.11 CLASSIFICATION OF LAMB PELTS OR SKINS

Class	Length of wool fiber, in. (mm)
Full wooled	More than 1.5 (38.1)
Fall shorn	1.0–1.5 (24.4–38.1)
Number 1	0.5–1.0 (12.7–25.4)
Number 2	0.25–0.5 (6.4–12.7)
Number 3	0.125–0.25 (3.2–6.4)
Number 4	0.125 (3.2)

only does the weight of the pelt reduce the carcass yield, but it may or may not increase the value of the lamb. Excessive dirt and manure may cause the pelts to be classified as "manure pelts" and discounted severely. The amount of wool on a lamb is used as a basis for classifying the pelts (or skins to use packer terminology). Table 13.11 gives the classification used in the industry. It has been estimated that it requires approximately 6 weeks for a lamb to grow a number 1 pelt after it has been sheared.

The value of the pelt fluctuates depending on wool prices, clothing styles, and other factors. During World War II when fleece-lined flight uniforms were being produced in large quantities, pelts were used extensively and, thus, were quite valuable. The same can be said when the fur trade is manufacturing mouton coats. Length of fleece on the pelt determines the weight and value of it. Prices quoted for pelts in 1981 and 1982 in the *National Provisioner,* a trade journal of the meat industry, show relative values for the various classifications of pelts (Table 13.12).

13.13.3. Value of other by-products By-products are usually classed as edible or inedible. Those classed as edible include the variety meats (heart, liver, sweetbreads, kidneys, fries, tongues, brains), cheek and headmeats, casings, gelatin, and edible fats. Inedible products besides the pelt include blood, inedible fats, bone and connective tissue, casings, and associated tissues and glands.

The value of the by-products, except for the pelt, are difficult to obtain. Usually the variety meats are the second most valuable products obtained from lamb slaughter, and all the rest, such as the inedible fats, bones, and sinews, are collectively worth very little.

TABLE 13.12 VALUE OF LAMB PELTS IN DOLLARS

Class of pelt	1981	1982	1983
Number 1	4.15–4.25	3.25	4.00
Number 2	2.75	2.00	2.80
Number 3	—	1.00	1.60
Fall shorn	4.50	3.75–4.00	5.00

Source: October 1981 and 1983 *National Provisioner.*

13.13.4. Transportation and processing costs Lambs must be moved greater distances from the farm, ranch, or feedlot to market and to the final point of consumption than most other meat animals. Moving live lambs and the meat always adds to the cost, which must ultimately be borne by the consumer. Slaughtering costs are higher for lambs per hundredweight than for cattle or hogs because of their light weight relative to salable meat and other products. If carcasses are shipped directly to retail stores, the fat, bones, and trimmings are of little value because the small volume of the products is insufficient to allow efficient processing. Transportation and processing costs have increased faster than the market price of live lambs from the 1960s to date.

13.13.5. Retailer's markup The retailer must realize a profit on lamb if it is to be carried in the store. Unlike many other grocery items, meat prices are often used to entice customers into the store. Sometimes cuts are sold at lower prices, even below costs, as a feature in advertising. These are called *loss leaders.* Beef and pork are frequently featured and lamb seldom. These activities are not beneficial to lamb sales. Thus, lamb can be expected to be priced at a level to realize a profit and not appear as attractive a value as other meats. In summary, the price of meat purchased by the consumer includes all the costs of preparation and delivery minus the value of by-products. If the consumer has a prejudice against lamb, does not have experience in preparing and serving it, or finds the price too high relative to other sources of protein, she or he may pass it by. It is interesting that the American Sheep Producers Council has conducted a number of promotions in various areas of the United States with positive results. These tests have demonstrated that consumption of lamb can be increased by educating the consumer. This represents an opportunity for improving the sheep industry in the future.

13.14 IMPORTANCE OF RELIABILITY OF SUPPLY ON CONSUMPTION

If an objective of the sheep industry is to maintain consistent consumption of lamb, it will be necessary to provide a steady supply of the product. In many areas, lamb is not readily available and, thus, it occupies only a small space in the meat counter. This space may also be located in an out of the way place.

Unfortunately, the American preference is for fresh meats as opposed to frozen products. It is interesting that many consumers insist on buying fresh meat and then go home and freeze it until it is to be cooked. If Americans could be persuaded to accept frozen lamb, it would make it possible to slaughter and store lamb in seasons of greater supply and sell it later. This should result in increased efficiency in slaughtering, processing, and distributing meat, and the costs should be decreased appreciably, thus making it more competitive with other meats.

13.15 LAMB AND SHEEP SLAUGHTER

In 1985, there were 831 establishments that slaughtered lambs and sheep under federal inspection. Many of the establishments were small butchers and locker plants. The 624 small plants (73% of the total) only slaughtered 9400 lambs and sheep. On

TABLE 13.13 SHEEP AND LAMB SLAUGHTER, 1970–1985 (1000 head)

	Commercial				
	Federally inspected	Other	Total	On the farm	Total
1970	10,010	542	10,552	249	10,801
1971	10,256	473	10,729	236	10,965
1972	9,905	396	10,301	224	10,525
1973	9,234	363	9,597	202	9,799
1974	8,556	291	8,847	217	9,064
1975	7,552	283	7,835	212	8,047
1976	6,474	240	6,714	197	6,911
1977	6,133	223	6,356	199	6,555
1978	5,169	200	5,369	174	5,543
1979	4,833	184	5,017	172	5,189
1980	5,363	216	5,579	163	5,742
1981	5,789	219	6,008	189	6,197
1982	6,273	175	6,449	194	6,643
1983	6,412	207	6,619	173	6,792
1984	6,549	210	6,759	141	6,900
1985	5,976	188	6,165	135	6,300

Source: Livestock and Meat Statistics, Supplement for 1983, ERS, Statistical Reporting Service, USDA, Bulletin 715.

the other end of the scale, 66 plants with volumes exceeding 10,000 head accounted for 91.9% of the total federally inspected kill. Table 13.13 shows the data on lamb and sheep slaughter from 1970 to 1985. Slaughter has decreased by approximately 40% during the 15-year spread.

Commercial lamb and sheep slaughter is restricted to a few areas in the United States. Table 13.14 shows the 10 leading states. Colorado, California, Iowa, and Texas account for 63% of the total. The 10 leading states accounted for 88% of U.S. slaughter. There has been a decided reduction in slaughter plants as sheep numbers have decreased.

13.15.1. Slaughtering process Although some small slaughtering plants and local butchers dress lambs on a hook, most of the volume plants use an endless chain. The steps are the same, but the latter is much more labor efficient. The steps in slaughter are:

1. Shackling and hoisting to the chain, usually by wheel hoist
2. Rendering insensible
3. Sticking and bleeding
4. Removal of front feet at the break joint
5. Removal of pelt

TABLE 13.14 COMMERCIAL SHEEP AND LAMB SLAUGHTER IN TEN LEADING STATES (1985)

State		Numbered slaughtered (000 head)	Percent of total slaughter
Colorado		1237	20.1
California		1173	19.0
Iowa		840	13.6
Texas		654	10.6
Minnesota		437	7.1
South Dakota		335	5.4
Michigan		296	4.8
Illinois		254	4.1
New Mexico		203	3.3
Pennsylvania		151	2.3
Total 10 leading states		5429	
All others		736	10.3
	Total US	6165	100

Source: Livestock and Meat Statistics Statistical Bulletin, Supplement for 1983, ERS, Statistical Service, USDA, Bulletin 715.

6. Removal of hind feet and head
7. Eviserating or opening of the carcass at the median line. Internal organs, windpipe, and gullet are removed, and the carcass is split at the breast bone.
8. The foreleg is bent at the knee and pinned to the shank by a skewer. The carcass is spread by a stick at the belly to shape it.
9. The carcass is washed, weighed, and then put in a cooler for chilling.

Cooler Shrink. Hot carcasses lose weight during chilling. This usually amounts to approximately 2.5%.

13.15.2. Kosher slaughter Meat for the Orthodox Jewish trade must be slaughtered under supervision of religious representatives, either a rabbi or a shochet. In kosher slaughter, the animal is hoisted without stunning and is killed by cutting across the throat by a very sharp knife called a chalaf. The lungs, stomach, and other organs must pass inspection by the shochet in order to be accepted in the kosher trade. There is a time limit of 216 hours on holding kosher carcasses before sale. To hold meat that long, it must be washed by religious representatives at 72-hour intervals. Most kosher meat moves rapidly to conform to the dietary laws.

13.16 PROCESSING LAMB

More than two-thirds of lamb sold in the United States is fresh and thus highly perishable. Most of the cuts are retailed on a fresh, bone-in basis. The lamb carcass can be reduced to wholesale cuts (called breaking) and then to retail cuts. But if retailers receive whole carcasses, they may process them into retail cuts directly.

13.16.1. Wholesale cuts of lamb Many lambs are sold in the carcass. In accordance with custom, some retail stores in areas where lamb is consumed in large quantities demand carcasses so that their butchers can cut or break them according to the demands of their clientele. Some store owners still produce as many as 43 different retail cuts of lamb.

Wholesale or primal cuts have also been used in selling lamb to retailers. Figure 13.12 shows the primal cuts of lamb or mutton carcasses. The carcass may also be broken into the fore and hind saddle. The fore saddle contains the shoulder, rack, shank, brisket, and breast; the hind saddle contains the loin, legs, and flank. Some other terminology used in merchandizing wholesale cuts are:

- *Front or yoke:* contains the shoulder, breast, and shank.
- *Bracelet:* contains the rack and breast.

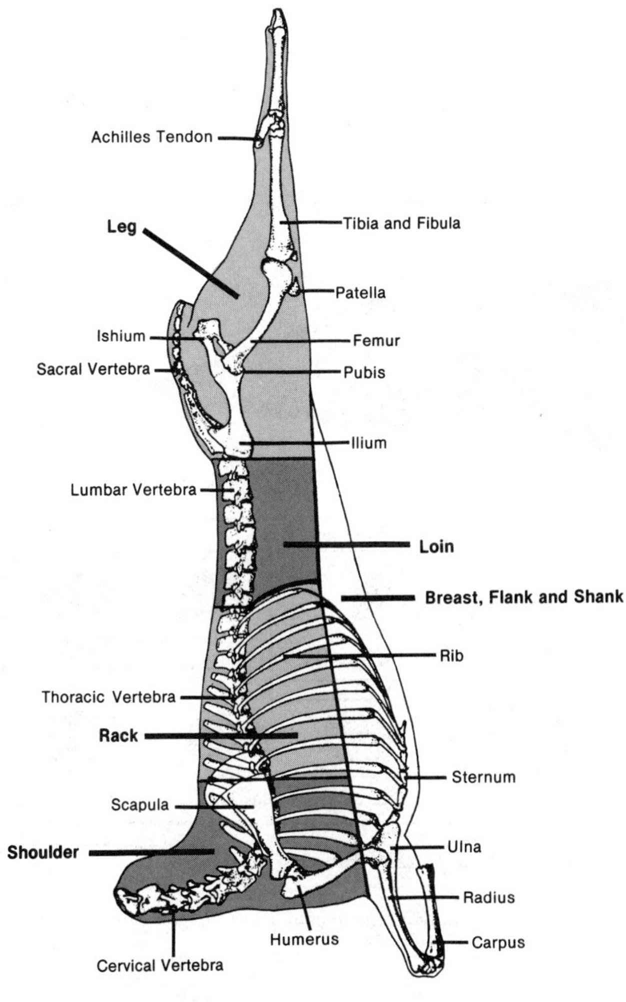

Figure 13.12 Wholesale cuts and major bones in a lamb carcass.

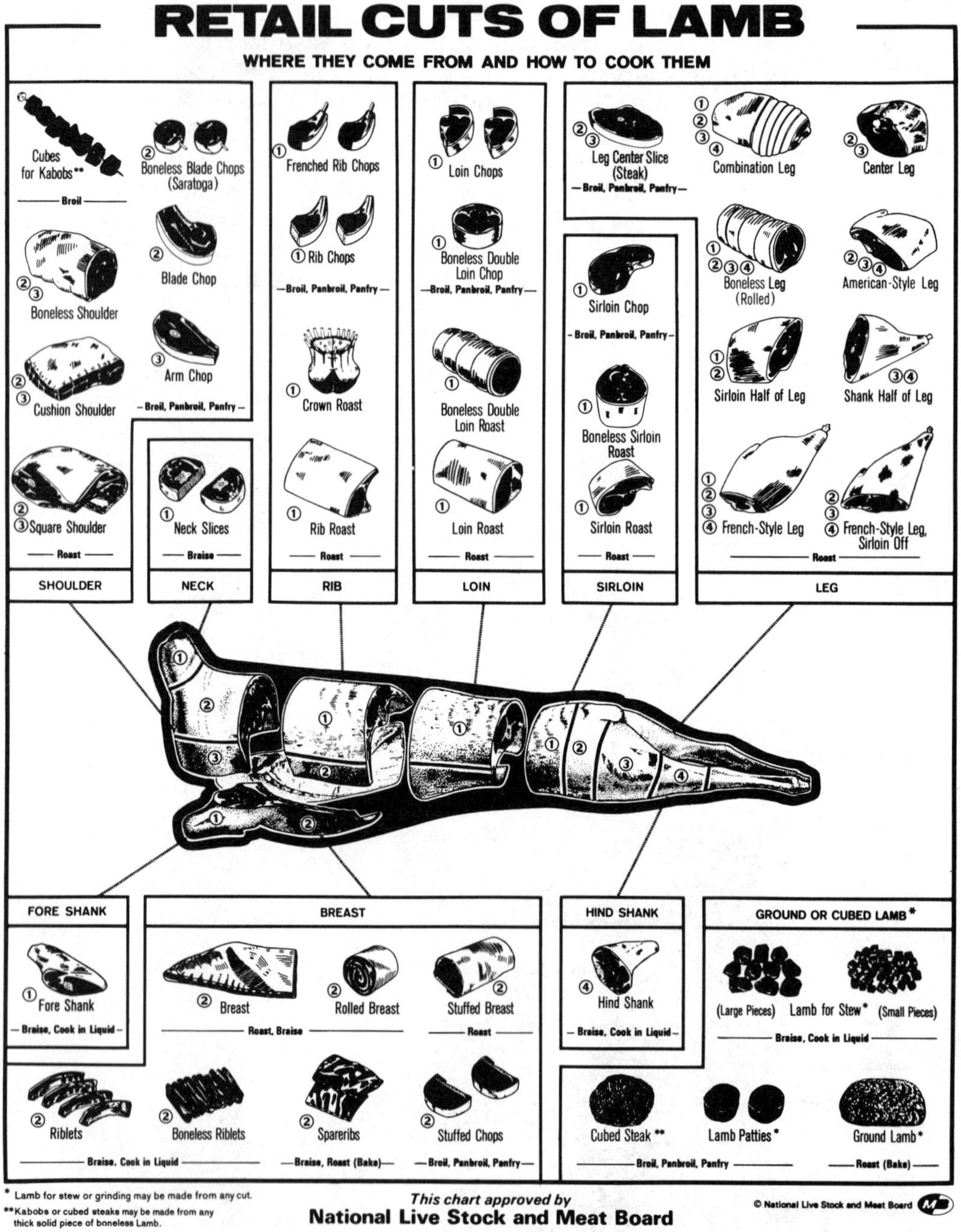

Figure 13.13 Retail cuts; where they came from and how to cook them. (Courtesy National Live Stock and Meat Board)

- *Long fore saddle or trunk:* contains the fore saddle plus the loin and flank.
- *Long hind saddle:* includes all the carcass except the yoke.

The carcass may also be broken into the yoke, middle, and legs, each part making up about one-third of the carcass.

13.16.2. Retail cuts of lamb Figure 13.13 shows standard retail cuts, the primal cuts they come from, and how the cuts should be cooked. Table 13.15 shows the expected yield of retail cuts from the various primal or wholesale cuts.

The most preferred cuts (loin, leg, and rib) comprise 45% of the whole carcass. In a 50-lb (22.7-kg) carcass, this only amounts to 22.5 lb (10.2 kg). The retail cuts from the loin, leg, and rack sell for higher prices than the others. Generally, loin chops bring the most per pound. Ground lamb and stew meat are nearly always as much as 40% lower per pound than the most preferred cuts. The preferred cuts must then sell at much higher prices in order for a profit to be made on the whole carcass.

13.16.3. Consequences of reduction in killing facilities There has been a major reduction in the number of packing plants slaughtering lambs during the 1970s and 1980s. Figure 13.14 shows the location of plants in the United States that slaughtered 100,000 or more lambs in 1980. The small number of packing plants is a matter of great concern to leaders of the sheep industry. The continued closing of plants reduces the options for sale of lambs and sheep. It makes competitive marketing very difficult, and it increases costs of procurement for the packers.

In addition to the reduction in numbers of buyers bidding on lambs as a result of the closing slaughtering plants, another factor seriously affects the lamb business. This is the high cost of slaughtering and processing the carcass. Since lambs are small, they require more labor relative to the weight of the carcass than heavier animals such as cattle and hogs. If they must be transported farther both to slaughter and then to the retailer, this also raises the price to the consumer or it reduces the price

TABLE 13.15 YIELD OF BONE-IN RETAIL CUTS FROM A CHOICE LAMB CARCASS

Wholesale cuts	Retail cuts	Percent of carcass
Leg	Roasts and Chops	24
Loin	Chops	8
Rack	Roasts and Chops	7
Breast, flank, and shank	Riblets, Stew Meat, Fore Shanks	17
Shoulder	Roasts and Chops, Neck Slices, Stew Meat	20
Total salable retail cuts		76
Bones, fat, and waste		24
	Total	100

Source: H. B. Hedrick (1984), Food Science and Nutrition Department, University of Missouri, Columbia, Missouri.

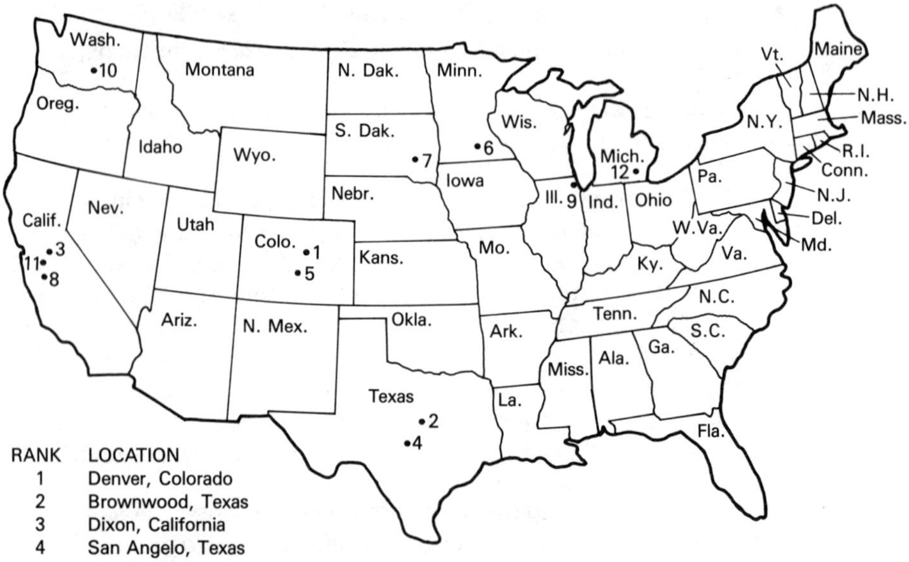

RANK	LOCATION
1	Denver, Colorado
2	Brownwood, Texas
3	Dixon, California
4	San Angelo, Texas
5	Denver, Colorado
6	Albert Lea, Minnesota
7	Sioux Falls, South Dakota
8	Chino, California
9	Chicago, Illinois
10	Ellensburg, Washington
11	Dixon, California
12	Detroit, Michigan

Figure 13.14 Location and rank of packing plants slaughtering more than 100,000 lambs and sheep in the United States in 1980.

to the producer. Since lamb is competing with beef and pork, lamb cannot get too far out of line in price and still be purchased. During a period from the early 1950s to the middle 1970s, the farm-retail price spread for lamb increased from \$22.6/100 lb (retail) to \$67.4, compared with changes of \$20.8 to \$50.5 for beef and \$21.0 to \$44.6 for pork (Holder, 1977). It is predicted that the spread may well become more unfavorable if lamb volume continues to decrease. Very few new plants are being built.

How and where lambs will be slaughtered and processed if production does increase is uncertain. Historically, facilities have been built when supplies of slaughter lambs have warranted the investment. But the changes have usually been slow to occur. Some producers and others are advocating cooperative producer-owned slaughter houses in the future.

13.17 PROBLEMS WITH MERCHANDIZING LAMB

If lamb carcasses are shipped to retail stores for breaking, store personnel must perform a great deal of expensive hand labor to prepare the meat for sale. It is estimated that at least 10% of the weight of the carcass ends up as waste at a store. Not only does the retailer lose the weight at the store, but the weight of those waste products had to be transported from the slaughter house to the retailer. It has been shown that for every 1 lb (0.45 kg) of retail cut sold, it required 2.4 lb (1.1 kg) of carcass weight to produce it. Generally, transportation and processing of lamb and mutton are inefficient and wasteful.

13.18 INNOVATIONS IN SLAUGHTER, PROCESSING, AND TRANSPORTATION

Although the acceptance of new technology for processing lamb and mutton appears to be lagging far behind that of beef and pork, progress has been made in the laboratory, which, if widely adopted, would very much improve the industry. Some examples are:

1. Use of larger lambs. Larger lambs are less expensive to process on the basis of body weight. Improved strains of sheep should be finished for market at heavier weights, but not too fat to meet trade demands. Improved boneless cuts from larger lambs have been shown to be acceptable to consumers.
2. Elimination of discrimination against young ram lambs. Research has shown that ram lambs under the age of 170 days are fully as acceptable as wethers and ewes as far as flavor is concerned, and they are usually not as fat.
3. Central cutting and processing is much more efficient than shipping whole carcasses to retail stores. Products such as fat, bones, and trim can be more efficiently used in the quantities obtained at a central cutting plant than at the various retail stores.
4. Improved packaging. Retail cuts ready for the counter can be vacuum packed for more efficient transportation and longer shelf life. If this is done in a central cutting plant, it is much more efficient than at a retail store.
5. Frozen lamb. This has not been popular, although most consumers buy the cuts fresh and store them in the deep freezer until needed. If this prejudice could be overcome, it would solve many problems of seasonal shortages, longer shelf life, and shrinkage.
6. Mechanical deboning. This process is much more efficient than removing the lean meat by hand. It would save much of the lean meat now poorly utilized and serve to reduce transportation on parts of the lamb that have low value.
7. Restructured cuts of lamb have been made and tested by taste panels. It is well accepted by persons who have eaten the meat. The restructured roasts and chops are prepared from muscles and parts of muscles that are bound together by muscle exudate. The roasts and chops can be manufactured to exact specifications as to size, shape, and composition. It is anticipated that the restructured meat should find acceptance in the food service industry and military, as well as the retail trade.
8. Hot boning of carcasses appears to be feasible now that electrical stimulation of the carcass soon after bleeding has been shown to speed up rigor mortis and reduce the aging time required to produce tender meat. Electrical stimulation of conventionally chilled carcasses has been demonstrated to increase tenderness, brighten the color, reduce the aging period, increase the USDA quality grade, and improve the appearance and shelf life of the cut. A further benefit of electrical stimulation immediately after bleeding is the reduced time required for chilling or freezing the carcasses.
9. Use of a pelt puller on the killing floor. Removing the pelt is a costly process. A pelt puller on the slaughter chain should decrease hand labor and reduce the cost of slaughter.
10. The radical reduction in the number of slaughter plants may very well limit

the resurgence of the lamb industry. Producers of the future may have to form cooperatives to slaughter and process their lambs and sheep. Some preliminary studies by Holder and Hogeland (1982) have demonstrated such steps may be feasible.

13.19 WHAT THE PRODUCER CAN DO TO IMPROVE LAMB MARKETING

Individually, a lamb producer has few options for improving lamb marketing over a wide geographical area, but collective action by many producers may get results. All lamb producers should be affiliated with organizations interested in maintaining the industry. Participation is important, because politics does affect business and politicians are sensitive to the feelings of business and farm leaders. Another good reason to become a part of a commodity group is to keep yourself informed of conditions of the industry. The following are some things a producer can do to improve marketing performance.

13.19.1. Study the markets If the producer does have marketing options, it is most important that they be thoroughly investigated. There is much difference in local markets depending on the amount of competition. It is doubtful that the top price for lambs will be received if there is only one buyer available. If more than one option exists, it is important to test them so that results can be compared. Splitting a group and comparing results may give valuable information. A visit to the markets to get acquainted is usually worth the time spent. A careful look at the expense of marketing may be crucial in making the decision. Transportation, shrinkage, yardage, commission, and other charges may make enough difference for the choice to be obvious. Compare market reports and expenses and see where the net advantage lies. Include in the costs your time if it is necessary to spend it on marketing. Finally, it is important to follow the markets by radio, newspaper, magazine, or special report.

13.19.2. Produce a quality product It may cost somewhat more to produce choice and prime high-cutting lambs, but it is surprising how little additional expense and effort is required to satisfy the market demand. Producers who take pride in their work and market top-quality lambs are soon spotted by the buyers. They will then associate the name with high-quality lambs. While it is true that buyers, as yet, do not discriminate as severely as they should, they do make a special effort to purchase the cream of the crop. It seems reasonable to expect that the industry will ultimately make use of the very good research on lamb yield, grades, and the like, and then it will be a greater advantage to have a reputation as a superior lamb producer.

13.19.3. Know when lambs are ready for market It is important for the producer to know when lambs are ready for market. Weight and finish are two of the most important factors used in determining market value. A farm scale is always a useful tool for the producer, not only from the standpoint of checking weights of products bought and sold, but also in helping to improve the judgment of live animals. With practice, most persons who have a good eye can learn to estimate weights and grades of lambs by visual inspection. Estimating weights and finish and then checking yourself by weighing and handling lambs is the best way to develop skill in grading. When estimating weights, it is well to remember that longer animals nearly always are heavier than shorter ones. Wider lambs invariably weigh more than narrow ones. Deeper lambs are usually heavier than shallower ones.

Eye appraisal for finish may also be developed. Lambs that are too thin may be too narrow and have a prominent backbone that slopes away on either side like an inverted V, a narrow loin and dock, and a narrow brisket. If wooled, the fleece will lack compactness and may even part down the back. Lambs that are too fat are usually wide and deep in the brisket and flat over the back, with very little difference in width of back, loin, and rump. The front flank may be full, and there is often a fold down over the elbow. The hind flank and twist of overfinished lambs are usually full and the fleece compact and oily.

Correctly finished lambs are wider at the shoulders and rump than the loin. The back is rounded on each side of the backbone. The brisket is wide, but trim, the foreflank is trim, and the elbow can be seen. The hind flank is trim and may even appear to be cut up. The twist will be trim, and the muscle shape is readily visible. The degree of finish of a lamb can be verified by handling it. Use the bulbs of your fingers to determine covering over the areas where the skeleton is near the surface. Keep fingers together, and don't press too hard.

The back ribs are slower to cover than those further forward. Move the fingers back and forth. A properly finished lamb will be covered so that the ribs do not feel bare, but still can be felt. If the backbone is sharp and the spinal processes are bare, the lamb is too thin. A lamb that has a perfectly smooth backbone when the hand is moved side to side and the spinal processes cannot be felt is probably too fat. Ideally, the backbone should have enough cover to prevent sharpness, but the processes can still be felt. If the width of the loin is spanned by a hand, the edges should have enough cover to prevent sharpness, but the spinal processes should still be felt. The depth of the loin edge indicates the degree of finish. Overfat lambs will be much deeper than properly finished lambs. The hipbones should have enough covering to prevent sharpness, but not enough so that they cannot still be felt. Thin lambs will have sharp hipbones, and overfat ones will be perfectly smooth. The shoulder vein should have sufficient covering so that it is not sharp, but can still be felt. The inside of the leg should be hard, and the muscle readily discernible. The twist of a properly finished lamb should not be filled with fat. Overfat lambs will have deep twists and feel soft, and also have a valley along the backbone and over the rump.

Long, wide rumps, heavy legs, and wide shoulders are the best indicators of meatiness in a lamb. These characteristics must be bred into them. Degree of finish plus conformation are important in a live lamb if it is to dress into a desirable carcass. Figure 13.15 shows excellent market lambs.

13.19.4. Make certain the lambs are marketed in good condition It is a tragedy that some very able lamb producers follow all the correct procedures in sheep husbandry and neglect perhaps the most important aspect, marketing. Regardless of how good the lambs are, their value may be diminished appreciably by handling

Figure 13.15 Top quality lambs like these are the ultimate objective of sheep production.

from the farm to market. Desirable practices for managing and transporting lambs to market are:

1. *Avoid heavy-concentrate feeding before shipping.* If lambs have eaten excessive quantities of grain, they may be expected to suffer digestive disturbances en route to market. It is usually much better to feed sparingly before shipping to avoid scouring by the lambs. Lambs should never be without water on the market. Bright, high-quality grass hay is probably the best feed to give lambs while they are in the holding pens awaiting sale. If lambs are brought in to the market several hours before sale time and allowed to rest and drink, they will look better, weigh more, and be more attractive to buyers.

2. *Load lambs without injury.* Any livestock farm should have efficient handling and loading facilities so that trucks can load and unload promptly and easily when they arrive at the farm. These need not be expensive to construct, and they should reduce the stress on sheep or lambs when they are loaded. Figure 13.16 shows satisfactory loading facilities. Some chutes use cleated ramps that can be raised or lowered for single- and double-decked trucks. Other loading facilities include a lower and higher ramp. Generally, stepped ramps are superior to cleats from the standpoint of footing for lambs. Cutting chutes and sorting pens are very beneficial not only in preparing lambs for market, but in other operations. Unless there is some compelling reason, these should be located near the loading chutes. Lambs are easily injured. They should not be forced through narrow openings or stressed so that they pile up in the chute. Patience and forethought will help to prevent problems. Never allow lambs to be picked up by their wool in loading. This not only stresses the lamb, but also causes bruises to the carcass. A tame, old ewe that can be led on a halter will often save much shouting and jostling of lambs to get them to move toward the chute.

Figure 13.16 Good loading facilities reduce stress in marketing lambs. (*left*) A stepped ramp for small numbers of sheep. Chute may be stationary or portable. (*right*) Electrically operated, adjustable chutes for single-, double-, or triple-decked trucks.

13.19.5. Make certain the truck is bedded properly Trucks should be cleaned and freshly bedded before hauling livestock. Sand is best for moderate to warm weather; 1 to 2 in. is recommended. In cool or cold weather, sand covered with several inches of straw is recommended.

13.19.6. Do not overcrowd sheep and lambs Lambs should not be jammed into trucks or railroad cars. Wooled sheep and lambs require more space than freshly shorn ones. Table 13.16 shows general space specifications for loading lambs.

13.19.7. Do not load lambs in the same compartment with mature sheep or other species Lambs and cattle do not ship well together. If they are loaded on the same deck, make certain there is a strong, solid partition between them. The same is true with lambs and mature sheep and hogs. It is much better for lambs to be loaded on a deck above hogs than on the same deck, unless partitions are very substantial. Never load hogs over sheep or lambs.

13.19.8. Ship when temperatures are most favorable In cool or cold weather, sheep suffer very little unless they have been recently shorn. It is not advisable to ship sheep or lambs in cold weather within 10 days of shearing.

In hot weather, it is preferable to load sheep or lambs late in the afternoon

TABLE 13.16 RECOMMENDED SPACE ALLOWANCE FOR TRANSPORTING LAMBS AND SHEEP

Weight (lb)		Space (ft²)	(m²)
55–60	Wooled lambs	2.0	0.19
55–60	Clipped lambs	1.8	0.18
70	Wooled lambs	2.3	0.21
75	Clipped lambs	2.2	0.20
90–100	Wooled lambs	2.6	0.24
100	Fed lambs, no. 1 pelt	2.6	0.24
110	Fed lambs, no. 1 pelt	2.8	0.26
120	Breeding ewes, no. 1 pelt	3.0	0.28
130	Breeding ewes, no. 1 pelt	3.2	0.30
140	Breeding ewes, no. 1 pelt	3.5	0.33
145	Pregnant ewes, full wool	3.7	0.34

or early in the morning. Generally, it is preferable to ship sheep and lambs in the afternoon before they are to be sold so they can rest and fill up on water and hay.

Make certain that the truck is ventilated without draft.

13.19.9. Know your truckers and use only those that have proved reliable Research done at the Sioux City Stockyards in the early 1950s demonstrated that the shrinkage and injury records of some truckers were far superior to others. The producer should personally know the trucker and be present to see the livestock loaded. It is also advisable to check periodically on the condition of sheep and lambs when they arrive at the market. The best drivers avoid excessive acceleration or abrupt stops or sharp turns, which may cause the livestock to lose their footing. The careful driver will avoid unnecessary stops and will unload his or her charges without injury. Good drivers are not common and should be asked for. It behooves the good producer to speak up if mistreatment of sheep and lambs occurs. It is important to make checks on shrinks to market and to keep records in order to judge the performance of truckers.

13.19.10. Understand shrinkage and how to deal with it In the lamb industry, the term shrink is often used. This refers to loss of weight when the live lamb, the carcass, or the meat is handled or processed. In the live lamb, it refers to loss of weight during handling or transporting. Two kinds of shrink occur in live lambs. They are excretory and tissue. By far the most important is excretory, which is mainly in the form of feces and urine. Tissue shrink occurs when the tissues of the body are broken down to provide energy to the animal. Tissue shrink is not great unless animals are deprived of feed and water for extended lengths of time. Cooler shrink is largely accounted for in loss of moisture when the carcass is chilled. This ranges from 1.5% to 3% with an average of 2.5%. Thus, hot carcasses should weigh approximately 2.5% more than they will after chilling.

Store Shrink. This occurs when the carcasses are broken down into retail cuts at the store and usually amounts to 5% to 5.5%.

The producer should understand the meaning of shrink, because it is used many times by buyers in trading on lambs. Shrinkage is an important factor in handling and transporting lambs. When lambs are sorted, driven, loaded, and transported, they lose weight. The more they are stressed, the more they lose. This reduces the amount they weigh when sold.

If lambs are sorted and driven some distance before they are loaded and then hauled to a scale, they may weigh several pounds less than they would if they were weighed immediately at the farm. Most buyers know this and thus may request an overnight stand in drylot. This is equivalent to approximately 3% shrink on fat lambs and more in feeders. Another method used is to agree on a shrink, say 4% and deduct it from the farm weight. This is called pencil shrink.

In an extensive study on shrinkage of lambs in transit, Roberts (1956) of the Utah Agricultural Experiment Station recorded the shrinkage in fat and feeder lambs for periods from 1 to 50 hours. These data are shown in Table 13.17.

Some general observations on shrinkage are:

1. The amount of shrink in transit depends on the previous handling of the lambs and sheep. If they have been sorted and driven before loading, it will be appreciably less than if they were loaded immediately.

TABLE 13.17 SHRINKAGE OF FAT AND FEEDER LAMBS AS RELATED TO HOURS IN TRANSIT (Utah Range Lambs, 1956)

	Percent of shrink[a]			
	Fat lambs[b]		Feeder lambs[c]	
Hours in transit	Total	(rate/hr)	Total	(rate/hr)
1	1.35	1.35	1.0	1.0
2	1.83	0.92	1.60	0.8
3	2.30	0.77	2.23	0.74
4	2.74	0.68	2.78	0.70
5	3.16	0.63	3.34	0.67
6	3.56	0.59	3.87	0.64
10	4.98	0.50	5.78	0.58
20	7.34	0.37	9.18	0.46
30	8.23	0.27	10.94	0.36
35	8.33	0.24	—	—
40	8.16	0.20	11.48	0.29
41	—	—	11.49	0.28
42	—	—	11.48	0.27
50	7.34	0.15	11.22	0.22

Source: Utah Agricultural Experiment Station Circular 141.

[a]Based on weights on and off transportation equipment.

[b]Data obtained on 141 shipments representing approximately 65,000 lambs.

[c]Data obtained from 55 shipments of approximately 30,500 lambs.

2. The feeding regime affects shrink. Lambs sold off pasture generally lose more weight than feedlot lambs.
3. Younger lambs shrink more than those that are older.
4. Rate of shrink in transit is highest immediately and decreases with time.
5. In the Utah studies, feeder lambs shrank less than fats for the first 4 hours; then their rate was higher.
6. Shrinkage is greatest during extremes of temperature, especially when it is hot.
7. Stress may cause excessive shrink.

A producer who sells lambs direct to a country buyer is at a distinct disadvantage unless a scale is available on the farm. It is also desirable to know how much weight is lost between the farm weight and the pay weight at a central market or auction. A study of shrinkage will usually pay good dividends to a producer.

In any dealing with a buyer, make certain the bid is thoroughly understood. Figure the pay weight and price offered each way before deciding. For example, if 110-lb (50-kg) lambs are bid at \$50.00 per hundred or with an overnight stand at \$51.00 per hundred, the producer would receive \$55.00 per head for those not shrunk versus \$54.42 for those with the overnight stand with a 3% shrink.

Careful management and supervision plus a knowledge of shrinkage should reduce it appreciably. Study the shrinkage involved in the marketing alternatives and take them into consideration before making a decision on how to sell lambs.

13.19.11. Study the situation and adopt a marketing system that meets your needs You as a producer should make every effort to improve conventional methods of marketing lambs as discussed previously, but should carefully examine some other possibilities. A few that have proved successful are listed next as an incentive for study and possibly the triggering of new ideas:

1. Some producers have been quite successful in contracting lambs to packers at specified prices and delivery dates. This assures profits and cash flow.
2. In some areas, producers have orders from restaurants for delivery of lamb on a year-round basis. The producer is paid a premium for furnishing the kind and number needed by the firm.
3. The same kind of agreement as that of the producer and restaurant may be made with a local butcher.
4. Producers in certain areas have formed a marketing–slaughtering cooperative to handle their lambs.
5. The producer has an agreement to furnish lamb to local clientele. The lambs are slaughtered at a locker plant or on the farm and delivered as scheduled.
6. The producer participates in a local farmers' market.
7. Some producers have a cooperative sausage plant that uses mutton from cull ewes.
8. In a number of areas near colleges and universities that have large enrollments of Moslem and other ethnic groups, lambs and sheep are sold and facilities for slaughtering are provided by local producers.

9. Cut down on the family grocery bill by eating your own produce. A yearling ewe that fails to lamb yields an excellent, economical carcass.

REFERENCES

Armstrong, J. H., H. V. Courtenay, and M. D. Judge, "Sheep and Lamb Marketing Publication," Purdue University Cooperative Extension Service, 1970.

Breimyer, H. F., "Demand Prices for Meat—Factors Influencing Their Historical Development," *Technical Bulletin No. 1254,* USDA, 1961.

Forrest, J. C., and others, *Principles of Meat Science.* San Francisco, Calif.: W. H. Freeman Co., 1975.

Haas, J. T. (chairman of committee), "The Future Role of Cooperatives in the Red Meat Industry," *Market Research Report No. 1089,* Economics, Statistics, and Cooperative Service, USDA, 1978.

Hedrick, H. D., Personal Communication, 1984.

Henneman, H. A., "Evaluation of Slaughter Lambs," *Sheepbreeder and Sheepman Magazine,* 102, no. 7, July 1982, pp. 106–11.

Holder, D., "Cooperative Marketing Alternatives for Sheep and Lamb Producers," *Agriculture Marketing Service Report No. 1081,* USDA, 1977.

______, and J. A. Hogeland, "Cooperative Lamb Slaughtering in the Northeast," *ACS Report No. 14,* Agriculture Cooperative Service, USDA, 1982.

Menzies, C. S. (chairman of task force), "The U.S. Sheep and Goat Industry: Products, Opportunities and Limitations," *CAST Report No. 94,* 1982.

Molpus, M., "Meat Facts," *American Meat Institute,* 1982.

National Provisioner, October 1981 and 1983. Quotation on Value of Lamb Pelts.

Roberts, N. K., "Shrinkage of Fat and Feeder Lambs in Transit," *Utah Agr. Exp. Sta. Cir. 141,* 1956

Scott, G. E., *The Sheepman's Production Handbook.* Denver, Colo.: Sheep Industry Development Program, Inc., 1982.

USDA, "Official United States Standards for Grades of Slaughter Lambs, Yearlings and Sheep," SRA 168, issued in 1951 and amended in 1960 and 1969.

______, "Packers and Stockyards Administration," Packers and Stockyards Résumé for various years.

______, "Statistical Reporting Service, Agriculture Marketing Service, Livestock and Meat Statistics," *Statistical Bulletins Nos. 230, 522, and 715.*

______, *"Yield Grades for Lamb Marketing," Bulletin No. 52,* 1969.

QUESTIONS FOR STUDY AND DISCUSSION

1. What has been the effect of change in lamb numbers since World War II on lamb markets and slaughtering?
2. Why is transportation so important to the lamb industry?
3. How do climatic conditions in the United States affect lamb marketing?
4. What has been the trend in lamb marketing during the last 20 years?
5. Why is lamb feeding important to the lamb industry?
6. Defend the role of the commission agent in lamb marketing.

7. What are the advantages and disadvantages of direct marketing to the producer?
8. Explain the difference between an auction, market, and tele-auction.
9. What economic law determines lamb prices?
10. Why are lamb prices becoming less and less elastic?
11. What has happened to lamb prices since 1954?
12. Why have retail prices of lamb risen faster than live prices?
13. What factors are considered in arriving at USDA grades of lambs?
14. What would be the benefits from yield grading lambs?
15. What areas of a lamb body do you handle for judging finish?
16. Why are retail cuts of lamb so much more expensive than carcass prices?
17. What possible recourse could producers have if live markets and slaughtering plants continue to disappear?
18. In what ways can a producer be kept informed on market prices?
19. Why is an understanding of shrinkage important in selling lambs?
20. In what ways would yield grading be beneficial to the lamb producer and the meat packer?

14 Production and Handling of Wool

14.1 NATURE OF WOOL

Wool is a natural fiber grown on the body of a sheep. It is largely a protein called keratin, which is basically a nitrogen, carbon, hydrogen, oxygen, and sulfer compound. These elements are contained in the 14 amino acids that are the building blocks of keratin. One particular amino acid, cystine, makes up 12.7% of wool fibers.

14.1.1. Why wool is valuable to humans Long before written history, wool was used by people, first as sheepskins and later as clothing. Now it is used for apparel, upholstery, carpets, and other industrial purposes. The reasons why wool is in much demand as a textile fiber are:

1. Wool is an excellent insulator. It prevents body heat from escaping and cold from entering. It also may protect against heat in summer.
2. Wool absorbs moisture. At 25% of its weight, it still feels dry. It may absorb 50% of its weight without saturation. This prevents perspiration from accumulating in clothing near the skin with a chilling effect.
3. Wool is light, yet strong and durable.
4. Wool is elastic and will stretch and return to shape. This resilience causes wool to resist wrinkling or sagging and causes it to be warm in winter and cool in summer.
5. Wool can be dyed easily and permanently in a wide spectrum of colors.
6. It has felting quality not found in other fibers.
7. It is not easily burned. When exposed to fire, it burns with difficulty and has a very disagreeable odor. A spark will leave a charred spot, whereas a hole would appear in many textiles.

14.1.2. Growth of wool The wool fiber originates in a follicle, which is a tubelike sac located in the outer layers of the sheep's skin. The fiber is formed at the bulblike base by cell division and pushing the cells outward to form a fibrous filament, which is the wool fiber. Figure 14.1 shows the primary wool follicle. The follicle includes sudiferous and sebaceous glands that secrete suint (sweat) and grease, which together form yolk in the fleece.

14.1.3. Structure of wool Wool fibers may vary in diameter from fine at 18.1 microns to 39.7 microns for coarse wool. They usually have many closely spaced waves called *crimp* that provide the elasticity of the fiber. A fine-wool fiber may be extended to 30% to 50% of its length without breaking and return to its original dimension without damage.

A cross section of a wool fiber under a microscope shows two distinct layers of cells, the cuticle on the outside and the cortex on the inside (Figure 14.2). The scales of the cuticle overlap each other in much the same way as the slates on a roof, except that the free end projects toward the tip of the fiber. The fibers present a ser-

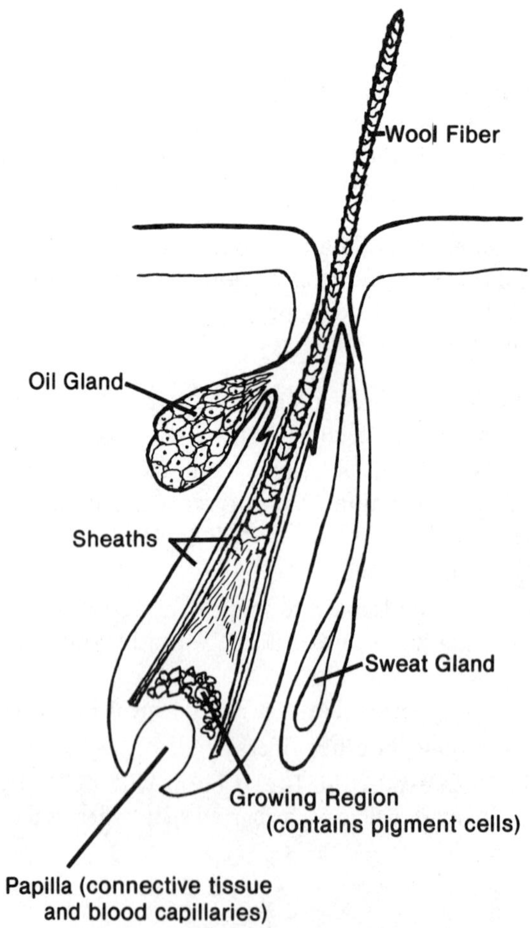

Figure 14.1 Diagram of wool follicle.

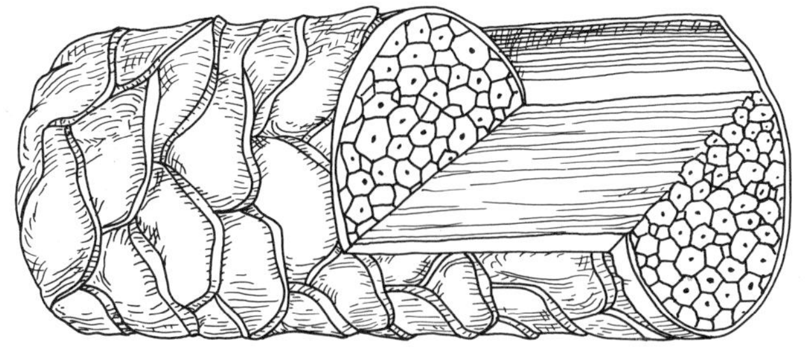

Figure 14.2 Cross section of a wool fiber. (Courtesy Sheep Industry Development Program, Denver, Colorado)

rated appearance due to the scalelike arrangement of the cells. These scales, called scabs, attach themselves to each other and adhere to form a strong bond. This bond is called *felting quality*. In fine wool, there are as many as 2400 scales per inch (2.54 cm) compared with 1800 for long wool. Likewise, the number of crimps per inch is usually, but not always, associated with the fineness of fibers. Coarse wool may only have gentle waves or even none. There is a general relationship between length of fleece and diameter of the fibers. The longest fleeces are generally the coarsest. The length of the fiber is called the staple.

14.1.4. Variation of wool on a sheep Ideally, the wool would be identical all over the body of a sheep. Unfortunately, this is not the case. There is a difference in the length and fineness of the fleece at various locations of the body (see Figure 14.3).

Fineness is greatest on the neck and shoulder intermediate on the side and coarsest on the britch or outside of the hind leg. The length of fiber is just the reverse. The fleece is cleanest on the shoulder, then on the side and belly, and most soiled on the head and twist.

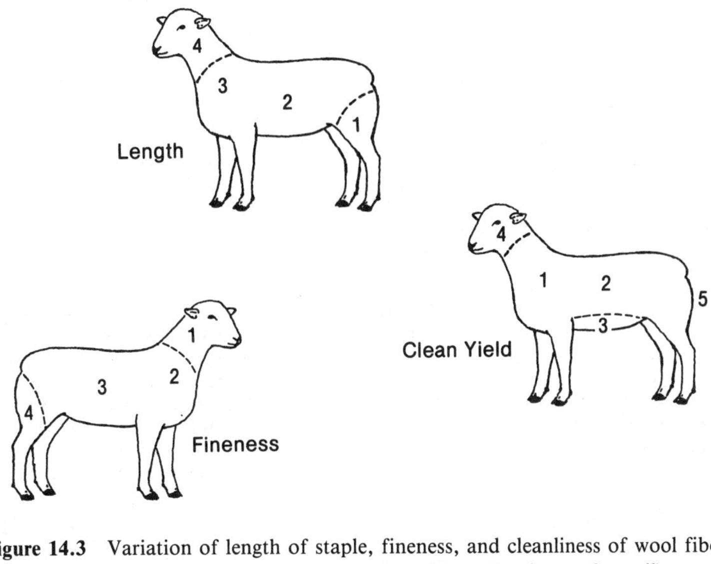

Figure 14.3 Variation of length of staple, fineness, and cleanliness of wool fiber on the body of a sheep. (Courtesy of American Sheep Producers Council)

14.1.5. Classification of wool Wool grown in the United States is classified according to origin and use. The United States is divided into areas according to the kind of sheep that are raised there and the amount of extraneous material that may contaminate the wool. Figure 14.4 shows the geographic distribution by wool type. The fleece wool area differs from the territory wool states because it has many more medium-wool sheep and they have much less dust and dirt in the fleece. Texas wools may be finer, but also may be shorter than territory wool, especially if they are shorn twice each year. California wool differs from territory wool in time of shearing.

Classification by Use. Fine-wool fleeces can be used in different ways according to length of staple and the way the yarn is spun. The two general classes of apparel wool are combing and clothing. Combing wool yarn is made from fiber at least 2¾ in. (7 cm) long called *staple* wool. The wool is processed by a machine with teeth on a roller that passes through the wool and removes the short fibers and aligns the rest in a continuous strand of parallel fibers called the *top*. The short fibers removed are called the *noils*. The top is then spun into yarn and woven into worsted cloth, which has a hard finish and a dense texture. The worsted yarns are usually very small in diameter and tightly twisted. Another process called French combing (more than 1¾ in. or 4.4 cm long) uses a different comb and can process shorter-staple wool.

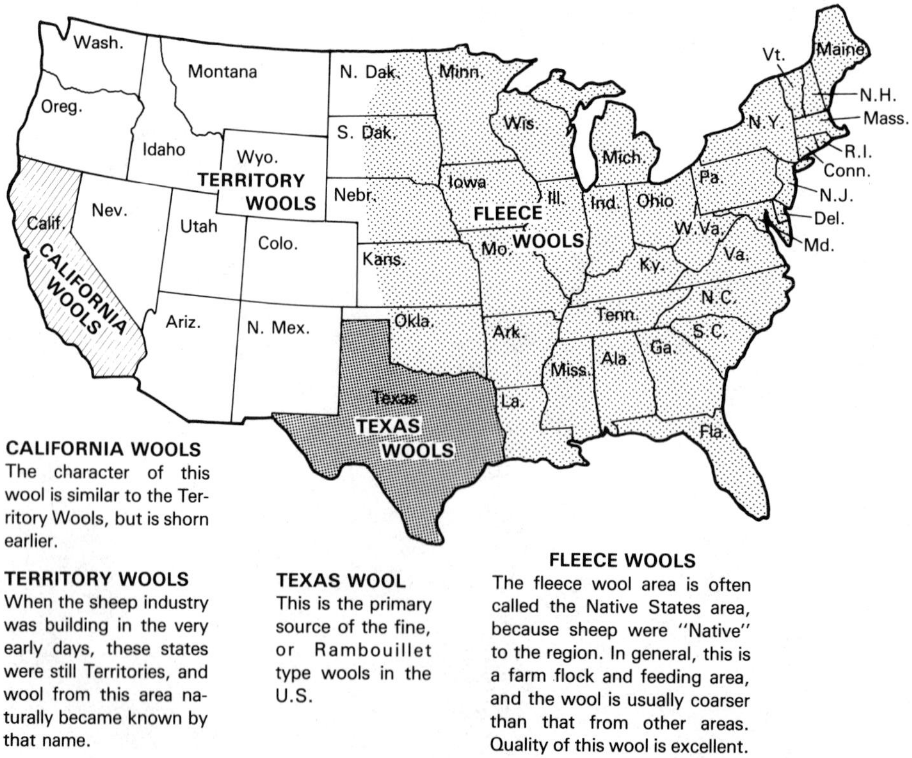

Figure 14.4 General geographic distribution of types of wool produced in the United States.

Woolen yarns are shorter in staple and are not combed, but carded, which results in the fibers being laid in every direction with only slight twisting.

Worsted cloth has a more clearly defined pattern than woolens. They hold their shape and wrinkle less than woolens. Examples of worsted cloths are serge, gabardine, and tropical worsteds. Tweed, jerseys, flannels, sweaters, and blankets are all woolen cloth, as is the felt used in hats.

The classification of apparel wool according to use would be *worsted* and *woolens.* Worsted yarns are made from longer-staple wool than woolen yarns.

14.2 GRADES OF WOOL

Wool is graded according to fineness of the fiber. There are three systems of grading wool. The oldest grading system was developed in the United States and is called the blood or American system. The grades were developed using the Merino fleece as the standard for fine wool, and other grades are based on wool from crosses with Merinos. Thus, the grades were based on the percentage of Merino blood. Based on fiber fineness, the grades are fine, one-half, three-eights, one-quarter, and low-quarter blood and common and braid. While the percentage of Merino blood has long since lost its meaning, the names of the grades and the standards continue to be used.

The spinning count or Bradford system was developed in England and is also based on fineness. Finer fibers are capable of being spun into finer yarn and, thus, more length of yarn can result. The spinning count is expressed in the number of hanks of yarn that can be spun from 1 lb of clean wool.

A hank of yarn is 560 yd (511.8 m) in length. The finest wool may yield 120 hanks of yarn to the pound and be graded as 120's. Other fine wool could grade 70's or 64's.

Comparable grades of the two systems are:

Blood or American system	Bradford or English system
Fine	80's, 70's, 64's
Half-blood	62's, 60's
Three-eighths blood	58's, 56's
Quarter-blood	50's, 48's
Low quarter-blood	46's
Common	44's
Braid	40's, 36's

The third system, called the micron count system, is based on actual measurement of the diameter of fibers. The micron count system uses the micron (1/25,400 of an inch) as the unit of measurement.

Fine wool with a spinning count of 80's has an average diameter of 18.1 microns. Common or braid with a spinning count of 36's has an average diameter of 39.7 microns. Wool grades vary by breeds of sheep, as shown in Table 14.1.

TABLE 14.1 BREEDS OF SHEEP IN THE UNITED STATES AND THEIR AVERAGE GRADE OF WOOL ACCORDING TO THE SYSTEMS OF GRADING

	Blood or American	Bradford or English	Range in diameter (microns)
Merino	Fine	64's to 80's	17.7–20.6
Rambouillet	Fine	62's to 70's	19.2–22
Corriedale	¼–⅜	50's to 60's	24.9–29.3
Columbia	¼–⅜	50's to 60's	24.9–29.3
Panama	¼–⅜	50's to 58's	26.4–31
Targhee	⅜–½	50's to 60's	23.5–31
Southdown	⅜–½	58's to 60's	23.5–26.4
Shropshire	¼–⅜	48's to 56's	26.4–32.7
Hampshire	¼–⅜	48's to 56's	26.4–32.7
Suffolk	¼–⅜	48's to 56's	26.4–32.7
Oxford	Low ¼–¼	46's to 50's	29.3–34.4
Dorset	¼–⅜	48's to 56's	26.4–34.4
Cheviot	¼–⅜	48's to 56's	26.4–34.4
Lincoln	Braid to low ¼	36's to 46's	34.4–40.2
Cotswold	Braid to low ¼	36's to 46's	34.4–40.2
Romney	Braid to ¼	40's to 48's	31.0–38.1

14.3 FACTORS AFFECTING THE VALUE OF WOOL

The class and grade of wool reflect the use of the fiber, but are not the only important factors determining the value of a fleece. Other characteristics that are important are character, purity, soundness, strength, cleanliness, shrinkage, grease content, uniformity, and color. These are discussed separately.

14.3.1. Character Character in wool refers to the appearance of the grease wool with regard to brightness, luster, white or creamy color, and a crimp pattern characteristic of the grade. The term denotes the recognizable characteristics that contribute to the attractiveness of a fleece.

14.3.2. Purity Purity relates to the kind and color of the wool fibers. *Kemp* is a chalky white, weak fiber. It has no crimp and detracts from the fleece. Dark fibers are also considered to be impurities that detract from the value of wool.

14.3.3. Soundness A sound fiber is one that has no weak spots in it. Tender spots caused by inadequate nutrition or sickness cause the fibers grown during that time to be small and weak. This is refered to as a *break* in the fleece.

14.3.4. Strength The fibers must be strong if the cloth made from it is to be durable. It is said that the fine-wool fibers are actually stronger than particles of steel of the same diameter.

14.3.5. Cleanliness The amount of foreign material in wool has a very decided effect on value. Manure, dust and dirt, plant matter, or manufactured materials in wool can all cause a downgrading in the value of a fleece.

Before wool is spun into yarn and woven into cloth, it must be scoured by running it through warm water that contains soap or detergent and rinsing to remove the soap or detergent. Then, after drying the shorter, finer wool, it is run through burr-picking machines or carbonized with acids to change vegetable matter to carbon compounds that can be removed by mechanical action. Wool that is longer is combed to remove vegetable matter.

14.3.6. Shrinkage This term refers to the loss of weight due to removal of all material in the wool except the clean fibers. Shrinkage is calculated as follows:

$$\text{shrinkage (\%)} = \frac{\text{grease weight} - \text{clean weight}}{\text{grease weight}} \times 100$$

It varies widely in wools from more than 60% in fine-wool fleece to as little as 30% in some long wools. The amount of yolk, dirt, water, vegetable, animal, and mineral matter all contribute to losses of weight when the wool is scoured. Territory wools uniformly shrink more than fleece wools. Wool from sheep produced in regions of high rainfall also have a lower shrinkage because much of the grease may have been lost by water falling on the sheep.

The amount of adhering material in wool is subject to wide variation depending on the circumstances and environment where the sheep are being produced. The following indicates the ranges in the amount of fiber and adhering material found in grease wool:

- Wool fiber, 15% to 75%
- Yolk, 7% to 40%
- Suint, 5% to 15%
- Foreign material, 5% to 40%
- Moisture, 3% to 20%

14.3.7. Grease content The grease content of wool affects shrinkage. It is usually higher in fine-wool fleeces than coarse wool and greater in ram fleeces than in the ewe. The grease is important, because it protects the scales on the wool fiber, prevents the fleece from becoming cotted or matted, and holds the fibers together to protect the sheep from the penetration of moisture. Excessive grease detracts from the value of a fleece, because it increases shrinkage.

14.3.8. Uniformity Fleeces usually are not uniform in length and fineness, and thus there may be a wide variation in the wool on an individual sheep. However, it is desirable for the fleece to be as uniform in fineness and length of fiber as possible.

14.3.9. Color The color of the wool is important, because it may limit its usefulness. From a manufacturing standpoint, almost any color other than white is undesirable. Only white wool can be dyed the many colors dictated by the wool market. A common defect is canary yellow, a bright-yellow stain that occurs in white wool due to moisture and heat or light.

The most desirable wool is bright and lustrous without dark fibers or stains. Care should be taken during shearing not to include black fleeces, dirty or wet wool, or dunglocks in the same wool sacks with top-quality fleeces because they will detract from the color and the value of a wool clip.

14.4 PRODUCTION OF SUPERIOR WOOL

High-quality wool that is in demand on the market only occurs after careful planning and care by the producer. Steps can be taken that are relatively inexpensive yet should result in much improvement in the volume and quality of wool produced and more income.

14.4.1. Breed for superior wool production The fineness and length of the fiber are largely controlled by genetics and are therefore dependent on the breeding program being followed. Careful selection of rams should result in high-quality wool if the breeding program is combined with proper care and management.

14.4.2. Feed properly and use superior management practices The size and strength of the wool fiber can be altered by nutrition. It has been determined that the protein requirement for a mature sheep on a maintenance diet is approximately 8%. This is so low that very few sheep should ever be deficient, but when a ewe is pregnant, suckling a lamb or lambs, or under physiological stress, the drain on her system may be so great that she will not be able to properly form the wool fiber, especially if she is also receiving insufficient protein in her ration. This inadequacy is usually only temporary and then the fiber returns to normal size and strength. But the fiber grown during the stress will be small and weak and will show up as a line across the fleece when it is parted. This is called a tender spot or *break* and very much detracts from its value. The same thing can happen if the sheep is sick. Thus, proper nutrition and care to prevent sickness are important in the protection of the fleece. Lack of energy seldom affects the growth of the fiber, but may cause a reduction in the amount of grease in the wool. Generally if ewes are fed and managed so that they raise a good lamb crop, they will also produce an adequate fleece.

14.4.3. Prevent contamination of fleeces As previously discussed, wool contains the natural compounds of yolk and suint, which are of value to the sheep. It is the extraneous matter that adheres to the wool that causes most of the discounts.

Such foreign material as burrs, seeds, and chaff and artificial substances must all be removed by the manufacturer before the wool can be processed into yarn. All are expensive to remove and thus, when they are present, the wool must be discounted.

Weeds that have burrs or seeds that may get into wool can be mowed, sprayed, or cut to remove them from pastures; if this is not practical, it may be desirable to use such pastures only at certain times of the year. Poor loft floors that allow chaff to sift down on the backs of sheep or hay racks that extend over the backs of feeding sheep can cause contamination of the fleece. Tight loft floors and properly designed feeding equipment should be used.

Binder twine made of sisal should never be used to tie fleeces because the small pieces of fiber that break off adhere to the wool and are dificult to remove. If not removed, they will not assume a dye.

Plastic twine used in baling hay also is a very serious impurity because it must be removed by hand at a high cost. It is especially troublesome when the hay is chopped or ground. Sheepraisers would do well to use wire for tying their bales. Only paint that is scourable should be used for branding or marking. Another artificial material that may get into wool is grease from machinery. *Sheep should not be housed where machinery is stored.* Manure may stain the wool if it is wet and sticky. Generally, sheep do not need much bedding unless the pens are damp; but if they are, bedding should be used. Outside lots may become quite muddy and cause the wool to become dirty. This can be avoided to some extent by using mounds and covering them with lime and a layer of straw. Sheep should not be allowed to accumulate dunglocks called tags. Pregnant ewes should be either crutched or sheared before they lamb. Tags or dirty wool should never be sacked with clean fleeces. The same should be the rule with burry or off-color wool.

14.5 SHEARING AND CARE OF WOOL

14.5.1. Preparation for shearing Since there is a great difference in the quality of the work done by different shearers, it is desirable to book a good one early enough to be sure you get the one you want when you desire to shear. Order wool sacks and paper twine so they will be on hand when needed.

The sheepraiser owes it to the sheep and shearer to provide a satisfactory place to work. The area used for shearing should be well lighted and clean. There should be holding or catch pens close to the shearing floor (Figure 14.5) so that the shearer

Figure 14.5 Properly designed facilities increase the efficiency of shearers and reduce stress on sheep.

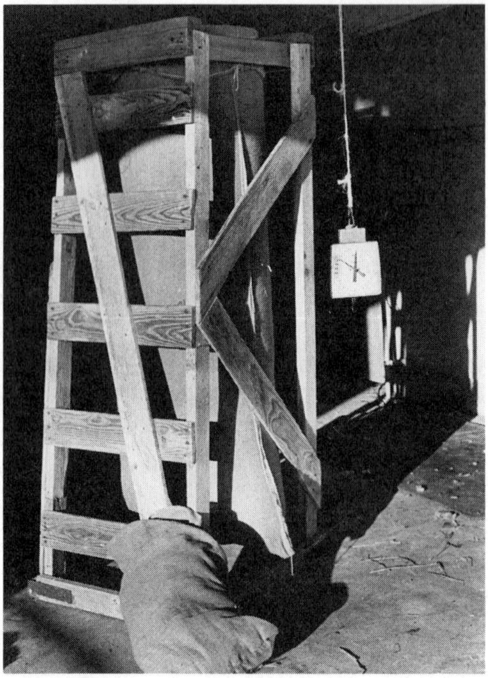

Figure 14.6 A holding stand for sacking wool with scale attached and a bag for holding tags.

will not have to spend valuable time and effort bringing the sheep to the floor. The holding pens and the shearing floor should never be bedded. Perhaps the ideal shearing floor is constructed of solid wood 6 ft × 8 ft (1.8 × 2.4 m) that can be swept frequently. Less comfortable, but adequate, is a clean, concrete floor. A tarpaulin or an old rug may be used to cover the floor. Electrical outlets should be convenient to the floor. It is desirable to think ahead and plan the sheep shearing layout so that the sheep can be moved easily and efficiently to and from the floor. Room should be provided for a sacking frame or frames (Figure 14.6) near the shearing area. If more than one kind of wool will be harvested (i.e., fine and medium wool), it may be desirable to have more than one sack open so that fleeces can be kept separate. Ram and black fleeces should be kept separate from those of ewes. A bag should also be available for tags, facings, and leg wool. This could be an ordinary burlap or paper bag.

14.5.2. Keep sheep dry before shearing Sheep should be dry when shorn. Wet wool may mildew in storage and stain the fibers. This will decrease the value of the clip. Special precautions should be taken to provide shelter for sheep prior to shearing. During wet, cold weather, sheep that have been thoroughly wet lose moisture slowly. It is usually necessary to put them outside in sunshine to dry them. Prevention is much to be preferred.

14.5.3. Insist on a good shearing job Either be present or have an experienced, responsible person present to supervise the operation. Emphasize the following points:

1. *Removal of the fleece in one piece.* This is standard practice for a good shearer. There are a number of ways of shearing sheep that are satisfactory, but perhaps the most widely used is the "long blow" or Australian procedure described by the Sunbeam Corporation (now called the Oster Corporation) (Figure 14.7). The sheep is held in 12 different positions that allow the shearer to maintain control but, at the same time, to efficiently remove the fleece. Shearing schools are conducted in most states by the agricultural extension service. The schools usually last for 2 to 7 days, but that is sufficient to learn the basic principles (Figure 14.8).

2. *Avoid second cuts.* Second cuts occur when a shearer cuts the fleece too far from the skin and must go over the same area again. This is most undesirable, because second cuts result in very short fibers (noils) that have little value. These may be avoided by working at a speed in accordance with the experience of the shearer. If the handpiece is kept close to the skin at every stroke, second cuts will be minimized. One of the best features of a good procedure is that the skin is kept taut at all times.

3. *Prevent contamination of the fleece on the shearing floor.* Keeping the fleece free of contamination can be accomplished by using clean pens without bedding, sprinkling dusty lots, keeping the shearing floor swept after each sheep is sheared, and separating out the sweat and dunglocks, facings, and leg wool before tying the fleece. In large operations where many sheep are to be sheared, the waiting pens are on slotted floors to prevent the wool from being soiled.

4. *Avoid injury of the sheep.* Shearing carefully to prevent injury can be done by skilled shearers. Careless workers may cut off the ends of teats and the tips of vulvas of ewes and damage the penis or scrotum of rams or injure the ligaments of

Figure 14.7 A skilled shearer removes the fleece in one piece without injury to the sheep.

Figure 14.8 Shearing schools are one of the best methods of learning the fundamentals of shearing.

the legs. Almost all shearers will occasionally nick a sheep, sometimes severely. The worst injuries should be treated with antibiotics and stitched. These sheep should be marked and cared for until they heal. Remember that skilled shearers can shear a pregnant ewe as close as 3 to 4 weeks before the end of gestation without injuring her. Watch the shearer and make certain he or she handles the sheep gently. It is important to have pens and equipment in good repair and properly set up to avoid injuries.

5. *Tie the fleece properly.* This will enhance the value of the wool. Generally, lamb wool is too short to tie and is therefore sacked loose. Fleece wool that is at least 2¼ in. (5.7 cm) in staple should be tied. When the fleece is removed, it should be spread out on the floor with the skin side down. Remove the tags, facings, and leg wool and put in a tag bag. The sides are then folded in and the fleece rolled from the rear to the front. The fleece should always be tied with special paper twine (Figure 14.9). For the usual fleece, 8 ft (2.4 m) of twine passed around and across the fleece twice should be adequate. A properly tied fleece will only show the clean skin side of the fleece and will be firmly tied, but without being solidly packed as it is possible to do with a tying box. Tying boxes are no longer recommended.

6. *Sack wool correctly.* Proper sacking is much easier to accomplish if a good rack or stand is available (Figure 14.10). Wool bags are 7 ft (2.1 m) long and thus some device is needed to hold the bag upright so the fleeces can be stored inside. A metal ring with another that can be slipped inside with the sack between is quite handy, or the bag can be pinned to one ring with nails. The bag should not touch the floor. Some cleats or a ladder will facilitate entry into the bag. Fleeces should only be bagged in special wool sacks. Using any other kind of bag may result in contamination of wool.

The first layer of fleece should be formed with a fleece in the center and one on each side. From there on, there should be five fleeces per layer, with one in the center and a fleece on each of the four sides. The second and third fleeces in a layer should not be next to the seams. The first three fleeces in a layer should be tromped

Figure 14.9 Correctly tying a fleece. Paper twine is passed across and around the fleece twice.

Figure 14.10 Correct sacking of wool includes placing five fleeces per layer (one on each corner and one in the middle) and tramping around the edges of the wool bag.

lightly into position, then the final two at the seams will require more effort. Once the five fleeces are in place, tramp around the edge with the sole of the shoe against the bag. Two or three round trips should be sufficient per layer. The fleeces should be packed tightly so that 30 fleeces are in each bag. Filled bags should weight 275 to 300 lb (125 to 136 kg). When the bag is full, it is necessary to lift it up so that the steel hoops can be removed. Then the bag should be closed. Stitching with cotton twine using a lock stitch is preferable so that, even though one stitch is cut or broken, the bag will stay closed. The bag should be distinctly marked with the kind of wool and the owners's name.

14.6 STORING WOOL ON THE FARM

Wool is usually stored for a period of weeks or months on the farm before it is sent to market. Properly protected, it can be held safely for months. The following should be done with stored wool:

1. Choose a storage area that can be kept clean and dry and not subject to contamination with manure from birds or invasion by rodents.
2. Protect the wool from moths by spraying empty wool sacks with 2.5% premium grade Malathion or an equivalent insecticide and allow to dry before filling. If wool is already sacked, it can be fumigated by covering with a tarpaulin and using paradichlorobenzine or a similar compound.
3. Thoroughly clean the area to be used for storage.
4. Spray walls and floors to the point of runoff using either 2.5% methoxychlor or a 2.5% solution of Malathion. One gallon of either solution should cover 500 to 1000 ft^2 (46.5 to 93 m^2) of surface area.
5. Store wool on wooden floors. Wool stored on dirt or concrete floors may draw moisture that causes molding and staining.

14.7 MARKETING WOOL

Unlike most farm products, there are no open markets for wool. Most wool is bought and sold by private treaty. Buyers may be representatives from woolen mills or brokers who represent the mills. In areas where sheep are produced, local dealers buy wool and resell it to brokers. Some shearers will purchase wool at the time they shear. Some will accept wool as payment for shearing.

Although it is not true of all local or county wool dealers, many buy wool for cash at a flat rate regardless of quality. This penalizes the better product and may result in a higher price than deserved for inferior-quality fleece. Many of these buyers are in the feed or grain business, or they may buy hides and junk.

State Dealers or Central Market Dealers. Dealers usually buy on a cash basis. They often have local buyers throughout a sheep-producing area. Since they are usually large, they may perform services such as furnishing bags and supplies needed by the sheepraiser. They also may grade fleeces and provide information to the grower.

Commission Merchants. Some agencies will take wool on a consignment basis and warehouse it until it is sold. They sell for a commission with the approval of the owner. Some of these houses also buy wool on a cash basis, which may or may not limit their usefulness.

Wool Marketing Cooperatives. Wool co-ops have been formed to protect the producer. These may be local, state, or national in scope. The cooperatives pool the wool from many producers, grade and sort it, and sell it in large quantities. They usually pay an advance on the wool. Then, when the pool has been closed, the wool is sold and, after paying expenses, the remainder is distributed to the growers based on quality and weight. Co-ops usually retain a portion of the value of the wool for operating capital. Thus, the consigner is part owner of the organization. The producer becomes a member of the co-op by doing business with it. The advantage claimed by such organizations is that they have highly skilled warehouse people and graders who sort the wool into uniform lots in large enough quantity so that manufacturers will be interested in bidding on it. They may sell it directly or ask for sealed bids. By handling a large volume, costs per unit may be reduced. Cooperatives also may furnish information on grade, classification, and shrinkage to the grower. Most co-ops also have service organizations that sharpen and repair shearing equipment and furnish drugs, implements, and a full line of supplies to growers.

It is important that the grower of wool learn the technique of marketing it. The price varies widely and is very difficult to establish. Unless the operation produces a large quantity of wool, it usually is not possible to make direct contact with woolen mills. Then, it is a good policy to investigate the possible outlets beginning with the local dealers and including the most sophisticated markets. If you produce a high-quality product, sell to people who buy on a grade and yield basis. Talk to the buyers and ascertain how well informed they are. Split some lots of wool and send it to more than one place and compare returns. Consider the advantages of pooling wool or joining a cooperative. Know your buyer and deal with the one who is willing to pay for a quality product.

14.8 THE NATIONAL WOOL ACT

In 1954, the Congress of the United States passed legislation to encourage increased production of sheep and wool, which was recognized as an essential and strategic commodity. Funding for an incentive program is derived from tariff on imported wool.

The National Wool Act authorized the secretary of agriculture to support the price of wool at a level sufficient to reach an objective of 300 million pounds (133,929 metric tons) of grease wool production annually. The incentive program of the National Wool Act is administered by the USDA agriculture and stabilization programs. Incentives are paid for wool produced and on the weight of unshorn lambs marketed. Payments are made by county ASCS offices to qualifying sheep producers.

Each year, the secretary of agriculture establishes a support level for the next calendar year. This is expressed at a price per pound (0.45 kg). During the year and until January 31 of the following year, producers file authorized sales reports with their county ASCS offices. These sales data are then transmitted to the national office and used to determine the average price received in the United States. The support or incentive price and average sales price are used to figure the incentive (%) to be paid the producer. If the support or incentive price was \$1.65/lb (0.45 kg) of wool (the level for 1984) and the average price throughout the United States for that year was \$0.795/lb (0.45 kg), the incentive level would be calculated as follows:

National incentive level: \$1.65/lb (0.45 kg)
Average U.S. sales price for the year: \$0.795

National incentive level − average sales price = \$0.855/lb (0.45 kg)

$$\frac{\$0.855}{\$0.795} \times 100 = 107.5\% \text{ per lb (0.45 kg) incentive to be paid}$$

The grower then multiplies the price received per pound (0.45 kg) by 107.5%. The following example illustrates the method.

If grower 1 sells wool at \$1.20/lb and the incentive is 107.5%, the payment from ASCS would be \$1.29/lb (\$1.20 × 1.075% = \$1.29). The grower would realize \$2.49/lb of fleece produced (\$1.20 + \$1.29 = \$2.49). Grower 2 receives \$0.76/lb for wool: \$0.76 × 1.075% = \$0.817/lb incentive. The sale price of \$0.76/lb plus \$0.817/lb incentive = \$1.577 realized per pound for wool produced. Producer 1 receives \$0.913 more per pound than producer 2. This is an incentive to produce high-quality wool and do a good job of marketing it. It should be pointed out that in 1973 the incentive price was \$0.72/lb and the average U.S. price was \$0.82. No incentive was paid that year. During the time the National Wool Act has been in effect, wool payments have contributed 20% to 70% of producers' return above cash costs for sheep production. Payments made for unshorn lambs in 1981 were at the rate of \$3.42/100 lb (45.4 kg) of lamb marketed.

The National Wool Act also implemented the establishment of the American Sheep Producers Council (ASPC), which is the promoting and advertising organization of the American sheep industry. This self-help program has the overall objective of increasing the stability, efficiency, and profitability of the sheep industry. Much of the money is spent advertising lamb and wool. Each time the Wool Act is extended, sheep producers are given the opportunity by referendum to approve or disapprove of the programs of the ASPC.

ASPC activities are directed by elected representatives of sheep producers and receive financial support by deductions from wool incentives paid to producers. In 1981, deductions were 2.5 cents/lb of wool and 12.5 cents/100 lb (45.4 kg) of lamb sold.

Lamb feeders can collect wool incentive on the gain on unshorn lambs. For example, if they purchase 80-lb feeder lambs and market 115-lb finished lambs, they may collect incentive on 35 lb of live weight or 35 × \$0.0342/lb = \$1.197 per lamb. If they buy the lambs and shear them, they are required to deduct the incentive paid on the live feeder lamb before calculating the incentive on the shorn wool. Thus, if the feeder lambs weighed 80 lb, the rancher would receive \$2.734 per lamb as incentive (80 × 0.0342). If the lambs sheared 5 lb of wool that sold for \$0.60/lb, the incentive would be figured as follows: 5 lb × \$0.60 = \$3.00 received for the wool. Then \$3.00 × 107.5% = \$3.225 minus \$2.734 incentive paid to grower = \$0.49 per lamb incentive.

REFERENCES

Botkin, M. P., Ray A. Field, and C. LeRoy Johnson, *Sheep and Wool.* Englewood Cliffs, N.J.: Prentice Hall, 1988.

Ensminger, M.E., *Sheep and Goat Science*. Danville, Ill.: Interstate Printers and Publishers, 1987.

McFadden, W. D., *Wool Science.* Boulder, Colo.: Pruet Press, 1967.

Onions, W., *Wool: An Introduction to Its Properties, Varieties, Uses and Production.* New York: Interscience-Wiley, 1962.

Ryder, M. L., and S. K. Stephenson, *Wool Growth.* New York: Academic Press, 1968.

Scott, George E., *The Sheepman's Production Handbook.* Denver, Colo.: Abegg Printing Company, Inc., 1982.

Von Bergen, W., *Wool Handbook.* New York: Interscience-Wiley, 1963.

QUESTIONS FOR STUDY AND DISCUSSION

1. Describe the chemical composition of wool.
2. In what ways is wool superior to artificial fibers?
3. What is felting quality, and how is this related to the structure of the wool fiber?
4. The highest quality wool is grown on which area of the body of a sheep?
5. The fleece from what area of the body is usually most soiled?
6. What is meant by territory wool, and how does it differ from wool grown in the fleece states of the United States?
7. Why doesn't grade determine the value of a fleece?
8. Describe the three different methods of grading wool.
9. What methods should be used in determining soundness of a fleece?
10. What is meant by shrink in wool?
11. Why are particles of plastic so discriminated against in wool?
12. Outline a program for producing superior wool.
13. What is the Wool Incentive Program and how does it work?
14. What is the function of the ASPC?

15 Maintaining Healthy Sheep

There is a great volume of literature on the diseases and parasites of sheep. Any study on the pathology of the *Ovis aries* species will indicate that there are many diseases in existence, but in total not many are problems in the United States. Sheep are hardy animals that have evolved over many centuries under primitive conditions with only rare outbreaks that have decimated their numbers comparable to such a disease as cholera in swine. They are responsive to treatment, and if managed intelligently and protected from exposure to diseased animals, they may be maintained in healthy, productive condition without a great deal of effort.

The old saying "a sick sheep is a dead sheep" is true only if the shepherd is blind to symptoms and insensitive to the basic needs of the flock. A skilled shepherd observes abnormalities and takes immediate action to correct them and to prevent their reoccurrence. Keeping sheep healthy involves planning to minimize risks and an understanding of the steps to be taken if, in spite of protection and prevention, the physical well-being of the flock is threatened. In planning a health program, it appears that the following fundamental assumptions are necessary.

15.1 ASSUMPTIONS REGARDING HEALTH OF SHEEP

1. Prevention of diseases and other health problems is generally preferable to cures.
2. A sheep flock that is properly nurtured and kept in a comfortable environment is much more resistant to diseases and parasites than one that has been neglected or mistreated.
3. Disease outbreaks are usually the result of failure to select healthy, disease-resistant replacements or to isolate and observe new purchases until they are proven to be healthy.
4. Spread of disease can be checked and losses reduced by maintaining a hospital pen where sick animals can be quarantined and their condition diagnosed and treated.

5. The assistance of a trained veterinarian is valuable in planning and maintaining a satisfactory health program.

15.2 ROLE OF THE VETERINARIAN

Colleges of veterinary medicine and surgery teach principles that, if followed, will allow their graduates to minister to domestic animals. Whether or not the students are given much exposure to sheep depends largely on the area of the country where the institution is located. The clinical experience is limited to the animals that are submitted for treatment by the clients. Thus, it may be that the local practitioner may have had little experience with sheep. This often leads to the statement; "Our vet doesn't know anything about sheep." To some extent this may be true initially, but it does not have to stay that way. Many of the best sheep doctors had very little experience with the species before they set up practice in sheep country. If the producers are patient and give the veterinarian an opportunity to work with them, they are nearly always repaid by developing an interested, valuable asset to their business.

Not many shepherds have the training to diagnose causes of unexplained deaths. Few of them can do surgery or prescribe drugs for unusual maladies or treatments for other than routine problems. The worst relationship is only to seek the assistance of the veterinarian when the sheep is nearly dead. Then it is usually necessary to pay a bill for a dead animal, while the veterinarian leaves the farm frustrated and fuming because the odds were against saving the victim. Nobody likes to lower their batting average. If a condition is observed that is unfamiliar and serious, medical assistance is indicated. If your local practitioner does not know what is wrong, the case can usually be referred to a state diagnostic laboratory, which requires time and money. If the local veterinarian can diagnose the disease or problem, correction is nearly always easier and quicker. Also, a health program can be worked out much more easily if your veterinarian knows the situation on your farm and assists with the formulation. While it is true you should have a program specific for your farm, the following general practices should be considered and possibly incorporated into it.

15.3 PRACTICES FOR MAINTAINING HEALTH OF SHEEP

1. Practice sanitation.

a. Provide clean water to sheep in waterers that cannot be contaminated with feces. If ponds are used, fence them and pipe the water to a tank. If cisterns are used, make sure surface water does not contaminate it. Plan waterers so that mudholes don't develop around them. Use gravel to good advantage. Fill in mudholes in lots and pastures. Fence out swamps and sloughs.

b. Provide adequate feeding facilities. Feeders should be built so feed will not be contaminated by feces or by animals walking in them. Hay should never be fed on the ground. Don't use refused hay for bedding pens.

c. Keep the barn clean and bedded. Barns should be cleaned at least once each year. When they are cleaned, walls and ceilings should be swept down and sprayed with disinfectant. Concrete floors should be scrubbed and disinfected. Bedding should be used in barns if the manure is wet and sticky.

d. Dispose of bedding used by sick animals where healthy sheep will not come in contact with it. Dispose of dead sheep by burning or burying them or using a dead animal service.

e. Provide your own shearing floor, canvasses, and the like, for shearing or require shearers to disinfect theirs before using.

f. Require shearers to disinfect shears before using on your sheep.

g. Sterilize all equipment used for docking and castrating before use. Sterilize needles used for vaccines administered between sheep.

h. Sweep hallways and passages to keep down dust.

2. Keep down traffic on your farm.

a. If a dead wagon is used, meet it at the road and don't allow it to come on the farm.

b. Provide a shallow metal vat with disinfectant in it for people who visit your barns. A 5% cresol solution soaked into foam rubber in the mat will kill bacteria. Another measure is to keep a pair of boots on hand for visitors.

3. Prevent exposure to infected animals.

a. Purchase replacement sheep only from healthy flocks. Quarantine all newly purchased livestock for 3 to 4 weeks. Observe them closely, and only turn with the flock if they are healthy.

b. Have good perimeter fences so livestock from neighboring farms don't come in contact with yours.

c. Do not allow sheep from other farms on your place.

4. Observe sheep at least once each day. The time to see them best is early in the morning when they are just leaving their beds. If you are quiet and make no sudden moves, they will allow you to approach them close and note the ones that are sick. Sheep should be observed often when they have just been turned on a pasture if it contains legumes, which may cause bloat. Vigilance is also required if predators are a problem. A good shepherd should constantly be on the lookout for abnormalities or changes in the behavior of the flock. These will often provide a clue to a developing problem.

5. When sheep show symptoms of disease, parasites, or other abnormalities, remove them from the flock and treat immediately if the problem is understood. If not, obtain a diagnosis with a suggested treatment.

6. Keep a record of animals treated, death losses, and those that are culled with the reasons.

7. At least once a year, study the records and decide on strategy for the future. If, for example, losses of newborn lambs are excessive, it usually indicates incorrect management or facilities not adequate for the job. Change the health program when records indicate that an alteration is needed. Be wary of sweeping changes. A program that has evolved is generally better adapted to local conditions than a revolutionary one, which may ignore all the good features that were successful in the past. Don't be disappointed if you are still making changes in your health program 20 years later. If changes are only made for valid reasons, it shows that the regimes are keeping pace with new technology and natural shifts in the industry.

The following is a discussion of parasites and diseases that affect sheep and suggested treatment and control.

15.4 INTERNAL PARASITES OF SHEEP

Of all the losses that occur in sheep, internal parasites may be the most universal. Few sheep are entirely free of worms of some kind and most are hampered in their production, to some degree, by the parasites. Losses from internal parasites are usually not as spectacular as a disease that causes sudden death, but they may be greater overall, because the performance of the whole flock may be lowered and some individuals may be stunted so that they never do well.

15.4.1. Stomach worms Throughout the world, the parasite that lives in the abomasum (true stomach) and derives its nutrition from sucking blood from the walls of the organ is known as the stomach worm or by its scientific name, *Haemonchus contortus.* This adult roundworm is approximately 1 in. (2.5 cm) long with a diameter of a common pin. It is called a barber pole worm, because the ovarian tubes are coiled around the intestine of the worm. When the intestine is full of blood, it is bright red; thus the red-and-white contrast.

Although the loss of blood ingested by the parasite is harmful to the sheep, there is a further loss due to continued internal bleeding after the parasite detaches itself. When many thousands are present, a sheep may become anemic in a very short time. The best visible symptom of stomach worm infection is anemia, as shown by paleness of the mucous membranes of the eyes and mouth. If the eyelid is white on the inside and the lips are white, the sheep is probably wormy. In advanced stages, the sheep will often have edema (swelling of tissues) between the jaws called "bottle jaw," off-character of the wool, and loss of condition (see Figure 15.1). Scouring is not a symptom of stomach worm infection, although it may occur. Lambs with a heavy load of parasites may, if exercised, have difficulty breathing.

Figure 15.1 A lamb with a severe infection of stomach worms (*Haemonchus contortus*).

A very good test to determine stomach worm infection is to draw a blood sample and do a hematocrit using capillary tubes and a centrifuge to determine the percent of packed red cells. Normal sheep have values of 30% or more. Infected sheep may be as low as 15%. Any values in the low twenties indicate a severe infection. Worm eggs counts of the feces will indicate the kind of parasites present, but will not necessarily tell the degree of infection. In very serious cases, the egg counts may be very low because the population is so great that reproduction has been curtailed.

It is a serious error to wait until lambs are unthrifty in appearance before treating them. Many lambs may have such a rapid buildup of parasites that they still are in good condition and yet are so heavily loaded that they are near death.

Some facts about *Haemonchus contortus* (stomach worms) are worth remembering. The stomach worms mate in the abomasum, and their eggs pass out with the feces. The eggs hatch in a few hours if the temperature is above 40 °F (5 °C) and there is adequate moisture. If the temperature is above 70 °F (21 °C), and there is sufficient moisture, the embryo develops within 2 to 4 days into sheathed larvae that are very active and resistant to weather. This is the infective stage of the parasite. The larvae crawl up the blades of grass and are eaten by grazing lambs. They mature in the abomasum within 15 days and are ready to reproduce (see Figure 15.2). Ideal conditions for the propagation of parasites are warm, humid weather, lush, rapidly growing forage, and a heavy concentration of sheep on an area. Under those conditions, the life cycle outside of the body of the sheep may require as little as 5 days. Sheep develop varying degrees of immunity to stomach worms after they have been con-

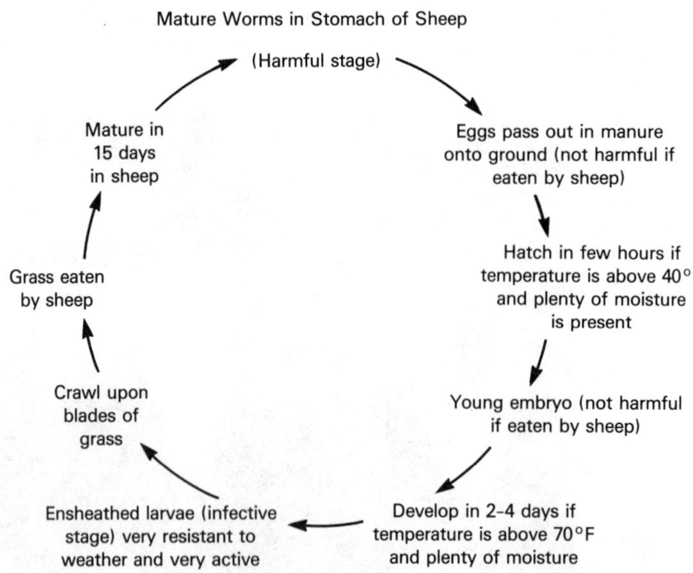

Note: That portion of the life cycle completed outside the sheep is very dependent on temperature and moisture. It may be completed within 5 to 6 days or it may take many months.

Figure 15.2 Life cycle of the stomach worm (*Haemonchus contortus*).

tinuously exposed to them. Young lambs have no immunity and, thus, are much more severely affected than mature sheep.

The stomach worms in the digestive system of a sheep go into a dormant or arrested stage of development during the winter months and thus are able to survive until spring. Research at the Ohio Agricultural Research and Development Center has demonstrated that the dormant worms develop into very active adults when the ewe approaches parturition in the spring so that, if not checked, the ewes can soon seed the pastures with worm eggs and cause the lambs to be infected. Long, continued use of an anthelmintic may cause the development of resistant strains of parasites. Work by LeJambre (1976, 1978) and LeJambre, Southcott, and Dash (1977, 1978) demonstrated the development of resistance to drugs. The alternating of drugs was not beneficial, but actually developed strains that were resistant to all the drugs. They suggested that one effective drug be used and continued until there was a decrease in efficacy.

It has been suggested by various authorities that a systematic drenching program combined with proper grazing management is the best strategy for controlling the parasites.

Drugs approved for worming sheep in the United States are rather limited due to the extensive tests required by the Food and Drug Administration. Many drugs that have been routinely used in Australia, New Zealand, and other sheep-producing nations are illegal in the United States. Of those available for use without a prescription, Tramisol (riperol-levamisole hydrachloride) appears to be most effective. It has a wide spectrum and appears to destroy any stage of development of the parasites. It is also relatively safe to use.

Thiabendazole, also called thiabenzole, has a wide spectrum of activity and a low toxicity to sheep. However, parasites appear to develop a tolerance for the drug with continued use. Other drugs have been used reliably for many years. Phenothiazine (micronized), used as a drench, is not as effective over as wide a range as the newer drugs, but in combination with arsenic is useful for drenching about the first of July to control nodular and tapeworms. Copper sulfate, also called bluestone or blue vitrol, was for many years the standard anthelmintic for sheep. It still works, but not as well as the newer drugs. It has a further disadvantage that sheep should be fasted overnight before treatment and 3 to 4 hours afterward. It is mixed in a 1% solution, being careful to use only blue crystals and to thoroughly dissolve them. Dosage rate is 1 ml/20 lb (9 kg) body weight with a maximum of 4 oz (118 ml) per sheep. Nicotine sulfate may be added to the copper sulfate mixture at the rate of 1 oz per gallon (3.8 liters) of water for tapeworms.

Phenothiazine and copper sulfate are mentioned because they are relatively inexpensive. If prices are so high that the newer drugs cannot be purchased, these two drugs still work. Since new drugs are continually coming on the market, it is important to check periodically with your state sheep extension specialist for the newest recommendations. Effectiveness of anthelmintics is influenced by date of administration. Based on data from extensive studies, Herd and Parker (1977) of the Ohio Agricultural Research and Development Center (OARDC) recommend the following treatments.

Strategic Anthelmintic Treatments Recommended by OARDC

1. *Winter treatment:* kills arrested worms and prevents reactivation and seeding of pastures.
2. *Prelambing:* kills arrested worms and adults, thus preventing the lactation surge of sexual activity of the worms. If ewes and lambs are moved to uncontaminated pastures, control is easier to accomplish.
3. *Weaning:* treat lambs at weaning and move to a clean pasture.

In research conducted at the Missouri Experiment Station by McDonald, Thompson, and Ross (1978), it has been shown that, if ewes are not allowed to seed down lamb pastures after the last hard freeze in the spring, there is very little advantage to drenching lambs grazing the clean pastures. Research has demonstrated that most eggs and larvae do not survive cold winters; thus it is important to drench mature individuals before they are put on pasture in the spring.

Two other worms found mostly in the western states are *Ostertagia,* known as the brown stomach worm, and *Trichostrongylus,* the stomach hair worm. These are no more than half the size of *Haemonchus* and difficult to find in postmortem. If parasitism is a problem, it is advisable to take some infected sheep to a diagnostic laboratory to determine the cause of the problem. The life cycles of all the stomach worms are similar. Since it is possible that the sheep are infected with more than one kind of stomach worm, it is advisable to use a broad-spectrum anthelmintic for their treatment.

15.4.2. Worms of the intestines The intestinal hair worm (*Trichostrongylus*), the thread-necked intestinal worm (*Nematodirus*), and the hookworm (*Brunostomum*) are parasites of the small intestine. In large numbers, they may cause severe damage to the walls of the small intestine and severe weight losses, generally in late summer or early fall when pastures are poor.

Nodular worms (*Oesophagostomum columbianum*) have a similar life cycle to other round worms, but they attach themselves to the walls of the large intestine. They burrow through the mucosa and into the muscle. Whitish cysts form around the parasite. Later, as the worm develops, the area around it is filled with whitish-greenish materials that become hard and raised into nodules, which remain when the mature worm leaves to mate and continue the life cycle. Young lambs may grow slowly or actually die from heavy infections. The usual damage is caused by tissue destruction, interference with digestion, and infections.

15.4.3. Other internal parasites The thread lungworm (*Dictyoaulus filaria*) is the most important lungworm affecting sheep in North America. It is most prevalent in the cooler regions.

The worm causes severe irritation and coughing. With heavy infections, there may be blockage of the air passages, causing rapid and labored breathing. Heavily infected sheep usually become thin and listless and may die of pneumonia after several months. The adults deposit eggs in the air passages and are moved up the air passage by coughing by the sheep. They are swallowed and hatch in the intestine, and the first-stage larva pass out with the feces. They molt twice and reach an infective stage in approximately 10 days. They are ingested by grazing sheep and pass through the digestive system to the small intestine.

The larvae penetrate the walls of the small intestine and enter the lymphatic system, where they molt in the mesenteric lymph nodes. Then the larvae migrate to the heart and the pulmonary system and finally the alveoli, bronchi, and trachea, where they develop into adults in 2 weeks. Lungworms are resistant to most drugs, but adults may be removed by treatment with L. tetramasole. If the parasite is a problem, do not graze sheep on low, moist areas.

Tapeworms (*Monezia expansa*) commonly infest sheep. Segments may be observed in manure from time to time. The parasite is white and approximately ½ in. (12.7 mm) wide. Within the small intestine, they may grow to a length of as much as 10 ft (3 m). They can cause severe scouring in lambs, and the parasite to some extent interferes with the action of the intestine. It also uses nutrients in competition with the host. Generally, the tapeworm is considered to be less damaging than other internal parasites. The life cycle begins with the adult worm, whose body consists of many segments filled with eggs. When segments are detached from the main parasite, they pass out with the feces. The segments disintegrate, leaving the eggs exposed. The eggs are picked up by mites, where they develop into an infective stage. When the mites are ingested with forage, the immature tapeworm passes through the digestive system and attaches itself to the walls of the small intestine. Tapeworms are seldom found in sheep older than a year. Apparently, they develop an immunity after being infested as a lamb. Sheep can be drenched with nicotine sulfate or lead arsenate to remove tapeworms.

Sheep are intermediate hosts to the dog tapeworm (*Taenia ovis*). The eggs are ingested with grass and develop into cysts chiefly in the liver and lungs, but may also be found inside muscle tissues. When these cysts are found in slaughter lambs, they require extensive trimming or even condemning of the carcass. In the meat industry, the condition is called "measles." Control of measles consists of preventing the infection of dogs by disposing of all dead sheep so that dogs cannot eat the carcasses. Offal or scraps of uncooked tissue should never be fed to dogs or allowed to be consumed by any canine. Dogs should be wormed periodically to rid them of tapeworms.

Liver flukes (*Fasciola hepatica*) are a serious problem in certain parts of the West and Southwest in areas of low, wet pastures. The mature flukes mate in the bile ducts and lay eggs. The eggs pass through the digestive system and are deposited with the feces. The eggs hatch, and a free-living ciliated organism emerges. The intermediate host is a snail. They leave the body of the snail and encyst themselves on vegetation. When the cyst is consumed with the vegetation, the fluke escapes from the cyst and burrows through the wall of the small intestine and into the peritoneal cavity of the sheep. Once in the abdominal cavity, the fluke locates the liver and embeds itself in the tissues. The immature fluke migrates through the liver, developing as it goes, until it is mature in 6 to 13 weeks. When it reaches the bile ducts, it mates and completes the life cycle. Mature flukes may live for many years in the large bile ducts.

Fluke attacks may be either acute or chronic. In the acute form, affected sheep suffer extensive damage to the liver due to invasion of many immature flukes. The sheep may die from hemorrhage. These attacks usually occur in the late summer and early fall. The infected animal may be in good condition, but usually has a distended belly and congested mucous membranes of the eye. Postmortem examination may show excess fluid in the abdomen, and the fluid may be stained with blood. The liver is enlarged, dark, and mottled, and if the surface is carefully examined, the scars left by entry of the immature flukes may be seen. The texture of the liver is soft and friable. In the chronic form, symptoms are caused by irritation and even blockage of the bile ducts, which causes the ducts to thicken and reduces the flow to the intestine. This causes the sheep to go off feed. Also, there is a loss of blood from the hemorrhages of the liver. Anemia is one of the best clues of infection. Bottle jaw may also be noted. The sheep loses condition and becomes unthrifty and may waste away and die in 1 to 2 months. Liver flukes are controlled most effectively by destroying the snails that are intermediate hosts. Copper sulfate at concentrations of 1 to 1.5 million parts water will kill the snails. Pastures may also be sprayed with $CuSO_4$ using 1 part to 100,000 to 1 million parts of water. An important step in controlling the disease is to drain swampy land and fence out sloughs and mudholes. Individual sheep can be treated by using carbon tetrachloride, hexachloroethane, or hexachloraphene. It is advisable to consult with the local veterinarian before treating sheep with drugs. The three drugs recommended for treatment for flukes are only effective against adults. For that reason, it may be necessary to treat the sheep several times to rid them of the parasite.

Nose bots are the larval stage of the sheep nasal fly (*Estrus ovis*). The parasite is also called "grub in the head" and may be a source of much annoyance to sheep from the warm summer days when the eggs or immature larva are deposited on the moist mucous of the nostrils through their development in the sinuses to their exit 10 months later. The life cycle includes a pupa or nymph stage in the soil for a month to 6 weeks and then the emergence of the adult. The cycle is complete when the adults mate and produce eggs and larvae. The bot reaches a length of approximately 1 in. (2.54 cm) and is about one-third as wide. The adult fly is somewhat smaller than the blow fly, a grayish-yellow in color and hairy. Sheep under sttack by the rapidly flying flies stamp their feet, rub their noses in the dust, run with their noses close to the ground, and seek shelter in a dark barn if it is available. The larvae cause much irritation, with heavy mucous discharge and frequent sneezing, with the head drawn back in an attempt to discharge the grub. The principal damage by the grub is in annoying the sheep and thus causing them stress. Treatment with ivermectin by a veterinarian may be effective in removing the parasite. Ivermectin is not approved by the FDA for general use. Otherwise, allowing the sheep to rest in a dark barn during midday will prevent some harassment by the flies.

15.5 PROGRAM FOR CONTROL OF INTERNAL PARASITES

1. Know what parasites are a problem on your farm. (Use fecal examinations or postmortem to identify parasites.)
2. Practice strategic drenching or treatment:
   a. Either prior to or after lambing to destroy very active adults.
   b. Before turning to pasture.
   c. When weather conditions are most favorable for parasite buildup or when

symptoms indicate infection (in warm, wet weather, it may be as often as biweekly).
   d. Approximately July 1 for tapeworms.
   e. In the fall after the first hard freeze.
3. Provide plenty of nutritious pasture.
4. Rotate pastures.
5. Control growth of forage by clipping or grazing with cattle.
6. During periods when pastures are low in protein, feed a low level of protein supplement, such as 0.25 lb (114 g) of soybean meal per lamb daily.
7. Provide clean pastures for lambs by not grazing with ewes in the spring.

15.6 EXTERNAL PARASITES OF SHEEP

The two most common external parasites of sheep are keds and lice. Both spread by contact between sheep. Spread is quite rapid when sheep are closely confined.

Keds and lice irritate sheep and cause damage to wool because of excessive rubbing. Infested sheep also fail to graze or eat well and thus become unthrifty. Both parasites appear to be most damaging to young lambs. If a sheep flock is doing excessive rubbing and scratching, examine them carefully in a strong light by parting the wool in several places. Keds are easily seen, but lice are as small as a timothy seed and more difficult to see.

15.6.1. Keds Sheep keds, which are actually wingless flies and not true ticks, spend their entire life on a sheep living off blood from the skin. The life cycle of the ked begins with the mating of two adults. The female develops one larva in her body at a time. She deposits the larva in a sheath glued to a wool fiber. The sheath hardens and covers the pupating parasite. Development occurs in the pupa, and it emerges in 4 to 6 weeks in the winter, but in as little as 3 weeks in the summer. From the date of mating, the female ked can produce larvae in as little as a week, especially in warm weather. Keds appear to be more harmful to lambs than to older sheep. When ewes are sheared, keds immediately migrate to the lambs. Dipping, spraying, or dusting soon after shearing very effectively controls the pest. Once the flock is rid of keds, special efforts should be made not to expose it to sheep that may be infested.

A number of excellent insecticides are available for removing the keds. The agriculture extension representative in the area should be contacted for recommendations.

15.6.2. Lice Sheep lice may belong to any of three species. Two are sucking lice and one is a biting louse. The most common is the red-headed biting louse. The adult louse is small, approximately 1/25 in. (1 mm) long, and may resemble a timothy seed. The life cycle begins with eggs deposited on the hairs or wool fibers with a gluelike substance. The eggs hatch in 5 to 18 days. They pass through a nymph stage and are mature and ready to reproduce within 2 weeks after hatching. The parasites irritate the skin, causing scratching and rubbing off of wool. Sheep with lice may be restless and lose condition.

15.6.3. Control of keds and lice Control of keds and lice is relatively simple. Dip, spray, or dust sheep with Toxophene or other preparations recommended by the state agriculture extension service as soon after shearing as possible. If there are still parasites on the sheep, treat them again between September and November.

15.6.4. Scabies Sheep scab or scabies is caused by a small mite that spreads on contact. Infected sheep scratch excessively and bite their wool, causing bare areas. The bare areas have a characteristic texture that resembles parchment and are greatly thickened and furrowed. Mites are very small and difficult to see. They may be seen in edges of bare spots that appear greasy, glistening white, and glossy. There is a program to eliminate scabies in the United States. Federal law requires the sheep to be dipped under the supervision of a federal veterinarian. Sheep that are scratching excessively, biting their wool, and have rubbed off patches of skin should never be purchased without veterinary inspection.

15.6.5. Screwworm and wool maggots The larvae of flies may infest sheep in the region of the dock, the anus, and the twist. Sheep on lush pasture or in the feedlot that are loose in the bowels are most likely victims. Infested sheep are nervous, become feverish and depressed, and are unthrifty in condition. Generally, sheep with "fly strike" have a peculiar twisting, squirmish action as they try to rub and bite the affected areas. Screwworm infestation is more serious than the maggots of blowflies, because the larvae of screwworms burrow into the *living* tissues of the lamb and cause much damage and even death in heavy infestations. Open sores caused by irritation, biting, and rubbing may provide entry for bacteria and subsequent infection. Blowfly maggots may be found in wool that has been wet for an extended period. They feed on the organic material in the wool, and the area may become putrid. The adult flies lay eggs on damp or soiled wool. The eggs hatch within a few days and the maggots begin feeding. It has been estimated that a generation turnover can occur within 3 weeks in the summer.

A federal program to control screwworms has been quite effective, but they may still be a factor. Anytime an open wound is exposed, there is danger from flies. Therefore, injured sheep should be treated and the wound covered with a repellent smear or spray. During fly season, it is best to use clamps to castrate lambs to avoid an open wound. If sheep have ear tags, look at their ears to see that they are not infested. Also, watch newly sheared lambs that may have wounds.

If sheep are loose or scouring, they should be tagged or crutched. In extreme cases, they may be sheared completely. Infested sheep should be treated with a 2% cresol solution to kill the maggots and then painted with a repellent.

15.7 DISEASES OF ADULT SHEEP

15.7.1. Abortion Ewes may abort because of disease, injury, severe malnutrition, poisons, or from reasons related to genetics. When a ewe gives birth to a lamb that is not completely developed, she should be isolated and the fetus examined by a veterinarian to determine whether or not it is an infectious disease. Care should be taken to clean up any contaminated bedding, hay, feed, or watering facilities. The ewe should not be returned to the flock until the trouble is diagnosed. Bang's disease caused by *Brucella ovis* is not nearly as serious in sheep as cattle. It does sometimes cause ewes to abort, but is most often found in rams and will be discussed under epididymitis.

The two principal diseases that cause ewes to abort are enzootic abortion of ewes, called EAE, and vibriosis. Both diseases are spread by infected ewes contaminating feed, water, or bedding with the placenta, fetus, or fluids of parturition. Both may result in significant losses of production.

15.7.2. Enzootic abortion of ewes (EAE) Enzootic abortion of ewes is widespread over the United States, but appears to occur most often in the mountain areas of the West. It is caused by an organism called *Chlamydia* and is characterized by premature births and abortions. Lambs born alive are weak and may die soon thereafter. Many full-term lambs are born dead. Ewes appear sick before and after they abort. They usually have a brownish vaginal discharge, and the placenta may be retained. It is important to preserve both the placenta and the fetus for laboratory diagnosis. Losses may be as low as 1% to 2%, but they may also be as high as 30% per year in subsequent years. Ewes that abort are apparently immune for life and may be kept. A vaccine is available that is effective for 3 years to life, but special permission must be obtained to use it. Preventive treatments with tetracyclines at the rate of 300 mg per ewe daily for 3 weeks at breeding and again 3 weeks preceding lambing will protect first lambing ewes. Strict sanitation to prevent contamination of feed and water and burning or destroying placenta, fetus, and contaminated bedding is advisable.

15.7.3. Vibriosis Vibriosis is a bacterial disease caused by the organism *Vibrio fetus*. The disease is often confused with EAE, but is usually more severe. Abortions occur late in pregnancy, but full-term lambs may be born dead or weak. The ewe may show no symptoms before she aborts or, if any are present, they may only be a slight swelling of the vulva and some discharge. Aborted lambs have swollen abdomens, and sometimes the liver upon autopsy will have white areas on it. The disease is spread by contamination of feed and water by infected ewes. Some veterinarians have been successful in preventing abortions by use of a *V. fetus* vaccine (5 ml) plus 5 ml of penicillin streptomycin, with a day of no treatment followed by 5 ml Penstrep injection each day for 3 days. Prevention can be accomplished by giving two annual vaccinations to all ewes, the first vaccination to be given at breeding time, and the second at the beginning of the last third of gestation. Another treatment to control abortions is to add tetracycline to the ewe rations at a level of 250 to 400 mg per ewe daily during the last 6 weeks of gestation. It has also been recommended that all replacement ewes be vaccinated before they are added to the flock.

15.7.4. Big head (light sensitization) A disease of the summer months may be a serious problem in certain areas of the United States. In midwestern states, it is sometimes called "stubble field disease." It is a noninfectious and noncontagious disease characterized by swelling of the ears, eyelids, face, and lips, which are bare or lightly covered with wool. According to Uren (1938), who described the disease in Missouri Extension Circular 386, it is thought to be caused by a combination of eating certain weeds or plants in the pasture that cause the formation of toxins in the blood and exposure of the sheep to direct sunlight, which activates the toxins and causes the edema.

Symptoms usually observed are restlessness, rubbing of heads on stationary objects, or scratching of heads with the hind feet. In a short time, the ears, eyelids, face, and lips begin to swell. Animals with the disease usually have a body temperature of 105° to 108°F (40° to 43°C). As the swelling continues, the animals show extreme irritation and are continually rubbing their heads against stationary objects. The disease progresses rapidly. There is a discharge from the nostrils and a profuse flow of tears out over the lower eyelid. The swollen parts of the head are hot and painful and there is usually an oozing of serum out onto the surface. The serum dries, forming a crust. The frantic scratching causes mechanical injury and bleeding. The mixture of tears and the yellow serum dries and forms a brown or black crust over the surface. The swelling usually crests within 24 to 48 hours after the first symptoms are noted. If the animals do not die during the high-fever stage, the affected skin may become leatherlike and dry with crackling and sloughing. Under extreme conditions, the ears may be sloughed off, dry up, and curl. Flies may then blow the injured areas and they may be infested with maggots.

Not all affected sheep show the swollen heads. In some flocks, there is only a jaundice, which shows up as a yellowish discoloration of the eye and a depressed appearance of the sheep. On postmortem, the body fat is usually yellow in color.

Treatment consists of removing the flock from the pasture immediately, confining them to a dark barn, and feeding a dry laxative feed, such as wheat bran and legume hay, plus plenty of fresh, clean water. This simple treatment removes them from the toxic weeds and prevents them from becoming sensitized by the sun. Some relief from the irritation may be obtained by application of a mixture of olive oil and lysol [1 pint (0.47 liter) olive oil and 1 teaspoon of lysol] or bathing the affected areas with an Epsom salts mixture [2 lb (0.91 kg) per gallon (3.8 liters) of water]. A repellent applied to the areas where the skin is broken will prevent fly strike.

Usually the confinement for 5 to 10 days with feed and water is sufficient to control the disease. But in the most severe cases when the victims refuse to eat, it may be necessary to drench the sick sheep with 1 pint (0.47 liter) of warm milk 4 to 5 times a day. On the first day, mix 0.5 pint (0.24 liter) of molasses with the milk early in the morning and late in the afternoon. If the bowels, which are often inactive with the disease, are not active at the end of the first day, repeat the molasses treatment the second day. If the sheep is still constipated the third day, administer an Epsom salts solution [3 oz (85 g) per pint of water] until the bowels move. As soon as the sheep begins to have bowel movement and is eating, discontinue the treatment.

Usually the plants causing the disease are restricted in area, and the toxins are not present for long. Always remove the sheep from the field where the condition occurs.

Another practice that may be desirable is to allow the sheep to have access to a darkened barn during the heat of the day. Not only will it protect them from the most intense irritation of the sun, but it will allow refuge from flies. One further observation is that big head usually occurs when pastures are dry and sparse. Sheep instinctively refuse to eat most harmful species of plants until they are forced to do so. This is another reason why a temporary pasture of sudan or some such forage may be useful.

15.7.5. Bloat Bloat occurs when an abnormal accumulation of gases occur in the rumen. Sheep naturally eructate frequently to relieve themselves of gases formed by digestion of feed. If a physiological condition is present that prevents eructation, gases will accumulate in the rumen, causing the sheep extreme discomfort and death in acute cases. A bloated sheep will be observed to have distension of the abdomen on the left side in front of the hindflank. In the most severe cases, the abdomen will be as tight as a drum and actually be raised above the back. Unless the pressure is relieved promptly, the animal will soon die. The most common cause of bloat is grazing a straight stand of certain lush legume pastures, such as ladino and alfalfa, when the sheep are exceptionally hungry. This type of bloat is caused by gases, but they are trapped in a frothy mixture of the vegetation and stomach contents so that the gases are not free to pass out. Frothy bloat is much better prevented than treated.

Gaseous bloat may be caused by eating too much grain or by mechanical means, such as lying with the back to the ground, which sometimes happens when short-legged sheep lie down in a ditch or on a hillside. These bloats may be relieved by passing a stomach tube to the rumen and releasing the gas. In extreme cases, if a tube is not available, the rumen may be punctured at the highest part of the swelling between the hip and last rib, using a trocar and canula if available or, if they are not, which is normal, a sharp, disinfected knife. Care should be taken to see that a hole made by a knife blade is kept open, and the gas contents not allowed to get between the rumen and skin.

The question that is frequently asked is how can you graze excellent legume pasture without having bloat problems. The answer is you can not always be sure, but you can minimize the odds by doing some things to prevent the condition. Some sheep are apparently natural bloaters and should not be kept in the flock. There is enough circumstantial evidence indicating a hereditary involvement to cause breeders to avoid using rams that are chronic bloaters.

Straight alfalfa or ladino pasture is not usually safe for grazing sheep. It would be much more satisfactory to have a grass–legume mixture or some coarse hay fed in addition to the pasture. Hungry sheep should not be turned on wet legumes. Always fill the sheep up with hay before turning them on pasture. Some very successful graziers routinely keep their sheep in drylot on hay all day and then in the late afternoon turn them to the pasture. They watch the sheep carefully, and if none bloat they leave them on the pasture continually. In Australia, a procedure followed is to fill the sheep up with hay and then turn on pasture for 20 minutes. The sheep are removed for 1 hour, followed by another 20 minutes on pasture and 1 hour in drylot, and the procedure is continued until they are full.

Poloxalene is a compound that appears to protect sheep from bloat. It may be provided in molasses block free choice, or if concentrates are fed, it may be added to the mixture at a level to provide 3 g/100 lb (45.4 kg) body weight.

Sometimes very high quality alfalfa hay causes bloat. If sheep are not accustomed to such high quality hay, they should be fed grass or mixed hay and the quantity of the alfalfa gradually increased. It may be necessary to routinely feed some coarser hay with the high quality hay.

15.7.6 Bluetongue Bluetongue is caused by a virus. As many as 8 to 12 strains of the virus have been identified. Not all the strains cause immunity to others. The disease is infectious, but not contagious. It is spread by a tiny biting midge (*Culicoides variopennis*). It is most often observed from midsummer to the first frosts of the fall, but it may occur all year in very mild climates. Symptoms observed are excessive salivation and reddening and swelling of the muzzle, lips, and mouth, which progresses to a deep red or bluish color. Swelling may interfere with blood circulation, causing the skin to turn white and to slough off. There may be a yellowish discharge from the nostrils. Eructation is often interfered with so that rumen contents appear on the muzzle and nostrils. The throat may become so swollen that the sheep may not be able to eructate so that the contents of the rumen cannot be swallowed, and they may pass into the lungs and cause pneumonia. Lameness may occur, and muscular weakness and emaciation are common with the disease. But not all the symptoms are observed on a single sheep. Since a number of symptoms are similar to other diseases, it is necessary to obtain a positive diagnosis.

Prevention consists of controlling the insect that carries the virus. The midge breeds in the mud or sandy areas around water tanks, water troughs, and streams. Spraying the breeding grounds with insecticide will destroy the adults and reduce spread of the disease. Modified live-virus-type vaccines may be used to protect the flock. In areas where bluetongue is a problem, it is recommended that all ewes be vaccinated at least a month before the breeding season and lambs not until they are 3 to 4 months old. Vaccinating pregnant ewes should not be encouraged, because the reaction may cause lamb losses due to head malformation. Bluetongue appears to be a disease of the western states to date.

Duration of bluetongue is usually 1 to 3 weeks. Death losses may be as high as 5%. To date, no drug treatment has been successful, but isolation in a comfortable, shady area with good-quality hay and clean, fresh water will reduce losses.

15.7.7 Coccidiosis Coccidiosis is caused by a minute, one-celled protozoal parasite. Protozoa are common inhabitants of the digestive tract where they do no apparent harm except under conditions of bad hygiene, overcrowding, and contamination of the feed or water by feces. The symptom of the disease first noticed is usually violent scours, which may be very dark or bloody. Later, the lambs appear unthrifty, are very gaunt, and may have a stiff gait as though paralyzed in the hindquarters. The organisms attack the lining of the intestines, causing bleeding and sloughing off of the epithelial tissue.

It may affect any age sheep, although it commonly causes more damage to lambs. Death losses may be as high as 20% to 30%, with many unthrifty lambs among the ones that recover.

The best way to cope with coccidiosis is to prevent infection by sanitation. Never feed hay on the ground or concentrates in troughs contaminated by manure. Make certain that the water supply is protected against fecal contamination. If Ruelene is available, it is very effective against the disease. Rumensin at the rate of 15 to 20 g/ton of feed is also useful in preventing infection.

15.7.8. Epididymitis Epididymitis is characterized in the ram by an inflammation of the epididymis, orchidis, and impaired fertility. It is an infectious, bacterial disease that may also be the cause of placentitis and abortion in ewes. The organism *Brucella ovis* is considered to be the primary cause of the disease, although there is evidence that other bacteria may cause it or contribute to the problem.

If the disease is caused by *Brucella ovis* (Bang's disease), it may spread to the ewes and cause abortion. The bacteria are spread by sexual contact between rams and ewes or even with other rams.

Detection of the disease consists of palpation of the testicles of all rams at least once and preferably twice a year. Rams with a distinct enlargement of the epididymis should be eliminated from the flock. It has been recommended that ram lambs, yearlings, and aged rams be kept in separate groups. Sheepraisers should learn to palpate the testicles of a ram so that they can do it themselves. Great care should be taken to make certain testicles of newly purchased rams are normal. While the rams are under quarantine, they may be blood tested for *Brucella ovis.* If the disease has been diagnosed and the causative organism is *Brucella ovis,* a vaccination program should be considered. Mature rams should be vaccinated 2 to 3 months prior to each breeding season and ram lambs at weaning, with another injection 4 weeks later.

15.7.9. Foot Rot In any survey asking sheepraisers to list the most serious disease of sheep, foot rot will usually rank first or second. The cause of the disease is bacteria, *Fusaformis nodosus* and *Bacillus necrophorus,* which act synergestically. *Fusaformis nodosus* lives only about 14 days in the soil. However, *Bacillus necrophorus* is common in manure and can survive in the soil for years. Foot rot is highly contagious and is spread by carrier sheep, which are chronic cases that may or may not show symptoms of the disease. This is why it is very difficult to eradicate once it is established in the flock. It has been estimated that if 5% of the sheep in a flock are limping, as many as 30% to 40% have the disease. The disease is seldom seen in areas of sandy, well-drained soil and in arid regions. Generally, predisposing factors precede the disease. Some injury or lowering of resistance of the skin or the feet, especially between the toes, is necessary for the organisms to enter the foot. The organisms attack the tissues beneath the horn of the hoof below the hairline. They progressively invade the sensitive laminae of the foot, destroying the tissues. One of the surest indications that a sheep has true foot rot is the smell. Once experienced, it will never be forgotten. It is surely one of the rottenest, most putrid odors anywhere. Severely affected feet have a foul discharge that is loaded with disease organism.

In the initial stage of the disease, the skin between the claws is reddened, with a small, moist area forming on the inner face or the bulb of the heel. The infection spreads under the thin, soft horn of the foot. During the next 2 weeks, the organisms continue to spread, undermine the sole, and extend under the walls of the hoof. By then, the horn is partly separated from the soft tissue beneath. Under the loose horn, there may be a layer of dirty gray, moist, slimy tissue that stinks. The sheep will not be able to walk on the foot. In the final stages, the disease destroys the tissues of the foot, and the horn is attached only at the coronet. This is the reason for the weird, distorted growth of feet of sheep with the disease.

The Sheep Industry Development Program, Inc., convened a group of veterinarians and sheep scientists in 1985 to make recommendations for the control of foot rot in sheep. The Foot Rot Panel (1985) recommended the following program:

Prevention of Contagious Footrot. Prevention is much superior to treatment. It is not only easier and less expensive but it is more effective. Recommended prevention practices are:

1. Never buy sheep infected with foot rot. Avoid purchasing apparently clean sheep from foot-rot-infected flocks. Avoid buying sheep at sale yards where clean and infected sheep are penned together.
2. Avoid common use trails and corrals where infected sheep have traveled or have been penned during the preceding 2 weeks.
3. Insist on proper cleaning and disinfection of commercial vehicles prior to transporting sheep.
4. Assume all new additions to your flock are infected with foot rot. Isolate new purchases for a mimimum of 2 weeks and trim all feet immediately upon arrival; treat feet of new sheep following trimming and reexamine periodically during the quarantine period (see treatment section).

Treatment of Infected Sheep

1. Vaccination: Vaccinate to prevent spread of foot rot and as a treatment for infected sheep. Two doses should be given 4 to 6 weeks apart. The first dose should be given approximately 2 weeks before a seasonal outbreak is anticipated. The vaccine is administered subcutaneously behind the ear.

2. Trimming: Feet must be properly trimmed if medication is to be effective. All diseased, dead, and undermined hoof areas must be pared away to allow medication and air to reach causative organisms. An effective trimming usually results in some bleeding. Topical medications are sprayed or painted on feet just after trimming or when foot bath facilities are unavailable. Use one of the following:

a. Ten percent zinc sulfate in water. Mix 0.25 lb (114g) zinc sulfate in 1 qt (0.95 liter) of water. Spray or paint on.

b. Ten percent copper sulfate in vinegar. Dissolve 0.25 lb (114 g) copper sulfate in 1 qt (0.95 liter) of vinegar. Spray or paint on.

c. Copper sulfate in pine tar. Mix two parts copper sulfate and one part pine tar. Paint on.

d. Oxytetracycline solution in alcohol. Add 25.69 g (one packet of Terramycin soluble powder to ½ cup (118 ml) of water. Add alcohol to bring solution to 2 qt (1.9 liters). Spray or paint on.

e. Penicillin solution in alcohol (10,000 units/ml). Mix 5 million units of potassium penicillin G with 10 mil of water and add solution to 1 pt (0.47 liter) of alcohol. Spray or paint on.

3. Foot baths: Solutions recommended for walk-through foot baths are:

a. Ten percent zinc sulfate in water. Mix 8 lb (3.6 kg) of zinc sulfate (hepta hydrate, $ZnSO_4 \cdot 7H_20$) in 10 gal (37.8 liters) of water.

b. Ten percent copper sulfate in water. Mix 8 lb (3.6 kg) copper sulfate ($CuSO_4$ blue vitrol or bluestone) in 10 gal (37.8 liters) of water. The $CuSO_4$ dissolves more rapidly in hot water. If the water is hard, addition of vinegar will assist in dissolving the crystals. *Caution:* Do not store this solution in metal containers. Unused solution should be stored in areas inaccessible to animals.

c. Five percent formalin in water. Mix 1 gal (37.8 liters) of 36% formaldehyde in 19 gal (71.9 liters) of water. This solution should be only mixed and used outdoors. It is the least effective of the foot baths.

Foot baths may be placed in a chute so that sheep are forced to walk through it or between food and water. The zinc sulfate solution seems to be most satisfactory for the voluntary walk-through baths.

4. *Foot soaks*: These are similar to foot baths but sheep are held in them for prolonged periods for more effective treatment. The 10% zinc sulfate solution used for foot baths is most suitable if a laundry detergent is added. *Never use formalin for foot soaks.*

5. *Dry chemicals:* When weather conditions are freezing, foot baths or soaks may be impractical. Then a mixture of 10% zinc sulfate in lime may be placed in a box between feed and water so that sheep must walk through. It is ineffective in the treatment of diseased animals but does reduce spread of the disease.

6. *Antibiotics:* These can be used for animals not to be slaughtered. They should be administered only after consulting your veterinarian. Use penicillin–streptomycin (Penstrep) at 1 ml/8 lb (3.6 kg) of body weight.

7. *Dry Pens:* Prolonged exposure to moisture is necessary for foot rot to develop. Sheep that have been treated should be placed in a dry clean area that has not been occupied by sheep for two weeks. A dry area will increase the effectiveness of topical treatments 60% to 80%.

Eradication of Foot Rot. The SID Foot Rot Panel developed a procedure for eradication of the disease. Eradication uses the same methods just outlined, but requires more commitment, perseverance, and willingness to cull. An initial increased expense in labor and facilities will be required, but will pay off many times over in the long run. Even dedicated use of the preceding control methods will not eliminate all carriers from a flock. Therefore, to eradicate foot rot, it is essential to have facilities for keeping infected animals isolated and to cull animals with recurring foot rot infection. Carrier animals keep the disease from being eradicated from the flock during treatment. Animals that periodically become infected or are the first ones to develop foot rot are carriers and must be identified. Once the animal is known as a probable carrier, one can keep track of the relapses. The use of an ear punch with a hole for each recurrence works well. Some carriers can be cured of the disease by dedicated treatment, while others do not respond to treatment. If, after employing vaccination, trimming, foot bathing or soaking, an animal still shows signs of foot rot, it should definitely be culled. Recent studies in the United States have shown that the incidence of foot rot can be reduced through selection. This genetic vulnerability to the disease further supports the culling of animals that fail to respond to successive treatments.

Producers with small flocks may wish to wait until lambs are weaned and for dryer weather conditions to begin an eradication program. Range flocks are often inaccessible during the ideal season for treatment, and therefore eradication measures must be introduced during confinement and lambing. This makes the process much more difficult and expensive, but the same principles must be employed. The following steps are suggested for range flock producers interested in eradication:

1. The herder must be able to recognize the disease from other causes of lameness.
2. The herder needs to be supplied with a means of permanently identifying any animal showing evidence of foot rot while on the range.
3. These animals should be removed from the flock at every opportunity and taken to an isolated area for treatment.
4. If there are relatively few infected animals, especially ewes, cull them!
5. If it is not feasible to cull, infected feet should be trimmed and treated. Vaccination may be used, but other treatments are recommended in addition for best results. If trimming cannot be done immediately, two soaks (1 hour each) 5 to 7 days apart in a zinc sulfate solution would be the treatment of choice. These sheep are not to be returned to the flock until completely cured of foot rot and, even then, they should be permanently identified. Cull those not showing signs of healing.
6. Foot bathing of all animals should be done if a large number of lame individuals are being removed from the band.
7. When confinement of animals for feeding or lambing is necessary, vaccination should be considered using two doses 4 to 6 weeks apart, or a single booster if vaccinated previously.
8. Zinc sulfate or other chemical foot bath solutions should be used every 5 to 7 days during confinement.
9. Bands should continually be monitored for lameness. Infected individuals should be removed immediately.
10. It is important not to mix infected animals with clean ewes during lambing or after, in spite of the inconvenience.
11. Cull any permanently marked sheep in which foot rot healed but recurred more than twice within a 2 to 3 month period.
12. When sheep go back to the range, all sheep that have had foot rot should be kept in a separate band. If there are not enough for one band, they should be combined with aged or other less desirable ewes. They definitely should not be combined with young animals, and their offspring should not be kept for replacements.

It will be impossible to eradicate the disease unless other users of common facilities are also working to eradicate the disease from their bands. Common users should set up cooperative programs. Animals lame due to foot rot should never be allowed use of common trails or facilities.

For help in implementing a cooperative drive against foot rot in your area, see your veterinarian, state animal disease control officials, state sheep specialist, state extension veterinarian, your county agent, and organizations of sheep producers.

Owners of smaller farm flocks should plan on starting their eradication program after weaning. Sheep should be tipped up and their feet thoroughly examined every 2 to 4 weeks; those with any suspicious signs can be removed and their feet trimmed. Foot bathing the entire flock whenever a new case is discovered will reduce the spread to normal animals. Treatment of infected animals should proceed as described previously. In this manner, carriers will be identified and eradication accomplished much quicker.

Treatment Schedule. Trim, treat (vaccine, topical foot bath), isolate, cull.

During First Week:

- Set up, trim, and examine all sheep.
- Separate flock into two groups: infected group and clean group.
- Hand treat infected feet with topical solution at time of trimming.
- Put both groups through foot bath or foot soak.
- Isolate clean sheep on clean ground that has been free from all sheep for two weeks.
- Isolate infected sheep for further treatment.

During Second Week:

- Foot bathe or foot soak infected sheep for second time.
- Examine sheep in clean group and trim, treat, and isolate any newly infected sheep (note Treatment and Eradication Sections).
- If any new cases are found, foot bathe clean sheep and move into another clean area if possible.

During Fourth Week:

- Reexamine all sheep.
- Foot bathe clean sheep. If any new cases are found, try to return to a clean area.
- Trim and treat infected group and establish a group of potentially clean sheep, foot bathe and return to clean ground (note Treatment and Eradication Sections).
- Foot bathe or foot soak infected group.
- Begin thinking about culling sheep that still show chronic signs of foot rot.

During Sixth Week:

- Reexamine all sheep.
- Move potentially clean sheep to clean group if no recurrence of disease.
- Always foot bathe clean sheep if a new case is discovered among them; try to put in another clean area. If no new cases, foot bathing and changing areas can be discontinued.
- Cull chronic cases of foot rot or repeat treatment procedure.
- *Note:* Clean sheep should be examined frequently for at least 3 months after the last case was discovered.

Further suggestions relative to foot rot treatment are:

1. Trimming feet must be thorough to be certain that diseased tissue is removed. It will probably be necessary to cut through healthy tissue to get to hidden pockets if infection. If a proper job is done, there is almost always some bleeding. Have a cord handy in case it is necessary to apply a tourniquet on the pastern. Don't leave a tourniquet on too long or it will permanently injure the foot. Topical treatments will not be effective if the foot is bloody. Use a sharp pocket or hoof knife, pruning shears, or nippers to trim the feet. A cradle for holding the sheep will result in reduced labor and increase the precision of trimming and treating feet.

   Begin trimming at the heel of the foot and then around the outside of the hoof. If the horn is separated, remove it. Next, trim the inside, leaving the toe till last. Probe for hidden pockets of infection with the point of the knife or locate them by tapping on the sole. Remove all pockets of infection and apply topical treatment.
2. Disinfect trimming equipment between feet and sheep.
3. Save all trimmings and burn them.
4. Construct a foot bath of concrete or wood so that copper or zinc sulfate can be used in it. Locate the foot bath outside. A water bath ahead of the foot bath will be beneficial in cleaning the feet before they go through the bath.
5. It is recommended that the foot bath be 16 to 20 ft in length, no more than 20 in. wide and 8 to 10 in. deep.
6. Sheep should spend 3 to 4 minutes in the solution when they go through the foot bath.

15.7.10. Other foot and leg problems These may be confused with foot rot and cause lameness. Anytime sheep are limping, they should be caught and examined. In most cases, very little is required to correct the trouble. Examples of causes of limping are bruises, sprains, cuts, scratches, stones, or mud wedged between the claws, and sometimes the interdigital gland may be plugged. Bruises and sprains should recover with time if the sheep are not stressed. Open wounds should be treated with a disinfectant and fly repellent. The plug in the gland may be removed by working it out with the fingers. Other causes may be associated with diseases such as arthritis, navel ill, tetanus, and white muscle disease. Generally, foot rot does not start in lambs although they may contract it from the ewes.

Foot abscess is a condition caused by the same bacteria, *Speropherous necrophorus,* that causes foot rot in cattle. The infection also affects the soft tissues of the foot and may cause severe lameness and swelling about the hoof. Joints may become involved in advanced stages. Abscesses may form and ooze pus when they break. But this disease can be differentiated from contagious foot rot by the absence of the putrid smell and the separation of the horn from the foot, which will not be noted. Opening up the abscesses and treating with disinfectants and antibiotics will usually control the problem. If conditions are unfavorable, such as very wet weather, and abscesses are numerous, it may be necessary to segregate and treat the infected animals and run the rest through a foot bath similar to the treatment given for contagious foot rot.

Foot scald is a contagious disease caused by the same organism or possibly a different strain of the organism (*Fusiform nodosus*) that is associated with contagious ovine foot rot. The disease occurs during extremely wet weather and in the initial stages resembles foot rot. It begins with inflammation between the claws and progresses to the rear portion of the heels, producing separation from the hoof. It usually affects the front feet only and may cause moderate to severe lameness. Treatment with chemicals used for foot rot is usually effective. Walking the flock through the foot bath at weekly intervals during wet weather will control the disease.

15.7.11. Listeriosis (circling disease) Listeriosis is a serious and often fatal disease of both mature sheep and lambs. The disease is caused by the bacterium (*Listeria monocytogenes),* which invades the brain, causing nervous symptoms. The organism is found in the nostrils and intestines of a large number of sheep. It can live in the soil, manure, straw, silage, or hay.

The disease is sometimes called silage disease. But if silage is well preserved with a high acid content (less than 5.6 pH), the organisms cannot survive. However, the organism often live in the top layer of a silo or in pockets of spoilage.

Outbreaks of the disease usually occur 3 to 4 weeks after silage feeding begins. The outbreaks also seem to be associated with blizzards and stress conditions. Sheep with listeriosis lose their appetite, are sluggish, have fevers up to 106 °F (41 °C), lag behind others, and may be found off by themselves. They may show congestion of the mucous membrane of the eye, paralysis of the tongue and jaw, blindness, and excessive salivation. They may also have a nasal discharge and only partly chew their feed. The name "circling disease" is used to describe their behavior. They usually hold their head to one side with one ear lower than the other and walk in circles, always in the same direction, because the brain on one side is damaged. Others may tremble excessively and stand with their head pushed up to a fence, feeder, or wall. Progressive paralysis usually causes them to collapse, and most are dead within 48 to 72 hours.

Abortion near the end of pregnancy may occur in ewes afflicted with the disease. Generally, circling and abortion do not occur in the same ewe. Although the number affected in the flock is usually not more than 2% to 3%, sheep with listeriosis seldom survive. The following have been recommended as measures that may help to prevent the disease:

1. Don't feed silage from the top of an upright silo.
2. Don't feed moldy or spoiled silage or hay.
3. Remove silage not consumed daily or that has been exposed to sunlight, causing it to heat up.
4. Provide plenty of clean, fresh drinking water that is not too cold.
5. Provide adequate quarters during cold weather.
6. Maintain a healthy, properly fed flock, including proper vaccinations and parasite control.

7. Avoid adding replacements from a *Listeria*-free area into a flock that has had cases of listeriosis.
8. Isolate ewes that have aborted with listeriosis. Burn aborted fetuses and fetal membranes. Exercise care in handling fetus, membranes, and contaminated bedding because the disease can be transmitted to humans.

Treatment is of doubtful value, although there is some evidence that antibiotics, particularly penicillin, if given early may be at least partially effective.

15.7.12. Mastitis (garget) In studies of the attrition rate of breeding ewes, udder problems are usually the principal reason for culling ewes. Mastitis is an infectious disease that may be caused by staphylococcus, streptococci, coliform or pasturella, and possibly other bacteria. It may be either gangrenous or nongangrenous. The gangrenous form is the more serious and sometimes is fatal.

Bruises, shear cuts, teats that are sore from nursing, and soremouth that has been transferred to the teats by lambs may provide entry for the organism. In the acute form, the disease causes a swelling and heat in the affected side or sides. Redness of the skin and feverishness of the ewe, lack of appetite, stiffness, and tenderness are common symptoms. The ewe may have no milk, or the teats and udder may be so sore that she will not allow her lambs to nurse. Within 24 hours, the skin of the udder may become moist and weeping and turn a dark-bluish color. This is the reason it is called "blue bag." The ewe may have a high fever and if not treated immediately may die in 1 to 2 days. In the chronic form, the udder may become abscessed and rupture, leaving a heavy scar. Usually the milk secretory tissue has been damaged and will never function again. Ewes that have had mastitis should be eliminated from the flock.

Prevention of mastitis consists of:

1. Careful selection of additions to the flock. Make certain that ewes with caked or lumpy udders are not selected.
2. Cull ewes that have had mastitis.
3. Isolate ewes with the disease to prevent their lambs, which are usually undernourished, from spreading the disease to healthy ewes by carrying the organisms in their mouths when they attempt to nurse.
4. Maintain strict sanitation in sheds, corrals, and lambing quarters.
5. Prevent injuries to the udder by removing high doorsills or other obstructions and by careful handling at shearing or when working or hauling sheep.
6. Treat all teat and udder sores promptly.

Treatment of gangrenous mastitis is usually not very effective, but some ewes will respond to wide-spectrum antibiotics. Sulfamethazine bolus (15 g) may be given immediately on discovery of the disease, followed by another within 24 hours. Streptomycin also may be effective. The swollen udder may be reduced by hot and cold packs and anti-inflammatory drugs. Drain and disinfect surface abscesses, being careful not to allow infected discharges to contaminate equipment or bedding used by healthy sheep. Ewes in close confinement appear to be more affected by the disease than those that are pastured or managed on the range. Thus, strict sanitation and segregation of diseased ewes must be followed when ewes are managed in close quarters.

15.7.13. Pregnancy disease (ketosis) Pregnancy disease is also known as lambing paralysis, ketosis, or twin-lamb disease. It is related to the metabolism of carbohydrates in the body of a ewe.

Pregnancy disease most often affects ewes that are either too thin or too fat during the last few weeks of pregnancy. It generally occurs in ewes that are carrying twins or triplets. Symptoms include the following: Ewes go off feed, have no energy, and get separated from the flock. They may be nervous at first and then later are dull and listless and have a staggering, unsteady gait. The ewe may fall and be unable to rise. Characteristically, she will lie with her head turned toward the rear as shown in Figure 15.3. She may appear blind and often will grind her teeth. The ewes have low blood sugar levels and ketone bodies in the blood and urine. The liver contains little glycogen, but is infiltrated with fat.

Prevention is simple and effective. These steps are recommended:

1. Furnish good pasture during pregnancy.
2. Feed hay in winter so that ewes will be required to walk at least 1 mile (1.6 km) daily to get it.
3. Don't allow ewes to get too fat before or during the first part of pregnancy.
4. Feed ewes concentrates during the last third of gestation so that they will gain weight.
5. Feed a balanced ration to pregnant ewes, including adequate vitamin A, minerals, and protein.
6. Furnish plenty of clean water in a convenient location.
7. Protect ewes from stress by providing shelter from rain, sleet, or snow. Keep handling to a minimum.

(a) (b)

Figure 15.3 (a) A ewe in early stages of pregnancy disease may be listless and lazy. (b) Later her limbs are usually uncoordinated and she lies with her head turned back.

Treating affected ewes must be done as soon as they show signs of the disease. Otherwise, the losses may be great. Treatment with 200 ml of glycerol (propyleneglycol) per ewe as a drench twice daily is recommended. Injections of insulin in addition to the glycerol enhances the treatment, but ewes should then be drenched with glycerol at 8-hour intervals.

An even more effective treatment consists of intravenous administration of 100 ml of 40% glucose solution, plus glycerol as a drench and insulin injections according to a veterinarian's recommendations.

15.7.14. Vaginal prolapse This condition is also called eversion. It occurs most often in overfat ewes, heavy in lamb, carrying twins or triplets or in pregnant ewes consuming large quantities of roughages. It also seems to be associated with lack of exercise. Ewes that are too fat and carrying twins have difficulty urinating if they are full of feed. When their bladder is distended, they must exert so much pressure to release the urine that they may force the vagina to break loose and appear outside the vulva. When the condition first occurs, the vagina may return to the normal position during the time the ewe is standing and walking, but it usually becomes worse and may remain outside the body at all times. The prolapse may block the opening of the urinary tract and prevent urination. Then the bladder may rupture and cause the death of the ewe. Attention to the ewe should be immediate to be effective.

First, cleanse the exposed organ with soap and water and dust it with an antibiotic powder. Elevate the hindquarters and gently return the prolapsed portion to the normal position. Two sutures across the vulva usually will prevent a recurrence. The sutures must be removed when the ewe goes into labor.

A device called a bearing retainer is available commercially that can be inserted into the vagina against the cervix and the arms attached to the wool by cords, as shown in Figure 15.4. If properly placed, it will not interfere with the ewe and she can lamb normally past the device.

There appears to be some hereditary basis for prolapse, and ewes tend to repeat in subsequent gestations. Therefore, it is advisable to cull the ewe from the flock. Docking replacement ewe lambs too close also may be a predisposing factor.

15.7.15. Scrapie Scrapie is an insidious and uncontrollable disease that results in a high mortality. It was first observed in the United States in 1947. It has been a problem in Europe for 200 years.

The cause of scrapie is not certain. Some authorities believe the condition is hereditary. Indeed, the work of Parry (1975) with extensive numbers over a period of years appears to justify his conclusion that the scrapie is caused by a natural autosomal recessive gene.

But there is a volume of research that indicates it is a disease caused by an agent similar to a virus because it is transmissible and the infection is amplified in the host, but antigenecity has not been demonstrated. The infection is persistent and degenerative rather than inflammatory.

The incubation period is long, with symptoms rarely apparent before the age of 2 to 3 years. The disease has been described as a relentlessly fatal neuropathy of

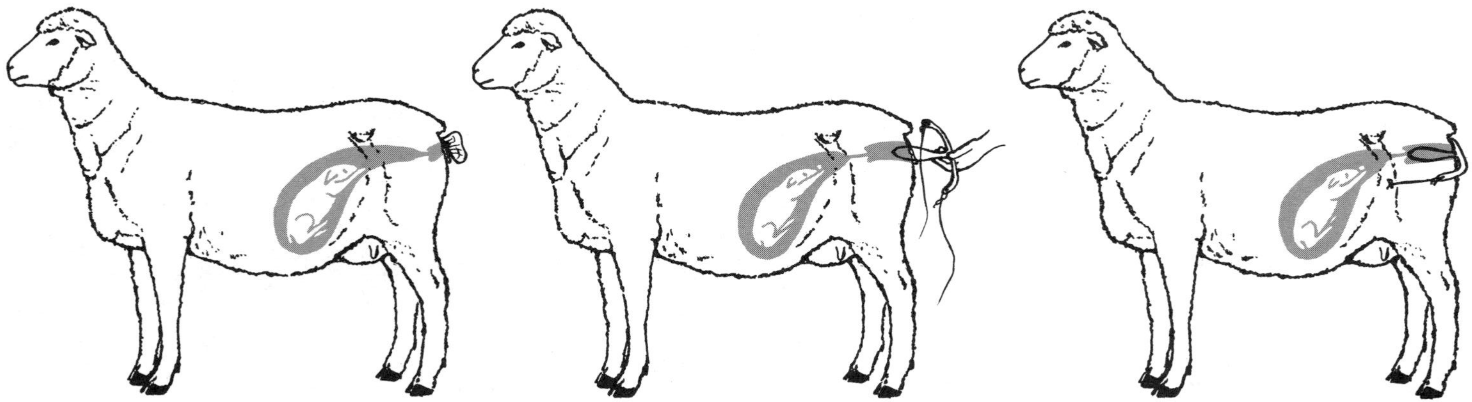

Figure 15.4 A bearing retainer in use to prevent vaginal prolapse.

sheep characterized by intense pruritis (itching), altered gait, and debility. Affected sheep are initially excitable, with fine tremors in the head and neck, and may have convulsions. As the disease progresses, the pruritis (itching) causes them to rub and scratch the affected area, with a resulting loss of fleece. Some areas are invariably rubbed raw. A common test for the disease is to rub or scratch over the loin and observe the sheep. Infected animals hold their heads high and make nibbling motions with their lips.

There is a federal eradication program for scrapie. The disease must be reported to state and federal veterinarians so the flock can be isolated and destroyed. Owners of the sheep are eligible to receive an indemnity based on a fair market value for the sheep.

Sheep from families that have been infected should neither be introduced into the flock nor allowed to remain.

15.8 DISEASES AFFECTING LAMBS

The greatest losses of lambs occur at birth and during the first week of life. Disease may be a major cause of death if proper precautions are not taken in managing growing lambs.

The shepherd must appreciate that the newborn lamb has to develop the ability to withstand radical environmental changes and resistance to disease and parasites and to fit into the social structure of the flock in a very short period. Prevention and protection are the two key words to disease control. Well-nurtured lambs that have started life in favorable environmental conditions are more resistant to disease than weak, poorly fed ones that have been exposed to unsanitary conditions and subjected to stress. Following are some common diseases of lambs and discussions of how to cope with them.

15.8.1. Diarrhea In a study by the USDA Sheep Experiment Station at DuBois, Idaho, it was shown that 46% of deaths at the station were caused by diarrheal disease. Whether or not the USDA experience is a true indicator of the extent of the problem, diarrhea, scours, or dysentery (as the disease is known) does account for heavy losses. Some authors have stated that scours is merely a symptom of a disease. Lambs may scour for a variety of reasons, such as sudden changes of feed, an oversupply of milk, coccidiosis, and hemorrhagic enterotoxemia. The most common cause of outbreaks in the United States appears to be caused by *Escherichia coli,* a bacteria that thrives in filth. Outbreaks are usually associated with chilling, cold, damp lambing quarters or with improper or erratic feeding of the ewes. Lambs with the disease are feverish, depressed, weak, colicky fail to nurse, and have a very fluid diarrhea, which may be tinged with blood. It may appear any time from 3 days to 6 weeks after birth. The younger the lamb, the more severe the effects of the disease. Dehydration and failure to eat are serious problems with young lambs and thus early treatment should be given. If many lambs are scouring, it is important to obtain a diagnosis of the disease. Always segregate scouring lambs and treat with antibiotics recommended by your veterinarian. After doctoring the lamb, wash your hands before working with another lamb. When the sick lamb is removed from the pen, clean it well and disinfect it. Feeding antibiotics to ewes seems to diminish problems with scours among other diseases.

15.8.2. Hemorrhagic enterotoxemia Hemorrhagic enterotoxemia is an acute, highly fatal disease of young lambs caused by the bacteria *Clostridium perfringens* (type C). The disease may affect young lambs, feeder lambs, and older sheep. Most commonly, vigorous, single lambs that are getting plenty of milk are affected when they are 1 to 4 weeks old. Small, weak lambs seldom have the disease. Symptoms may include the sudden death of apparently healthy lambs or the lambs may be depressed, refuse to eat, shiver, bleat, and show evidence of colic. Later the lamb may appear to be straining and black or bloody scours may be observed. The lamb may have a high fever and die before it begins to scour. Because the disease moves so rapidly, treatment is seldom successful. This disease is not widespread, but it does occur. If a positive diagnosis is made, vaccinate ewes twice the first year with enterotoxemia type C or C and D. The vaccinations should be separated with an interval of 2 to 6 weeks prior to lambing. Type C antitoxin has been useful as a preventive.

15.8.3. Enterotoxemia Enterotoxemia is also known as overeating disease, pulpy kidney, and apoplexy. All ages of sheep are susceptible to enterotoxemia. Lambs in the feedlot may sustain heavy losses. It is caused by a bacteria, *Clostridium perfringens* (type D), that is commonly found in the bowels of most sheep. The bacteria appear to cause no difficulties until the sheep is given high levels of concentrates. Then they may produce powerful toxins that are absorbed into the bloodstream and result in death. One of the best indications of the disease is the sudden death of some of the largest, fastest-gaining lambs. Other symptoms include nervous disturbances, such as circling, frothing at the mouth, grinding of teeth, butting, or throwing the head from side to side or backward. The animal becomes steadily weaker, finally collapsing and often going into convulsions before dying. Treatment for the disease is seldom successful because it occurs so quickly. Prevention in the feedlot is effective and not difficult by using good management and vaccinating.

Prevention of Enterotoxemia in Lambs

1. Management practices
   a. Follow recommended procedures for adjusting lambs to feedlot conditions.
   b. Treat lambs for internal parasites if needed.
   c. Follow proper procedures for maintaining lambs on feed.
2. Vaccination
   a. Vaccinate feeder lambs as soon as rested after they arrive at the feedlot, usually 2 to 3 days. Use 2 ml of toxoid. It requires 9 to 10 days to develop immunity.
   b. Consult with a veterinarian regarding the type and number of vaccinations to give.
   c. Vaccinate ewes twice if they have never been vaccinated, the second one 2 weeks after the first and 2 to 4 weeks before lambing. Give ewes an annual booster shot 2 to 4 weeks before lambing.

d. Vaccinate suckling lambs when they are 4 weeks old and again when they are weaned if your veterinarian recommends it.
e. To avoid troublesome abscesses, do not shear within 10 days of vaccination.

Sometimes a violent outbreak of enterotoxemia occurs with very serious losses. This usually happens in lambs that have not been vaccinated. Under those conditions, the lamb feeder is usually desperate to take action to cut losses. The following measures are suggested:

1. If tetracylines have not been included in the ration, do so immediately.
2. Reduce the concentrate levels to 50% or below and vaccinate larger lambs with type D antitoxins if they are fleshy and do not have long to go before they are finished.
3. Increase concentrate levels slowly to top levels to finish lambs.
4. Smaller lambs should be vaccinated with toxoid and held on rations containing less than 50% concentrates for 10 days and then worked back up to top levels.

15.8.4. Entropion Entropion is a condition in which eyelids, either the upper but more frequently the lower lid, are turned so that the eyelashes make contact with the eyeball. This irritates the eye and if allowed to continue will result in infection and blindness. It may occur in either one or both eyes. Eyes of newborn lambs should be examined and corrective action taken if the lids are turned under. In some cases, an ointment applied for a few days is all that is needed.

Bathe the eye with a mild solution of boric acid and either use a piece of thread through the eyelid and the cheek to hold it open, as shown in Figure 11.22 in Chapter 11, or use suture clips in the lower eyelid and skin below it to hold it open. If the eye appears to be infected, treat it with an antibiotic ointment. Since this is hereditary, make note not to save the lamb as a replacement.

15.8.5. Polyarthritis (naval Ill) Polyarthritis occurs in lambs as a result of improper sanitation and failure to disinfect the naval at birth or improper precautions when lambs are docked and castrated. Pathogens enter the cord or wound and cause infection. The infection often spreads to the joints, causing them to swell, and results in stiffness and inability to move freely. The infection may be so severe that lambs die within a few days after birth or they may linger on for weeks. The disease is caused by an agent (chlamydia) belonging to the psittacosis lymphogranuloma venerium trachoma (PTL) group. Treatment with antibiotics is sometimes effective, but not usually satisfactory. Prevention consists of clipping the naval cord to a length of approximately 2 in. (5 cm) and immersing it in iodine (7% solution) at birth. Sanitation in the lambing quarters and premises is imperative. Lambing pens should be kept clean and bedded with clean straw. In feedlot lambs, tetracyclines at 250 mg per lamb per day in the feed for 1 to 3 days is recommended.

15.8.6. Pneumonia Some authorities claim that respiratory diseases are the most serious problem with young lambs. In the farm flock states, it is often the greatest cause of deaths. In certain flocks of sheep, it may reach disastrous proportions. Mortality can be as great as 20% in lambs 24 to 48 hours old. It affects all ages of sheep, but its worse in young lambs.

Symptoms of the disease include labored breathing, refusal to eat, and fever. As the case progresses, lambs may become depressed and weak, and there may be discharges from the nostrils and eyes. Older sheep may cough and sneeze or may die without any visible signs that they are sick. Sheep that recover may have the disease in a chronic form and never do well.

Causes of the disease include bacteria, usually *Pasteurella hemolytica* or *Pasteurella multocida,* but viruses and other pathogens may be involved. Other causes may be dust, smoke, weed seeds, and lungworms. Stress also seems to trigger the occurrence. Chilling, exposure to wind and rain, and exhaustion from shipping and handling appear to be contributing factors. Crowding sheep in closed barns also increases the risk of pneumonia.

Prevention is the most satisfactory solution to a pneumonia problem. Steps that may be taken to prevent the disease are:

1. Thoroughly clean and disinfect lambing quarters well ahead of the lambing season.
2. Keep the moisture down in the lambing quarters by changing bedding frequently. Slaked lime on floors may assist in this.
3. Shear or crutch ewes before lambing.
4. Properly feed ewes so lambs will be strong at birth.
5. Reduce stress on ewes during lambing. Pregnant ewes should have dry shelter with a windbreak and run in a clean dry lot.
6. Eliminate ewes with respiratory problems from the flock.
7. Do not overuse heat lamps.
8. Move ewes and their lambs out of closed quarters as soon as possible.
9. In situations where pneumonia is a continuing problem, ewes may be treated with antibiotics before they lamb.

A regime for treating ewes with sulfonamide in the drinking water has been very effective in areas where pneumonia is a problem. The sulfonamide is added at the rate of 1 lb. (0.45 kg) or 1 gal (3.8 liters) of 12.5% solution to 128 gal (485 liters) of water. Pregnant ewes are then given access to the medicated water on the following schedule:

Days	Treatment
1–5	Medicated water
6–7	Nonmedicated water
8–9	Medicated water
10–11	Nonmedicated water
12–13	Medicated water

The treatment is most effective when it is administered 2 weeks before lambing. Ewes must therefore be grouped according to lambing date for treatment.

Bailey (1982), an eminent sheep veterinarian from Roseburg, Oregon, strongly recommends an intranasal vaccine called Nasalgen for prevention of the disease. It is manufactured by Burroughs-Wellcome Co. He recommends vaccinating ewes with 1 ml of Nasalgen before they lamb, and lambs with 0.5 ml^3 at birth. In situations where there is much difficulty wtih pneumonia, he also recommends vaccination semi-annually with 2 ml of pasturella bacterin.

15.8.7. Rectal prolapse Eversion of the rectum, also called piles, is a condition in which the rectum protrudes outside the body of the sheep. It is quite often seen in creep fed or feedlot lambs, but it is also seen in lambs on excellent pasture. The causes of the problem may include such varied ones as heredity, dusty feed (coughing), use of estrogenic compounds, internal parasites, injuries, and docking the lamb too close. Some authorities have stated that any one of these can cause the sheep to prolapse, but it is usually a combination that causes it. There is some evidence that certain lines of breeding are more susceptible to prolapse than others. Heavy coughing from any cause may be a contributing factor. It occurs in ewes as well as rams and wethers. Ewes that have prolapses may also suffer a vaginal prolapse when they lamb.

Treatment should be immediate. If the lamb is ready for market, it should be sent without delay. That, too, is a gamble, since if the lamb prolapses at the slaughter plant it may not pass inspection. Consider slaughtering it for home use if there is doubt about getting it passed by the inspectors.

Sometimes it is possible to treat the condition by washing it with soap and water and reinserting it. A purse-string suture will sometimes hold it in place. If the lamb is not close to being finished, it can be put on a higher roughage ration until the condition is cured. Some shepherds successfully amputate the prolapse by using a 3-in. length of garden hose inserted inside the rectum. The hose has a groove cut around the outside so that it may be held in place by putting an elastrator over the gut end and over the grooved end of the hose at the point where the gut leaves the body of the sheep. The hose is tapered on the inside so the feces can pass through it. Fecal material must be cleansed from the hose twice daily by flushing it out with a syringe. This operation can be performed by a veterinarian if the shepherd has had no experience and does not wish to try it.

15.8.8. Shipping fever (hemorrhagic septicemia and pasteurellosis) Shipping fever is usually associated with transportation and/or stress. Lambs with the disease may have high fever, pleuritis, pneumonia, weepy eyes, discharges from the nostrils that harden and cause difficult breathing, and a weak, feverish appearance. Some lambs may die suddenly 1 to 3 weeks after arrival at the feedlot or farm. This is usually the first sign that an outbreak is occurring. Postmortem examination may show lung congestion, lung adhesions, small subcutaneous and intramuscular hemorrhages, and subserol and other internal hemmorhages.

Although outbreaks may happen infrequently, they can cause a mortality of 10% to 25%, with many other lambs making only a partial recovery. The disease is infectious and contagious, and thus sick lambs should be segregated and treated separately. Causes are not clear, although it appears that pasteurella organisms along with pleuro-pneumonia-like organisms (PPLO), viruses, and possibly salmonella may be implicated.

It is generally agreed that long hauls without proper feed and water and exposure to inclement weather contribute to the conditions favorable to the appearance of shipping fever. Careful handling in transit and upon arrival, with clean water and bright-green grass hay, then plenty of rest, usually result in very little problem. Sick lambs may be treated with sulfonamide (1.5 g/lb of body weight), plus additional treatments of 0.5 g/lb of body weight per day on the second through fourth days. The rest of the shipment may also be drenched with sulfonamide to prevent spread of the disease. Vaccines are available that can be administered before shipping.

15.8.9. Soremouth (contagious ecthyma) Ecthyma is a virus disease that primarily affects the lips and mouth of sheep, but it is frequently transmitted to the teats of ewes by suckling lambs. It can also affect the nostrils, eyelids and mouth, vulva, and feet. First noticeable signs of the disease may be small red spots in the corners of the mouth. Blisters usually are followed by pustules, which break and scab over in 2 to 3 days. The scabs are dark and crusty. Sores inside the mouth may prevent the lamb from eating. Lambs are feverish and lose weight during the 1 to 3 weeks the scabs are present. The disease is spread by direct contact and may affect all the lambs in the flock. Death may occasionally occur due to starvation, but the greatest losses are due to setback and failure to gain while the lambs are infected.

Soremouth can be prevented by vaccination. If the disease has been a problem, lambs should be vaccinated routinely at a young age. Vaccination consists of scarifying the skin on the inside of the thigh and brushing the vaccine into the wound.

Humans are susceptible to the disease (''orf''), and thus care should be taken in handling the live virus. Rubber or plastic gloves are recommended. Sheep that are to be taken to shows should always be vaccinated at least a month prior to the show. Treating the sores is not usually very successful, but some sheep raisers do rub off the scabs and treat the wound with a combination of iodine and glycerine to prevent secondary infections. Some authorities discourage the removal of scabs and recommend working a vasoline–phenol mixture (3% phenol) into the lesions. Treatment of sore teats with an antibiotic ointment may prevent entry of mastitis organisms.

15.8.10. Tetanus (lockjaw) Tetanus is caused by an anaerobic bacteria, *Clostridium tetani*, an organism commonly inhabiting the soil and usually found in large numbers in horse manure. The spores live for years in the soil and may be a continuing problem on some farms. This is the reason for not choosing a sheep barn building site that was once a horselot or converting a horse barn for sheep housing.

The bacteria gain entry through wounds from shearing, docking, castrating, eartagging, and vaccinating. They may also infect the reproductive tract when the ewe requires assistance to lamb. Stiffness of limbs, difficulty in getting up and walking, and walking with a straddling gait are symptoms of the disease. In later stages the tail, jaw, and back may become rigid. Sudden noises may cause spasmodic jerking of the muscles.

There appears to be no effective treatment once the disease has developed. Prevention is the only practical solution. All surgical procedures should be done under strictly sanitary conditions. All instruments should be sterilized by immersing them in boiling water for approximately 30 minutes. On premises where the disease has been a problem, the use of elastrators for docking and castrating is discouraged. Immunizing the flock by two injections of the toxoid 30 to 60 days apart is recommended. In addition, an annual booster may be given prior to lambing. If tetanus is responsible for losses, it is advisable to work with the local veterinarian to develop control measures.

15.8.11. Urinary calculi (gravel, kidney stones, water belly) Formation of stones in the kidney, ureters, bladder, and urethra in male sheep occurs over a wide area of the United States. It is serious because it may cause a blockage in the urinary tract. When this occurs, the sheep will stand rigidly with its back arched and strain to pass urine. The animal will be in obvious pain, may kick at the belly, elevate one hind leg, lie down, and get up again. If the pressure is not relieved, the bladder may rupture, with the development of uremia and usually death.

Although the urinary calculi condition has been studied at length, there are still a number of unanswered questions regarding the reasons it occurs. It is known that certain factors appear to contribute to the condition. Rations high in sorghums plus cottonseed meal and cottonseed hulls have been observed to be calculogenic. Rations high in beet pulp have also been implicated. An imbalance between calcium and phosphorous, an excess of vitamin D, and the use of diethylstilbestrol are other factors that may cause the condition. The most satisfactory solution is to pay close attention to the calcium to phosphorous ratio, making certain that it is near 1.6 to 1 and adding ammonium chloride or ammonium sulfate to the ration at the rate of 0.5% (8 to 10 lb or 3.6 to 4.5 kg) per ton (909 kg) of mixed feed or 0.25 oz (7g) per head per day. If mixed diets are used, ammonium chloride may be mixed with molasses to increase palatability.

Treating a sheep with calculi is usually a problem for veterinarians. If the stones are lodged at the end of the penis at the urethral process and the surgery is done soon enough, the filiform appendage can be removed. This may save the life of the sheep, but it is not very satisfactory because of the difficulty of healing.

15.8.12. White muscle disease (stiff lamb disease, muscular dystrophy) White muscle disease results in the degeneration of skeletal and cardiac muscles. It is a nutritional condition associated with a vitamin E and/or selenium deficiency. It is much more prevalent in some areas than others. It is seen most often in lambs 1 to 8 weeks old. The disease appears to be related to selenium, vitamin E, and sulfur-bearing amino acid deficiency.

Ewes on selenium-deficient diets may be expected to produce lambs that develop white muscle disease. Some areas of the United States have soils very low in selenium. Feeds grown on such soils have a low selenium content. Regions known to be deficient are northern California, western Oregon and Washington, states east of the Mississippi River and north of the Ohio River, and possibly eastern Iowa, Minnesota, and much of Wisconsin. But there are pockets in the United States in areas assumed

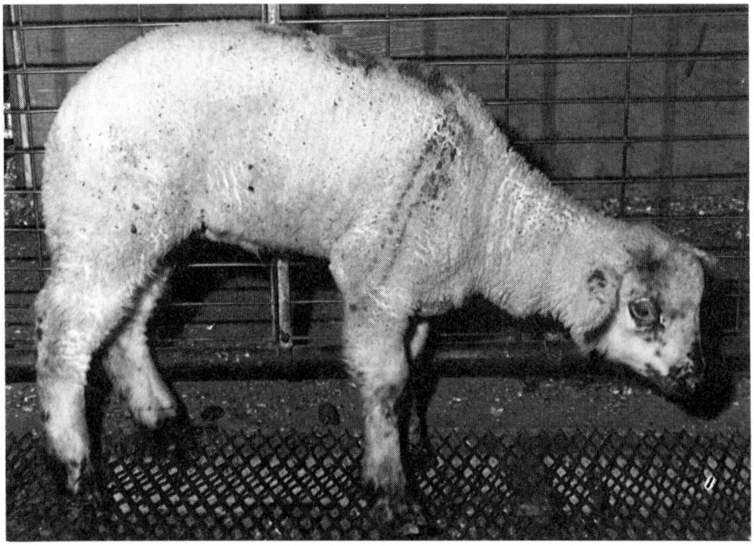

Figure 15.5 A lamb with white muscle disease due to selenium or vitamin E deficiency. (Courtesy Animal Science Department, Ohio Agricultural Research and Development Center, The Ohio State University, Wooster)

to be adequate in selenium. It also is possible that vitamin E in grains and forages may have been destroyed by overheating during drying or by storage for extended periods.

Except in the selenium-deficient areas, rations containing wheat, wheat bran or middlings, linseed oil meal, good pasture, high-quality legume hay, and distillers dried grains should be sufficient to prevent white muscle disease. If the condition is a problem, a number of steps can be taken. Use of a trace-mineral salt containing 39 ppm of selenium is usually sufficient except in very deficient areas. Caution must be exercised not to use a mixture formulated for other species, such as swine, which contains toxic levels of copper for sheep. Pregnant ewes may be given injections 1 month before lambing with Bo-Se, a product of Burns-Biotec, at the rate of 2.5 ml/100 lb (45.4 kg) of weight. This will protect the lambs at birth. In severe cases, it may be advisable to also treat lambs at birth with Bo-Se. If lambs develop the disease, immediate injection of Bo-Se appears to be an effective treatment. If sheep are on grain rations, a supplement providing 0.1 ppm selenium to the total diet should be fed. A lamb with white muscle disease is shown in Figure 15.5.

REFERENCES

Bailey, Don, Prevention of Pneumonia in Lambs by Vaccinating with Nasalgen. Personal communication with author, 1982.

Belschiner, H. G., *Sheep Management and Diseases.* Sydney, Australia: Angus and Robertson Ltd., 1965.

Blood, D. C., O. M. Radostits, and J. A. Henderson, *Veterinary Medicine, a Textbook of Diseases of Cattle, Sheep, Pigs, Goats and Horses*, 6th ed. London: Bailliere, 1983.

Coop, I. E., *Sheep and Goat Production.* New York: Elsevier Scientific Publishing Company, 1982.

Fraser, Allan, and J. T. Stamp, *Sheep Husbandry and Diseases.* London: Crosby Lockwood and Son Ltd., 1961.

Herd, R. P., and C. F. Parker, "Strategic Treatment for Stomach Worms in Sheep." Sheep Day Report, Ohio Agricultural Research and Development Center, 1977.

Jensen, Rue, and Brinton L. Swift, *Diseases of Sheep,* 2nd ed. Philadelphia: Lea and Febiger, 1982.

LeJambre, L. F., *Anthelmintic Resistance in Gastrointestinal Nematodes in Sheep. The Epidemiology and Control of Gastrointestinal Parasites of Sheep in Australia,* A. D. Donald, ed. Melbourne, Australia: Commonwealth Scientific and Industrial Research Organization, 1978.

______, "Resistance of Certain Lines of Haemonchus Contortus to Thiabendazole, Morantel Tartrate and Levamisole," *Int. J. Parasitology,* 6, 1976, pp. 217–222.

______, W. H. Southcott, and K. M. Dash, "Resistance of Certain Lines of Ostertagia Circumcincta to Thiabendazole, Morantel Tartrase and Levamisole," *Int. J. Parasitology,* 7, 1977, pp. 473–479.

______, ______, and ______, 1978, "Development of Simultaneous Resistance in Ostertagia Circumcincta to Thiabendazole, Morantel Tartrate and Levamisole," *Int. J. Parasitology,* 8, 1978, pp. 443–447.

McDonald, Cynde, J. M. "Thompson, and C. V. Ross, "Effect of Worming Lambs Grazed on Relatively Uncontaminated Pasture," *Missouri Sheep Day 1978,* pp. 37–40.

Merck Veterinary Manual, 5th ed. Rahway, N.J.: Merck and Company, Inc., 1979.

Sheep Industry Development Program, Inc., Foot Rot Panel, 1985, "Control of Foot Rot in Sheep," *S.I.D. Research Digest,* Vol. 1 No. 2, Winter 1985.

Uren, A. W. "Bighead or Light Sensitization in Sheep and Lambs," *Missouri Agr. Ext. Cir. 386,* 1938.

Yearbook of Agriculture, "Keeping Livestock Healthy," Washington, D.C.: U.S. Department of Agriculture, 1942.

Yearbook of Agriculture, "Animal Diseases," Washington, D.C.: U.S. Department of Agriculture, 1956.

QUESTIONS FOR STUDY AND DISCUSSION

1. Why should you reject the generalization that a sick sheep is a dead sheep?
2. Why are services of a veterinarian important even though the shepherd is an expert at working with sheep?
3. Discuss the importance of practicing sanitation on a sheep farm.
4. What steps can be taken to prevent the spread of infections in a flock of sheep?
5. Although parasites seldom cause heavy death losses, why are they considered more serious than most diseases?
6. What steps other than drenching are important in controlling internal parasites?
7. Outline the life cycle of the stomach worm and indicate at what points it can be attacked with best results.
8. What is an anthelmintic?
9. What is the site of damage of the *Haemonchus contortus* parasite?

10. Justify stringent efforts to prevent infestation with external parasites.
11. How can sheep lice and keds be identified?
12. What is the difference in the feeding habits of screwworms and wool maggots?
13. What is the difference between abortion by EAE and vibriosis?
14. What two factors must be present for sheep to have big head?
15. Discuss the kinds of bloat and how they differ.
16. What forages in a pure stand will likely cause bloat in sheep?
17. What drug can be used to protect sheep from bloat?
18. What is the surest indication that a sheep has foot rot?
19. What conditions are favorable to the spread of foot rot?
20. Discuss reasons for eliminating a ewe from the flock after she has mastitis.
21. Outline a program for preventing pregnancy disease.
22. Why are newborn lambs more susceptible to disease than adults?
23. What steps may be taken to prevent pneumonia in lambs?
24. What is the cause of white muscle disease, and how can it be prevented?

16 Protection of Sheep from Poisons and Poisonous Plants

16.1 POISONOUS CHEMICALS AND DRUGS

In modern agriculture, it is necessary to use various chemicals for controlling weeds and plant and animal parasites and for other purposes around the farmstead. Many of the chemicals and drugs are harmful if ingested by animals. If care is not exercised, they may be eaten and cause difficulties if not death to sheep. A junkpile or a dump, as shown in Figure 16.1, in a pasture may have empty cans and cartons that could be the source of compounds dangerous to sheep. When a container has been emptied, it should be disposed of in a safe place where livestock cannot be exposed to it. If landfills are available, these types of containers should be taken there for disposal or a fenced-off area should be used for the purpose.

Directions on the label of drugs or chemicals will state if compounds are dangerous. If they are, exercise special care in storing and handling them. Poisons used to rid the premises of rodents and predators are usually lethal to other animals too. They should be stored in a safe, dry place and used according to directions.

Errors in administering or carelessness in handling or storing drugs are frequently the cause of unnecessary deaths. For example, copper sulfate in concentrations suitable for treating sheep for foot rot if used for drenching sheep for internal parasites is usually disastrous. This problem can be eliminated by proper labeling of all containers. Sheep can also be poisoned by copper sulfate if they are very thirsty before they go through a foot bath. Arsenic is a poison that accumulates in the tissues of an animal. Too frequent drenching or dipping with arsenicals may result in lethal doses. Thus, it is recommended that arsenic be used in a drench for tapeworms only once each year. Lead may also accumulate in the body of a sheep. This is a good reason for never painting any sheep equipment with lead-based paints. Chemicals used to control flies in a barn may accidentally be sprayed on feed, or seed grains that have been treated for weevil may contain lethal doses of a poison. Lead, zinc, mercury, and phosphorous compounds are all poisons under certain conditions. If

Figure 16.1 Dumps such as this should be fenced to prevent sheep from being poisoned.

animals die unexpectedly of unknown causes, this is another reason for getting a post-mortem examination to diagnose the cause of death. Such compounds as phenol if improperly mixed in a dip may poison by absorption. Nicotine sulfate or carbon tetrachloride used as anthelmintics in overdoses may be lethal. Salt may poison sheep if they have been deprived of it. Then care must be taken to limit intake until they are accustomed to salt again. Any limitation of access to plenty of fresh water may contribute to poisoning of other species as well as sheep. A salt brine mixture may be drunk by very thirsty sheep with disastrous results.

Nitrates are poison to sheep and may occur in lethal quantities in drinking water or feeds contaminated with fertilizer. Natural sources will be discussed later. Healthy, well-fed animals are much more resistant to nitrate poisoning than starved, weak ones. If cisterns are used for the water supply, it is advisable to make certain that surface runoff does not contaminate the well. But it is also possible to have high nitrates without surface runoff. Periodic tests for nitrates and bacterial counts of the farm water supply are good procedures to follow.

16.2 POISONS IN DECAYING PLANT AND ANIMAL MATTER

Toxins may be produced by organisms grown on decaying plant and animal matter. They may only cause minor discomfort, but they could be highly lethal. Feeding moldy silage, hay or grain may be the means by which sheep ingest the organisms that produce listeriosis. Other problems from feeding damaged feed may be the result of the action of molds and other fungi. Mycotoxins are the result of mold action on grain, hay, or grass and may cause sickness and even death.

16.2.1. Mycotoxic lupinosis This is a disease caused by eating lupines that have been damaged by fungus-causing stem blight. The disease is not the same as lupine poisoning caused by alkaloids in the plant. The fungus causes linear stem lesions containing black stromatic masses and also affects seeds and pods. It grows on decaying matter, especially after a rain, and thus may be present in stubble fields. In acute cases, sheep are lethargic and have no appetite, and the rumen ceases to function. Within a few days, they may show severe jaundice and become constipated. Postmortem lesions are fatty degeneration of the liver. In chronic cases, almost complete cirrhosis has been observed.

Ergotism is also caused by mycotoxins produced by a fungus that is parasitic on seeds of rye and other small grains or bromes, bluegrasses, and ryegrass. Sheep do not appear to be as severely affected as cattle, but the toxin may cause mouth ulcers, gastroenteritis, and a convulsive syndrome. Control consists of changing the diet to feeds that do not contain ergot.

16.2.2. Fescue lameness Sheep do not show the classical symptoms of fescue foot observed in cattle, but they are affected by the same toxin. Although they may or may not show lameness, they usually lose condition, and the lambing percentage may be reduced as much as 60% to 70%. Older stands of fescue tend to have greater infestation with fungi and thus may be more dangerous. Severe droughts are often followed by increased problems with fescue foot. Removal from the fescue and providing high-quality forage is usually all that is required to alleviate the situation.

16.2.3. Botulism In certain areas of the world where phosphorus is quite deficient in the soil, sheep may chew on the skeletons of dead animals that have died of botulism, a bacterial disease, and contract the disease. The disease also has been known to be spread through stagnant water in pastures if the water has drained through spoiled vegetation or if decayed carcasses of animals have been near the water supply.

16.3 POISONOUS PLANTS

Of all the plants that exist in the world, there are relatively few that are dangerous to sheep. In fact, sheep can graze 90% of all weeds without harm. But there are poisonous plants everywhere. Some plants contain compounds that interfere with the physiology of sheep. The effect may vary from an irritation to death. The poison may vary with the stage of growth of the plant or certain conditions such as the rainfall, frost, cutting, curing, and storage. Sheep are discriminate grazers and usually will not eat poisonous weeds or shrubs if they are accustomed to the range. But they may graze toxic plants if they are exceptionally hungry or mineral starved (see Figure 16.2). Then it is a matter of survival, and their usual judgment is not functioning. Poisonous plants are discussed according to the toxic substances they contain.

16.3.1. Alkaloids Alkaloids affect the nervous system, causing such disorders as convulsions and paralysis. Some alkaloids produce gastroenteritis and damage the liver. The following are common names of plants containing alkaloids: belladonna, jimson weed, rattle box, groundsel, fiddleneck, poison hemlock, Indian tobacco, wild and domestic tobacco, fitweed, Dutchman's breeches, poppy, ergot, Carolina jasmine,

Figure 16.2 During droughts sheep may eat poisonous weeds such as nightshades if nothing else is available. (Courtesy University of Missouri, Columbia)

golden chain, lupine, mescalbean, tomato, nightshades, potato, staggergrass, false hellebore, death camas, monkshood, larkspur, onion, boxwood, fritillaria, gloriosa lily, Star of Bethlehem, yew, colchicum, and locoweed.

16.3.2. Amines and peptides Compounds in this class of poison may cause sudden death or may inhibit metabolism so that chronic disability occurs. These include one kind of bluegreen algae, some mushrooms, a Mexican mistletoe, ergot, and sweet pea.

16.3.3. Toxic proteins (phytotoxins) Phytotoxins have properties similar to those of bacterial toxins. They contain proteolytic enzymes that cause the loss of vital protein and the accumulation of ammonia. Animals suffering from these poisons have severe gastroenteritis and edema in several organs. Plants containing phytotoxins are precatory bean, tung tree, barbados nut, castor bean, and black locust.

16.3.4. Glycosides Some of the sorghums, Johnson grass, black cherry, choke cherry, arrow grass, velvet grass and Christmas berry contain substances called cyanogenic glycosides. When glyocides are acted on by enzymes, they release the compound prussic acid, which is quite poisonous to sheep. Wilting or bruising the plant containing glycosides may cause the enzymes to become active and release the prussic acid. The enzymes in the digestive tract of a sheep also can cause the release of the prussic acid. The poisonous property in the plant usually develops in dangerous amounts only when the growth of the plant is checked by lack of moisture, frost, cutting, or processing of the forage. If the forage is cut and cured for hay or silage and stored for several months, it is usually safe. Pure stands of sudan rarely are affected with prussic acid. Thus, special care should be exercised to obtain seed that has no contamination with other sorghums when sudan is to be used for pasture.

Animals suffering from prussic acid poisoning have increasing weakness or may have a rapid collapse. Muscular tremors, incoordination of gait, increased respiration, bloating, convulsions, and death from respiratory failure are also observed. Bright-red venous blood is considered the most reliable diagnostic sign.

16.3.5. Goitrogenic glycosides There are species of the mustard family that have the capability of blocking the uptake of iodine by the thyroid and thus changing the metabolism of the animal. This may be quite severe in areas of acute iodine deficiency. Lambs may be born dead, bare, or with goiter. The following belong to the class: chard, cabbage, kohlrabi, mustard seed, rape seed or meal, kale, broccoli, brussel sprouts, rutabaga, turnips, soybean, and flax.

16.3.6. Glycosides as irritant oils These compounds cause gastroenteritis. Plants containing them are wild mustard, horseradish, wild radish, and fan weed. Buttercups and marsh marigolds contain protoanemonin, which is a potent irritant.

16.3.7. Coumarin Sweet clover and lespedeza contain coumarin derivatives that are capable of being changed to dicumarol by molds and fungi in poorly cured hay. Dicumarol competes with vitamin K in the formation of prothrombin, which is necessary for blood to clot. Sheep fed toxic silage for long periods may die of hemorrhages. It should be a rule never to feed spoiled silage or hay to sheep.

16.3.8. Cardiac glycosides Action of these compounds is on the heart. They are capable of slowing the beat and increasing the force. The best-known plant containing the compound is digitalis, also known as foxglove. When consumed in large quantities, there may be vomiting and gastrointestinal irritation. Not all have dangerous levels, but the substance is toxic. Some plants containing the substance are dogbane, lily of the valley, and oleander.

16.3.9. Saponins This group of toxic plants contains glycosides that form a colloidal solution in the body and act to destroy red blood cells. Some also are irritating to the digestive tract of sheep. Plants that secrete saponins are corn cockle, tung tree, English ivy, and pokeweed.

16.3.10. Oxalates Oxalates act to precipitate serum calcium and cause hypocalcemia. The crystals also form in the kidney and block action of the organ. High levels of alfalfa hay or inclusion of dicalcium phosphate in the mineral mixture will usually prevent difficulties with the compounds.

Halogeton is the most dangerous plant in the group. Plants may contain as much as 20% of the oxalates in the early spring in arid or semiarid regions. Sheep can adjust to consumption of halogeton with time, but if they have been shipped and are very hungry or if they have been confined for shearing, it may be dangerous for them to eat it. Protection has been provided by feeding pellets composed of 90% alfalfa and 10% dicalcium phosphate until the sheep have had time to adjust to the weed. Other plants containing oxalates are sorrels, docks, rhubarb, and Russian thistle. Greasewood is sometimes the cause of large losses on the western ranges. Other plants contain calcium oxalate, a substance very irritating to the mouth and throat. They are Jack-in-the-pulpit, dumb cane, philodendron, and skunk cabbage.

16.3.11. Resinoids Most common of the plants that contain resins are milkweeds. The toxic substance affects the nervous system and may cause violent convulsions. Other plants containing the substance are Chinaberry, marijuana, and water hemlock.

16.3.12. Photosensitization Certain plant compounds, when digested and absorbed by sheep, are transported to the skin where they absorb ultraviolet rays and cause an irritating reaction. The skin becomes reddened and swollen, and the capillaries just beneath the skin rupture and cause fluid to accumulate. In later stages, the skin may slough off. The sheep will be very irritable and rub itself and possibly cause infections and invasions of maggots.

Some photosensitizing agents, such as those found in buckwheat and St. John's wort, pass directly through the liver and to the skin with no damage to the liver. Others damage the liver so that it can no longer eliminate the chlorophyll derivatives that cause the photosensitivity. The plants that secrete the substance are Lechuguilla (agave), rape, lantana, beargrass, horsebrush, and puncture vine. Photosensitivity was discussed previously in Chapter 15 (see Big Head in sheep).

16.3.13. Miscellaneous poisonous plants Other poisonous compounds in plants and common names of the plants are thiaminase (horsetails and bracken fern); trematol (white snake root and rayless goldenrod); gossypol (cotton); hydroquinone (cocklebur, especially at the two-leaf stage); and circutoxin (water hemlock).

16.3.14. Toxic material from soil concentrated in plants In areas where the selenium content is high and the soil is alkaline, some plants absorb and store the element and thus provide a high level to livestock grazing the forage. These plants are divided into two categories. The obligate accumulators require high levels of selenium to grow. Woody aster, goldenweed, and prince's plume belong to that group. The facultative group does not require high levels of selenium, but can accumulate it in toxic amounts. Examples of the facultative group are asters, saltbrush, paintbrush, beardtongue, and ironweed. Areas with high levels of selenium are found in South Dakota, Montana, Wyoming, Utah, and Kansas. There are also soils in some of the other Great Plains and Rocky Mountain states that may have high levels of the element.

Much more is known about selenium poisoning in cattle than sheep. The symptoms in sheep are anorexia, emaciation, weakness, anemia, and giving birth to weak, blind lambs with bone deformaties. Postmortem examination indicates subserous hemorrhages in the peritoneum, endocardium, and epicardium, necrosis in the abomasum, some enteritis, and swelling of the mesenteric lymph nodes. The liver may be swollen with areas of necrosis and the kidney swollen and hemorrhagic. The sheep with chronic poisoning may have a flabby heart and a liver with necrosis and fatty degeneration.

Nitrates may accumulate in pasture species of oats, barley, or wheat that are cut for hay. Pigweed and lamb's quarters are two weeds that may have high levels of the element. Generally, sheep have the ability to handle nitrates if they are fed concentrates in addition to the forage. Use of herbicides may also cause an accumulation of nitrates. The most common problem with nitrates occurs during droughts, particularly with corn that has been heavily fertilized with nitrates. The stalk tends to have high levels of the substance, especially in the part close to the ground. Nitrates are converted to protein in the grain, and therefore ears are not dangerous. Many farmers use drought corn for green chop or silage when the grain crop has been severely damaged by lack of moisture. Research has shown that the nitrate level is much reduced if the corn is not chopped before ears have formed. It is also recommended that cutting the stalks 18 in. (45.7 cm) above the ground will appreciably decrease the nitrate content. It is suggested that cutting the corn when it is dryer than usual is also helpful. Storage in a silo for a period of months appears to lower the level also.

During very dry years, it may be desirable to test some of the corn to determine the nitrate levels. Most agriculture extension services have the equipment and know-how to make the tests.

When silos filled with drought corn silage are opened, it is advisable to test the silage and very carefully observe sheep when they are first put on it. Any evidence of dizziness, accelerated breathing, and pulse rate may indicate poisoning. Blueness of the whites of the eyes and inside the lips are additional signs. Dead sheep may have chocolate-colored blood. Nitrate poisoning can cause abortion and stoppage of milk flow in ewes. Nitrates cause the distress in sheep by combining with the hemoglobin in the blood to form methemoglobin, which does not have the ability to transport oxygen. Thus the animal dies of lack of oxygen. Injections with methylene blue have been effective in treating the disease. Most sheep will recover if they are changed to a ration that does not contain nitrates.

16.4 METHODS FOR DECREASING LOSSES FROM POISONS AND POISONOUS PLANTS

1. Follow directions in the use of all chemicals and drugs.
2. Dispose of used cartons or containers and any leftover material in a safe place.
3. Label all dangerous substances so that they will not be used mistakenly.
4. Obtain a diagnosis of animals that die of unknown causes.
5. Provide adequate feed and pasture for sheep. Toxins affect well-fed animals much less than those that are starved.
6. Feed only wholesome grain and forage. Avoid using moldy, spoiled, or damaged feed unless there is assurance that it is safe.
7. Provide plenty of wholesome water so that sheep are watered daily. If not, never extend waterless periods longer than 2 days.
8. Know the poisonous plants and how to control or live with them.

9. Make certain that adequate forage is available in pastures so that sheep will not have to eat poisonous weeds to exist.
10. Control poisonous plants, if possible, or avoid using heavily infested pastures.
11. Always provide adequate minerals free choice.
12. Schedule grazing so that sheep are not on ranges when they are most dangerous.
13. Don't put mineral-starved sheep on a pasture. Feed hay and allow sheep to become used to minerals before turning them out.
14. Watch sheep diligently, and take immediate action when problems appear.

REFERENCES

Belschiner, H., *Sheep Management and Diseases.* Sydney, Australia: Angus and Robertson Ltd., 1965.

Ensminger, M. E., *Sheep and Goat Science.* Danville, Ill.: Interstate Printers and Publishers, Inc., 1986.

Scott, G., *The Sheepman's Production Handbook.* Denver, Colo.: Sheep Industry Development Program, Inc., 1982.

Wagstaff, D. J., *A Manual of Poisonous Plants.* Columbia, Mo.: College of Veterinary Medicine, University of Missouri, 1973.

QUESTIONS FOR STUDY AND DISCUSSION

1. Outline do's and don'ts for handling farm chemicals.
2. How should cartons, bottles, or cans containing poison chemicals be disposed of to prevent harm to livestock?
3. Why shouldn't farm equipment used for livestock be painted with lead-based paint?
4. How can you avoid salt poisoning in your sheep?
5. What is botulism? How may animals come in contact with the toxins?
6. What causes lameness from fescue consumption?
7. If there are many poisonous plants available, why don't sheep die from their consumption?
8. Why should hedge and shrubbery clippings not be fed to sheep?
9. What is the principle of poisoning by the glycosides found in black cherry, sorghums, and Johnson grass?
10. What are symptoms of nitrate poisoning?
11. Why should moldy feed be fed sparingly?
12. How would you find out what killed an animal on pasture if you suspected poisonous plants?

17 Controlling Predator Losses in Sheep

17.1 LOSSES TO PREDATION

In surveys of producers throughout the United States, one of the most serious constraints to sheep operations is losses to predators. In some regions, the damage is sufficient to force sheepraisers to cease operations and turn to alternative livestock enterprises. The problem is by no means new. Since people first began keeping sheep, they have always had to contend with raiding carnivores. In every age, people have managed to thwart the aggressors, and the present generation must develop the technology to do so also. The methods used in Biblical times are still practiced in the region and are effective, but are economically impractical in modern agriculture in the United States. Conservationists, ecologists, and groups interested in animal rights have captured the attention of the public and have been responsible for restrictions on the activities of sheepraisers that have, in many instances, swung the balance in favor of predators. Now it appears that some of the unfavorable regulations may be removed and the never-ending battle may be somewhat more in favor of the sheep producer. Even under the best of conditions, the sheep producer must exercise vigilance or severe losses may occur.

17.2 COST OF PREDATION TO THE SHEEP INDUSTRY

Various studies have been made of losses in the sheep industry due to predation. Terrill (1980) presented the results of a study made by the U.S. Department of Agriculture, Science and Education Administration, on data from 33 states from 1958 through 1979 to the Third Symposium on Test Methods for Vertebrate Pest Control and Management Materials by the American Society for Testing and Materials. He stated that the area of predator losses was enlarged by movement of the coyote both eastward and southward. In general, adult sheep losses were about one-third as great as lamb losses in the range states and two-thirds as great in farm states. Total losses to predators was 1.5 million in 1979. This was during a period of declining sheep numbers and represented an increase from 4% in 1958 to 6.8% in 1977. The value of sheep and lamb losses increased from $25 million in 1961 to $103 million in 1979. Terrill concluded that predation appeared to be the major reason for the decline in sheep numbers in the United States.

17.3 KINDS OF PREDATORS

Animals that prey on sheep include coyotes, dogs, bears, mountain lions, bobcats, foxes, eagles, and wild hogs. All are capable of injuring or killing lambs and all but foxes and eagles may destroy adult sheep. Shelton (1977) reported on the losses of sheep and lambs in the United States to various predators. He noted that coyotes accounted for 84.7% of the losses, dogs, 3.9%, bobcats, 3.4%, bears, 2.6%, mountain lions, 0.5%, eagles, 0.5%, and others, 4.2%.

17.3.1. Coyotes The coyote, *Canis latrans*, is closely related to the domestic dog, *Canis familiaris*. They are much like dogs in many of their habits and will readily crossbreed with them. Coyotes are found as far south as Central America, in all western states, in most states in the East and South and over much of Canada. They have existed with people for so long that they are familiar with their scent and activities. They may even adapt to urban conditions and have become a problem in such metropolitan areas as Los Angeles. They are the most destructive to sheep of all predators on the ranges and are becoming increasingly a problem in farm states.

Coyotes range in size from 20 to 48 lb. (9 to 22 kg). Males are somewhat larger than females. They may live from 10 to 15 years. They breed once each year in late winter or early spring and have a gestation period of 63 days. Litter size may vary from 2 to 10. Pups usually run with the group until they are more than 1 year old. Coyotes first breed when they are 2 years old. According to biologists, rabbits and rodents are the main staple in the diet of a coyote. They also eat carrion, insects, grass, berries, seeds, and fruit. Not all coyotes kill sheep and lambs, but a few individuals may cause great damage. Wildlife conservation authorities urge producers who are suffering losses to trap the culprits, since others of the species may not be causing the damage. They urge anyone who is suffering losses to identify the offending animals by their tracks or the method of killing. Coyote tracks are similar to those of a dog, but are longer and narrower and fall more nearly in a straight line. Figure 17.1 shows tracks of the dog, coyote, and fox. Coyotes are active primarily at night, but may attack livestock in daylight if the terrain or cover offer adequate concealment. They are capable of jumping or climbing over high fences or burrowing under a fence, but they prefer to crawl through or under if possible. Individuals mate for life and range together over an area that may vary from 2 to 25 square miles (52 to 65 km^2). Most of the sheep and lamb kills by coyotes occur during the spring when the whelps are young and again in the fall when they are growing fast and have appetites more demanding than the parents can satisfy by killing game.

When coyotes attack sheep, they may kill one to three and feed on one or more. Usually the kill is made by puncturing the animal's throat (Figures 17.2 through 17.8). There may be very little marks on the throat. In experiments at the California Experiment Station and the USDA Sheep Experiment Station at DuBois, Idaho, by Timm and Connolly (1980), using captive coyotes, detailed pictures have shown that the

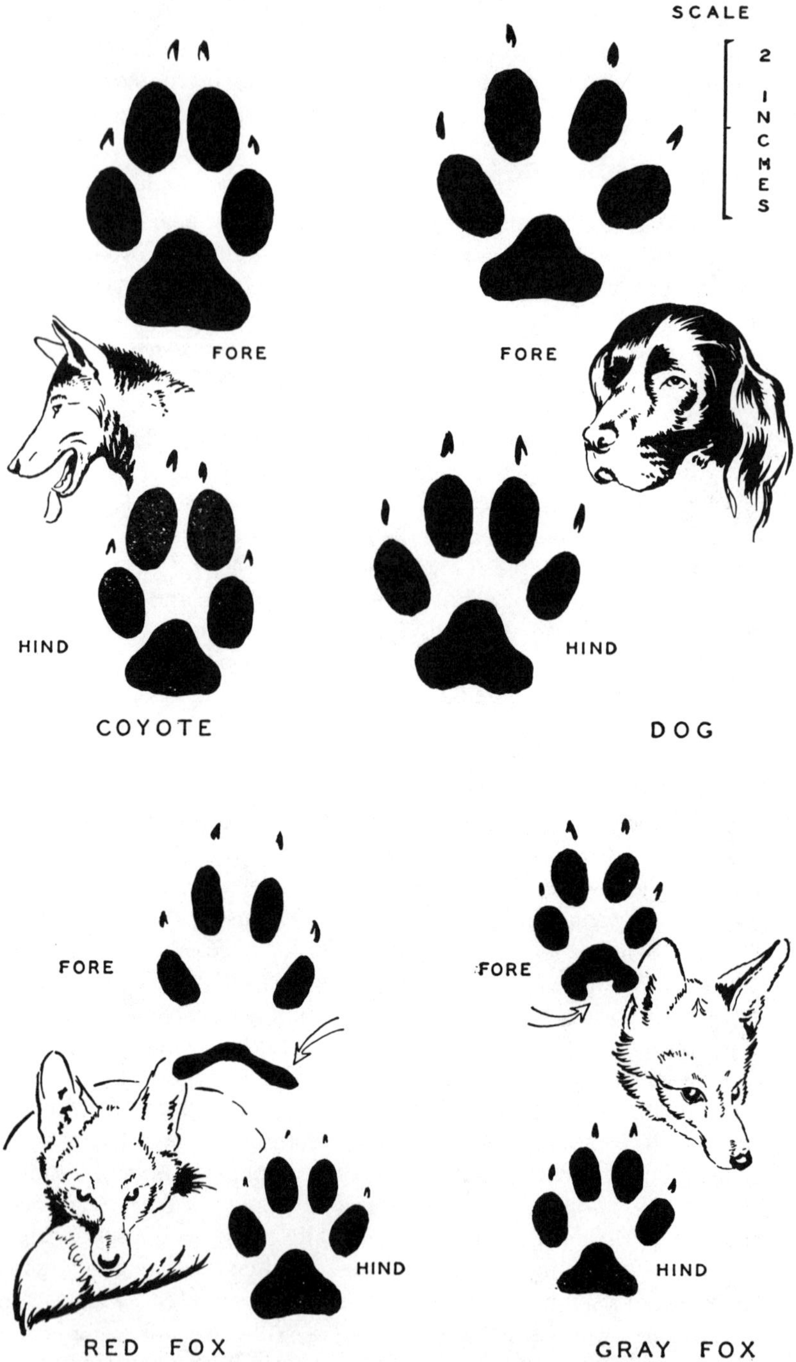

Figure 17.1 Tracks made by the domestic dog, coyote, and fox. (Courtesy Missouri Conservation Commission)

Figure 17.2 Adult male coyote prepares to attack a lamb in predation behavior studies at the University of California, Hopland Field Station. (Photo by Guy Connolly, University of California, Davis)

Figure 17.3 Coyote selects a lamb in a study of sheep-killing behavior of captive coyotes at the University of California, Hopland Field Station. (Photo by Guy Connolly, University of California, Davis)

Figure 17.4 Two captive coyotes attacking a 90-lb ewe at the University of California, Hopland Field Station. (Photo by Guy Connolly, University of California, Davis)

Figure 17.5 Coyotes usually attack sheep with a throat hold, clamping off the windpipe and causing death by suffocation. The death grip was maintained for an average of 13 minutes. (Photo by Guy Connolly, University of California, Davis)

predator runs parallel to the sheep and grabs the throat behind and below the ear. Young lambs may be killed by biting through the head. The method of killing may not provide absolute proof that a coyote was the aggressor. They have also been known to attack at the flank and hindquarters. In Montana studies (1977), coyotes attacked the head and neck region in 89% to 100% of the cases observed. But some dogs also have been known to attack the head and neck region. Thus, if the predator is to be

Figure 17.6 When coyotes prey on sheep, they usually begin to feed as soon as the prey stops struggling. Captive coyotes, on the average, fed for 25 minutes and ate about 4 pounds. (Photo by Mike Beaucage, USDA Agricultural Research Service)

Figure 17.7 This view shows the ewe in Fig. 17.8 prior to dissection. Note that little tissue damage is visible. (Photo by Guy Connolly, University of California, Davis)

identified, other evidence such as tracks, droppings, or hair may be needed. Most states with wildlife conservation departments employ experts on coyote control, who may be of assistance in identifying the predator.

17.3.2. Domestic dogs Dogs are second only to coyotes in predation in the United States. In certain localities, they may be highly destructive. Not only may the family dog be a lone killer, but packs of dogs may be formed from free-ranging

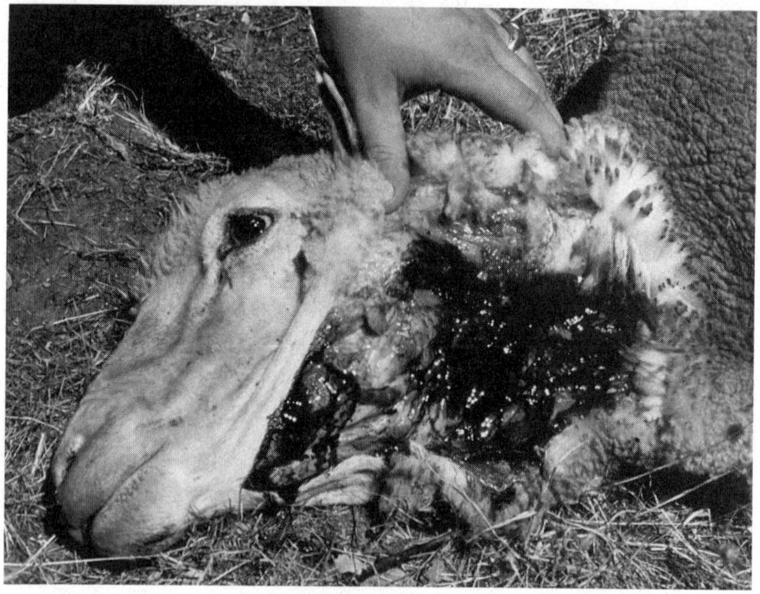

Figure 17.8 Severe tissue damage and hemorrhaging in the throat of this coyote-killed ewe is visible under the skin. See Fig. 17.7 for a view of the same sheep prior to dissection. (Photo by Guy Connolly, University of California, Davis)

pets in a neighborhood. Sometimes dogs that have been released or abandoned go wild and become a menace to livestock. This situation may occur where a city dump is not properly managed, especially if it is located in terrain providing cover. Dogs do occasionally crossbreed with coyotes to produce "coydogs," which have been known to cause much damage. These crossbreds are usually larger than a coyote and are sometimes mistakenly identified as a wolf.

Dogs are most troublesome near towns and cities and especially in the vicinity of trailer courts and housing developments that do not have leash laws or ordinances requiring the pets to be controlled. Attacks by dogs may occur at night or during the day and at any place on the farm. Dogs attack sheep more for sport than for food.They seldom eat much of the carcass. Often, the greatest damage is the mutilation of many sheep. Figures 17.9 through 17.11 show typical damage by dogs. The victims are usually bitten and torn in a number of places, especially in the hindquarters. Dogs may also stampede the sheep through fences or cause them to pile up in corners and suffocate. Or they may chase the sheep until they die of exhaustion. Dogs may attack singly, but ordinarily they run in packs. Once dogs kill sheep, they develop a mania for the activity and it is seldom possible to cure them of it. Additionally, sheep-killing dogs tend to lead others to do it.

17.3.3. Bears Bears may be a problem in the mountainous areas of the west, some northern states, and in New England. Not all bears kill sheep, and there are restrictions in most states on destroying them. Generally, losses to bears are restricted to certain areas, and it may be only one or a few animals doing the damage. Never-

Figure 17.9 This ewe exhibits the type of injuries that often result from attack by dogs. (Courtesy The Texas A&M University System)

Figure 17.10 These ewes exhibit the indiscriminate mutilation that commonly results from dog attacks. (Courtesy The Texas A&M University System)

Figure 17.11 This ewe was mutilated and partially disemboweled by dogs which killed her. (Courtesy The Texas A&M University System)

theless, bears do have the capability of destroying many sheep and thus the damage could be serious.

17.3.4. Cougars or mountain lions Cougars may kill sheep in certain areas in the mountains of the western United States. The damage is usually localized and may involve only a few predators.

17.3.5. Other predators Bobcats, foxes, wild hogs, and eagles may sometimes kill young lambs. No figures appear to be available on the extent of losses.

17.4 TREND IN NUMBERS OF PREDATORS

The U.S. Fish and Wildlife Service *Predator Abundance Survey* (1979) indicated that coyotes have been increasing in numbers and expanding their range in the United States. Bear, cougar, and red fox populations are increasing in the western states. Eagle population in the West is increasing, and bobcats are stable to increasing. The domestic dog population continues to grow.

17.5 NONLETHAL METHODS OF CONTROLLING PREDATORS

Nonlethal control of predators should always be considered first, because they do not incur liability for the property of other people. Some husbandry practices and equipment used by sheep producers to reduce losses to predators are:

1. Confinement rearing of lambs
2. Penning at night under lights
3. Close surveillance
4. Use of guard animals
5. Exclusion fencing
6. Control of carrion

17.5.1. Confinement rearing While it is true that a confinement system of production may make it possible to protect larger numbers of sheep with less effort, some thought must be given to preventing dogs from coming in contact with the sheep. A dog-proof perimeter fence is essential if the sheep are in a drylot. A board fence 5 ft (1.5 m) tall will usually turn most dogs or use a woven-wire fence 32 in. (0.81 m) high topped with five strands of barbed wire 5 to 7 in. (12.7 to 17.8 cm) apart, with another barbed wire offset 5 in. (12.7 cm) from the top of the post. A barbed wire stretched along the surface of the ground under the woven wire will discourage digging under.

17.5.2. Penning at night under lights In areas where there are many predators, lotting sheep at night in a lighted corral prevents attacks. The fence should be tight and have a gate as good as the fence.

17.5.3. Close surveillance Close surveillance is always important, especially if there is ground cover near the pastures. A rifle or shotgun should be handy in case predators are successful in getting into the pasture with sheep.

17.5.4. Guard animals Guard dogs are being used by some sheep producers. These dogs belong to breeds imported from Europe and the Middle East where they have been used to protect sheep. The puppies are raised with sheep, and they stay with the flock at all times. Some preliminary reports are encouraging, but it seems too soon to make a wise judgment on their effectiveness. There have also been reports on the use of other animals such as donkeys, billy goats, and llamas as protectors of a sheep flock. There is research underway at a number of experiment stations on the use of guard animals, but information is lacking.

17.5.5. Exclusion fencing Exclusion, if feasible, may be the method of choice of all means of controlling predator losses. Of course, on rangeland it may be impractical to fence out coyotes and dogs, but in many areas it may be done quite effectively. Coyotes and dogs have the ability to jump or climb over fences in excess of 60 in. (1.5 m), but they would prefer to crawl under or go through a hole in a fence. A well-built woven-wire fence 60 in. (1.5 m) high with barbed wire on top provides a formidable obstacle to an exploring animal. Most of the time, entry is gained by crawling under a gate or walking under a fence where the water has washed out the soil. If, in addition to a strong perimeter fence, an electric fence is used on the outside, the protection may be excellent. At the Missouri Experiment Station (1975), losses of lambs from coyotes and dogs was a yearly occurrence until an electric fence was strung around the outside of the perimeter fence. The simple fence, illustrated in Figure 17.12, consists of two wires 6 in. (15.2 cm) and 18 in. (45.7 cm) from the ground on insulated steel stakes set 20 in. (50.8 cm) from the woven-wire fence. The most serious problem with the electric fence is the control of vegetation.

Figure 17.12 An electric fence on the outside of a wall-built perimeter fence will deter most predators. (Courtesy University of Missouri, Columbia)

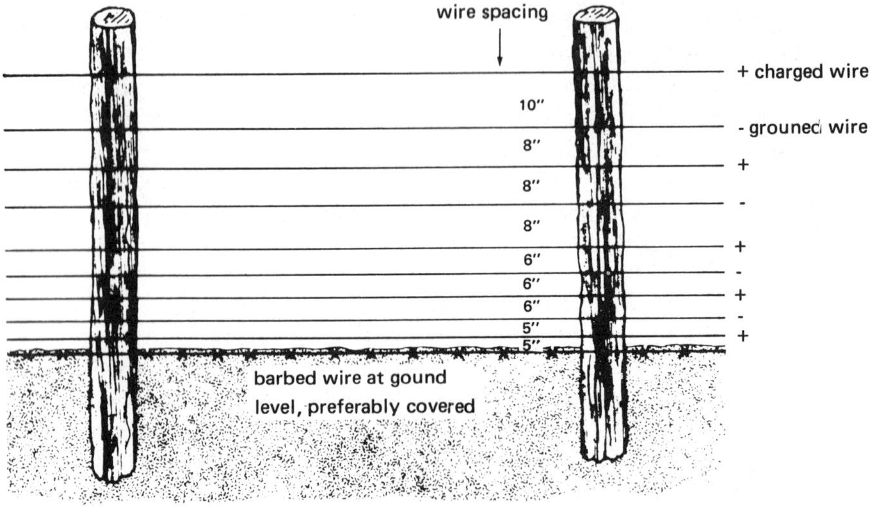

Figure 17.13 Plan for an electric fence used by the University of Illinois. (Courtesy of the University of Illinois)

However, that is solved by the use of herbicide twice during the summer. Other experiment staions have developed plans for electric fences as shown in Figures 17.13 and 17.14. Perhaps the most important feature of the fence is the kind of charger used. Scientists in New Zealand have developed energizers that operate with high voltage, but at a high pulse rate (3/10,000 of a second) that is safe to both humans and animals. The very rapid pulse rate also destroys vegetation and prevents grounding. A word of caution: In the interest of safety, don't use a charger unless it is approved by the Underwriters' Laboratory. Since the technology of electric fencing is improving rapidly, consult with your extension agricultural engineer or county agent before constructing one. If further information is desired, it can be obtained from the USDA, Agriculture Research Service, U.S. Sheep Experiment Station, DuBois, Idaho.

17.5.6. Additional methods of control Other nonlethal methods of preventing sheep depredation in an experimental stage include repellents sprayed on sheep, belling, propane or acetylene devices that explode periodically, and radio and tape recorders that broadcast music, voices, or the sounds of vehicles at intervals. Some of the methods appear to be promising, but their effectiveness may be only temporary until the predators become accustomed to them.

17.6 CONTROL OF CARRION

Disposal of dead animals is an important factor in the control of predators. Dead livestock should either be buried, burned, or sent to a rendering plant. Leaving them in the open is an invitation to predators to come on the property.

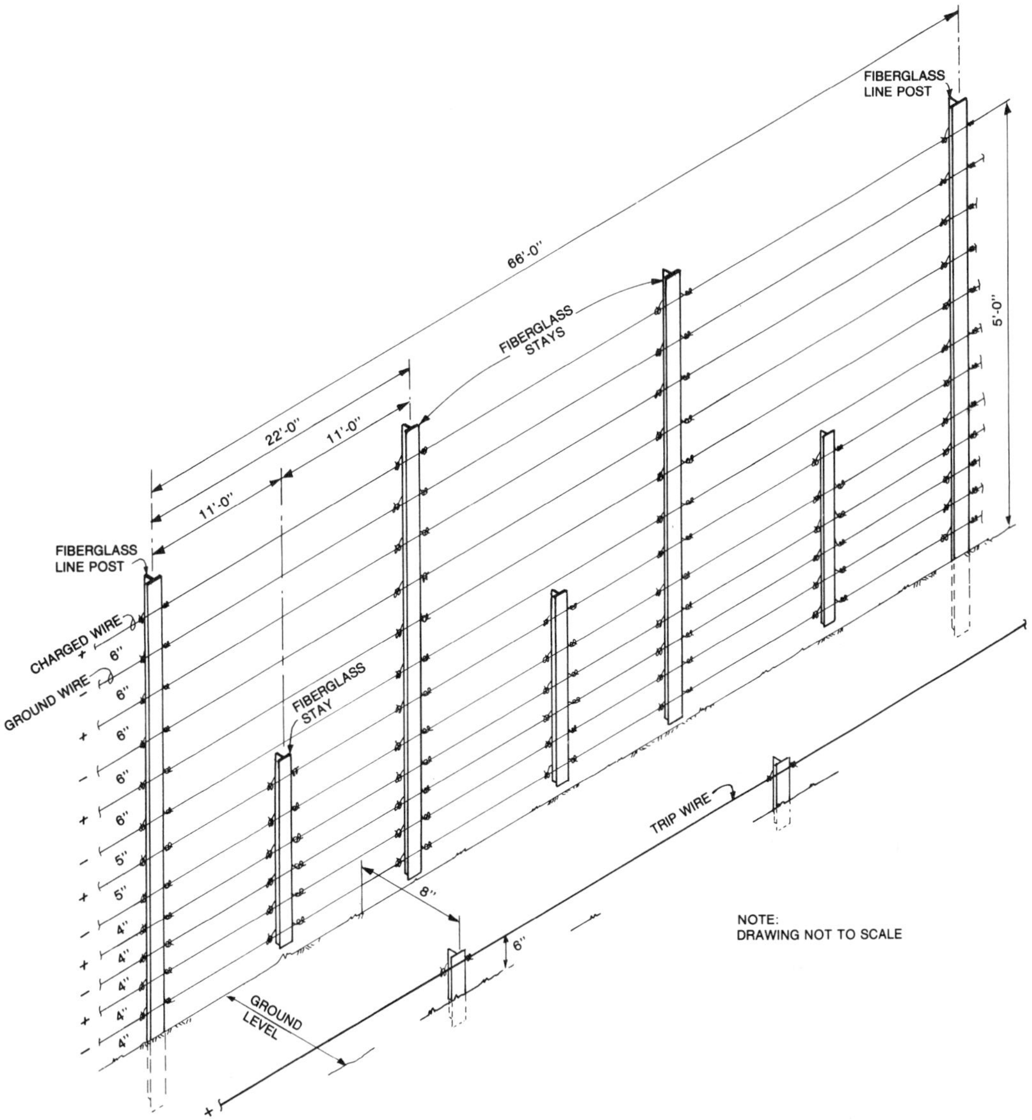

Figure 17.14 Plan for an electric fence used by USDA.

17.7 LETHAL METHODS OF CONTROLLING PREDATORS

Animal damage control (ADC) personnel of the U.S. Fish and Wildlife Service use lethal techniques to solve depredation problems. They have used aeriel gunning with much success in areas where light planes can operate, but it may be too expensive to operate the planes. Now they rely more on steel traps, snares, M-44's (a cyanide gun), and calling and shooting to eliminate predators. Poisons have been successful, but there is always the danger of destroying innocent animals. A thorough check should be made with the Wildlife Conservation Service in the state before they are used. Trouble caused by bears and cougars on public lands should be followed by an appeal for help from the ADC. Many states employ an expert trapper whose services are available to instruct groups of livestock producers in the techniques of eliminating coyotes by trapping them. If the sheep operation is located in a region with a high population of coyotes, trapping skills should be developed as insurance against the occasional predator that seems to be immune to most methods of control. In some localities, concerted action by sheep producers has appeared to at least temporarily alleviate a severe problem. Hunting with packs of hounds, chasing with greyhounds, use of special noisemakers, calling, denning, or drives have all been at least partially successful. But such methods usually are resisted by the wildlife conservation organizations, because so many animals not guilty of depredation are eliminated.

17.8 DOG LAWS

Legal files are loaded with records of lawsuits involving the killing of dogs. There is a great variation in dog laws in the United States. Knowing your legal rights is very important to the sheep producer regardless of where the operation is located. Proposing and passing fair and equitable laws regarding the control of dogs is the responsibility of all citizens, but may be most important to livestock producers. Unified efforts by the owners of all species can often bring results. Indiscriminate killing of a dog merely because it is on your property is not usually legal. Killing a dog that is chasing, worrying, or attacking sheep is legal in most states. Owners have the right to protect their property from damage, but there are all kinds of variations in the interpretation of protecting property and therein lies the basis for many bitter and costly lawsuits. Good dog laws, well enforced, prevent problems from occurring. The following are some provisions of workable dog laws:

1. The law must be just to command the moral support of the people.
2. Each dog should be registered and tagged and taxes paid to the state.
3. Dogs should be under control any time they are away from the premises of their owner. This should include confining dogs from sunset to sunrise.
4. Dogs found at large from sunset to sunrise are to be impounded, and the owner notified. The owner regains custody by paying a fine for violation of the law.
5. Any dog, licensed or unlicensed, may be killed by anyone when caught in the act of chasing, worrying, injuring, or killing sheep or other livestock. The dog owner should have no recourse from the killing of the dog.
6. Any dog found running at large on enclosed lands of a person other than the owner of the dog may be killed at the time of finding it by the owner of the land, his agent, or employee.
7. Livestock owners should be compensated for loss or injury of animals from funds collected for licenses.

8. Damages caused by dogs, including court costs, should be assessed against the owner and payable to the state.
9. It should be mandatory for the owner to destroy any dog that has attacked livestock.

REFERENCES

Arnsperger, D. R., T. Nabors, and J. D. Rhoads, "Experience with Electric Fences for Sheep at the University of Missouri," Missouri Sheep Day Report 1975.

Arthur, L. M., "Measuring Public Attitudes toward Natural Resource Issues: Coyote Control," *Tech. Bull. 1657*, Economic Research Service, USDA, 1981.

Beckhoff, Marc, ed., *Coyotes: Biology, Behavior and Management*. New York: Academic Press, 1978.

Bogges, E. K., F. R. Henderson, and Clifford Spaeth, "Managing Predator Problems: Practices and Procedures for Preventing and Reducing Livestock Losses," *Kansas Agr. Ext. Serv. Circ. 620*, 1980.

Brawley, K. C., Domestic Sheep Mortality during and after Tests of Several Predator Control Methods, M.S. Thesis, University of Montana, Missoula, 1977, p. 698.

Coppinger, R., and Lorna Coppinger, "Livestock Dog Project," Chestnuthill Road, RFD, Montague, Mass., 1978.

Fox, M. W., ed., *The Wild Canids—Their Systematics, Behavioral Ecology and Evolution*. New York: Van Nostrand Rheinhold Co., 1975.

Gates, Norman, "Constructing an Electric Anticoyote Fence," *Leaflet No. 656*, USDA, 1978.

Gee, C. K., and others, "Sheep and Lamb Losses to Predators and to Other Causes in the Western United States," *Agric. Res. Rept. No. 369*, Economic Research Service, USDA, 1977.

______, "Factors in the Decline of the Western Sheep Industry," *Tech. Bull. 377*, Economic Research Service, USDA, 1977.

Green Jeffrey, "Eliminating Coyotes—Where We Are," *Sheep Breeder and Sheepman*, 100, no. 5, 1982, p. 264.

Green, J. S., T. T. Tueller, and R. H. Woodruff, "Livestock Guarding Dogs, Economics and Predator Control," *Rangelands*, Dec. 1980.

Griffel, D. E., and J. V. Basile, "Identifying Sheep Killed by Bears," *Intermountain Forest and Range Station Research Note Int. 313*, Forest Service, USDA, 1981.

Gum, R. L., L. M. Arthur and R. S. Magleby, "Coyote Control: A Simulation Evaluation of Alternative Strategies," *Agr. Econ. Report No. 408*, Economics, Statistics, and Cooperative Services, USDA, 1978.

Larson, G. E., and others, "Coyote Predation in Sheep," *Circular No. 3*, Dixon Springs, Ill.: University of Illinois Agriculture, 1975.

Menzies, C., and Sheep Industry Task Force Members, "The U. S. Sheep and Goat Industry, Products, Opportunities and Limitations," *Report No. 94*, Council for Agricultural Science and Technology, 1982.

Sampson, F. W., and W. O. Nagel, "Controlling Coyote and Fox Damage on the Farm," *Missouri Conservation Comm. Booklet*, 1951.

Shelton, M., "Symposium on Livestock Predators: Nature of the Problem—Livestock," *Annual Meeting Proceedings*, American Association for the Advancement of Science, Feb. 1977.

Simmons, V. L., "Sheep Killing Dogs," *Farmers Bull. No. 1268*, USDA, 1935.

Terrill, C. E., "Trends of Predator Losses of Sheep and Lambs from USDA Mortality Statistics," *Third Symposium Proceedings on Test Methods for Vertebrate Pest Control and Management*, Fresno, Calif. American Society for Testing and Materials, March 7, 1980.

Timm, R. M., and Guy E. Connolly, "How Coyotes Kill Sheep," *National Wool Grower*, 1980.

U.S. Fish and Wildlife Service, *Predator Abundance Survey*, 1979.

QUESTIONS FOR STUDY AND DISCUSSION

1. What is meant by predation?
2. What is the extent of losses to sheep producers by attacks of predators?
3. Discuss the distribution of coyotes in the United States and neighboring countries.
4. Justify the statement that coyotes are the most serious predator problem for sheep owners.
5. How can a coyote kill be identified?
6. Describe wounds made by attacking dogs.
7. What is the trend in predator numbers in the United States?
8. List six steps that can be taken to reduce predator losses.
9. In what way should guard dogs be trained to protect the sheep? How should they be trained?
10. Describe an exclusion fence for sheep pastures. What are the advantages of using high-tensile wire electric fences?
11. Discuss the five most important features of a satisfactory dog law.
12. Who can lend assistance if coyotes are killing your sheep?

18 Housing and Equipment

18.1 GENERAL CONSIDERATIONS

In most areas where sheep are raised in the United States, optimum performance requires some protection from weather conditions. The benefits vary from a slight improvement to avoidance of disaster. Decisions must be made relative to the needs for and the benefits from housing and equipment. The operation should be analyzed to determine what is required and how to provide it economically. If the sheep are to perform their function of profitably producing meat and wool, the following are the basic requirements of buildings and equipment for sheep:

1. Protection from cold, rainy weather
2. Shade and sanctuary from flies in the summer
3. Well-drained lots to ensure dry conditions
4. Security from predators
5. Storage for feed, bedding, and equipment
6. Convenient, efficient working area for handling sheep

The overall objective of providing physical facilities is to enable the operator to perform, efficiently and effectively, the necessary production practices with the flock.

Because physical equipment may be very expensive, it is necessary to spend some time thinking about providing an additional facility beyond the bare essentials for each unit. Answering such questions as the following may be of benefit in reaching wise decisions: What are the results expected? Will it increase production, reduce losses, save labor, increase feed efficiency, increase flexibility, or are there esthetic or compelling personal reasons for procuring it? Can the cost be justified for business or tax purposes? Best of all, will the addition improve the operation sufficiently to pay for itself quickly? The author once worked for a very successful businessman who required that the arithmetic on any outlay of money for physical facilities be worked out. His criteria for making a decision were: If it will pay for itself in 3 years, buy it. If it requires 5 years, give it serious consideration. If it takes 10 years, it is doubtful and not usually worth the effort to get it.

The point of this discussion is to know why you want to make the improvement, what it costs, and what you expect from it. No doubt immediate profits should not be the only criterion for making a decision, but it should occupy a prominent spot in your priority list. All too often a person who is carried away with the craze to build facilities finds that the financial balance sheet shows a high net worth, but is shocked to discover that there is not sufficient money to operate the unit. Then it is necessary to talk to the banker, who may be expected to be very conservative and reduce the options available to the operator. Many bankers have very little knowledge or appreciation for the sheep business.

Generally, except for lamb finishing, sheep operations should not be classified as risky. They should be expected to return a steady but not spectacular income with modest expenses. High investments in physical facilities are seldom needed or justified. Many of them may be built by a person handy with tools and having some knowledge of construction.

18.2 HOUSING FOR SHEEP

The sheep barn is likely to be the most expensive of all the physical facilities for sheep. It is necessary to know what is needed, what is available, and whether or not it is most advantageous to remodel an existing building or to construct a new one tailored to the operation. Not only should the present needs be considered, but the building should be designed to permit expansion or alternate use if the situation warrants it. The specifications for sheep housing are known and available. Most are included in this chapter. Most states have extension agricultural engineers who are available for assistance in planning facilities. In addition, it may be advantageous to visit successful sheep operations to get ideas. A review of available publications may be helpful. One in particular, *The Sheep Housing and Equipment Handbook* (1982) (MWPS-3), published by the Midwest Planning Service, Iowa State University, Ames, Iowa, 50011, should be invaluable.

If barns or other buildings exist on the farm, think very carefully before wrecking them and building another. Not only is it costly to dismantle a building, but it may be much more expensive to construct a new one than to remodel an existing structure. Seldom is it possible to adapt a building such as a chicken house to use for sheep with completely satisfying results, but it should be pointed out that it has been done in many instances. Perhaps the only buildings that should not be converted to a sheep barn are former horse barns. They should be avoided because of the danger of tetanus, which tends to stay on the premises as spores and can affect the lambs for years. Dairy and beef cattle and swine barns, as well as machinery sheds and tobacco barns, have all been successfully converted into usable sheep barns. In instances where the building can not be remodeled into a sheep barn, it could be used for feed storage or for housing machinery. Since buildings are used for years, it is important to plan and construct them carefully or be penalized because they are inadequate. What features are desirable in a sheep barn? The following are considerations and suggestions for planning and constructing a barn.

18.2.1. Location of the sheep barn The building site should be chosen carefully where it is handy to the house yet not so near as to be a nuisance from the standpoint of odors and flies. It should be near feed supplies, where water and electricity are available, and on all-weather roads to allow constant access to machinery and trucks. The barn should also be located where there is adequate space for outside lots and working lots and facilities. Consideration should also be given to access to pastures for the sheep.

Ideally, the barn would be located where there is protection from the gales of winter on a south or east slope on well-drained soil. It should be northeast or east of the house to be downwind in the summer. The least desirable site would be in a valley with level, poorly drained soils and subject to flooding. There should also be room for expansion if, at a later date, more barn space is needed.

18.2.2. Type of construction At one time, a frame barn with a complete loft and lambing quarters in the middle was considered ideal. This type of construction was satisfactory when building materials were relatively inexpensive. Later, pole barns have proved to be so much less expensive that frame buildings are seldom used. Lofts have been superceded by pole barn hay storage or by building the barn bigger and using some of the pen space for storage. It is mandatory to keep the building costs low, but the building should be functional and durable, and require little maintenance. A pole barn for sheep housing, as shown in Figure 18.1, has given good service at the University of Missouri.

Figure 18.1 Pole barn used for housing sheep at the University of Missouri. (Courtesy University of Missouri, Columbia)

18.3 DESIRABLE FEATURES OF A SHEEP BARN

18.3.1. Flexibility During a production year, it may be desirable to use the barn in a number of ways. For example, it is usually desirable to vary the sizes of the inside pens. Larger pens are needed for open ewes in the summer, smaller ones for ewes with lambs, and various sizes for finishing lambs or breeding groups. If the building has clear-span construction, then by using panels and temporary posts the configuration of pens may be varied as needed. The area that can be closed up for the lambing quarters should also be available in other seasons for shearing and working with ewes, such as crutching, facing, or foot trimming. Locating water fountains on the perimeter of the building will allow you to plan for larger and smaller pens as needed.

18.3.2. Accessibility Plan so that it is possible to move sheep in and out of the barn to certain pens without disturbing the rest of the flock. A working alley along the back or front of the barn is useful in this regard. An alley to allow access for bringing in feed is desirable. Sometimes a feed alley down the center of the barn is most satisfactory. This should be wide enough for machinery (10 to 16 ft) (3 to 4.9 m). Some modern barns have fence-line bunks along the outside of the building for feeding. Those usually have a work alley running down the center of the barn. A width of 4 to 5 ft (1.2 to 1.5 m) is usually satisfactory for handling sheep.

In the planning stage, keep in mind that the barn may be used throughout the year. Ideally, in the fall and winter, it is desirable to house ewes in several groups according to stage of gestation and move them progressively nearer to the lambing quarters as they approach lambing. The ones that are springing heavily should be right next door to the lambing quarters. Then, after lambing, more than one pen for ewes and lambs is needed, the first pen for the newest lambs, and others for older ones. Later in the winter and spring when the lambs are weaned, there should be pens for finishing them. There should also be sheltered pens for holding newly shorn ewes. In the summer, it is useful to have access to pastures from the barn.

If the handling and working corrals are outside the barn, it should be possible to reach them easily.

18.3.3. Ventilation Sheep should not be housed in closed barns unless it is necessary to save lambs at parturition or after shearing. Warm buildings with high humidity are usually conducive to respiratory problems. If sheep are sheltered from cold rain, wet snow, and drafts, low temperatures are not harmful.

The reasons for ventilation of barns are to:

1. Remove moisture from inside the barn.
2. Provide fresh air for the sheep.
3. Remove excess heat in hot weather.
4. Remove odors and gases from the sheep and their offal.

Except for special conditions, a barn or shed with an open front to the south or leeside of the prevailing wind with some adjustable openings on the windward size should be satisfactory for air movement. Most modern designs of pole barns with gable roofs also have an open ridge to facilitate ventilation. Long barns may develop drafts inside the structure. It is recommended that solid partitions be constructed at 50-ft (15.2-m) intervals to prevent swirling of winter winds within the building.

In areas where ewes are lambed during severe winter weather, the lambing quarters should be closed to maintain temperatures above freezing. Then ventilation with a fan is needed. The fan can run continuously or be controlled by a thermostat. The fans should be used either to force fresh air in or to exhaust stale air. Some suggestions on the use of fans are as follows: Locate them in tight walls or in the ceiling and away from doors, windows, stairwells, or hay chutes. Face fans away from the prevailing winds.

It may be that some heat is also needed in the closed room. Almost always some insulation is required in the walls and ceiling.

Ventilation of closed buildings is a technical problem and warrants engineering assistance. If you are inexperienced in building construction, begin with the extension agricultural engineer or a consulting engineer and follow recommendations.

18.3.4. Electricity Sheep barns should be wired for electricity. Lights are needed in order to handle the sheep and do the chores, and power is needed for electrical equipment and machinery. The pens may be satisfactorily lighted using one 100-watt bulb per 400 ft^2 (37.2 m^2) of floor space. Outlets will be needed for water heaters and equipment and a nightlight outside. If outlets are spaced throughout the barn at intervals of 20 to 30 ft (6 to 9 m), they will be useful. Outlets are needed for shears and other equipment in the lambing quarters and also for heat lamps above lambing pens. They should be designed so that the cord hangs vertically and will come unplugged if the heat lamp is knocked down. Provide outlets for a refrigerator, lights, heaters, and so on, in the shepherd's quarters. Within animal quarters, the receptacles should be located above the height the sheep can reach. Use only type U-F (underground feeder) wire for barns. Most sheep barns should be wired for 220-volt or three-phase electrical service as well as 110–120 volts.

18.3.5. Water Too often not enough thought is given to furnishing water to sheep, even though an inadequate supply can adversely affect performance. Nonfreeze faucets or hydrants in the sheep barn are a valuable asset that will save much labor and problems, especially in very cold weather. Plan to furnish the requirements for sheep; then include such demands as sanitation, lagoon dilution, spillage, fire protection, and possible expansion of the operation. It is recommended that the following be available per day for sheep:

	Gallons/day (liters/day)
Rams	2 (7.5)
Dry ewes	2 (7.5)
Ewes with lambs	3 (11.3)
Small lambs	0.1–0.3 (0.4–1.1)
Feeder lambs	1.5 (5.7)

Underground lines should be buried deep enough to prevent freezing. Check with extension agricultural engineers for depth below the frost line in the area. Pay particular attention to pipes under compacted driveways and concrete floors. Where underground pipes are to be brought to the surface in an animal pen, protect them by insulation and enclosing in a larger-diameter, rigid plastic pipe to avoid damage by sheep and machinery and heaving of concrete slabs in winter.

A storage tank is invaluable as insurance against loss of water supply. A pond nearby or a lagoon may be instrumental in saving the barn in case of fire. Up to

50 gal (189 liters) per minute may be needed to fight fire. Water quality should be checked to make certain that it is wholesome. High mineral content, nitrates, or bacteria can be harmful to sheep.

18.3.6. Feed and bedding storage Some principles of storage are:

1. Move materials as little as possible, preferably storing near the point of use.
2. If possible, handle feed in bulk.
3. It is preferable to provide a continuous flow from storage to feeders.
4. Consider costs versus benefits of equipment.

Factors affecting location of storage are:

1. Available power for processing
2. Access roads (whether paved or graveled)
3. Appearance of the unit or units
4. Odor, dust, noise and traffic at the unit dictate that the storage area should be no closer than 100 yd (91 m) from the house on the downwind side.

Factors to consider in providing feed storage are:

1. The type of operation (extensive or confinement).
2. Form of rations to be fed: whether ground mixed, pelleted, or concentrate mixtures plus hay or silage.
3. Source of feed: whether home grown and stored at harvest time or purchased when needed.

To estimate the storage capacity needed, figure the composition and amount to be fed daily and multiply by number of days to feed. For example, on a concentrate plus corn silage ration, estimate 10 lb of silage (includes waste) plus 2 to 3 lb (0.9 to 1.4 kg) of concentrates per head daily for 13 weeks of gestation and lactation. If ewes are in confinement, their requirements are 20% to 30% less than when managed in a more extensive manner. If ewes have many twins, the requirements are considerably greater. If hay plus concentrates is fed, estimate 4 lb (1.8 kg) of hay plus 2 to 3 lb (0.9 to 1.3 kg) of concentrates daily. Using these figures for corn silage and concentrates, it would require 1300 lb (165 kg) of corn silage and 228 lb (103.5 kg) of concentrates per ewe. Using hay, it would require 364 lb (165 kg) of hay and 228 lb (103.5 kg) of concentrates per ewe. The sensible move is to calculate rations for gestation and lactation, and also for finishing lambs. In addition, it is practical to have a reserve of roughage in case of unusual circumstances such as blizzards or droughts. A reserve of 15% to 25% of hay or silage is good insurance.

In many operations, it is not practical to store the total supply of hay and bedding in the sheep barn, but some space is needed near the place of feeding or pens that will require bedding. Traditionally, a loft with a tight floor is excellent for storing a ready supply of baled hay or straw. If the barn does not have a loft, then an area in a pen or pens may be partitioned off for that purpose.

Bulk storage may be provided by pole-type barns near the sheep barn in climates where snow may be a problem, but they should be on all-weather driveways so they can be reached in inclement weather. Consider the possibility of locating the hay storage so that it can act as a partial windbreak for the barn. In dry climates, baled hay and straw may be stacked and covered with tarpaulins or plastic. In some places, it may be more practical to chop and stack the hay. Large and small round bales may sustain nutrient and waste losses from 10% to 30% if left outside in humid climates compared with 5% in covered storage. Over a period of years, those losses may be enough to pay for adequate storage.

18.3.7. Lambing quarters Climatic conditions affect the kind of quarters needed for lambing out ewes. In most places, it should be possible to close the room when temperatures are low. Many producers who have small flocks adapt existing buildings to the purpose. Unless the operation is located where winters are severe, small flocks of 100 ewes or less seldom require special ventilation or heat. Larger numbers may require forced air ventilation and extra heat as discussed previously.

The lambing quarters should be located near the holding pens for easy access, near the shepherd's quarters, handy to feed supplies, and with water available. The amount of space will depend on the rapidity with which the ewes lamb and the anticipated weather conditions during parturition. Some authorities recommend 8 ft^2 (0.74 m^2) per ewe in the quarters in order to set up a lambing pen for each eight ewes and to have space for others heavy with lamb to be closed up in the quarters prior to lambing.

Sanitation is important in the lambing quarters. Ceilings, walls, and floors should be constructed of materials that can be cleaned and disinfected. Concrete floors poured over coarse gravel with drains can be cleaned easily, scrubbed, and disinfected. These floors should be dry and warm if bedded with 6 in. (15 cm) of straw. A cabinet to store supplies and equipment (the shepherd's kit) should be located either in the lambing quarters or the shepherd's quarters if they are adjacent to each other.

The quarters should be wired as previously discussed and serviced with water. There should be space for storing panels used for lambing pens, buckets, feed boxes, straw for bedding and hay, and concentrate for feeding the ewes, unless they are kept in a handy feed and storage room nearby.

18.3.8. Shepherd's quarters It is highly desirable to have a heated room next to the lambing quarters so that records can be kept, chilled lambs revived, and many other jobs done in comfort. If a cot is provided, it may pay dividends for the shepherd to spend part of the night in the quarters to more closely observe the ewes and to render assistance when needed. The furnishings of the lambing quarters will depend on the amount of use it will receive. For large flocks, there should be a water heater, refrigerator–freezer, storage cabinets, space heater, desk, shower, milk replacer mixer, and an exhaust fan. Some such quarters have telephones and radios and are used as an office. Some producers who do not have shepherd's quarters have used closed-circuit TV to observe activity in the lambing quarters from the house.

18.3.9. Floors Dirt floors are very satisfactory for holding pens and outside lots. As discussed previously, there are definite advantages to using concrete in the lambing quarters. It is preferable for the floors in the inside pens to be built up higher than the outside ground level to prevent surface water from entering. If the inside floors are built up 12 to 18 in. (0.3 to 0.45 m) with clay and thoroughly packed down before the barn is built, they should be quite beneficial in keeping the barn dry and sanitary. The barn floor should be sloped toward the front. Dirt floors are, of course, the cheapest to build and are quite satisfactory if properly maintained each year when the barn is cleaned. If dirt floors are used, it is easy to drive steel posts and set up various pen arrangements. This also is an advantage when the barn needs cleaning. Then all the pens can be removed easily so machinery can work unobstructed.

There are some good reasons for using concrete floors. Concrete is more permanent, requires less maintenance, does not develop mudholes, and may be dryer. On the other hand, it must be bedded, and it is initially more expensive than dirt.

Other types of floors are being used, particularly in confinement buildings. Wooden, steel, concrete, aluminum, plastic (PVC), and steel slats have been developed for sheep floors. Steel, aluminum, and other materials fabricated into mesh or expanded metal have been compared by researchers at the Illinois Experiment Station at Dixon Springs. They looked at costs, durability, cleaning ability, footing under various conditions, and other factors. In a report at the National Symposium on Confinement Sheep Production in 1981 at St. Louis, Missouri, Lewis summarized their research work. He noted that many of the floors tested were satisfactory for sheep. He stated that the choice of materials depends on cost, design of existing facilities, life expectancy desired, availability of materials, labor, and the program planned. From their comparisons, it was concluded that ¾-in. (1.9-cm) no. 9 carbon steel Safe-T-mesh expanded metal floors were the most satisfactory. It is among the best cleaning floor under varying conditions, and it keeps the feet of sheep well worn and provides excellent footing for them. This flooring requires more support than slats and must be replaced in 5 to 6 years if used constantly. The lifetime of galvanized, plastic-coated or stainless steel Safe-T-mesh is longer, but it also is more expensive. Since the technology is changing rather rapidly, it is wise to check with agricultural engineering specialists before making a decision on floors.

If floors are slatted or expanded metal, they must be raised at least 2 to 3 ft (0.6 to 0.9 m) above the ground level in order to have room for the droppings to accumulate. If manure is only cleaned once or twice a year, it must be possible to remove the floor for cleaning. Some confinement barns are built over a pit and the manure flushed down periodically (see Figure 18.2). The Ohio Experiment Station has a scraper in their pit. The Illinois Experiment Station at Dixon Springs uses a deep pit on a hill sloping away from the barn (Figure 18.3). Overhead doors allow access to the pit. They have routinely cleaned out the pit once a year and have been able to control flies and odors.

18.3.10. Cleaning and sanitation Cleaning barns used in confinement setups requires special planning as indicated previously. Cleaning is also a problem in conventional barns, and it should receive attention in the planning stages. If clear-span construction is used and partitions are only temporary, it should be possible to take them down and clean large spaces mechanically. If tractors with scoops are to be used, 9 to 10 ft (2.7 to 3 m) of head room is needed. Doors should be 8 ft (2.44 m) wide to facilitate the movement of tractors, trucks, or wagons.

Figure 18.2 Front and rear view of the confinement barn at the Dixon Springs Experiment Station, University of Illinois. (Courtesy University of Illinois)

Figure 18.3 Exterior and interior of the confinement barn at the OARDC, Ohio State University. (Courtesy Animal Science Department, Ohio Agricultural Research and Development Center, The Ohio State University, Wooster)

18.3.11. Control of pests Pests such as rodents, flies, and birds should be considered in planning a barn. Mice and rats eat and spoil much feed. They may also be carriers of disease. Feed rooms should be constructed so that entry by pests is difficult. Doors should fit frames and not provide an entry under them. Holes made for plumbing and wiring should be closed or protected by hardware cloth collars. Nesting and roosting of birds in the rafters and ledges can be prevented by the use of poultry netting tacked to the overheads of the holding pens.

18.4 PROVIDE ADEQUATE SPACE FOR SHEEP

Requirements for space in a sheep operation have been determined by trial and error and research. These may vary depending on such factors as the system of production, rations fed, size of the sheep, climatic conditions, and the kind of buildings and equipment available. The following standards appear to be satisfactory.

18.4.1. Floor space Floor space recommendations are as follows:

	Dry ewes		Rams		Ewes w/lambs		Creep space		Finishing lambs	
	ft^2	m^2	ft^2	m^2	ft^2	m^2	ft^2	m^2	ft^2	m^2
Solid floors	12–16	1.1–1.5	20–30	1.9–2.8	15–20	1.4–1.9	1.5–2.0	0.14–0.19	8–10	0.74–0.93
Slotted or metal floors	8–10	0.74–0.93	14–20	1.3–1.9	10–20	0.93–1.9	1.5–2.0	0.14–0.19	4–5	0.37–0.46

18.4.2. Lambing pens One 4 ft (1.2 m) × 4 ft (1.2 m) lambing pen per six ewes for flocks of under 100 ewes; one pen for eight ewes in flocks of more than 100 ewes. Large ewes weighing 160 lb (73 kg) or more need 4 × 5 ft (1.2 × 1.5 m) pens or even 5 × 5 ft (1.5 × 1.5 m) pens.

18.4.3. Feeder space Feeder space is dependent on the method of feeding, size of the sheep, whether wooled or shorn, time of year, objective of feeding, and the feeds fed. Feeder space recommendations are as follows:

Feeder space/head	Rams		Ewes: Pregnant		Ewes: Lactating		Creep fed lambs		Finishing lambs	
	in.	cm	in.	cm	in.	cm	in.	cm	in.	cm
Self-feeders	6	15.2	6	15.2	6–8	15.2-20.3	2	5	5	12.7
Concentrate troughs (sheep fed daily)	12-16	30–41	12-16	30–41	16–20	41–51	2	5	9–12	23–30
Hay racks	12-16	30–41	12-16	30–41	16–20	41–51			12	30

18.4.4. Waterer space Water consumption is dependent on type of feed fed, weather conditions, size of sheep, function being performed by the sheep, and temperature of the water. Sheep tend to drink more water if it is above 35 °F and below 75 °F. They consume more in warm weather than cold and lactating ewes more than those that are pregnant; dry, open ewes need the least of all. Waterer recommendations per head of sheep are as follows:

	Rams	Ewes		Creep fed lambs	Finishing lambs
		Pregnant	Lactating		
No./bowl or nipple	10	40–50	40–50	Avail. near creep	50–75
No./foot (0.305 m) of tank or trough	5	15–20	15–20	15–25	25–40
Gallons (liters) needed per day	2–3 (7.5–11.3)	2 (7.5–11.3)	3 (11.4)	0.1–0.3 (0.4–1.1)	1.5 (5.7)

18.5 SHEEP EQUIPMENT

The kind of equipment needed will depend on a number of factors. Some of these are listed next.

1. Type of operation: whether ewe–lamb or lamb finishing
2. Time of year ewes lamb: fall, winter, spring, or all three in accelerated systems
3. Size of operation
4. Location of the operation
5. Kind and form of feeds to be fed
6. Labor availability and costs
7. Costs of power
8. Costs of building materials
9. Costs of manufactured equipment versus home construction

The type of operation affects the kind and amount of equipment required. A ewe–lamb operation may require much more than a simple lamb finishing lot, because the ewes must be lambed out and the lambs reared before final finishing. Ewes that lamb in the fall and spring require less equipment than those that produce winter lambs.

Larger operations may justify automation to save labor, while smaller ones may use more labor and be handled by members of the family. Labor may be in surplus in some areas, while it may be scarce and expensive in others. Cost of power may be prohibitive, but it could be relatively low and justify the use of more power and less hand labor. The skill and expertise of the operator may be an important factor. Much of the equipment for feeding and handling sheep can be constructed by someone who is good in a farm shop. If, in addition, the operation is such that there are slack periods then labor on building feeders, and the like can be justified.

There has been much progress in developing plans for feeding sheep mechanically. This should be studied carefully before such an installation is constructed, but it should be kept in mind that labor costs continually increase and the employment situation has been known to fluctuate radically. The more the operator can insulate against dependence on hired help, the greater security and stability the operation should have. Understanding the requirements for equipment and the options should contribute to better decisions in procuring them. Following is some information on the various kinds of equipment that may be needed.

18.5.1. Feeders These may include a number of types, including the following:

- Grain troughs
- Hay racks
- Combination grain and hay bunks
- Fenceline bunks
- Self-feeders

Grain Troughs. These should be 4 to 6 in. (10 to 15 cm) deep, 10 to 12 in. (25 to 30 cm) wide, and 12 to 15 in. (30 to 38 cm) from the ground. They should be built of sturdy material and treated to resist weather conditions. The box should be tight and easily cleaned. If wood is used, 1-in. (2.5-cm) material will result in a bunk light enough to be moved readily.

Hay Racks. These may be constructed to feed from one or both sides. Following are suggested dimensions: height, 12 to 15 in. (30 to 38 cm) from ground; width, 14 to 16 in. (36 to 41 cm) for a one-side feeder or 20 to 24 in. (51 to 61 cm) for both sides. These may be constructed of the least expensive, unplaned material. If slats are used, they should be 3½ in. (8.9 cm) apart. There is a distinct advantage to building a combination grain and hay feeder, because the leaves from the hay tend to fall into the trough and are more readily consumed by the sheep.

Combination Hay and Grain Feeders. There are many types of satisfactory combination feeders in use. Usually, they are 12 to 16 ft (3.65 to 4.9 m) long, so they can be moved. If they are built of native hardwood and are very heavy, they may be built on skids. A simple combination feeder consists of a rectangular box attached to 30-in. (0.76-m) legs, 6 in. (15.2 cm) from the ground, with an 8 to 9 in. (20 to 23 cm) opening and 1 × 10 in. (2.54 to 25.4 cm) boards around the top. These feeders usually have a hinged flap or removable board at each end so they can be opened to sweep the box (Figures 18.4 and 18.5).

Fence-line Bunks. The purpose of fence-line bunks is to enable the operator to feed the sheep without entering the pen. They also are well adapted to feeding mechanically (Figure 18.6). A concrete or gravel pad 4 ft to 6 ft (1.2 to 1.8 m) wide is usually constructed before the bunks are built. This will prevent the area near the bunk from becoming muddy during rainy weather. A hard surface drive is needed

Figure 18.4 Handy combination hay and grain feeder for ewes or lambs.

Figure 18.5 Combination bunks placed end to end make it possible to feed mechanically.

on the outside for the same reason. The bunk may be constructed of wood, concrete, or a combination of the two or even of sheet metal, depending on prices. The bunk should be 6 in. (15 cm) deep on the side from which the sheep eat, 12 in. (30 cm) wide at the bottom, and 24 in. (61 cm) wide at the top. Height from the bottom of the bunk to the top of the outside should be 16 to 18 in. (41 to 46 cm). The opening

Figure 18.6 Fenceline bunks reduce hand labor in feeding sheep.

for sheep to eat is 10 in. (25 cm) unless it is to be used for light lambs. Then it may be reduced by adding another strip. The total height of the fence may be adjusted to fit the temperament of the sheep, but it is usually 36 in. (0.9 m) at least.

If fence-line bunks are being contemplated, it is wise to visit some installations where they have been in use for a number of years to learn the strengths and weaknesses of the various types. There are a number of clever innovations that may be quite satisfactory and reduce the required investment.

Self-feeders. There are also many plans for self-feeders. If grain or pellets are to be fed, the problem of selection is much simpler since there is little or no difficulty with feeding down. Then it is wise to consider how large the feeder should be and how to clean it and to control feed wastage. Building stationary self-feeders has some advantages in construction costs, but there are great advantages to being able to move a feeder. Covers over the hopper to protect from rain are a distinct advantage. They also may prevent birds from getting into the hopper. Before building feeders, look at commercial products. Sometimes it is cheaper to buy them.

Self-feeders for ground mixed complete rations are much more difficult to construct because of the tendency of such feeds to bridge over and require the operator to punch down the feed. If the feeder has vertical sides, it will usually feed down better than those that slant. There should also be some provision for restricting the flow of feed into the trough; otherwise, many feeders provide too free a flow and much feed is wasted.

18.5.2. Waterers The concepts of watering livestock have changed somewhat during recent years. One of the latest innovations is the use of nipple waterers, which are generally satisfactory. This innovation appears to be well adapted to confinement production and may be useful in feedlots and possibly other operations, too.

Perhaps the simplest method of watering sheep is to use troughs or tanks. It has been said that the ideal setup would be for water to trickle into a shallow trough. This does not work well during cold weather, but may be unexcelled during the summer. Metal or concrete troughs are satisfactory provided they are cleaned frequently. Automatic waterers with heating elements usually save much time and trouble during cold winter months. A good selection of commercial waterers is available. Features to consider before selecting a waterer are reliability, ease of repair, ease of cleaning, adaptability to the situation, cost of operation, and safety if it is heated electrically.

If tanks are used in the summer, they should be shaded if possible so that the sheep will drink more. Also, if tanks are used, make certain that lambs that fall in can climb out of shallow tanks.

18.5.3. Mineral feeders Feeders should be located near the source of water. The feeder should be at least 18 in. (0.46 m) above the ground and built so sheep cannot stand in it. Covers will be needed for outside feeders. Feeders that have insertable wooden boxes are handier and can be cleaned when mineral cakes. Metal feeders should always have wooden boxes for holding the mineral. One foot (0.305 m) of space is sufficient for 20 to 25 ewes (Figures 18.7 and 18.8).

18.5.4. Lambing pens Depending on the part of the country you are from, these may be called jugs or jails, but the purpose is the same wherever ewes are lambed. A most satisfactory way of constructing lambing pens is to use hinged panels. The panels may be 4 × 4 ft (1.2 × 1.2m) or 4 × 5 ft (1.2 × 1.5m) or 5 × 5 ft (1.5 × 1.5 m) and 36 in. (0.9 m) high. They are usually constructed of 1-in (2.5-cm) material with a 1 × 12 in. (2.5 × 30.5 cm) or 1 × 10 in. (2.5 × 25.4 cm) board on the bottom and three 1 × 4 in. (2.5 × 10.2 cm) boards spaced 3 to 4 in. (7.6 to 10.2 cm) apart as shown in Figure 18.9. The panels can also be made solid with plywood or some people have made very serviceable panels out of snowfence with the slats horizontal. A few hinged panels 8 × 8 ft × 36 in. (2.44 × 2.44 × 0.9 m) are handy for segregating a ewe that is having lambing trouble. These can be constructed of 1 × 4 in (2.5 × 10.2 cm) material, but should probably have a diagonal brace for rigidity.

18.5.5. Fencing panels Pens inside the sheep barn can be easily constructed with panels that can be fastened to steel posts. These fences can be taken down with a minimum of effort to change the pen configuration or to clean the barn. They should be constructed of very durable material, such as wood, that has been treated with CCA (chromate copper arsenate) or other preservatives. They may be built of 1 × 4 in. (2.5 × 10.2 cm) slats, 12 to 16 ft (3.7 to 4.9 m) long. Height of the panel should be 36 to 42 in. (0.9 × 1.07 m). Diagonal cross braces will add to the strength and rigidity of the panels. Distance between slats should be no more than 4 in. (10.4 cm) for the lowest three slats, then 6 in. (15 cm) and even 8 in. (20.3 cm) between the top two. Panels can also be constructed with a frame of 1 × 4 in. (2.54 × 10.2 cm) with diagonal braces and woven wire stapled to the frame. Hog panels of electrically welded metal are also very useful, but may be too expensive for most sheep barns.

Figure 18.7 A metal mineral feeder with wooden box inside.

Figure 18.8 A wooden mineral feeder.

Figure 18.9 Hinged 8′ × 8′ panels have many uses, especially when a ewe needs help at lambing.

18.5.6. Corrals and outside lots Well-drained and protected corrals or lots adjacent to the barn will result in a healthier flock and will reduce labor in handling them. In many regions of the country, much of the feeding and working with the sheep can be done in outside facilities. If properly fenced, the outside lot may provide a safe sanctuary from predators during the night. The lot or corral should slope away from the barn with enough grade for the water to drain off promptly. The lots should be protected from drainage from other areas by terraces or diversion ditches.

In many states, laws require feedlots to catch the runoff and hold it in lagoons to prevent pollution to streams. These regulations should be considered before building a facility. If there are requirements that must be met, it is much easier to comply when the facility is being built than later.

The following are recommendations for corral or yard space:

	Open Ewes		Rams		Ewes w/lambs		Finishing Lambs	
	ft²	m²	ft²	m²	ft²	m²	ft²	m²
Unpaved lots	25–40	2.3–3.7	25–40	2.3–3.7	30–50	2.8–4.6	20–30	1.9–2.8
Paved lots	16	1.5	16	1.5	20	1.9	10	0.93

If the soil is light and there is plenty of slope (6 to 10%), there may be few problems with the lot. But if there is little slope and the soil is heavy, it may be very difficult to prevent the lot from becoming very muddy and unhealthy for the sheep. One solution is to pave the lot with concrete. This will provide solid footing, but it may be too expensive. It may also be unsatisfactory unless there is adequate slope to carry off the water. A second option is to build mounds in the lot.

Mounds in Corrals or Lots. The most satisfactory mounds appear to be the type that run parallel with the slope and are built 4 to 6 ft (1.2 to 1.8 m) high with a top width of approximately 6 ft (1.8 m) and rounded. Side slopes ratios away from the mound, should be 1 : 5 ft (0.3 : 1.5 m) to 1 : 6 ft (0.3 : 1.8 m), that is, 1 ft (0.3 m) of drop per 5 to 6 ft (1.5 to 1.8 m) horizontal distance. The longitudinal slope of the mound should be no more than 6%. Mounds should be started at feed bunks and slope away.

Mounds must be maintained yearly. The upper half of the mound should be stabilized by using fine-cut bedding or limestone. Further detail can be obtained from the Midwest Planning Service Publication (MWPS–3) of 1982.

The corral or lot should usually have a predator-proof perimeter fence around it. This may be wire or board and should be 5.5 to 6 ft (1.7 to 1.8 m) high. The west or north side of the lot may be enclosed in a taller-board fence used as a windbreak. A fence built of 36–in. (0.9–m) woven wire with three or four strands of barbed wire 4 to 6 in. (10 to 18 cm) apart will repel most predators. If there is an electric fence outside the perimeter, it will be virtually predator proof as long as it is maintained.

Shelter Belts or Windbreaks. In the more northern and western states, wind protection is valuable. Shelter belts with evergreens or conifers can be grown in only a few years and will last a long time. If they are well laid out, these may be quite attractive. Windbreaks may also be built of many materials, such as bales of straw, plywood, sheet roofing, or sawmill slabs, but care must be taken to anchor them securely. Windbreaks are usually 6½ ft (2 m) high.

18.6 SHEEP-HANDLING FACILITIES

Handling facilities should be planned with the following jobs in mind: gathering, sorting, grading lambs, weighing, tagging, crutching, shearing, foot trimming, vaccinating, worming, castrating and docking, treating sheep that are sick or have external parasites, handling rams during the breeding season including semen evaluation, pregnancy testing of ewes, and other necessary jobs.

These facilities should be located near the barn and easily accessible to it. They should be designed with the peculiarities of sheep in mind if they are expected to flow smoothly through the system. McNeal (1985), sheep and wool specialist at Utah State University, has suggested that sheep tend to react as follows when they are being handled:

1. To move toward freedom.
2. To move toward a friend or other sheep.
3. To avoid things that frighten them.
4. To flow better around curves or slight corners.
5. To move away from buildings.
6. To move uphill rather than down.
7. To follow the sheep in front of them.
8. To prefer lighted areas to dark barns, alleys, or chutes.
9. To move toward a horizon.
10. To move themselves at their own speed rather than to be crowded and intimidated.

It should be possible to handle large as well as small groups or individual animals through the facility.

Some desirable features of working chutes and pens are:

1. Solid walls with smooth surfaces on the inside. All posts and obstructions on the outside.
2. Sharp corners, projections, and sudden drops should be avoided.
3. Gates located near corners.
4. Gates in the perimeter 8 ft (2.44 m) or wider.
5. Latches, chains, or catches easy to work by people but not by sheep.
6. The layout should be as simple as possible.
7. The race in a sorting chute narrow enough to prevent turning around.
8. The race long enough to permit an unobstructed view of sheep before they reach the cutting gate.
9. Sorting gates must present a clear view ahead to the oncoming sheep.
10. The position of the operator should not repel the sheep flow.

Excellent steel or wood equipment is available commercially either as permanent or portable installations. Such equipment is usually on display at the larger fairs and sheep meetings. If the manufactured products are too expensive, it is possible to construct very satisfactory ones using plans from the *Sheep Housing and Equipment Handbook* (Midwest Planning Service, 1982).

Construct the working facilities of sturdy material. If plywood is to be used, it will usually be preferable to use marine plywood. Outside facilities have a much longer life if built of lumber treated with CCA or other preservative. Sides of chutes should be 36 in. (0.9 m) high for most sheep, but may be greater for exceptionally wild ones. Excellent handling facilities at Ohio State University are shown in Figure 18.10.

Two holding pens are about minimum. A large one for gathering the whole group should lead into a smaller pen equipped with gates to funnel them into the chute. A circular funneling pen with one of the gates attached near the end of the chute will allow the sheep to be squeezed into the funnel as the gate is rotated.

Some small pens opening off the sorting chute will allow sheep to be divided into a number of groups. If the working facilities are to be used extensively in all weather, the floor of chutes should be made of rough-finished concrete so that the sheep will have good footing when the pens are muddy. Cinders should never be used in pens, runways or chutes because they cut the feet and open them to infections.

A loading chute should be located near the working areas so that lambs can be loaded immediately after they are sorted. There are a number of types of loading chutes that are satisfactory. They must be located where trucks can reach them. A hard-surfaced driveway is mandatory for this facility, and there must be paved or graveled maneuvering room for trucks. If the terrain has much slope, it may be possible to fill dirt behind a restraining wall so the sheep can walk on the level into the truck. If it is necessary to walk up and down an incline, steps are more satisfactory

Figure 18.10 Good handling facilities increase efficiency and reduce stress on sheep. Facilities at the Ohio Agricultural Research and Development Center, Wooster, Ohio. (Courtesy Animal Science Department, Ohio Agricultural Research and Development Center, The Ohio State University, Wooster)

than a cleated chute. An adjustable chute that has counterweights for ease of raising or lowering will save time and simplify loading and unloading of double-decked trucks. Plans are also available for portable chutes. The most important consideration is to build chutes of strong durable material that can be kept in good repair at all times. Although it is not mandatory, lights over the working area are quite desirable so that loading, unloading, sorting, and weighing can be done when it is dark. A roof over the area also improves working conditions. It will keep the sun and rain off the operator, sheep, and handling equipment.

Equipment for Controlling External Parasites and Foot Rot. Depending on the size of the flock, it may be prudent to plan for an area where sheep can be dipped or sprayed. Dipping vats may be made of galvanized metal and portable or built permanently of concrete with adjacent drip pens. This unit can be connected to a crowding pen, which is also used for funneling sheep into the sorting and loading chutes and the scales. Midwest Planning Service (1982) has an excellent plan for such facilities, as well as spraying chutes. A metal or polyurethane insert can be put on the floor of the sorting chute if a foot bath is needed.

18.7 FENCES

Perimeter fences on a sheep farm *should be well constructed not only to restrain the sheep, but to prevent the entry of other animals.* Woven wire 36 in. (0.9 m) high with a fine enough mesh to prevent them from sticking their heads through should be topped with several strands of barbed wire, with the top strand 54 in. (1.37 m) from the ground. Fences should be built next to the ground, and all holes caused by erosion closed to prevent entry of predators. Cross fences should be at least 32 to 36 in. (0.85 to 0.9 m) high. Temporary fences 36 in. (0.9 m) high will turn most sheep. Cross fences 26 to 32 in. (0.7 to 0.85 m) high with a strand or two of barbed wire are usually satisfactory for sheep and cattle. A six- or seven-strand barbed wire fence is also satisfactory for cross fencing.

Gates should be well braced, have strong hinges, and swing close to the ground. They should have slats no wider than 6 to 8 in. (15 to 20 cm) and spaces between slats 4 to 6 in. (10 to 15 cm) to a height of 3 ft. (0.9 m) and then 8 to 10 in. (20 to 25 cm) at the top. They should fit close to posts and be braced sufficiently so that they will open and close easily. Very wide gates need special bracing.

Electric fencing technology has improved with the introduction of information from Australia and New Zealand. Use of high-tensile wire and high-energy fence chargers has resulted in much more effective electric fences. The key to the effectiveness of the chargers is the very short, but extremely high, energy pulse. Energizers are also capable of preventing weed growth from interfering with the installation.

Requirements for satisfactory function of an electric fence include taut wires and an electrically hot fence at all times. Wire recommended for electric fencing is 11 to 14 gauge, high-tensile steel, galvanized wire with a breaking strength of 1000 lb (454 kg). The wire must be flexible enough to tie knots. Tension of 200 lb (91 kg) should be maintained by the use of springs. For sheep, it is recommended that five strands spaced from the ground at 5, 10, 17, 27, and 38 in. (13, 25, 43, 68, and 97 cm) be used. Corner and end posts must be well anchored and of substantial strength. Line posts serve only to hold up the wire and to maintain spacing. Insulators may be of a conventional type, but must be capable of withstanding the tension on the wires and allow the wire to slide freely.

At the Missouri Experiment Station (1982), it has been possible to prevent predator attacks on the university sheep flock by using two-strand electric fence around a well-built, 54 to 60 in. (1.37 to 1.52 m) perimeter fence. The strands are 6 in. (15 cm) and 18 in. (46 cm) from the ground and both are charged.

18.8 STOCK GUARDS

Gates in busy driveways may be replaced with stock guards to save time and extra effort. On the perimeter of the farm, there should also be a gate if there are times it is desirable to control vehicular traffic. The guard should be built to withstand the weight of most trucks. It should be placed where the pit can be easily drained. Since most stock guards are 10 ft (3 m) wide, it is desirable to have a 10 to 12 ft (3 to 3.7 m) gate beside it to allow livestock and wider machinery to pass around it. Pipe, railroad rails, or hardwood 2 × 6 in. (5 to 15 cm) set on edge is satisfactory

Figure 18.11 Strategically located stock guards save time.

to prevent sheep from moving across the guard. A useful homemade stock guard is shown in Figure 18.11.

18.9 SHADES AND SHELTERS IN PASTURES

In the hot summer time, sheep will suffer if they do not have some protection from the sun. Many kinds of shade are used. Portable shed-type structures in which the sides and back can be raised are very satisfactory, as in a Quonset type that is open in the front and with a flap that can be raised in the back. Space under shades may vary from 4 to 6 ft^2 (0.37 to 0.56 m^2) per sheep, depending on the size of the sheep. An alignment with the long axis east and west and the shorter north and south with all sides open is preferable for summer. There does not appear to be complete agreement on how high a permanent shade should be from the ground. Four ft (1.2 m) is as low to the ground as most shades should be built if there is to be much air movement under it. There is some danger of sheep piling up underneath a low-lying shade on extremely hot days. There is also the disadvantage of it being difficult to work under the shade. Shades 6 to 8 ft (1.8 to 2.4 m) high tend to be dryer and much easier to clean under and have greater air circulation than those lower to the ground.

Building material for shades may vary from snow fence to straw over a woven wire frame to sheet metal roofing or plywood on a flat frame. All appear to be satisfactory if solidly constructed. Another option is to plant shade trees where they can provide protection.

REFERENCES

Black, C. A., "The US Sheep and Goat Industry. Opportunities and Limitations," *Council of Agricultural Science and Technology Report 94,* 1982.

McNeal, Lyle G., "Sheep Handling Concepts and Equipment," Sheep Industry Development Program Research Digest: 2 No. 1 p. 1–5, 1985.

Midwest Planning Service (MWPS-3), *Sheep Housing and Equipment Handbook,* 3rd ed. Ames, Iowa: Iowa State University, 1982.

Munoz, J. R., Causes of Sheep Mortality at the Cook Ranch, Florence, Montana, 1975–1976. M.S. thesis, University of Montana, Missoula, 1977.

Proceedings: A National Symposium on Confinement Sheep Production," September 11, 1981. Urbana, Ill.: University of Illinois, Animal Science Department.

Science and Education Administration, "Housing and Equipment for Sheep," *Farmers Bulletin No. 2242,* USDA.

Sheep Industry Development Program, Inc., "Proceedings of Symposium—Intensive Sheep Management," Paducah, Ky., June 20–23, 1972.

QUESTIONS FOR STUDY AND DISCUSSION

1. Why do sheep need shelter? When is the most critical time?
2. What are the basic requirements for a sheep barn?
3. Discuss the reasons for being cautious about investing money in buildings and equipment.
4. Why is it undesirable to renovate a horse barn for sheep?
5. Discuss the most desirable location of a sheep barn.
6. Explain how a sheep barn may be built so it is flexible.
7. What are the basic reasons for ventilation of sheep barns?
8. Discuss the benefits of having feed storage capacity adjacent to a sheep operation.
9. What are the requirements for satisfactory lambing quarters?
10. What is a main advantage to clear-span construction in a sheep barn?
11. Discuss the difference in floor space requirements for ewes on solid or metal floors.
12. How do feeder space requirements change with the method of feeding?
13. What are the possible benefits of mechanized feeding for sheep?
14. What can be done to increase the life of wooden feeders used on the outside of the barn?
15. What problems can be encountered with the design for self-feeders for complete ground mixed lamb rations? How can these be overcome?
16. What are the benefits of automatic waterers for sheep?
17. Where should mineral feeders be located?
18. What special provisions can be made for mud prevention in outside lots on heavy, poorly drained soil?
19. Why is it important to check legal restrictions before building a feedlot?
20. How are mounds protected against erosion in a sheep lot?
21. What are some characteristics of sheep that should be considered in constructing working facilities?
22. Describe a predator-proof fence.
23. What are some advantages of an electric fence?
24. What is the secret to the success of the Australian or New Zealand type of electric fences?
25. What are conditions that indicate the need of a stock guard?
26. What are some alternate methods of providing shade for sheep?

Appendix

APPENDIX TABLE 1 TABLE OF EQUIVALENTS: CUSTOMARY WEIGHTS AND MEASURES WITH METRIC EQUIVALENTS

Length	
1 inch	= 2.54 centimeters
1 foot	= 30.48 centimeters
	= 0.3048 meter
1 yard	= 0.9144 meter
1 mile	= 1609.34 meters
	= 1.609 kilometers
Area	
1 square inch	= 6.452 square centimeters
1 square foot	= 0.0929 square meter
1 square yard	= 0.8361 square meter
1 acre	= 0.4047 hectare
1 square mile	= 259.0 hectares
Capacity or volume	
1 cubic inch	= 16.387 cubic centimeters
1 cubic foot	= 0.0283 cubic meter
1 cubic yard	= 0.7646 cubic meter
1 fluid ounce (U.S.)	= 29.573 milliliters
1 liquid pint (U.S.)	= 0.4732 liter
1 liquid quart (U.S.)	= 0.9463 liter
1 gallon (U.S.)	= 3.7853 liters
Weight	
1 ounce (avdp.)	= 28.50 grams
1 pound (avdp.)	= 453.592 grams
	= 0.4536 kilogram
1 ton (short)	= 0.907 ton (metric)
1 ton (long)	= 1.016 ton (metric)
	= 1016.05 kilograms
Volume per unit area	
1 gallon (U.S.)/acre	= 9.354 liters/hectare
Weight per unit area	
1 pound (avdp.)/square inch	= 0.0703 kilogram/square centimeter
1 pound (avdp.)/acre	= 1.121 kilograms/hectare
Area per unit weight	
1 square inch/pound (avdp.)	= 14.22 square centimeters/kilogram

APPENDIX TABLE 2 TABLE OF EQUIVALENTS: METRIC WEIGHTS AND MEASURES WITH CUSTOMARY EQUIVALENTS

Length	
1 millimeter	= 0.03937 inch
1 centimeter	= 0.3937 inch
1 meter	= 39.37 inches
	= 3.281 feet
	= 1.094 yards
1 kilometer	= 0.6214 mile
Area	
1 square centimeter	= 0.155 square inch
1 square meter	= 1.196 square yards
	= 10.764 square feet
1 hactare (10,000 m^2)	= 2.471 acres
1 square kilometer	= 0.386 square mile
	= 247.1 acres
Capacity of volume	
1 cubic centimeter	= 0.061 cubic inch
1 cubic meter	= 35.315 cubic feet
	= 1.308 cubic yards
1 milliliter	= 0.0338 fluid ounce (U.S.)
1 liter	= 33.81 fluid ounces (U.S.)
	= 2.1134 pints (U.S.)
	= 1.057 quarts (U.S.)
	= 0.2642 gallon (U.S.)
1 kiloliter	= 264.18 gallons (U.S.)
Weight	
1 gram	= 0.03527 ounce (avdp.)
1 kilogram	= 35.274 ounces (avdp.)
	= 2.205 pounds (avdp.)
1 metric ton (1080 kg)	= 0.984 ton (long)
	= 1.102 tons (short)
	= 2204.6 pounds (avdp.)
Volume per unit area	
1 liter/hectare	= 0.107 gallon (U.S.)/acre
Weight per unit area	
1 kilogram/square centimeter	= 14.22 pounds (avdp.)/square inch
1 kilogram/hectare	= 0.892 pound (avdp.)/acre
Area per unit weight	
1 square centimeter/kilogram	= 0.0703 square inch/pound (avdp.)
Temperature conversion formulas	
Celsius (centigrade)	= $\frac{5}{9}$ (Fahrenheit − 32)
Fahrenheit	= $\frac{9}{5}$ Celsius (centigrade) + 32
Illumination	
1 foot-candle	= 10.764 lux
1 lux	= 0.0929 foot-candle

When conversions are made, the results should be rounded to a meaningful number of digits relative to the accuracy of the original measurements. Values for weights and volumes are based on pure water at 4 °C under 760 mm of atmospheric pressure.

For additional conversion factors, or for greater accuracy, refer to National Bureau of Standards Miscellaneous Publication 233 or to the *CRC Handbook of Chemistry and Physics,* Cleveland, Ohio, 1987/1989.

APPENDIX TABLE 3 DAILY NUTRIENT REQUIREMENTS OF SHEEP

Body Weight		Weight Change/Day		Dry Matter per Animal[a]			Nutrients per Animal: Energy[b] TDN				Crude protein					
(kg)	(lb)	(g)	(lb)	(kg)	(lb)	(% body weight)	TDN (kg)	TDN (lb)	DE (Mcal)	ME (Mcal)	Crude protein (g)	Crude protein (lb)	Ca (g)	P (g)	Vitamin A Activity (IU)	Vitamin E Activity (IU)
Ewes[c]																
Maintenance																
50	110	10	0.02	1.0	2.2	2.0	0.55	1.2	2.4	2.0	95	0.21	2.0	1.8	2,350	15
60	132	10	0.02	1.1	2.4	1.8	0.61	1.3	2.7	2.2	104	0.23	2.3	2.1	2,820	16
70	154	10	0.02	1.2	2.6	1.7	0.66	1.5	2.9	2.4	113	0.25	2.5	2.4	3,290	18
80	176	10	0.02	1.3	2.9	1.6	0.72	1.6	3.2	2.6	122	0.27	2.7	2.8	3,760	20
90	198	10	0.02	1.4	3.1	1.5	0.78	1.7	3.4	2.8	131	0.29	2.9	3.1	4,230	21
Flushing—2 Weeks prebreeding and first 3 weeks of breeding																
50	110	100	0.22	1.6	3.5	3.2	0.94	2.1	4.1	3.4	150	0.33	5.3	2.6	2,350	24
60	132	100	0.22	1.7	3.7	2.8	1.00	2.2	4.4	3.6	157	0.34	5.5	2.9	2,820	26
70	154	100	0.22	1.8	4.0	2.6	1.06	2.3	4.7	3.8	164	0.36	5.7	3.2	3,290	27
80	176	100	0.22	1.9	4.2	2.4	1.12	2.5	4.9	4.0	171	0.38	5.9	3.6	3,760	28
90	198	100	0.22	2.0	4.4	2.2	1.18	2.6	5.1	4.2	177	0.39	6.1	3.9	4,230	30
Nonlactating—First 15 weeks gestation																
50	110	30	0.07	1.2	2.6	2.4	0.67	1.5	3.0	2.4	112	0.25	2.9	2.1	2,350	18
60	132	30	0.07	1.3	2.9	2.2	0.72	1.6	3.2	2.6	121	0.27	3.2	2.5	2,820	20
70	154	30	0.07	1.4	3.1	2.0	0.77	1.7	3.4	2.8	130	0.29	3.5	2.9	3,290	21
80	176	30	0.07	1.5	3.3	1.9	0.82	1.8	3.6	3.0	139	0.31	3.8	3.3	3,760	22
90	198	30	0.07	1.6	3.5	1.8	0.87	1.9	3.8	3.2	148	0.33	4.1	3.6	4,230	24
Last 4 weeks gestation (130-150% lambing rate expected) or last 4-6 weeks lactation suckling singles[d]																
50	110	180 (45)	0.40 (0.10)	1.6	3.5	3.2	0.94	2.1	4.1	3.4	175	0.38	5.9	4.8	4,250	24
60	132	180 (45)	0.40 (0.10)	1.7	3.7	2.8	1.00	2.2	4.4	3.6	184	0.40	6.0	5.2	5,100	26
70	154	180 (45)	0.40 (0.10)	1.8	4.0	2.6	1.06	2.3	4.7	3.8	193	0.42	6.2	5.6	5,950	27
80	176	180 (45)	0.40 (0.10)	1.9	4.2	2.4	1.12	2.4	4.9	4.0	202	0.44	6.3	6.1	6,800	28
90	198	180 (45)	0.40 (0.10)	2.0	4.4	2.2	1.18	2.5	5.1	4.2	212	0.47	6.4	6.5	7,650	30
Last 4 weeks gestation (180-225% lambing rate expected)																
50	110	225	0.50	1.7	3.7	3.4	1.10	2.4	4.8	4.0	196	0.43	6.2	3.4	4,250	26
60	132	225	0.50	1.8	4.0	3.0	1.17	2.6	5.1	4.2	205	0.45	6.9	4.0	5,100	27
70	154	225	0.50	1.9	4.2	2.7	1.24	2.8	5.4	4.4	214	0.47	7.6	4.5	5,950	28
80	176	225	0.50	2.0	4.4	2.5	1.30	2.9	5.7	4.7	223	0.49	8.3	5.1	6,800	30
90	198	225	0.50	2.1	4.6	2.3	1.37	3.0	6.0	5.0	232	0.51	8.9	5.7	7,650	32
First 6-8 weeks lactation suckling singles or last 4-6 weeks lactation suckling twins[d]																
50	110	−25 (90)	−0.06 (0.20)	2.1	4.6	4.2	1.36	3.0	6.0	4.9	304	0.67	8.9	6.1	4,250	32
60	132	−25 (90)	−0.06 (0.20)	2.3	5.1	3.8	1.50	3.3	6.6	5.4	319	0.70	9.1	6.6	5,100	34
70	154	−25 (90)	−0.06 (0.20)	2.5	5.5	3.6	1.63	3.6	7.2	5.9	334	0.73	9.3	7.0	5,950	38
80	176	−25 (90)	−0.06 (0.20)	2.6	5.7	3.2	1.69	3.7	7.4	6.1	344	0.76	9.5	7.4	6,800	39
90	198	−25 (90)	−0.06 (0.20)	2.7	5.9	3.0	1.75	3.8	7.6	6.3	353	0.78	9.6	7.8	7,650	40

First 6-8 weeks lactation suckling twins																
50	110	−60	−0.13	2.4	5.3	4.8	1.56	3.4	6.9	5.6	389	0.86	10.5	7.3	5,000	36
60	132	−60	−0.13	2.6	5.7	4.3	1.69	3.7	7.4	6.1	405	0.89	10.7	7.7	6,000	39
70	154	−60	−0.13	2.8	6.2	4.0	1.82	4.0	8.0	6.6	420	0.92	11.0	8.1	7,000	42
80	176	−60	−0.13	3.0	6.6	3.8	1.95	4.3	8.6	7.0	435	0.96	11.2	8.6	8,000	45
90	198	−60	−0.13	3.2	7.0	3.6	2.08	4.6	9.2	7.5	450	0.99	11.4	9.0	9,000	48
Ewe lambs																
Nonlactating—First 15 weeks gestation																
40	88	160	0.35	1.4	3.1	3.5	0.83	1.8	3.6	3.0	156	0.34	5.5	3.0	1,880	21
50	110	135	0.30	1.5	3.3	3.0	0.88	1.9	3.9	3.2	159	0.35	5.2	3.1	2,350	22
60	132	135	0.30	1.6	3.5	2.7	0.94	2.0	4.1	3.4	161	0.35	5.5	3.4	2,820	24
70	154	125	0.28	1.7	3.7	2.4	1.00	2.2	4.4	3.6	164	0.36	5.5	3.7	3,290	26
Last 4 weeks gestation (100-120% lambing rate expected)																
40	88	180	0.40	1.5	3.3	3.8	0.94	2.1	4.1	3.4	187	0.41	6.4	3.1	3,400	22
50	110	160	0.35	1.6	3.5	3.2	1.00	2.2	4.4	3.6	189	0.42	6.3	3.4	4,250	24
60	132	160	0.35	1.7	3.7	2.8	1.07	2.4	4.7	3.9	192	0.42	6.6	3.8	5,100	26
70	154	150	0.33	1.8	4.0	2.6	1.14	2.5	5.0	4.1	194	0.43	6.8	4.2	5,950	27
Last 4 weeks gestation (130-175% lambing rate expected)																
40	88	225	0.50	1.5	3.3	3.8	0.99	2.2	4.4	3.6	202	0.44	7.4	3.5	3,400	22
50	110	225	0.50	1.6	3.5	3.2	1.06	2.3	4.7	3.8	204	0.45	7.8	3.9	4,250	24
60	132	225	0.50	1.7	3.7	2.8	1.12	2.5	4.9	4.0	207	0.46	8.1	4.3	5,100	26
70	154	215	0.47	1.8	4.0	2.6	1.14	2.5	5.0	4.1	210	0.46	8.2	4.7	5,950	27
First 6-8 weeks lactation suckling singles (wean by 8 weeks)																
40	88	−50	−0.11	1.7	3.7	4.2	1.12	2.5	4.9	4.0	257	0.56	6.0	4.3	3,400	26
50	110	−50	−0.11	2.1	4.6	4.2	1.39	3.1	6.1	5.0	282	0.62	6.5	4.7	4,250	32
60	132	−50	−0.11	2.3	5.1	3.8	1.52	3.4	6.7	5.5	295	0.65	6.8	5.1	5,100	34
70	154	−50	−0.11	2.5	5.5	3.6	1.65	3.6	7.3	6.0	301	0.68	7.1	5.6	5,450	38
First 6-8 weeks lactation suckling twins (wean by 8 weeks)																
40	88	−100	−0.22	2.1	4.6	5.2	1.45	3.2	6.4	5.2	306	0.67	8.4	5.6	4,000	32
50	110	−100	−0.22	2.3	5.1	4.6	1.59	3.5	7.0	5.7	321	0.71	8.7	6.0	5,000	34
60	132	−100	−0.22	2.5	5.5	4.2	1.72	3.8	7.6	6.2	336	0.74	9.0	6.4	6,000	38
70	154	−100	−0.22	2.7	6.0	3.9	1.85	4.1	8.1	6.6	351	0.77	9.3	6.9	7,000	40
Replacement ewe lambs[e]																
30	66	227	0.50	1.2	2.6	4.0	0.78	1.7	3.4	2.8	185	0.41	6.4	2.6	1,410	18
40	88	182	0.40	1.4	3.1	3.5	0.91	2.0	4.0	3.3	176	0.39	5.9	2.6	1,880	21
50	110	120	0.26	1.5	3.3	3.0	0.88	1.9	3.9	3.2	136	0.30	4.8	2.4	2,350	22
60	132	100	0.22	1.5	3.3	2.5	0.88	1.9	3.9	3.2	134	0.30	4.5	2.5	2,820	22
70	154	100	0.22	1.5	3.3	2.1	0.88	1.9	3.9	3.2	132	0.29	4.6	2.8	3,290	22

(CONTINUED)

APPENDIX TABLE 3 DAILY NUTRIENT REQUIREMENTS OF SHEEP (CONTINUED)

Body Weight		Weight Change/Day		Dry Matter per Animal[a]			Nutrients per Animal: Energy[b] TDN		DE	ME	Crude protein		Ca	P	Vitamin A Activity	Vitamin E Activity
(kg)	(lb)	(g)	(lb)	(kg)	(lb)	(% body weight)	(kg)	(lb)	(Mcal)	(Mcal)	(g)	(lb)	(g)	(g)	(IU)	(IU)
Replacement ram lambs[e]																
40	88	330	0.73	1.8	4.0	4.5	1.1	2.5	5.0	4.1	243	0.54	7.8	3.7	1,880	24
60	132	320	0.70	2.4	5.3	4.0	1.5	3.4	6.7	5.5	263	0.58	8.4	4.2	2,820	26
80	176	290	0.64	2.8	6.2	3.5	1.8	3.9	7.8	6.4	268	0.59	8.5	4.6	3,760	28
100	220	250	0.55	3.0	6.6	3.0	1.9	4.2	8.4	6.9	264	0.58	8.2	4.8	4,700	30
Lambs finishing—4 to 7 months old[f]																
30	66	295	0.65	1.3	2.9	4.3	0.94	2.1	4.1	3.4	191	0.42	6.6	3.2	1,410	20
40	88	275	0.60	1.6	3.5	4.0	1.22	2.7	5.4	4.4	185	0.41	6.6	3.3	1,880	24
50	110	205	0.45	1.6	3.5	3.2	1.23	2.7	5.4	4.4	160	0.35	5.6	3.0	2,350	24
Early weaned lambs—Moderate growth potential[f]																
10	22	200	0.44	0.5	1.1	5.0	0.40	0.9	1.8	1.4	127	0.38	4.0	1.9	470	10
20	44	250	0.55	1.0	2.2	5.0	0.80	1.8	3.5	2.9	167	0.37	5.4	2.5	940	20
30	66	300	0.66	1.3	2.9	4.3	1.00	2.2	4.4	3.6	191	0.42	6.7	3.2	1,410	20
40	88	345	0.76	1.5	3.3	3.8	1.16	2.6	5.1	4.2	202	0.44	7.7	3.9	1,880	22
50	110	300	0.66	1.5	3.3	3.0	1.16	2.6	5.1	4.2	181	0.40	7.0	3.8	2,350	22
Early weaned lambs—Rapid growth potential[f]																
10	22	250	0.55	0.6	1.3	6.0	0.48	1.1	2.1	1.7	157	0.35	4.9	2.2	470	12
20	44	300	0.66	1.2	2.6	6.0	0.92	2.0	4.0	3.3	205	0.45	6.5	2.9	940	24
30	66	325	0.72	1.4	3.1	4.7	1.10	2.4	4.8	4.0	216	0.48	7.2	3.4	1,410	21
40	88	400	0.88	1.5	3.3	3.8	1.14	2.5	5.0	4.1	234	0.51	8.6	4.3	1,880	22
50	110	425	0.94	1.7	3.7	3.4	1.29	2.8	5.7	4.7	240	0.53	9.4	4.8	2,350	25
60	132	350	0.77	1.7	3.7	2.8	1.29	2.8	5.7	4.7	240	0.53	8.2	4.5	2,820	25

[a]To convert dry matter to an as-fed basis, divide dry matter values by the percentage of dry matter in the particular feed.

[b]One kilogram TDN (total digestible nutrients) = 4.4 Mcal DE (digestible energy); ME (metabolizable energy) = 82% of DE. Because of rounding errors, values in Table 1 and Table 2 may differ.

[c]Values are applicable for ewes in moderate condition. Fat ewes should be fed according to the next lower weight category and thin ewes at the next higher weight category. Once desired or moderate weight condition is attained, use that weight category through all production stages.

[d]Values in parentheses are for ewes suckling lambs the last 4-6 weeks of lactation.

[e]Lambs intended for breeding; thus, maximum weight gains and finish are of secondary importance.

[f]Maximum weight gains expected.

Source: Nutrient Requirements of Sheep, 6th revised edition, © 1985 by The National Academy of Sciences.

APPENDIX TABLE 4 NUTRIENT CONCENTRATION IN DIETS FOR SHEEP (EXPRESSED ON 100% DRY MATTER BASIS[a])

Body Weight		Weight Change/Day		Energy[b]			Example Diet Proportions		Crude Protein	Calcium	Phosphorus	Vitamin A Activity	Vitamin E Activity
(kg)	(lb)	(g)	(lb)	TDN[c] (%)	DE (Mcal/kg)	ME (Mcal/kg)	Concentrate %	Forage %	(%)	(%)	(%)	(IU/kg)	(IU/kg)
Ewes[d]													
Maintenance													
70	154	10	0.02	55	2.4	2.0	0	100	9.4	0.20	0.20	2,742	15
Flushing—2 weeks prebreeding and first 3 weeks of breeding													
70	154	100	0.22	59	2.6	2.1	15	85	9.1	0.32	0.18	1,828	15
Nonlactating—First 15 weeks gestation													
70	154	30	0.07	55	2.4	2.0	0	100	9.3	0.25	0.20	2,350	15
Last 4 weeks gestation (130-150% lambing rate expected) or last 4-6 weeks lactation suckling singles[e]													
70	154	180 (0.45)	0.40 (0.10)	59	2.6	2.1	15	85	10.7	0.35	0.23	3,306	15
Last 4 weeks gestation (180-225% lambing rate expected)													
70	154	225	0.50	65	2.9	2.3	35	65	11.3	0.40	0.24	3,132	15
First 6-8 weeks lactation suckling singles or last 4-6 weeks lactation suckling twins[e]													
70	154	−25(90)	−0.06 (0.20)	65	2.9	2.4	35	65	13.4	0.32	0.26	2,380	15
First 6-8 weeks lactation suckling twins													
70	154	−60	−0.13	65	2.9	2.4	35	65	15.0	0.39	0.29	2,500	15
Ewe Lambs													
Nonlactating—First 15 weeks gestation													
55	121	135	0.30	59	2.6	2.1	15	85	10.6	0.35	0.22	1,668	15
Last 4 weeks gestation (100-120% lambing rate expected)													
55	121	160	0.35	63	2.8	2.3	30	70	11.8	0.39	0.22	2,833	15
Last 4 weeks gestation (130-175% lambing rate expected)													
55	121	225	0.50	66	2.9	2.4	40	60	12.8	0.48	0.25	2,833	15
First 6-8 weeks lactation suckling singles (wean by 8 weeks)													
55	121	−50	0.22	66	2.9	2.4	40	60	13.1	0.30	0.22	2,125	15
First 6-8 weeks lactation suckling twins (wean by 8 weeks)													
55	121	−100	−0.22	69	3.0	2.5	50	50	13.7	0.37	0.26	2,292	15
Replacement Ewe Lambs[f]													
30	66	227	0.50	65	2.9	2.4	35	65	12.8	0.53	0.22	1,175	15
40	88	182	0.40	65	2.9	2.4	35	65	10.2	0.42	0.18	1,343	15
50-70	110-154	115	0.25	59	2.6	2.1	15	85	9.1	0.31	0.17	1,567	15
Replacement Ram Lambs[f]													
40	88	330	0.73	63	2.8	2.3	30	70	13.5	0.43	0.21	1,175	15
60	132	320	0.70	63	2.8	2.3	30	70	11.0	0.35	0.18	1,659	15
80-100	176-220	270	0.60	63	2.8	2.3	30	70	9.6	0.30	0.16	1,979	15
Lambs Finishing—4 to 7 months old[g]													
30	66	295	0.65	72	3.2	2.5	60	40	14.7	0.51	0.24	1,085	15
40	88	275	0.60	76	3.3	2.7	75	25	11.6	0.42	0.21	1,175	15
50	110	205	0.45	77	3.4	2.8	80	20	10.0	0.35	0.19	1,469	15
Early Weaned Lambs—Moderate and rapid growth potential[g]													
10	22	250	0.55	80	3.5	2.9	90	10	26.2	0.82	0.38	940	20
20	44	300	0.66	78	3.4	2.8	85	15	16.9	0.54	0.24	940	20
30	66	325	0.72	78	3.3	2.7	85	15	15.1	0.51	0.24	1,085	15
40-60	88-132	400	0.88	78	3.3	2.7	85	15	14.5	0.55	0.28	1,253	15

[a]Values in Table 2 are calculated from daily requirements in Table 1 divided by DM intake. The exception, vitamin E daily requirements /head, are calculated from vitamin E/kg diet × DM intake.

[b]One kilogram TDN = 4.4 Mcal DE (digestible energy); ME (metabolizable energy) = 82% of DE. Because of rounding errors, values in Table 1 and Table 2 may differ.

[c]TDN calculated on following basis: hay DM, 55% TDN and on as-fed basis 50% TDN; grain DM, 83% TDN and on as-fed basis 75% TDN.

[d]Values are for ewes in moderate condition. Fat ewes should be fed according to the next lower weight category and thin ewes at the next higher weight category. Once desired or moderate weight condition is attained, use that weight category through all production stages.

[e]Values in parentheses are for ewes suckling lambs the last 4-6 weeks of lactation.

[f]Lambs intended for breeding; thus, maximum weight gains and finish are of secondary importance.

[g]Maximum weight gains expected.

APPENDIX TABLE 5 NET ENERGY REQUIREMENTS FOR LAMBS OF SMALL, MEDIUM, AND LARGE MATURE WEIGHT GENOTYPES[a](Kcal/d)

Body Weight (kg)[b]:	10	20	25	30	35	40	45	50
NE_m Requirements[c]:	315	530	626	718	806	891	973	1053
Daily Gain (g)[b]								
NE_g Requirements								
Small mature weight lambs[d]								
100	178	300	354	406	456	504	551	596
150	267	450	532	610	684	756	826	894
200	357	600	708	812	912	1,008	1,102	1,192
250	446	750	886	1,016	1,140	1,261	1,377	1,490
300	535	900	1,064	1,219	1,368	1,513	1,652	1,788
Medium mature weight lambs[e]								
100	155	261	309	354	397	439	480	519
150	233	392	463	531	596	658	719	778
200	310	522	618	708	794	878	960	1,038
250	388	653	771	884	993	1,097	1,199	1,297
300	466	784	926	1,062	1,191	1,316	1,438	1,557
350	543	914	1,080	1,238	1,390	1,536	1,678	1,816
400	621	1,044	1,234	1,415	1,589	1,756	1,918	2,076
Large mature weight lambs[f]								
100	132	221	262	300	337	372	407	439
150	197	332	392	450	505	558	610	660
200	263	442	524	600	674	744	813	880
250	329	553	654	750	842	930	1,016	1,099
300	394	663	785	900	1,010	1,116	1,220	1,320
350	461	775	916	1,050	1,179	1,303	1,423	1,540
400	526	885	1,046	1,200	1,347	1,489	1,626	1,760
450	592	996	1,177	1,350	1,515	1,675	1,830	1,980

[a]Approximate mature ram weights of 95 kg, 115 kg, and 135 kg, respectively.
[b]Weights and gains include fill.
[c]$NE_m = 56$ kcal $\cdot W^{0.75} \cdot d^{-1}$.
[d]$NE_g = 317$ kcal $\cdot W^{0.75} \cdot$ LWG, kg $\cdot d^{-1}$.
[e]$NE_g = 276$ kcal $\cdot W^{0.75} \cdot$ LWG, kg $\cdot d^{-1}$.
[f]$NE_g = 234$ kcal $\cdot W^{0.75} \cdot$ LWG, kg $\cdot d^{-1}$.

Source: Nutrient Requirements of Sheep, 6th revised edition, © 1985 by The National Academy of Sciences.

APPENDIX TABLE 6 NE_{preg} (NE_y) REQUIREMENTS OF EWES CARRYING DIFFERENT NUMBERS OF FETUSES AT VARIOUS STAGES OF GESTATION

Number of Fetuses Being Carried	Stage of Gestation (days)[a]					
	100	%[b]	120	%[b]	140	%[b]
	NE_{preg} Required (kcal/day)					
1	70	100	145	100	260	100
2	125	178	265	183	440	169
3	170	243	345	238	570	219

[a]For gravid uterus (plus contents) and mammary gland development only.
[b]As a percentage of a single fetus's requirement.

From *Nutrient Requirements of Sheep,* 6th revised edition, © 1985 by The National Academy of Sciences.

APPENDIX TABLE 7 CRUDE PROTEIN REQUIREMENTS FOR LAMBS OF SMALL, MEDIUM, AND LARGE MATURE WEIGHT GENOTYPES[a] (g/d)

Body Weight (kg)[b]:	10	20	25	30	35	40	45	50
Daily Gain (g)[b]								
Small mature weight lambs								
100	84	112	122	127	131	136	135	134
150	103	121	137	140	144	147	145	143
200	123	145	152	154	156	158	154	151
250	142	162	167	168	168	169	164	159
300	162	178	182	181	180	180	174	168
Medium mature weight lambs								
100	85	114	125	130	135	140	139	139
150	106	132	141	145	149	153	151	149
200	127	150	158	160	163	166	163	160
250	147	167	174	175	177	179	175	171
300	168	185	191	191	191	191	186	181
350	188	203	207	206	205	204	198	192
400	209	221	224	221	219	217	210	202
Large mature weight lambs								
100	94	128	134	139	145	144	150	156
150	115	147	152	156	160	159	164	169
200	136	166	170	173	176	174	178	182
250	157	186	188	190	192	189	192	195
300	179	205	206	207	208	204	206	208
350	200	224	224	224	224	219	220	221
400	221	243	242	241	240	234	234	234
450	242	262	260	256	256	249	248	248

[a]Approximate mature ram weights of 95 kg, 115 kg, and 135 kg, respectively.
[b]Weights and gains include fill.

Source: Nutrient Requirements of Sheep, 6th revised edition, © 1985 by The National Academy of Sciences.

APPENDIX TABLE 8 MACROMINERAL REQUIREMENTS OF SHEEP (PERCENTAGE OF DIET DRY MATTER)[a]

Nutrient	Requirement
Sodium	0.09-0.18
Chlorine	—
Calcium	0.20-0.82
Phosphorus	0.16-0.38
Magnesium	0.12-0.18
Potassium	0.50-0.80
Sulfur	0.14-0.26

[a]Values are estimates based on experimental data.

Source: Nutrient Requirements of Sheep, 6th revised edition, © 1985 by The National Academy of Sciences.

APPENDIX TABLE 9 MICROMINERAL REQUIREMENTS OF SHEEP AND MAXIMUM TOLERABLE LEVELS (ppm, mg/kg OF DIET DRY MATTER)[a]

Nutrient	Requirement	Maximum Tolerable Level[b]
Iodine	0.10-0.80[c]	50
Iron	30-50	500
Copper	7-11[d]	25[e]
Molybdenum	0.5	10[e]
Cobalt	0.1-0.2	10
Manganese	20-40	1,000
Zinc	20-33	750
Selenium	0.1-0.2	2
Fluorine	—	60-150

[a]Values are estimates based on experimental data.
[b]NRC (1980).
[c]High level for pregnancy and lactation in diets not containing goitrogens; should be increased if diets contain goitrogens.
[d]Requirement when dietary Mo concentrations are <1 mg/kg DM. See text for requirements under other circumstances.
[e]Lower levels may be toxic under some circumstances. See text.

Source: Nutrient Requirements of Sheep, 6th revised edition, © 1985 by The National Academy of Sciences.

APPENDIX TABLE 10 VITAMIN E REQUIREMENTS OF GROWING–FINISHING LAMBS AND SUGGESTED LEVELS OF FEED FORTIFICATION TO PROVIDE 100% OF REQUIREMENTS

Body Weight		α-Tocopheryl Acetate		Feed Intake per Lamb		Amount of Vitamin E Added to Concentrate			Amount of Vitamin E Added to Protein Supplement[b]		
(kg)	(lb)	(mg/lamb/day)[a]	(mg/kg diet)	(kg)	(lb)	(mg/kg)	(mg/lb)	(mg/ton)	(mg/kg)	(mg/lb)	(mg/ton)
10	22	5.0	20	0.23	0.50	20	9.1	18,200	133	60	120,000
20	44	10.0	20	0.45	1.00	20	9.1	18,200	133	60	120,000
30	66	15.0	15	0.96	2.10	15	6.8	13,600	100	45	90,000
40	88	20.0	15	1.30	2.86	15	6.8	13,600	100	45	90,000
50	110	25.0	15	1.60	3.50	15	6.8	13,600	100	45	90,000

[a]Rounded values based on approximate diet intake containing recommended vitamin E levels.
[b]Assumes the concentrate diet contains 15 percent protein supplement.

Source: Nutrient Requirements of Sheep, 6th revised edition, © 1985 by The National Academy of Sciences.

APPENDIX TABLE 11 SUGGESTED CREEP DIETS

Ingredient	Amount (as-fed basis, %) Diet A	Diet B	Diet C
Simple Diets (grind for lambs under 6 weeks of age; feed whole thereafter; hand- or self-feed)[a]			
Barley, grain	38.5	—	—
Corn, dent yellow, grain, ground, grade 2 US, min 54 lb/bu	40.0	60.0	88.5
Oats, grain	—	28.5	—
Wheat, bran, dry milled	10.0	—	—
Linseed meal, soybean meal, or sunflower meal	10.0	10.0	10.0
Limestone, ground, min 33% calcium	1.0	1.0	1.0
Trace mineralized salt with selenium	0.5	0.5	0.5
Total	100.0	100.0	100.0
Alfalfa hay, sun-cured, early bloom should be fed free choice in conjunction with any of the above diets.			
Commercially Mixed Diets (hand- or self-fed as meal, but usually as pellets)[b]			
Alfalfa, sun-cured, early bloom or dehydrated alfalfa	—	10.0	20.0
Barley, grain	20.0	—	—
Corn, dent yellow, grain, grade 2 US, min 54 lb/bu	54.5	34.5	44.5
Oats, grain	—	30.0	10.0
Linseed, soybean, or sunflower meal	10.0	10.0	10.0
Bran, wheat	10.0	10.0	10.0
Beet or cane molasses	4.0	4.0	4.0
Limestone, ground, min 33% calcium	1.0	1.0	1.0
Trace mineralized salt with selenium	0.5	0.5	0.5
Total	100.0	100.0	100.0
Chlortetracycline or oxytetracycline	15.0-25.0 mg/kg	15.0-25.0 mg/kg	15.0-25.0 mg/kg
Vitamin A, IU/kg	500.0	500.0	500.0
Vitamin D, IU/kg	50.0	50.0	50.0
Vitamin E, IU/kg	20.0	20.0	20.0

[a]Limestone will separate from whole grain, so a combination of protein supplement with 10% limestone may be top dressed on the whole grain. Equal parts of trace mineralized salt and limestone is an additional way to maintain adequate calcium intake and prevent urinary calculi.

[b]The addition of 0.25 to 0.50% ammonium chloride will minimize urinary calculi. Corn may be substituted for all the barley and oats. Weight gains are depressed when barley or oats exceed 40% of the ration.

Source: Nutrient Requirements of Sheep, 6th revised edition, © 1985 by The National Academy of Sciences.

APPENDIX TABLE 12 COMPOSITION OF SOME SHEEP FEEDS: DATA EXPRESSED ON AN AS-FED AND DRY BASIS (100% DRY MATTER)

Feed Name Description	International feed number	Dry matter (%), DM	DE sheep (mcal/kg)	ME sheep (mcal/kg)	NE_m (mcal/kg)	NE_g (mcal/kg)	TDN sheep (%)	Crude protein (%)	Cell walls (%)	Cellulose (%)	Hemicellulose (%)	Lignin (%)	Calcium (%)	Cobalt (mg/kg)	Phosphorus (%)	Provitamin A (carotene) (ng/kg)	Vitamin E (mg/kg)	Vitamin D_2 (IU/g)
Alfalfa, *Medicago sativa,* hay	1-00-054	90	2.29	1.88	1.12	0.61	52	17.9	36	21	7	6	1.38	0.08	0.26	181	—	—
sun-cured, late vegetative		100	2.56	2.10	1.24	0.68	58	20.0	40	23	8	7	1.54	0.09	0.29	202	—	—
Hay, sun-cured, early bloom	1-00-059	90	2.22	1.82	1.06	0.55	51	16.2	38	22	8	7	1.27	0.15	0.20	126	23	1796
		100	2.47	2.03	1.18	0.61	56	18.0	42	24	9	8	1.41	0.16	0.22	140	26	1996
Hay, sun-cured, midbloom	1-00-063	90	2.22	1.82	1.06	0.55	51	15.3	41	23	9	8	1.27	0.32	0.22	30	—	1389
		100	2.47	2.03	1.18	0.61	56	17.0	46	26	10	9	1.41	0.36	0.24	33	—	1544
Hay, sun-cured, full bloom	1-00-068	90	2.10	1.72	0.96	0.47	47	13.5	45	25	10	9	1.13	0.29	0.20	59	—	—
		100	2.34	1.92	1.07	0.52	53	15.0		28		10	1.25	0.33	0.22	65	—	—
Hay, sun-cured, mature	1-00-071	91	2.17	1.78	1.01	0.50	49	11.7	53	26	12	13	1.03	0.08	0.17	11	—	1287
		100	2.38	1.95	1.11	0.55	54	12.9	58	29	13	14	1.13	0.09	0.18	12	—	1411
Meal dehydrated, 15%	1-00-022	90	2.27	1.86	1.09	0.58	52	15.6	46	26	—	11	1.24	0.17	0.22	74	82	—
protein		100	2.51	2.06	1.21	0.64	57	17.3	51	29	—	12	1.37	0.19	0.24	82	91	—
Meal dehydrated, 17%	1-00-023	92	2.43	1.99	1.23	0.71	55	17.3	41	22	—	10	1.40	0.30	0.23	120	111	—
protein		100	2.65	2.17	1.34	0.77	60	18.9	45	24	—	11	1.52	0.33	0.25	131	121	—
Silage	3-00-212	41	1.08	0.88	0.52	0.29	24	7.4		—	—	—	0.62	0.04	0.12	41	—	120
		100	2.60	2.13	1.28	0.71	59	17.8		—		—	1.50	0.09	0.28	99	—	289
Silage, 30% dry matter	3-08-149	26	0.73	0.60	0.37	0.22	17	4.9		—	—	—	0.37	—	0.08	—	—	—
		100	2.78	2.28	1.41	0.83	63	18.7		—	—	—	1.40	—	0.32	—	—	—
Silage, 30% to 50% dry	3-08-150	43	1.11	0.91	0.53	0.29	25	7.8		—	—	—	0.60	—	0.12	62	—	—
matter		100	2.56	2.10	1.24	0.68	58	17.9		—	—	—	1.39	—	0.27	144	—	—
Silage, 50% dry matter	3-08-151	57	1.42	1.17	0.69	0.36	32	9.8		—	—	—	0.54	—	0.17	61	—	—
		100	2.51	2.06	1.21	0.64	57	17.3		—	—	—	0.95	—	0.30	107	—	—

Barley, *Hordeum vulgare,* grain	4-00-549	88	3.35	2.74	1.87	1.28	76	11.9	17	4	—	2	0.04	0.09	0.34	2	22	—
		100	3.79	3.11	2.12	1.45	86	13.5	19	5	—	2	0.05	0.10	0.38	2	25	—
Grain, Pacific coast	4-07-939	89	3.46	2.84	1.94	1.36	79	9.6		—	—	—	0.05	0.09	0.34	—	26	—
		100	3.88	3.18	2.18	1.50	88	10.8	—	—	—	—	0.06	0.10	0.39	—	30	—
Beet, sugar, *Beta vulgaris altissima* pulp, dehydrated	4-00-669	91	2.96	2.43	1.60	1.04	67	8.8	49	—	—	2	0.63	0.07	0.09	0	—	577
		100	3.26	2.68	1.76	1.14	74	9.7	54	—	—	2	0.69	0.08	0.10	0	—	637
Pulp, wet	4-00-671	11	0.37	0.30	0.20	0.13	8	1.2	—	—	—	—	0.10	—	0.01	—	—	—
Bonemeal, steamed	6-00-400	97	—	—	—	—	—	—	—	—	—	—	29.8	—	12.49	—	—	—
		100	—	—	—	—	—	—	—	—	—	—	30.7	—	12.86	—	—	—
		100	3.35	2.75	1.82	1.19	76	11.2	—	—	—	—	0.87	—	0.10	—	—	—
Pulp with molasses, dehydrated	4-00-672	92	3.11	2.55	1.70	1.12	71	9.3	40	—	—	3	0.56	0.21	0.09	0	—	—
		100	3.40	2.78	1.85	1.22	77	10.1	44	—	—	3	0.61	0.23	0.10	0	—	—
Brewers grains, dehydrated	5-02-141	92	2.84	2.33	1.50	0.95	65	27.1	41	—	—	6	0.30	0.08	0.51	0	26	—
		100	3.09	2.53	1.63	1.03	70	29.4	42	—	—	6	0.33	0.08	0.55	1	29	—
Brome, *Bromus* spp., hay, sun-cured	1-00-890	91	2.20	1.80	1.04	0.53	49	8.8	—	—	—	4	0.31	—	0.17	31	—	948
		100	2.43	1.99	1.14	0.58	55	9.7	—	—	—	5	0.35	—	0.19	34	—	1407
Citrus, *Citrus* spp., pomace without fines, dehydrated	4-01-237	91	3.38	2.77	1.87	1.27	76	6.1	21	—	—	3	1.67	0.14	0.11	0	—	—
		100	3.70	3.04	2.06	1.40	84	6.7	23	—	—	3	1.84	0.16	0.12	0	—	—
Clover, alsike, *Trifolium hybridum,* hay, sun-cured	1-01-313	88	2.25	1.84	1.09	0.60	51	13.1	—	—	11	—	1.13	—	0.23	164	—	—
		100	2.56	2.10	1.24	0.68	58	14.9	—	—	13	—	1.29	—	0.26	187	—	—
Clover, crimson, *Trifolium incarnatum,* hay, sun-cured	1-01-328	87	2.12	1.73	0.99	0.50	48	16.1	—	—	—	—	1.22	—	0.19	20	—	—
		100	2.43	1.99	1.14	0.58	55	18.4	—	—	—	—	1.40	—	0.22	23	—	—
Clover, ladino, *Trifolium repens,* hay, sun-cured	1-01-378	90	2.61	2.14	1.36	0.82	59	19.7	32	—	—	6	1.21	0.15	0.28	75	—	—
		100	2.91	2.39	1.51	0.91	66	22.0	36	—	—	7	1.35	0.16	0.31	83	—	—
Clover, red, *Trifolium pratense,* hay, sun-cured	1-01-415	89	2.34	1.92	1.17	0.66	53	14.2	50	23	8	9	1.35	0.14	0.22	18	—	1694
		100	2.65	2.17	1.31	0.74	60	16.0	56	26	9	10	1.53	0.16	0.25	20	—	1914
Hay, sun-cured, early vegetative	1-01-394	87	2.50	2.05	1.28	0.77	57	18.7	—	—	—	—	1.35	0.20	0.32	—	—	—
		100	2.87	2.35	1.47	0.88	65	21.4	—	—	—	—	1.55	0.23	0.37	—	—	—
Hay, sun-cured late vegetative	1-01-397	87	2.56	2.10	1.34	0.82	58	17.0	—	—	—	—	1.29	0.15	0.29	—	—	—
		100	2.95	2.42	1.54	0.94	67	19.6	—	—	—	—	1.49	0.17	0.33	—	—	—
Hay, sun-cured, early bloom	1-01-400	86	2.44	2.00	1.24	0.74	55	16.0	—	—	—	—	1.24	0.14	0.30	—	—	—
		100	2.82	2.31	1.44	0.86	64	18.5	—	—	—	—	1.44	0.16	0.35	—	—	—
Hay, sun-cured, midbloom	1-01-401	88	2.41	1.98	1.21	0.70	55	16.0	—	—	—	—	1.48	—	0.26	—	—	—
		100	2.73	2.24	1.38	0.80	62	18.1	—	—	—	—	1.67	—	0.29	—	—	—
Hay, sun-cured, full bloom	1-01-403	86	2.38	1.96	1.21	0.71	54	13.1	—	—	—	—	1.25	0.11	0.23	—	—	1643
		100	2.78	2.28	1.41	0.83	63	15.3	—	—	—	—	1.46	0.13	0.27	—	—	1914

APPENDIX TABLE 12 (Continued)

Feed Name Description	International feed number	Dry matter (%), DM	DE sheep (mcal/kg)	ME sheep (mcal/kg)	NE_m (mcal/kg)	NE_g (mcal/kg)	TDN sheep (%)	Crude protein (%)	Cell walls (%)	Cellulose (%)	Hemicellulose (%)	Lignin (%)	Calcium (%)	Cobalt (mg/kg)	Phosphorus (%)	Provitamin A (carotene) (ng/kg)	Vitamin E (mg/kg)	Vitamin D_2 (IU/g)
Corn, dent yellow, *Zea*	1-28-234	90	2.03	1.66	0.96	0.41	46	2.8	80	25	—	6	0.11	0.12	0.04	1	—	—
mays identata, cobs, ground		100	2.25	1.84	1.07	0.45	51	3.2	89	28	—	7	0.12	0.13	0.04	1	—	—
Distillers grains, dehydrated	5-28-235	94	3.59	2.94	2.02	1.39	82	21.5	40	—	—	—	0.10	0.08	0.40	3	—	—
		100	3.84	3.15	2.12	1.48	87	23.0	43	—	—	—	0.11	0.09	0.43	3	—	—
Distillers grains with	5-28-236	92	3.52	2.89	1.98	1.36	80	23.0	40	13	—	4	0.14	0.17	0.65	3	40	551
solubles, dehydrated		100	3.84	3.15	2.15	1.48	87	25.0	44	14	—	4	0.15	0.18	0.71	3	43	600
Distillers solubles,	5-28-237	93	3.48	2.85	1.94	1.33	79	27.6	21	6	—	1	0.33	0.20	1.27	1	46	—
dehydrated		100	3.75	3.07	2.09	1.43	85	29.7	23	6	—	1	0.35	0.21	1.37	1	49	—
Ears, ground (corn	4-28-238	87	3.17	2.60	1.77	1.19	71	7.8	—	—	—	—	0.06	0.27	0.24	3	18	—
and cob meal)		100	3.66	3.00	2.03	1.37	83	9.0	—	—	—	—	0.07	0.31	0.27	4	20	—
Gluten, meal	5-28-241	91	3.54	2.90	1.98	1.37	80	42.7	34	7	—	1	0.15	0.08	0.45	16	31	—
		100	3.88	3.18	2.18	1.50	88	46.8	37	8	—	1	0.16	0.08	0.50	18	34	—
Gluten with bran (corn	5-28-243	90	3.29	2.70	1.83	1.23	75	23.0	—	—	—	—	0.33	0.09	0.74	6	12	—
gluten feed)		100	3.66	3.00	2.03	1.37	83	25.6	—	—	—	—	0.36	0.10	0.82	7	14	—
Grain, grade 2 (54 lb/bushel)	4-02-931	88	3.39	2.78	1.89	1.30	77	8.9	—	—	—	—	0.02	0.03	0.31	2	22	—
		100	3.84	3.15	2.15	1.48	87	10.1	—	—	—	—	0.02	0.04	0.35	2	25	—
Grain, grade 3 (52 lb/bushel)	4-02-932	86	3.30	2.71	1.85	1.27	75	8.7	—	—	—	—	0.02	—	0.25	—	—	—
		100	3.84	3.15	2.15	1.48	87	10.1	—	—	—	—	0.02	—	0.29	—	—	—
Silage, aerial part without	3-28-251	31	0.72	0.59	0.33	0.16	16	1.9	21	8	—	2	0.12	—	0.09	5	—	—
ears without husks (stalklage) (stover)		100	2.34	1.92	1.07	0.52	53	6.3	68	25	—	7	0.38	—	0.31	15	—	—
Silage, well eared	3-28-250	33	1.03	0.85	0.54	0.34	23	2.7	17	—	—	—	0.08	0.02	0.07	15	—	40
		100	3.09	2.53	1.63	1.03	70	8.1	51	—	—	—	0.23	0.06	0.22	45	—	119
Cotton, *Gossypium* spp.,	1-01-599	91	1.96	1.60	0.85	0.35	44	3.7	82	53	—	22	0.13	0.02	0.09	—	—	—
hulls		100	2.16	1.77	0.93	0.39	49	4.1	90	59	—	24	0.15	0.02	0.09	—	—	—
Seeds, meal mechanically	5-01-617	93	3.06	2.51	1.66	1.08	70	41.0	26	12	—	6	0.19	0.16	1.08	0	32	—
extracted, 41% protein		100	3.31	2.71	1.79	1.16	75	44.3	28	13	—	6	0.21	0.17	1.16	0	35	—

Seeds, meal solvent extracted, 41% protein	5-01-621	91	2.86	2.34	1.52	0.96	65	41.2	—	11	—	—	0.17	0.15	1.10	—	16	—
		100	3.13	2.57	1.67	1.06	71	45.2	—	12	—	—	0.18	0.17	1.21	—	17	—
Fescue, Kentucky 31, *Festuca arundinacea,* hay, sun-cured, early bloom	1-09-186	91	2.49	2.04	1.26	0.73	56	18.4	54	26	—	3	0.35	—	0.24	—	—	—
		100	2.73	2.24	1.38	0.80	62	20.2	59	29	—	3	0.38	—	0.26	—	—	—
Hay, sun-cured, midbloom	1-09-187	92	2.44	2.00	1.21	0.68	55	15.2	58	28	—	3	0.42	—	0.27	—	—	—
		100	2.65	2.17	1.31	0.74	60	16.4	63	30	—	4	0.45	—	0.29	—	—	—
Hay, sun-cured, full bloom	1-09-188	92	2.34	1.92	1.14	0.63	53	11.1	61	29	—	5	0.39	—	0.29	—	—	—
		100	2.56	2.10	1.24	0.68	58	12.1	67	32	—	5	0.43	—	0.32	—	—	—
Fish, Menhaden, *Brevoortia tyrannus,* meal mechanically extracted	5-02-009	92	2.83	2.32	1.50	0.95	64	61.1	—	—	—	—	5.18	0.15	2.89	—	12	—
		100	3.09	2.53	1.63	1.03	70	66.7	—	—	—	—	5.65	0.17	3.16	—	13	—
Flax, common, *Linum usitatissium,* seeds, meal mechanically extracted (linseed meal)	5-02-045	91	3.32	2.72	1.85	1.25	75	34.3	23	—	—	6	0.41	0.41	0.87	0	8	—
		100	3.66	3.00	2.03	1.37	83	37.9	25	—	—	7	0.45	0.46	0.96	0	9	—
Seeds, meal solvent extracted (linseed meal)	5-02-048	90	3.14	2.58	1.72	1.14	72	34.6	23	—	—	5	0.39	0.19	0.80	—	14	—
		100	3.48	2.86	1.91	1.27	79	38.3	25	—	—	6	0.43	0.21	0.89	—	15	—
Grass–legume, aerial part, molasses added, silage	3-02-309	28	0.74	0.61	0.36	0.20	17	3.4	—	—	—	—	0.30	—	0.10	—	—	—
		100	2.60	2.13	1.28	0.71	59	11.8	—	—	—	—	1.07	—	0.34	—	—	—
Silage	3-02-303	29	0.81	0.66	0.40	0.23	18	3.3	—	—	—	—	0.25	0.04	0.08	68	—	85
		100	2.73	2.24	1.38	0.80	62	11.3	—	—	—	—	0.85	0.13	0.27	230	—	289
Lespedeza, common, *Lespedeza striata,* hay, sun-cured	1-08-591	88	1.95	1.60	0.85	0.37	44	14.8	—	—	—	—	0.78	—	0.25	44	—	—
		100	2.21	1.81	0.97	0.42	50	16.8	—	—	—	—	0.88	—	0.29	50	—	—
Hay, sun-cured, early vegetative	1-20-886	89	2.40	1.97	1.19	0.69	54	14.3	—	—	—	—	1.09	—	0.26	—	—	—
		100	2.69	2.21	1.34	0.77	61	16.0	—	—	—	—	1.22	—	0.29	—	—	—
Hay, sun-cured, late vegetative	1-20-881	90	2.18	1.79	1.03	0.52	50	14.4	—	—	—	—	—	—	—	—	—	—
		100	2.43	1.99	1.14	0.58	55	16.0	—	—	—	—	—	—	—	—	—	—
Hay, sun-cured, midbloom	1-02-554	92	2.30	1.89	1.11	0.59	52	13.8	—	—	—	—	1.08	—	0.22	—	—	—
		100	2.51	2.06	1.21	0.64	57	15.1	—	—	—	—	1.18	—	0.24	—	—	—
Hay, sun-cured, full bloom	1-20-887	89	2.28	1.87	1.10	0.61	52	12.8	—	—	—	—	1.02	—	0.19	—	—	—
		100	2.56	2.10	1.24	0.68	58	14.3	—	—	—	—	1.14	—	0.21	—	—	—
Limestone, dolomitic	6-02-633	99	—	—	—	—	—	—	—	—	—	—	22.1	—	0.04	—	—	—
		100	—	—	—	—	—	—	—	—	—	—	23.3	—	0.04	—	—	—
Limestone, ground	6-02-632	99	—	—	—	—	—	—	—	—	—	—	33.7	—	0.02	—	—	—
		100	—	—	—	—	—	—	—	—	—	—	34.0	—	0.02	—	—	—

APPENDIX TABLE 12 (Continued)

Feed Name Description	International feed number	Dry matter (%), DM	DE sheep (mcal/kg)	ME sheep (mcal/kg)	NE_m (mcal/kg)	NE_g (mcal/kg)	TDN sheep (%)	Crude protein (%)	Cell walls (%)	Cellulose (%)	Hemicellulose (%)	Lignin (%)	Calcium (%)	Cobalt (mg/kg)	Phosphorus (%)	Provitamin A (carotene) (ng/kg)	Vitamin E (mg/kg)	Vitamin D_2 (IU/g)
Meadow plants, intermountain,	1-03-181	95	2.43	1.99	1.18	0.65	55	8.3	—	—	—	—	0.58	—	0.17	32	—	—
hay, sun-cured		100	2.56	2.10	1.24	0.68	58	8.7	—	—	—	—	0.61	—	0.18	33	—	—
Hay, sun-cured, early	1-08-464	90	2.78	2.28	—	—	63	—	—	—	—	—	—	—	—	—	—	—
vegetative		100	3.09	2.53	—	—	70	—	—	—	—	—	—	—	—	—	—	—
Hay, sun-cured, late	1-09-176	94	1.90	1.56	0.78	0.27	43	7.6	—	—	—	—	—	—	—	—	—	—
bloom, cut I		100	2.03	1.66	0.83	0.29	46	8.1	—	—	—	—	—	—	—	—	—	—
Hay, sun-cured, mature	1-08-465	91	1.64	1.35	0.58	0.10	37	5.1	—	—	—	—	—	—	—	—	—	—
		100	1.81	1.48	0.64	0.11	41	5.6	—	—	—	—	—	—	—	—	—	—
Molasses, *Beta vulgaris*	4-00-668	78	2.64	2.16	1.44	0.95	60	6.6	—	—	—	—	0.13	0.36	0.03	—	4	—
altissima, beet, sugar,		100	3.40	2.78	1.85	1.22	77	8.5	—	—	—	—	0.17	0.46	0.03	—	5	—
molasses, more than 48%																		
invert sugar, more than																		
79.5 degrees Brix																		
Molasses, *Saccharum*	4-04-695	94	3.16	2.60	1.71	1.12	71	9.7	—	—	—	—	1.04	1.15	0.14	—	5	—
officinarum, sugarcane,		100	3.35	2.75	1.83	1.19	76	10.3	—	—	—	—	1.10	1.21	0.15	—	6	—
molasses, dehydrated																		
Sugarcane, molasses > 46%,	4-04-696	75	2.60	2.13	1.43	0.95	59	4.4	—	—	—	—	0.75	0.90	0.08	—	5	—
invert sugars > 79.5°		100	3.48	2.86	1.91	1.27	79	5.8	—	—	—	—	1.00	1.21	0.11	—	7	—
Brix (blackstrap)																		
Oats, *Avena sativa,*	4-03-309	89	3.02	2.47	1.65	1.09	68	11.8	28	10	13	2	0.07	0.06	0.33	0	14	—
grain		100	3.40	2.78	1.85	1.22	77	13.3	32	11	15	3	0.07	0.06	0.38	0	15	—
Grain, grade 1, heavy,	4-03-312	89	3.18	2.61	1.75	1.17	72	12.6	—	—	—	—	—	—	—	—	—	—
46.3 kg/hl		100	3.57	2.93	1.97	1.32	81	14.2	—	—	—	—	—	—	—	—	—	—
Grain, grade 2, 41.2 kg/hl	4-03-316	89	2.98	2.44	1.62	1.06	68	11.4	—	—	—	—	0.06	—	0.27	—	—	—
		100	3.35	2.75	1.82	1.19	76	12.8	—	—	—	—	0.07	—	0.30	—	—	—
Grain, light, less than	4-03-318	91	2.64	2.16	1.37	0.83	59	11.9	—	—	—	—	—	—	—	—	—	—
34.7 kg/hl		100	2.91	2.39	1.51	0.91	66	13.1	—	—	—	—	—	—	—	—	—	—

Grain, Pacific coast	4-07-999	91	3.13	2.56	1.71	1.13	71	9.1	—	—	—	—	0.10	—	0.31	—	20	—
		100	3.44	2.82	1.88	1.24	78	10.0	—	—	—	—	0.11	—	0.34	—	22	—
Groats	4-03-331	90	4.03	3.30	2.12	1.49	91	15.8	—	—	—	—	0.08	—	0.43	—	15	—
		100	4.50	3.69	2.35	1.65	102	17.7	—	—	—	—	0.08	—	0.48	—	16	—
Hay, sun-cured	1-03-280	91	2.13	1.75	1.04	0.53	49	8.5	—	—	—	5	0.22	0.07	0.20	25	—	1410
		100	2.34	1.92	1.14	0.58	53	9.3	—	—	—	6	0.24	0.07	0.22	28	—	1544
Silage	3-03-298	31	0.84	0.69	0.43	0.25	19	3.0	—	—	—	—	0.10	0.02	0.07	14	—	—
		100	2.73	2.24	1.38	0.80	62	9.6	—	—	—	—	0.34	0.06	0.24	45	—	—
Straw	1-03-283	92	1.91	1.57	0.73	0.23	44	4.1	64	37	—	13	0.22	—	0.06	4	—	609
		100	2.07	1.70	0.79	0.25	47	4.4	70	40	—	14	0.24	—	0.06	4	—	662
Orchardgrass, *Dactylis glomerata,* hay, sun-cured, early bloom	1-03-425	89	2.04	1.68	0.93	0.44	47	13.4	54	26	24	4	0.24	0.38	0.30	33	—	—
	286	100	2.29	1.88	1.04	0.49	52	15.0	61	29	27	5	0.27	0.43	0.34	38	—	—
Hay, sun-cured, full bloom	1-03-427	93	2.00	1.64	0.86	0.36	45	11.5	—	—	—	4	—	—	—	—	—	—
	288	100	2.16	1.77	0.93	0.39	49	12.4	—	—	—	5	—	—	—	—	—	—
Oystershell (ground)	6-03-481	99	—	—	—	—	—	—	—	—	—	—	37.6	—	0.07	—	—	—
		100	—	—	—	—	—	—	—	—	—	—	38.0	—	0.07	—	—	—
Peanut, *Arachis hyogaea,* Kernels, meal mechanically extracted (peanut meal)	5-03-649	93	3.84	3.15	2.19	1.53	87	48.1	13	4	—	—	0.19	0.11	0.57	0	2	—
		100	4.14	3.40	2.35	1.65	94	52.0	14	5	—	—	0.20	0.12	0.61	0	3	—
Kernels, meal solvent extracted (peanut meal)	5-03-650	92	3.12	2.56	1.70	1.12	71	48.1	—	—	—	—	0.27	0.11	0.62	—	3	—
		100	3.40	2.78	1.85	1.22	77	52.3	—	—	—	—	0.29	0.12	0.68	—	3	—
Phosphate, defluorinated from phosphoric acid	6-26-336	100	—	—	—	—	—	—	—	—	—	—	32.0	—	18.0	—	—	—
Phosphate, Diammonium	6-00-370	97	—	—	—	—	—	—	—	—	—	—	0.57	—	20.0	—	—	—
		100	—	—	—	—	—	—	—	—	—	—	0.59	—	20.6	—	—	—
Phosphate, Dicalcium	6-00-080	96	—	—	—	—	—	—	—	—	—	—	26.0	—	18.5	—	—	—
		100	—	—	—	—	—	—	—	—	—	—	27.1	—	19.3	—	—	—
Phosphoric acid, feed grade	6-03-707	75	—	—	—	—	—	—	—	—	—	—	—	—	23.7	—	—	—
		100	—	—	—	—	—	—	—	—	—	—	—	—	31.6	—	—	—
Prairie plants, Midwest, hay, sun-cured, midbloom	1-07-956	95	2.26	1.86	1.05	0.52	51	6.7	—	—	—	—	—	0.07	—	—	—	—
	298	100	2.38	1.95	1.11	0.55	54	7.0	—	—	—	—	—	0.07	—	—	—	—
Hay, sun-cured, full bloom	1-03-184	89	2.03	1.67	0.93	0.40	46	5.4	—	—	—	—	0.34	—	0.12	—	—	879
	300	100	2.29	1.88	1.04	0.45	52	6.1	—	—	—	—	0.38	—	0.14	—	—	992
Hay, sun-cured, mature	1-03-187	91	1.92	1.58	0.82	0.32	44	4.6	—	—	—	—	0.34	—	0.14	9	—	—
	302	100	2.12	1.74	0.90	0.35	48	5.1	—	—	—	—	0.38	—	0.16	10	—	—

APPENDIX TABLE 12 (Continued)

Feed Name Description	International feed number	Dry matter (%), DM	DE sheep (mcal/kg)	ME sheep (mcal/kg)	NE_m (mcal/kg)	NE_g (mcal/kg)	TDN sheep (%)	Crude protein (%)	Cell walls (%)	Cellulose (%)	Hemicellulose (%)	Lignin (%)	Calcium (%)	Cobalt (mg/kg)	Phosphorus (%)	Provitamin A (carotene) (ng/kg)	Vitamin E (mg/kg)	Vitamin D_2 (IU/g)
Redtop, *Agrostis alba,*	1-03-885	92	2.19	1.80	1.02	0.51	50	7.4	—	—	—	—	0.39	0.13	0.20	4	—	—
hay, sun-cured	320	100	2.38	1.95	1.11	0.55	54	8.1	—	—	—	—	0.43	0.15	0.22	4	—	—
Hay, sun-cured,	1-03-886	94	2.33	1.91	1.11	0.57	53	11.0	—	—	—	—	0.60	—	0.33	5	—	—
midbloom	326	100	2.47	2.03	1.18	0.61	56	11.7	—	—	—	—	0.63	—	0.35	5	—	—
Rye, *Secale cereale*	5-04-023	92	2.59	2.13	1.32	0.79	59	21.6	—	—	—	—	0.05	—	0.48	—	—	—
distillers grains, dehydrated	334	100	2.82	2.31	1.44	0.86	64	23.5	—	—	—	—	0.16	—	0.52	—	—	—
Grain	4-04-047	88	3.28	2.69	1.84	1.26	75	12.1	—	—	—	—	0.06	—	0.32	0	15	—
	336	100	3.75	3.07	2.09	1.43	85	13.8	—	—	—	—	0.07	—	0.37	0	17	—
Safflower, *Carthamus*	5-04-109	91	2.50	2.05	1.26	0.73	57	20.2	54	—	—	—	0.25	—	0.71	—	1	—
tinctorius, seeds, meal		100	2.73	2.24	1.38	0.80	62	22.1	59	—	—	—	0.27	—	0.78	—	1	—
mechanically extracted																		
Seeds, meal solvent	5-04-110	92	2.28	1.87	1.09	0.56	52	23.4	53	—	—	13	0.34	—	0.75	—	1	—
extracted		100	2.47	2.03	1.18	0.61	56	25.4	58	—	—	14	0.37	—	0.81	—	1	—
Sorghum, *Sorghum bicolor,*	5-04-374	94	3.51	2.88	1.96	1.34	79	32.2	—	—	—	—	0.15	—	0.69	—	—	—
distillers grains, dehydrated		100	3.75	3.07	2.09	1.43	85	34.4	—	—	—	—	0.16	—	0.74	—	—	—
Grain	4-04-383	90	3.48	2.85	1.96	1.35	79	11.1	21	3	16	—	0.03	0.16	0.29	1	10	26
		100	3.88	3.18	2.18	1.50	88	12.4	23	3	18	—	0.04	0.18	0.33	1	12	29
Grain, less than 9% protein	4-08-138	89	3.36	2.75	1.89	1.29	76	8.9	—	—	—	—	0.03	0.07	0.27	—	2	—
		100	3.79	3.11	2.12	1.45	86	10.1	—	—	—	—	0.03	0.08	0.31	—	2	—
Grain, 9% to 12% protein	4-08-139	89	3.36	2.76	1.89	1.29	76	10.2	—	—	—	—	0.03	0.07	0.28	—	1	—
		100	3.79	3.11	2.12	1.45	86	11.5	—	—	—	—	0.03	0.07	0.31	—	1	—
Grain, > 12% protein	4-08-140	89	—	—	—	—	—	11.6	—	—	—	—	0.03	—	0.29	—	—	—
		100	—	—	—	—	—	13.0	—	—	—	—	0.30	—	0.32	—	—	—
Silage	3-04-323	30	0.74	0.61	0.36	0.12	17	2.2	—	—	—	2	0.10	0.09	0.06	5	—	196
		100	2.51	2.06	1.21	0.64	57	7.5	—	—	—	6	0.35	0.30	0.21	15	—	662
Sorghum, milo, *Sorghum*	4-04-444	89	3.44	2.82	1.94	1.34	78	10.0	—	3	—	—	0.04	0.47	0.30	0	12	25
bicolor subglabrescens, grain		100	3.88	3.18	2.18	1.50	88	11.3	—	3	—	—	0.05	0.53	0.34	0	13	29
Heads	4-04-446	90	3.33	2.73	1.85	1.26	76	9.0	—	—	—	—	0.12	—	0.22	1	—	—
		100	3.70	3.04	2.06	1.40	84	10.0	—	—	—	—	0.13	—	0.25	1	—	—

Sorghum, sorgo, *Sorghum*	3-04-468	27	0.73	0.60	0.30	0.20	16	1.7	—	—	—	—	0.09	—	0.05	10	—	—
bicolor saccharatum, silage		100	2.65	2.17	1.31	0.74	60	6.2	—	—	—	—	0.34	—	0.17	36	—	—
Soybean, *Glycine max,*	1-04-558	89	2.25	1.84	1.08	0.57	51	14.3	—	—	—	—	0.15	0.08	0.25	41	26	947
hay, sun-cured		100	2.51	2.06	1.21	0.64	57	16.0	—	—	—	—	1.29	0.09	0.28	45	30	1059
Seed coats (hulls)	1-04-560	91	2.17	1.78	1.01	0.50	49	11.0	61	42	16	2	0.45	0.11	0.19	—	6	—
		100	2.38	1.95	1.11	0.55	54	12.1	67	46	18	2	0.49	0.12	0.21	—	7	—
Seeds	5-04-610	92	3.79	3.11	2.16	1.52	86	39.2	—	—	—	—	0.25	—	0.60	1	33	—
		100	4.14	3.40	2.35	1.65	94	42.8	—	—	—	—	0.27	—	0.60	1	37	—
Seeds, meal mechanically	5-04-600	90	3.37	2.77	1.88	1.29	77	42.9	—	—	—	—	0.26	0.18	0.61	0	7	—
extracted		100	3.75	3.07	2.09	1.43	85	47.7	—	—	—	—	0.29	0.20	0.68	0	7	—
Seeds, meal solvent	5-04-604	90	3.48	2.85	1.96	1.35	79	44.8	—	—	—	—	0.30	0.09	0.63	0	3	—
extracted		100	3.88	3.18	2.18	1.50	88	49.9	—	—	—	—	0.34	0.10	0.70	0	3	—
Straw	1-04-567	88	1.66	1.36	0.63	0.16	38	4.6	61	33	—	14	1.40	—	0.05	—	—	—
		100	1.90	1.56	0.72	0.18	43	5.2	70	38	—	16	1.59	—	0.06	—	—	—
Sugarcane molasses																		
(see Molasses)																		
Sunflower, common,	5-09-340	90	1.79	1.46	0.68	0.23	41	23.3	36	—	—	11	—	—	0.93	—	—	—
Helianthus annuus, seeds,		100	1.98	1.63	0.75	0.25	45	25.9	40	—	—	12	0.23	—	1.03	—	—	—
meal solvent extracted																		
Seeds without hulls,	5-04-739	93	3.12	2.56	1.69	1.11	70	46.3	—	—	—	—	0.41	—	0.91	—	11	—
meal solvent extracted		100	3.35	2.75	1.82	1.19	76	49.8	—	—	—	—	0.44	—	0.98	—	12	—
Sweetclover, yellow,	1-04-754	87	2.04	1.67	0.93	0.45	46	13.7	—	—	—	—	1.11	—	0.22	86	—	1636
Melilotus officinalis,		100	2.34	1.92	1.07	0.52	53	15.7	—	—	—	—	1.27	—	0.25	99	—	1874
hay, sun-cured																		
Timothy, *Phleum pratense,*	1-04-881	89	2.56	2.10	1.31	0.78	58	15.2	49	25	23	3	0.59	—	0.30	112	—	—
hay, sun-cured, late vegetative		100	2.87	2.35	1.47	0.88	65	17.0	55	28	26	3	0.66	—	0.34	125	—	—
Hay, sun-cured,	1-04-882	90	2.21	1.81	1.06	0.55	51	13.4	55	28	26	4	0.48	—	0.22	47	12	—
early bloom		100	2.47	2.03	1.18	0.61	56	15.0	61	31	29	4	0.53	—	0.25	53	13	—
Hay, sun-cured,	1-04-883	89	2.35	1.93	1.17	0.66	53	8.1	60	29	28	4	0.43	—	0.20	47	—	1764
midbloom		100	2.65	2.17	1.31	0.74	60	9.1	67	33	31	5	0.48	—	0.22	53	—	1985
Hay, sun-cured	1-04-885	88	2.14	1.76	1.00	0.51	49	6.9	62	30	26	6	0.34	—	0.16	40	—	—
late bloom		100	2.43	1.99	1.14	0.58	55	7.8	70	34	29	7	0.38	—	0.18	45	—	—
Trefoil, birdsfoot,	1-05-044	92	2.36	1.93	1.14	0.63	53	15.0	43	22	—	8	1.57	0.10	0.25	173	—	1421
Lotus corniculatus,		100	2.56	2.10	1.24	0.68	58	16.3	47	24	—	9	1.70	0.11	0.27	188	—	1544
hay, sun-cured																		
Turnip, *Brassica rapa,*	4-05-067	9	0.35	0.29	0.19	0.13	8	1.1	4	—	—	1	0.05	—	0.02	—	—	—
roots, fresh		100	3.79	3.11	2.12	1.45	86	11.8	44	—	—	10	0.59	—	0.26	—	—	—

APPENDIX TABLE 12 (Continued)

Feed Name Description	International feed number	Dry matter (%), DM	DE sheep (mcal/kg)	ME sheep (mcal/kg)	NE_m (mcal/kg)	NE_g (mcal/kg)	TDN sheep (%)	Crude protein (%)	Cell walls (%)	Cellulose (%)	Hemicellulose (%)	Lignin (%)	Calcium (%)	Cobalt (mg/kg)	Phosphorus (%)	Provitamin A (carotene) (ng/kg)	Vitamin E (mg/kg)	Vitamin D_2 (IU/g)
Vetch, *Vicia spp.*,	1-05-106	89	2.16	1.77	1.01	0.52	49	18.5	43	—	—	7	1.05	0.32	0.29	411	—	—
hay, sun-cured		100	2.43	1.99	1.14	0.58	55	20.8	48	—	—	8	1.18	0.36	0.32	461	—	—
Wheat, *Triticum aestivum*,	4-05-190	89	2.78	2.28	1.49	0.94	63	15.2	46	9	30	3	0.11	0.10	1.22	3	18	—
bran		100	3.13	2.57	1.67	1.06	71	17.1	51	11	34	3	0.13	0.11	1.38	3	21	—
Flour by-product, less than	4-05-205	89	3.21	2.63	1.78	1.20	72	16.4	—	—	—	—	0.11	0.09	0.88	3	23	—
9.5% fiber (wheat middlings)		100	3.62	2.97	2.00	1.35	82	18.4	—	—	—	—	0.13	0.10	0.99	3	25	—
Grain	4-05-211	89	3.41	2.80	1.91	1.32	78	14.2	—	7	—	—	0.04	0.12	0.37	0	15	—
		100	3.84	3.15	2.15	1.48	87	16.0	—	8	—	—	0.04	0.14	0.42	0	17	—
Grain, hard red spring	4-05-258	88	3.48	2.85	1.97	1.36	78	15.1	—	7	—	—	0.03	0.12	0.38	0	13	—
		100	3.97	3.25	2.24	1.55	90	17.2	—	8	—	—	0.04	0.13	0.43	0	14	—
Grain, hard red winter	4-05-268	88	3.42	2.81	1.92	1.32	78	12.7	—	—	—	—	0.04	0.14	0.38	—	11	—
		100	3.88	3.18	2.18	1.50	88	14.4	—	—	—	—	0.05	0.16	0.43	—	12	—
Grain, soft red winter	4-05-294	88	3.42	2.81	1.92	1.32	78	11.5	—	—	—	—	0.04	0.10	0.38	—	16	—
		100	3.88	3.18	2.18	1.50	88	13.0	—	—	—	—	0.05	0.12	0.43	—	18	—
Grain, screenings	4-05-216	89	2.95	2.42	1.59	1.03	67	14.1	—	5	—	7	0.13	—	0.35	—	—	—
		100	3.31	2.71	1.79	1.16	75	15.8	—	6	—	8	0.15	—	0.39	—	—	—
Straw	1-05-175	89	1.60	1.32	0.57	0.10	36	3.2	75	35	—	12	0.16	0.04	0.04	2	—	587
		100	1.81	1.48	0.64	0.11	41	3.6	85	39	—	14	0.18	0.05	0.05	2	—	662

Source: Nutrient Requirements of Sheep, 6th revised edition, © 1985 by the National Academy of Sciences.

APPENDIX TABLE 13 SHEEP BREED ASSOCIATIONS

Sheep breed	Name of association	Secretary	Address
Blackfaced Highland	Scottish Blackfaced Sheep Breeders Association	R. J. Harward	39282 River Dr. Lebanon, OR 97355
Border Leicester	American Border Leicester Association	Beverly Tiffany	7594 S.R. 534 W. Farmington, OH 44491
Border Leicester	North American Border Leicester Association	R. J. Harward	39282 River Dr. Lebanon, OR 97355
Cheviot	American Cheviot Sheep Society	Ruth Bowles	RR 1, Box 100 Clarks Hill, IN 47930
Columbia	Columbia Sheep Breeders Association of America	Richard L. Gerber	P.O. Box 272 Upper Sandusky, OH 43351
Cormo	American Cormo Sheep Association	Travis Jones	18106 Woodgate Rd. Montrose, CO 81401
Corriedale	American Corriedale Association, Inc.	Russell E. Jackson	Senaca, IL 61360
Cotswold	American Cotswold Record Association	Patricia Frisella	282 Meadowboro Rd. Rochester, NH, 03867
Debouillet	Debouillet Sheep Breeders Association	Mrs. J. R. Skeen	300 South Kentucky Ave. Roswell, NM 88201
Delaine Merino	American and Delaine Merino Record Association	Mrs. Elaine A. Clouser	1193 Twp Rd. 346 Nova, OH 48859
Delaine Merino	Texas Delaine Merino Record Association	Mrs. G. A. Glimp	Route 1, Box 26 Burnet, TX 78611
Delaine Merino	Blacktop and National Delaine Merino Sheep Association		290 Beech St. Muse, PA 15350
Dorset	Continental Dorset Club	Marion Meno	P.O. Box 577 Hudson, IA 50643
Finnsheep	Finnsheep Breeders Association, Inc.	Claire Carter	P.O. Box 34303 Indianapolis, IN 46244
Hampshire	American Hampshire Sheep Association	Jim Cretcher	P.O. Box 345 Ashland, MO 65010
Karakul	American Karakul Fur Sheep Registry	Karen Koch	8756 Concord Rd. Powell, OH 43065
Lincoln	National Lincoln Sheep Breeders Association	Teresa M. Kruse	RR 6, Box 24 Decatur, IL 62521
Montidale	Montidale Sheep Breeders Association	Mildred Brown	P.O. Box 44300 Indianapolis, IN 46244
North Country Cheviots	American North Country Cheviots Sheep Association	Ann Trimble	717 Fall Creek Rd. Longview, WA 98632
Oxford	American Oxford Down Record Association	Wally Watts	Route Four Ottawa, IL 61350
Panama	American Panama Registry Association		Route 1 Jerome, ID 83338
Perendale	American Perendale Breeders Association	Norlaine Shulze	1811 New Hampshire Ashton, MD 20861

APPENDIX TABLE 13 (Continued)

Sheep breed	Name of association	Secretary	Address
Polypay	American Polypay Sheep Association	Linda Wick	Rt. 2, Box 2172 Sidney, MT 59270
Rambouillet	American Rambouillet Sheep Breeders Association	Laverne McDonald	2709 Sherwood Way San Angelo, TX 76901
Romney	American Romney Breeders Association	John J. Landers, Jr.	4375 Northeast Weslinn Dr. Corvallis, OR 97333
Scottish Blackfaced Highlands	Scottish Blackfaced Sheep Breeders Association	R. J. Harward	39282 River Dr. Lebanon, OR 97355
Shropshire	American Shropshire Registry Association	Kathy Buckham	6508 West R. Ave. Schoolcraft, MI 49087
Southdowns	American Southdown Breeders Association	Mrs. Florence Strouse	Rt. 4, Box 14B Bellefonte, PA 16823
St. Croix	St. Croix Sheep Breeders Association	Neil and Helen Simpson	P.O. Box 200 Russel, AR 72139
Suffolk	American Suffolk Sheep Society	Annette Jenson	P.O. Box 256 Newton, UT 84327
Suffolk	National Suffolk Sheep Association	Kathy Kafka	P.O. Box 6175B Columbia, MO 65205
Targhee	U.S. Targhee Association	Vickie Arnold	Box 40 Absarokee, MT 59001
Tunis	National Tunis Sheep Registry	Leona Fitzpatrick	Wayland, NY 14572

APPENDIX TABLE 14 SHEEP PERIODICALS

Name	Editor/Publisher	Address
California Sheepman's Quarterly		3382 El Camino Ave. N 06 Sacremento, CA 95821
Colorado Sheepman	Sandy Brinkman, editor	211 Livestock Exchange Bldg. Denver, CO 80216
Corriedale Extra		Box 29C Seneca, IL 61360
Dorset Journal	Douglas P. Gillespie and Denise W. Gillespie, editors and publishers	Box 1617 Champaign, IL 61820
Hampshire Journal	Douglas P. Gillespie and Denise W. Gillespie, editors and publishers	Box 1617 Champaign, IL 61820
Idaho Wool Growers	Marvin Cox and Stan Boyd, editors	P.O. Box 2596 Boise, ID 83701
Marker (Natural Colored Wool Growers Association)	Clifford Cain, editor	Star Route Box 48 Brooks, CA 95606
Montana Wool Grower		P.O. Box 1693 Helena, MT 59601
National Wool Grower	Monica Cronberg, editor	8 East Broadway, Suite 415 Salt Lake City, UT 84111
Purple Circle	Fred Bean, managing editor	P.O. Box 458 Brownwood, TX 76801
Ranch	Scott Campbell, editor and publisher	P.O. Box 2678 San Angelo, TX 76902
Sheep Breeder and Sheepman Magazine	Larry E. Mead, editor and publisher	P.O. Box 796 1701 West Worley Columbia, MO 65205
Sheepherder Sheet		Wyoming Wool Growers Assoc. P.O. Box 115 Casper, WY 82602
Sheep! Magazine	Doris Thompson, publisher	120 S. Highland Jefferson, WI 53549
Shepherd Magazine	Guy Flora and Pat Flora, editors	5696 Johnson Road New Washington, OH 44854
Southern Sheep Producer	Grant Laird, editor and publisher	Box 277 Puryear, TN 38251
Speaking of Columbias	Richard L. Gerber, editor	Box 272 Upper Sandusky, OH 43351
Suffolk Banner	Greg Deakin, editor and publisher	Box A. A. Cuba, IL 61427
Washington Wool Growers	Gwenyth P. Caldwell, editor	Rt. 1, Box 266 Goldendale, WA 98620
The Wool Sack	Dick Boniface, editor	315 Fifth Street, Box 328 Brookings, SD 57006

Glossary of Sheep and Wool Terms

Abattoir: A slaughter house.

Abcess: A localized collection of pus surrounded by inflamed tissue.

Abomasum: The fourth compartment of the ruminant stomach. Similar to the non-ruminant stomach where enzymatic digestion occurs.

Abortion: Premature expulsion of a fetus before it has matured enough to survive.

Absorption: Passage by osmosis or other means through membranes, as in passage of nutrients from the gastrointestinal tract to the bloodstream.

Acclimatization: The process of becoming adapted to different climatic conditions.

Acute: A rapid and often severe development of a disease. Compare with chronic condition, which is continuous and long term.

Adaptation: Adjustment to a different environment.

Additive: A substance included in a diet or ration, usually in small quantities to fortify them with nutrients, stimulants, or medicines.

Adipose tissue: Fatty tissue.

Ad libitum: Feeding without limitation; also called free choice feeding.

Adsorption: The process of being taken up and held on the surface of a substance.

Aerobe: A microorganism that requires the presence of oxygen to live and reproduce.

Afterbirth: The placenta and fetal membranes expelled from the uterus after parturition.

AI: Abbreviation for artificial insemination.

Air dry: Feed dried by natural air movement is termed air dry. Most feeds are air dry and contain approximately 90% dry matter (DM).

Alkali: A soluble salt or salts present in arid or semiarid soils in quantities detrimental to plant growth.

Alleles: Genes that occupy the same location on homologous chromosomes but affect the same trait in a different or alternative manner.

Allergy: A hypersensitivity to a particular substance that causes body reactions such as hives, sneezing, and labored breathing upon exposure to the substance.

Ambient temperature: Temperature surrounding the animal at any given time. The same as environmental temperature.

American Feed Manufacturers Association (AFMA): National organization of feed manufacturers organized to improve quality and promote the use of commercial feeds, to encourage high standards, and to protect the best interests of feed manufacturers and livestock producers in legislative programs.

Amino acids: Organic compounds containing nitrogen that are the building blocks in the formation of proteins.

Anabolism: The buildup of complex substances such as body tissues from simple compounds.

Anaerobe: A microorganism that can exist without free oxygen.

Anatomy: The structure and arrangement of the parts of an animal.

Anemia: A condition in which the blood is deficient in red blood cells, hemoglobin, or total volume.

Anestrus: The period of sexual quiescence between periods of sexual activity in cyclically breeding mammals.

Anorexia: A lack or loss of appetite for feed.

Anoxia: A deficiency of oxygen in the blood from anemia or other causes.

Anthelmintic: A drug or chemical agent used to destroy or control internal parasites.

Antibiotic: A substance that destroys or inhibits the growth and activity of microorganisms.

Antibody: A protein substance (modified type of blood serum globulin) developed or synthesized by lymphoid tissue of the body in response to an antigenic stimulus. It provides protection against specific diseases by interacting and neutralizing a disease-causing agent.

Antigen: An enzyme, toxin, or other "foreign" protein to which an animal's body reacts by producing antibodies specific to the invading antigen.

Antioxidant: A compound that prevents oxidative rancidity of polyunsaturated fats. Used in feeds and foods to prevent rancidity.

Antiseptic: A chemical substance that prevents growth and development of microorganisms.

Apparel wools: All wool suitable for manufacture into apparel fabrics.

Appetite: The craving for food; the desire to eat.

Artificial insemination (AI): Depositing semen into the female reproductive tract by other than natural means (i.e., by a technician).

As fed: Describes the analysis of feed as it is fed to livestock. The moisture may vary with conditions of harvest.

Ash: That mineral matter derived from incinerating organic matter.

Aspirate: To remove liquid by suction.

Assay: To determine purity or potency of a substance. To determine the level of a substance in a feed.

Assimilation: Process by which the body digests, absorbs, distributes, and metabolizes the nutrients in feed.

Astringent: A substance that causes shrinkage or contracting of localized blood vessels to reduce blood flow and any discharge.

Atresia: The absence of a normal body orifice, as in atresia ani (born without an anus).

Atrophy: Wasting away or decrease in size, usually muscular.

Attenuated: Weakened or made less virulent. Used to describe vaccines that can cause development of antibodies and protection from a disease without causing the disease.

Autosomes: All chromosomes except sex chromosomes.

Average daily gain (ADG): Average daily increase in body weight.

Avirulent: Incapable of causing disease.

Avoirdupois weights and measures: English system of weights and measures customarily used as standard in the United States, as opposed to the metric system.

Axillary space: The armpit; medial or ventral space or cavity at the junction of the foreleg and shoulder.

Baby combing wool: Same as French combing wool; short, fine wool usually manufactured in the French system of worsteds.

Backcross: Result of mating a crossbred to one of the parental breeds.

Bacteria: Single-cell plant microorganisms widely found in nature. Many are beneficial; others are pathogenic.

Bactericide: A product that destroys bacteria.

Bacterin: A suspension of killed bacteria injected into an animal to increase disease resistance.

Bag: Same as udder.

Bagging out: To inspect the udders of ewes for soundness or before lambing to group by estimated lambing dates.

Balanced ration: A daily allowance of feed that provides proper amounts and proportions of nutrients.

Bale: A method of compacting wool into large packages of 300 to 1000 lb.

Band: An operational unit of sheep in range country, which may number from 1000 to 3500 head.

Barren: Incapable of reproduction.

Basal metabolic rate (BMR): The energy required to maintain the vital functions of an animal (i.e., cellular activity, respiration, and circulation). BMR is measured in a thermoneutral environment, resting, postabsorbtive state, consciousness, quiescence, and sexual repose.

Bellies, or belly wool: Short and often less desirable wool from the belly of the sheep.

Bellwether: An animal equipped with a collar and bell that has been trained to lead groups of sheep when they are moved through working alleys, pens, and chutes. Judas goats serve the same purpose.

Biological control: Control of undesirable insects, other animals, or plants by the use of natural enemies of the pests.

Biosynthesis: The production of a chemical compound by an organism.

Black wool: Any fleece from sheep that is not completely white. It may contain fibers that are black, gray, or brown in color.

Bleat: The sound or cry of a sheep.

Bloat: An abnormal condition found in ruminants characterized by distension of the rumen on the upper left side by excessive accumulation of gas.

Block: Trimming the fleece of a show sheep to enhance its good qualities and to minimize its faults.

Blocky: A term indicating a conformation that is deep, wide, short, and low set.

Blood: A method of grading wool with the standards based on the amount of Merino breeding in the animal. Thus, fine for a pure Merino, half-blood, etc. It also is used to denote the amount of improved breeding in an animal.

Bolus: The undigested feed regurgitated during rumination. Also the name of a large pill.

Bomb calorimeter: An instrument for measuring the gross energy produced by incinerating a sample of feed.

Botany wools: A term applied to Australian fine wool.

Bradford system: A system of grading wool based on the number of hanks of yarn from a pound of clean fleece.

Braid wool: The coarsest of U.S. grades of wool using the blood system. It has long, coarse, lustrous fibers and is used primarily for carpets. Synonymous with luster wool.

Brand: Numbers or letters stamped on animals using special paint.

Break in wool: A clearly defined line across fleeces caused by a week spot in the fibers due to stress, nutrition, or illness.

Break joint: The cartilage at the lower end of the shank bone, which can be snapped when butchering lambs. It is used as an indication of age of the lamb.

Bred: Synonymous with pregnant. Also used to indicate mating.

Breech birth: Presentation of the fetus at birth with the hind feet and buttocks first instead of the front feet and head.

Breed: A group of individuals that as a result of selection and breeding have similar characteristics that distinguish them from other groups of the same species.

Breed character: Those characteristics that distinguish the breed from other breeds.

Breed type: Those characteristics that are commonly accepted as ideal for a particular breed.

Bright wool: Wool produced in areas where little contamination from windblown dust or sand occurs. It is lighter in color and has lower shrinkage than semibright wool from the southwest United States.

Britch wool: Wool from the hindquarters of a sheep. It is usually the coarsest wool on the body.

Broad tail: The pelt from the slaughter of newborn Karakul lambs.

Broken mouth: A mouth of an aging sheep with broken or missing teeth.

Browse: Tender shoots, twigs, or leaves of trees or shrubs fit for grazing livestock.

Buck fleece: Fleeces from rams, which tend to shrink heavily due to excessive yolk.

Bucks: A term used to designate ram fleeces.

Bulk grade: The largest percentage of a grade in an original bagged consignment of wool.

Burry wool: Wool heavily contaminated with vegetable matter, including burrs, seeds, leaves, and twigs. Such wool is discounted because of the expense of removing the foreign matter.

Caesarian section: Removal of the fetus by surgical means at or near term. Caesarean sections are necessary if the pelvic opening is too small or when the fetus is hopelessly malpositioned.

Calcification: Hardening of tissues by deposits of calcium salts.

Caloric: Relating to heat energy.

Calorie (cal): The amount of heat at a pressure of 1 atmosphere to raise the temperature of 1 gram of water 1 degree Celsius. Called small calorie. A kilocalorie or large calorie is the amount of heat to raise the temperature of 1 kilogram of water 1 degree Celsius.

Calorimeter: An instrument for measuring energy.

Caracul: The pelt from newborn Karakul lambs.

Carbonizing: Removing vegetable matter from raw wool by treating with sulfuric acid, drying, baking, and neutralizing.

Carcass: The dressed body of a lamb or sheep after the offal, pelt, feet, and head are removed.

Carcass weight: Weight of the carcass on the rail, either hot or chilled.

Carcass yield: The percent of live weight in carcass (carcass weight $\div$ live weight $\times$ 100). Also called dressing percent.

Card: A machine for processing wool before spinning. Also a hand instrument for disentangling wool when trimming show sheep.

Carding: A process in wool manufacture in which loose, clean scoured wool is converted into continuous untwisted strands ready for spinning.

Carding wool: Fine wool with 1¼-in. staple and coarse wool of 2 in. may be carded for the production of woolen yarn. Synonymous with clothing wools.

Carotene: Substance occurring naturally in green plants that is the precursor of vitamin A.

Carpet wool: Long staple, strong, coarse wool suitable for carpet manufacture. Little carpet wool is produced in the United States.

Carrying capacity: A measure of the number of animal units that can be maintained for a year on an area. It includes land grazed plus that used to produce harvested feed.

Caruncles: The maternal "buttons" of the uterus that attach to the cotyledons or "buttons" of the fetal placenta.

Castrate: (a) To render sterile by removal or blocking the activity of testicles or the ovaries. (b) An animal that has been castrated.

Catabolism: The degradation of a complex substance into simple compounds by living cells.

Catch pen: A small enclosure, usually in the corner of a large field, where animals can be driven or enticed by feed so they may be handled.

Cattle guard: A structure placed in lanes, driveways, or roads that prevents livestock from crossing. It is usually constructed with concrete, steel, or wood rails spaced over a pit so that the livestock cannot walk across.

Caudal: Toward the tail or posterior of the animal as opposed to cranial, which is toward the head or front end.

Cauterize: To burn away or seal with a hot instrument or caustic substance to prevent infection and supress blood flow.

Celsius (C): The metric expression of temperature (formerly, centigrade). Conversion to Fahrenheit may be done by multiplying the Celsius reading by $\frac{9}{5}$ + 32.

Cereal: A plant of the grass family which produces seeds used to feed humans and livestock.

Cervix: The structure located between the vagina and uterus that protects the uterus during gestation.

Character of wool: Denotes characteristics of grease wool that contribute to the attractiveness of the fleece. Included are brightness, luster, color, and crimp pattern.

Chronic: A disease condition that continues over a long period of time.

Clean content of wool: Clean scoured wool after removal of vegetable and other material and corrected for moisture and alcohol-soluble material.

Cleaning: The act of casting the afterbirth or placenta at parturition.

Cleanliness of wool: Describes the amount of foreign matter in the fleece.

Clip: Total annual production of wool produced by a flock or band, a state or nation.

Clitoris: The reproductive organ in the female that corresponds to the penis in the male.

Clothing wool: Short staple wool suitable for woolen yarn production. Also called carding wool.

Cloudy wool: Tinged or discolored wool in some areas of the fleece.

Coarse wool: Wool grading one-quarter or common using blood grades or 46's to 48's under the Bradford system.

Coccidiosis: An intestinal disease caused by protozoan coccidium parasites.

Coefficient of digestibility: The amount of nutrient absorbed by an animal. Derived by the following formula: nutrient in feed minus nutrient in feces ÷ by nutrient in feed × 100 = coefficient of digestibility (%).

Color: The actual color of the wool. Color may be affected by climatic conditions, breeding, and feed. Sheep with much yolk in fleeces have yellower wool.

Colostrum: The first milk secreted by mammals after parturition. It is rich in antibodies and is laxative.

Combing wool: Fine wool fibers 1½ to 6 in. in staple used to produce worsted yarn.

Common: A coarse grade of wool using the blood grading system corresponds to 44's under the Bradford system.

Complete ration: A portion of a feed or combination of feeds that satisfies the nutritional needs of an animal for one day.

Concentrates: Classification of feedstuffs for those high in energy and low in fiber. Includes grains and other seeds and oil meals.

Conception: (a) The union of sperm and ovum to form a zygote. (b) To become pregnant.

Condition: A descriptive term indicating the degree of finish an animal carries.

Conformation: The structure or shape of an animal.

Contagious: A disease transmissible by direct or indirect contact.

Core sampling: A method of drawing a sample from bales or bags of wool with a coring tube.

Corpus luteum: The yellow body on the ovary that replaces the follicle and produces progesterone to maintain pregnancy.

Cortical cells: Inner section of the wool fiber.

Cotyledons: Buttonlike structures on the placenta that attach to the maternal caruncles in the uterus of a ewe.

Creep: An area within a pen, corral, or pasture constructed so that it excludes ewes and allows lambs access to special feed.

Crimp: The natural waves in wool fiber that allows it to have resilience and elasticity.

Crossbreeding: The practice of mating parents of two or more breeds.

Crude fat: Ether extract from moisture-free feeds. It contains 2.25 times the energy of carbohydrates. It consists mostly of fats and oils, with small amounts of waxes, resins, and other organic materials.

Crutching: The trimming of wool from around the vulva and udder of a ewe before lambing.

Cryptorchid: A male animal in which one or both testicles have failed to descend into the scrotum. A unilateral cryptorchid (one testicle descended) may be fertile. Bilateral (neither descended) is invariably sterile.

Cud: The bolus of regurgitated feed that the ewe masticates during rumination.

Cuticle: Outer scalelike surface of the wool fiber as seen under the microscope.

Cutting: (a) Castration. (b) Sorting sheep into groups.

Cutting chute: A narrow chute with gates that allows individuals to be sorted into groups.

Dam: Female parent or mother of an animal.

Dead wool: Wool pulled from dead sheep.

Defecation: Evacuation of feces. Voiding of dung.

Deficiency disease: Disease resulting from lack of an essential nutrient. Usually it is a mineral or vitamin.

Dehydration: Loss of water from the body of an animal due to sickness or lack of drinking water.

Density: Number of wool fibers per unit area. Fine-wool sheep have greater density of wool than breeds with coarser wool.

Dental pad: That cartilaginous part of the front of the upper jaw that the lower incisors press against when the sheep grazes.

Depilatory: Chemical solution applied to the flesh side of a sheep pelt to loosen the fibers for pulling.

Depraved appetite (pica): Tendency of animals on a nutrient deficient diet to eat unnatural feed.

Dermatitis: Inflammation of the skin.

Desiccate: To dry completely.

Diarrhea: Abnormal frequency of intestional discharge. Also called scours.

Diet: (a) Feed ingredients including water consumed by an animal. (b) A mixture of feed ingredients fed ad libitum to animals.

Digestibile nutrients: That part of each nutrient digested by an animal. Obtained by subtracting fecal nutrient content from amount fed.

Digestion coefficient: *See* Coefficient of digestibility.

Disinfectant: A strong chemical used to kill microorganisms.

Disposition: The temperment or personality of an animal.

Dock: (a) To sever the tail from a lamb. (b) To reduce the price because of a defect. (c) That part of a sheep where the tail was cut.

Domestic wool: (a) Wool grown in the United States. (b) Wool grown east of the Mississippi River.

Dominant: The gene of a pair of alleles that controls the expression of the phenotype. The gene that prevails is dominant.

Dorsal: Pertaining to or located on the back of an animal as opposed to ventral, which is the lower part or belly.

Drench: (a) To mechanically apply oral medication. (b) The medication applied orally.

Dressing percentage: The percentage of the live weight of an animal that is carcass at slaughter (Carcass weight/live weight) × 100 (also called carcass yield).

Dry: Nonlactating.

Drying off (up): The process of causing cessation of laction when lambs are weaned.

Dry matter basis: Expression of nutrients on a moisture-free basis.

Dung: Fecal matter.

Dung locks: Wool from the area of the britch or twist that is contaminated by feces.

Dystocia: Difficult parturition.

Early lambs: Lambs born early enough to be marketed that spring, usually before March 15.

Early maturing: Sexually mature at a young age.

Ear notching: Marking an animal by cutting triangular notches in the ear according to a specified system.

Ecto parasite: A parasite that lives on the exterior of the host, for example, flies and mosquitoes.

Edema: Swelling caused by accumulation of water in tissues of the body.

Elasticity: The ability of a wool fiber to return to its original length after being stretched.

Elastrator: A bloodless means of docking and castrating by use of rubber bands. The band blocks circulation and the member sloughs off.

Electroejaculation: Use of a rectal probe and electrical impulses to stimulate ejaculation.

Electrolytes: Body acids, bases, or salts necessary for proper functions of nerves, respiration, digestion, and muscle contraction.

Emasculatome (burdizzo): An instrument used for bloodless castration by severing the spermatic cord without injury to the scrotum.

Emasculator: An instrument used for docking and castrating lambs. It has cutting blades for severing testes and tails and crushing jaws to prevent bleeding.

Endocrine: Ductless gland that secretes hormones necessary for the function of an animal. The hormones pass directly into the bloodstream from the glands.

Endogenous: Secretions originating within the body of an animal.

Enteritis: Inflammation of the intestines.

Enterotoxemia: A toxemia or blood poisoning caused by the toxin of *Clostridium perfringens;* also called overeating disease.

Entropion: An abnormality of the eye in which the lid and lashes are turned inward and are against the surface of the eyeball.

Environment: The aggregate of all the external conditions that affect an animal.

Epididymis: The convoluted tube at the posterior end of the testicle that serves as a passageway from the seminiferous tubules to the vas deferens.

Epithelium: Covering of the interior and exterior of the body.

Ergosterol: A plant sterol that, when activated by ultraviolet rays, becomes vitamin D_2.

Ergot: A fungus disease of plants.

Eructation: Belching to eliminate gas by way of the esophagus and mouth.

Essential amino acids: Amino acids required by an animal that cannot be synthesized in the body.

Estrous cycle: The physiological events occurring in the reproductive system of a ewe between one estrus and another. Average length is approximately 17 days.

Estrus (noun); estrous (adjective): The period when a ewe is physiologically and psychologically receptive to sexual advances of a ram. Also known as the heat or estrous period.

Ether extract (EE): Substances in feed soluble in ether. Included are fats, oils, waxes, resins, etc. Energy content is 2.25 times that of nitrogen-free extract or crude protein.

Euthanasia: Painless killing of animals.

Eviscerate: To remove the viscera when butchering or slaughtering an animal.

Ewe: A female sheep.

Excreta: Substances excreted by an animal, primarily feces and urine.

Exogenous: Substances introduced into the body of an animal.

Expectorant: A medicine that promotes the discharge of mucus from the respiratory tract.

External anatomy: The parts of an animal.

Fall shorn: Wool shorn in the fall in Texas or California or other areas where ewes are shorn twice a year. Also a pelt of a market lamb that has a staple of 1 to 1.5 in. (2.56 to 3.8 cm) in length.

Fattening: Feeding animals so they will store energy in the form of fat. Also called finishing.

Feces: Waste material of the gastrointestinal tract. Also called dung.

Fecundity: Ability to produce offspring.

Feed: Material capable of nurturing livestock.

Feed additive: An ingredient or substance added to a basic feed, usually in small quantities, to fortify it with nutrients, medicines, or stimulants.

Feed efficiency: The relationship between feed consumed per day and increase in weight or other measure of production. Usually expressed as feed per unit of production. For example, a lamb feeder will be much interested in weight of feed required per unit of gain in live weight.

Feed grain: Grains commonly used for feeding livestock.

Feeder lambs: Lambs under one year of age that need to grow and fatten before being suitable for slaughter.

Feeder's margin: The difference between the price per 100 lb of a feeder lamb and the price per 100 lb of finished lamb.

Feedstuff: A product suitable for nurturing livestock.

Felt: A dense, lightly woven, heavily fulled wool cloth.

Felting quality: The ability of wool fibers to lock together when subjected to heat, moisture, and pressure.

Feral: Wild and untamed animal.

Fertilization: The union of the sperm and ovum to form a zygote or independent individual.

Fetus: The developing organism in the uterus of the pregnant female.

Fiber content of feed: The content of complex carbohydrates (cellulose) and lignin in feed. These are difficult to digest and tend to lower feeding value.

Fill: The fullness of the digestive tract. In market terms, the weight increase at the market due to feed and water.

Fine wool: The finest grade of wool using the blood system. Fine corresponds to 64's or finer using the spinning count system.

Finish: Refers to the degree of fatness of the animal. Also used synonymously with fattening an animal.

Fitting: Preparing an animal for show, which includes special feeding, exercising, training, and grooming.

Fleece: The wool from one sheep.

Fleece wool: Wool produced east of the Rocky Mountains. So designated because of lower dirt content.

Flock: A small group of sheep.

Flushing: The practice of feeding a higher than normal level of energy prior to and during the breeding season to increase ovulation rate.

Fly strike: A seasonal situation in which flies irritate or attack livestock blowing sores, open wounds, incisions, or damp wool.

Food and Drug Administration (FDA): The federal agency in the Department of Health and Human Service that has the responsibility of safeguarding consumers against injury from unsanitary or unsafe food and fraud. New drugs must be approved by this agency.

Forage: Vegetable matter consumed by livestock as pasture, silage, or hay.

Fortify: To add nutrients, medicines, or stimulants to feed.

Fostering: Process of transferring of a lamb from its mother to another ewe. Also called grafting.

Founder: An inflammation of the foot caused by overeating grain, green grass, or cold water. It also may be caused by severe concussion to the foot.

Free choice: Allowing uninhibited access to feed; same as ad lib feeding.

French combing wool: Wool intermediate in length between woolen and worsted. Must be fine and 1.25 to 2.5 in. (3.2 to 6.4 cm) in length.

Full feed: A condition in the feedlot when the animal is consuming concentrates to capacity without digestive upset. A full feed of ensilage or hay would be all the forage an animal would consume daily or ad libitum feeding.

Fulling: Shrinking and felting a woolen fabric to make it thicker and denser.

Fumigant: A substance that is vaporized to destroy pathogens, insects, or rodents.

Fungi: Saprophytic, parasitic lower plants such as molds, rusts, mildews, and mushrooms.

Garneting: Reducing woven or knitted fabric to a sufficiently fibrous state so it can be carded and reused.

Gastrointestinal: The digestive tract of an animal.

Gene: The element on the chromosome that transmits a hereditary character of an animal.

Generation interval: The time span between the same stage in the life cycle of two successive generations.

Gestation (pregnancy): Period between fertilization and parturition. Time the fetus is carried in the uterus. Average gestation for sheep is approximately 147 days.

Get: Progeny of a male animal.

Grade: (a) Measure of how well an animal or product meets the standards of a class. (b) Animals that have characteristics of a breed but cannot be registered because a parent or parents are not purebred. Offspring of a purebred and one of inferior breeding. (c) The classification system used to describe grease wool.

Grading up: Continued use of purebred sires of one breed on a flock of females of another breed.

Grafted: Process of transferring the offspring of one ewe to another. Also called fostering.

Grain: Seeds from cereal plants.

Graze: To consume standing vegetation.

Grease wool: Wool shorn from live sheep.

Gregarious: The tendency to stay in groups.

Grind: The process of reducing grain to small segments by impact, cutting, or crushing.

Groats: Hulled grain.

Grow out: To feed for development rather than finishing. Usually done by pasturing or feeding harvested forage.

Growth: The increase in size of the muscles, bones, and internal organs.

Growthy: An animal large and well developed for its age. Usually large-framed animals.

Gummer: A sheep with no incisor teeth. A sign of old age.

Half-blood wool: The grade immediately below fine using the blood system. Corresponds to 62's and 60's using the spinning count system.

Hand feeding: A method of feeding lambs in which they are fed grain twice daily plus hay.

Handle or hand: The feel of the wool. Wool with a desirable hand is soft with great resiliency and pleasant to touch.

Hand mating: A method of mating in which the ram occupies a pen and ewes in estrus are brought to him for one service each.

Hank: A standard length of yarn wound on a reel. Length 560 yards (512 m).

Heat (estrus): The period when a ewe is physiologically and psychologically ready to accept service from a ram.

Heat increment (HI): Also called specific dynamic effect. The increase in heat production following feeding an animal in a thermoneutral environment.

Heavy metals: Metals such as lead and mercury that have a high specific gravity and are poisonous if they contaminate feed and water.

Heavy wool: Wool with a high shrinkage due to a high grease and dirt content.

Hematoma: A swelling or tumor that is filled with blood.

Hemoglobin: The iron-containing protein in red blood cells responsible for oxygen transport in respiration.

Hemorrhage: Loss of blood from blood vessels. Bleeding.

Herder: A person charged with the responsibility of taking care of a band of sheep on the range.

Hernia: The protrusion of an organ or intestine through a rupture of the surrounding structures (e.g., muscle walls, mesentery).

Heterosis (hybrid vigor): Increased vigor of the offspring over that of parents when unrelated individuals are mated.

Hoggett: A foreign term for a sheep from the age of weaning till first sheared as a yearling.

Homologous chromosomes: Chromosomes having the same size and shape and containing genes affecting the same characters.

Hormone: Substance secreted by endocrine glands that regulate the activity of organs in the body.

Hot-house lambs: Fall or very early winter lambs marketed at a young age (usually under 16 weeks) for Christmas to Easter trade.

Hulls: Fibrous covering of grains and seeds. Used as roughage component of feeds.

Hypocalcemia: Below normal levels of calcium in the blood, which may result in convulsions or parturient paresis as in milk fever.

Hypoglycemia: Below normal levels of blood glucose.

Hypomagnesemia: Below normal levels of magnesium in the blood.

Hypothalamus: That part of the brain that regulates body temperature, appetite, hormone releases, and other functions.

-icide: A suffix meaning to destroy or kill; for example, a fungicide destroys fungi.

Immunity: The ability of an animal to resist or overcome a disease by preventing development of pathogens or by counteracting effects of its products.

Immunization: Creating immunity by such means as vaccination.

Impaction: Stoppage of movement of the digestive tract by partially digested or indigestible material wedged into a segment of the system.

Inbreeding: Mating of related individuals.

Incisors: The sharp cutting or biting teeth located in the front of the lower jaw.

Infection: Invasion of the body by pathogenic microorganisms.

Inflammation: Reaction of tissues to injury characterized by pain, heat, redness, and swelling.

Infusion: The therapeutic introduction of a substance into the body by gravity.

Ingestion: The act of eating or drinking.

Ingredient: A constituent of a mixture.

Inguinal (ring): In a male, the space or opening through which the spermatic cord passes from the interior of the body to the scrotum.

Instinct: A natural ability to accomplish certain tasks without benefit of prior experience. It is probably hereditary.

Intradermal: Within or between layers of the skin.

Intramuscular: Being within or going into a muscle.

Intravenous: By way of a vein.

In vitro: Observable in a test tube or other artificial environment.

In vivo: Occurring within a living body.

Involution: Return of an organ to normal size and condition from an enlargement. For example, the involution of the uterus after parturition.

Irradiation: Exposure to rays of ultraviolet light or other forms of radiant energy.

IU: Abbreviation for international unit. A standard of measure potency of a biologic agent as specified by the International Conference for Unification Formulae. Potency is based on a bioassay that produces a particular effect. Potency is also expressed as USP units.

Jail: Synonomous with jug or lambing pen. A small pen for isolating a ewe and her lamb or lambs during and after parturition.

Joint ill: Also called navel ill. The result of invasion of pathogenic bacteria through the navel cord at birth. May cause umbilical abscesses, diarrhea, stiff joints, peritonitis, and convulsions and death.

Jug: Same as jail or lambing pen. Term more often used by western herders.

Ked: An external parasite of sheep. It is an insect that lives on blood from the skin of the sheep. Often incorrectly called the sheep tick.

Kemp: A straight, chalky white, weak, heavy medulated fiber that will not take a dye. It is highly objectionable in wool.

Keratin: A complex protein substance that is the principal constituent of wool.

Ketosis: Also known as lambing paralysis, pregnancy disease, or twin-lamb disease. A preparturient disease in ewes that affects their nervous system. Usually observed in ewes carrying twins late in gestation. Probably caused by inadequate carbohydrate metabolism, as indicated by ketone bodies in the blood and urine.

Kosher slaughter: Slaughter and handling of meat in accordance with the laws of the Jewish Orthodox religion by a rabbi or a specially trained representative called shohet or shochet.

Lactation: The period of milk production. The period a mother nurses her young.

Lactose: The sugar in milk.

Lamb: (a) A sheep under one year of age. (b) the act of giving birth by the ewe.

Lambing paralysis: *See* ketosis.

Lamb's wool: Wool shorn from a lamb 7 to 8 months old. It is softer and has spinning qualities superior to wool of the same quality from an older sheep.

Lanolin: Refined wool grease. Used in medicines, cosmetics, and paints.

Laxative: A drug or feed used to relieve constipation.

Lesion: The change in structure or form of an animal's body caused by disease or injury.

Lethal gene: A gene, or genes, that cause the death of an individual possessing them.

Libido: Sexual drive or desire by the males of an animal species.

Linebreeding: A form of inbreeding designed to concentrate the blood of an outstanding ancestor, but by mating individuals no more closely related than cousins, half-sibs, etc., to avoid loss of vigor.

Line cross: Crossing inbred lines.

Linkage: Two or more genes located on the same chromosome so they tend to be transmitted together.

Litter: (a) Number of young produced in breeds that normally have multiple births. (b) Bedding in barns.

Live weight: Weight of an animal on foot.

Lock: A tuft or small finger-size bit of wool that tends to cling together when shorn from the sheep.

Lofty wool: Open, springy, and bulky wool in comparison to its weight.

Long wool: Wool with a staple of 12 to 15 in (30.5 to 38 cm) that has a large diameter and very little crimp. Produced by Lincoln, Leicester, and Cotswold breeds of sheep.

Lot: The entire offering of wool for sale.

Low set: An animal with short legs.

Luster: The natural gloss or sheen of wool from long-wooled sheep.

Macrominerals: The major minerals: calcium, phosphorus, sodium, chlorine, potassium, magnesium, and sulfur.

Macroscopic: Large enough to be visible to the naked eye.

Maintenance requirement: Nutrients required solely to maintain body weight.

Malnutrition: A ration deficient in required nutrients.

Manure: Usually a mixture of feces, urine, bedding, and waste feed.

Margin (spread): The difference in price per 100 lb between feeders and finished lambs.

Market class: Classification of livestock according to the use for which it is bought.

Market grade: Grouping of animals within a market class based on the degree to which they satisfy specifications for quality.

Mastication: Chewing feed.

Mastitis: Inflammation of the mammary gland.

Meal: (a) A mixture of ground feed ingredients. (b) A particle size larger than flour.

Meats: (a) Animal tissue used as food. (b) Edible portions of fruits and nuts.

Medicated feeds: Feeds containing drugs for the treatment or prevention of diseases.

Medium wool: Wools grading by blood one-quarter, three-eighths, and one-half blood (50's to 62's by count).

Merino: A breed of sheep known for producing high-quality wool.

Metritis: Inflammation of the uterus.

Microbe: Same as microorganisms.

Microingredient: A feed ingredient added to a mixture in quantities of milligrams or micrograms or parts per million.

Microorganisms: Organisms of microscopic size such as bacteria.

Milk let-down: Release of milk from the alveola into the cisterns of the udder by the action of oxytocin.

Mill by-products: By-products of milling grain for flour. Examples are bran, shorts, and screenings.

Minerals: Inorganic elements required by animals for their normal function.

Moisture: The water content of a feed, expressed as a percent.

Moisture free (MF): Oven-dried feed subjected to 212 °F (105 °C) temperatures until weight becomes static. This is known as dry matter.

Molars: The grinding jaw teeth.

Molds: *See* fungi.

Morbidity: The condition of being diseased. The sick rate.

Mortality: Death or death rate.

Mouton: Fur made from sheep pelts.

Mucous membranes: The soft membranes lining the passages and tracts of the body.

Mutation: A change in a gene that may change the phenotype.

Mutton: The meat of a mature sheep.

Mycotoxins: Toxic substances produced by some molds.

Nano-: A prefix meaning one-billionth or 10^{-9}.

National Research Council (NRC): A division of the National Academy of Sciences that promotes the effective utilization of scientific and technical resources. For example, a subcommittee of the NRC calculates and complies nutrient requirements for animals.

Native wool: Wool grown east of the Mississippi River. Same as domestic wool.

Natural immunity: Inherited immunity to disease or parasites; not acquired immunity.

Navel ill: Same as joint ill.

Necropsy: Postmortem examination of internal organs.

Necrosis: Death and decay of body tissue that is still in contact with living tissue.

Nephritis: Inflammation of the nephrons in the kidney.

Neoplasm: A new and abnormal growth of tissue; a tumor.

Nick: A mating where the desirable characteristics of each parent are expressed and thus the offspring is superior to either parent.

Nitrogen fixation: The ability of legumes and bacteria to convert atmospheric nitrogen into organic nitrogen compounds and thus enrich the soil.

Nitrogen-free extract (NFE): In proximate analysis, the sum of the contents of moisture, crude protein, crude fat, crude fiber, and ash are subtracted from 100 and the result is called nitrogen-free extract. It contains sugars, starches, pentoses, and nonnitrogenous organic acids.

Noils: Short fibers removed during combing wool.

Nonprotein nitrogen (NPN): Simple organic compounds containing nitrogen, which can be utilized by ruminants to satisfy part of their protein requirements. Examples are urea and biuret.

Nulliparous: Females that have never given birth to young.

Nutrient requirements: Values of nutrients considered necessary for maintenance, optimum production, and prevention of signs of nutritional deficiencies of animals.

Nutrients: Chemical substances present in feeds that are used by animals for maintenance, growth, and production.

Offal: Organs and tissues removed at slaughter.

Off sorts: Shorts, britch wool, kemp, gray, stained, and other wool removed when wool is sorted at the warehouse.

Omasum: The third compartment of the ruminant stomach. It is located between the reticulum and abomasum.

Open faced: A sheep with very little wool on its face.

Open wool: A wool that lacks density and may part down the back.

Ossification: Hardening of the bone tissues by calcification as the animal ages.

Osteomalacia: Softening of the bones caused by vitamin D deficiency or lack of calcium or phosphorus or an improper ratio between the two.

Outcross: Introduction of unrelated individuals into a breeding program. Mating of an individual to another of the same breed which is not closely related to it.

Out of: Mothered by.

Ovary: The reproductive organ of the female.

Ovum: The female reproductive cell (gamete).

Oxytocin: The pituitary hormone that causes milk let-down and contractions of the uterus by causing contractions of smooth muscles.

Palpate: To investigate or examine by feeling or touching with the hands.

Parrot mouth: A defect of the mouth caused by a shortness of the lower jaw that does not permit the incisor teeth to come in contact with the dental pad.

Parts per million (ppm): Equals milligrams per kilogram or milliliters per liter.

Parturition: The act of giving birth.

Pathogenic: Causing or capable of causing disease.

Paunch: Same as rumen.

Pedigree: A written record of an animal's ancestry.

Pelt: The skin of a sheep as removed at slaughter.

Performance test: The evaluation of an animal by recording performance under specified conditions.

Peritonitis: Inflammation of the peritoneum, the thin membrane that covers the organs of the abdomen.

Per os: To administer a substance by way of the mouth.

Persian lamb: Fur made from pelts of young Karakul lambs.

pH: A measure of the acidity or alkalinity of a solution with values from 0 to 14. Most acid is 0 and most alkaline is 14, with 7 as neutral.

Phenotype: Those characteristics of an animal that may be observed and measured.

Photoperiod: Day length or period of light.

Photosynthesis: The formation of carbohydrates by the action of sunlight on chlorophyll in a plant.

Physiological: Healthy or normal function of a living organism.

Pinning: Interference with bowel movement of a young lamb by a collection of dung around the anus and between it and the tail.

Pizzle: The penis of a ram.

Placenta: The vascular organ that surrounds and protects the fetus and provides the means by which nourishment is transferred from the mother to the young.

Pleura: Delicate membranes that line the thorax of an animal.

Pneumonia: A disease of the lungs.

Polled: A naturally hornless animal.

Polyestrus: Animals showing regularly occurring estrus throughout the years.

Pool: A collection of wool clips from a group of producers offered for sale in one or more lots.

Posterior: Refers to the back portion of an individual.

Postnatal: Event occurring to the offspring after birth.

Postpartum: Event occurring to the mother after birth.

Pot bellied: A defect of conformation characterized by an exceptionally large abdomen.

Preconditioned: Management before shipping, weaning, etc., to reduce stress. It includes vaccination, use of antibiotics, and controlled feeding.

Precursor: A compound converted by the body to one needed by the animal. For example; carotene is a precursor to vitamin A.

Prehension: Grasping grass or feed by the tongue lips and teeth and conveying it to the mouth.

Premix: A thoroughly mixed collection of the micronutrients in the correct amount to be added to the formula in a mixed diet.

Prenatal: Event occurring to the fetus before birth.

Prepotency: The ability of an individual to transmit its characteristics to its offspring.

Produce: The offspring of a female.

Progeny: The offspring of animals.

Prognosis: An estimate of the progression and final outcome of a disease.

Prolapse: Abnormal protrusion of an organ or part of organ.

Prolific: Ability to produce numerous offspring.

Protein: A molecule consisting of chains of amino acids.

Provitamin: A precursor of a vitamin.

Proximate analysis: Chemical evaluation of feedstuff. These include dry matter (DM), crude protein (CP), total digestible nutrients (TDN), ether extract (EE), ash, crude fiber (CF), and nitrogen-free extract (NFE).

Puberty: The time an animal reaches sexual maturity.

Pulled wool: Wool pulled from the pelts of sheep after slaughter.

Purebred: An animal eligible for registration in the herd book of the breed to which it belongs.

Purity: A wool term indicating the kind and quality of wool fibers. Dark fibers and kemp are impurities.

Purulent: Containing or consisting of pus.

Pus: Thick, opaque, yellow-white fluid containing leucocytes, tissue debris, and microorganisms.

Putrefaction: Decay as a result of enzyme or bacterial action or both.

Qualitative traits: Traits with a sharp contrast between phenotypes that are controlled by the action of a single or only a few pairs of genes.

Quality: A descriptive form pentaining to the refinement of the head, hide, and bone of a live aniamal.

Quantitative traits: Traits that do not vary much in phenotypic expression but include most of those economic importance. Inheritance is through many pairs of genes.

Quarantine: Segregating of sick, exposed, or susceptible animals to prevent the spread of infectious disease.

Quarter-blood wool: A grade in the blood-grading system of wool. Same as 46's in the count system.

Radioactive: Emitting atomic energy as in the form of alpha, beta, and gamma rays.

Ram: A sexually intact male sheep.

Rancid: A foul odor or taste as in decomposed fat.

Range wool: Wool produced by sheep on the range lands of the West and Southwest.

Rangy: A descriptive term denoting an animal with long legs and a long, lightly muscled body.

Rate of passage: Time required for undigested feed to pass through the digestive tract.

Ration: The amount of feed supplied to an animal in one day.

Raw wool: The same as grease wool.

Recessive gene: A gene whose phenotypic expression is covered up (masked) by its own dominant allele.

Reclaimed wool: Wool salvaged by reworking new or old fabrics.

Rectum: The terminal part of the colon.

Registered: Designating animals recorded in the herdbook of their breed association.

Regurgitation: The movement of undigested fibrous feed from the rumen to the mouth for further chewing.

Rejects: Wool that for any reason is not acceptable for milling and is rejected from the clip. Examples are black, tender, cotted, kempy, dead, dungy, seedy, chaffy, and burry wool. Such wool is penalized in price.

Replacements: Animals selected to go into the breeding flock.

Retained placenta: A placenta not expelled at birth or soon afterward.

Reticulum: The second compartment of the rumen stomach where pregastric fermentation occurs. It is the site of hardware disease.

Ridgeling: A male with testicles that have not descended into the scrotum. Same as cryptorchid.

Rigor mortis: Temporary rigidity of muscles after death.

Rolag: The neat roll of wool that comes from the hand card.

Roughage: High-fiber, bulky green or dry feeds or by-products.

Roving: Slightly twisted roll or strand of wool produced in the step before the wool is drawn and spun into yarn.

Rumen: The first and largest compartment in the ruminant stomach. Also called the paunch.

Rumen flora: Microorganisms in the rumen.

Rumenitis: Inflammation of the rumen wall.

Saccharides: Sugars. Mono-, di-, tri-, and poly- are prefixes indicating the number of sugars contained in the molecule.

Sacking: Woven fiber bags used for storing and transporting fleeces. Usually 7 ft long and 30 in. in diameter.

Saliva: The clear liquid solution secreted by glands in the mouth that contains enzymes used in digestion of carbohydrates.

Salmonella: Aerobic, rod-shaped pathogenic bacteria that are the cause of severe diarrhea.

Scouring: (a) Severe diarrhea. (b) The washing of dirt, grease, and foreign matter from grease wool.

Scrotum: The bag or sack on a male that contains the testicles.

Scurs: Small, rounded, loose, horny growths attached to the skin at the hornpits of a sheep.

Secondary infection: An infection following an infection but with different organisms.

Second cuts: Short fibers in a fleece caused by failure to hold the shears next to the skin and thus having to go over the cut a second time; also called fribs.

Seedy wool: Wool contaminated by seeds or other vegetable matter.

Selection: The decision whether or not to keep an animal in the flock.

Self-fed: Feeding free choice or ad libitum.

Self-feeder: A storage container that allows animals access to feed at will or ad libitum.

Semen: The ejaculate of a ram that contains fluids from the accessory glands and spermatazoa.

Sensitivity test: A laboratory procedure for determining the sensitivity of microorganisms to antibiotics.

Septicemia: Blood poisoning caused by bacteria in the circulatory system.

Serrations: The scaly surface of the outer layer of cells of a wool fiber that causes fibers to cling together in felting. Fine wool generally has more serrations than coarse wool.

Serum: The colorless fluid component of blood.

Service: Mating of male and female.

Settled: Becoming pregnant.

Shearing: Clipping the fleece either by hand or power shears.

Shearling: (a) Pelts with length of fiber from 0.25 to 1.0 in. (0.6 to 2.5 cm). (b) A one-year-old sheep.

Shives: Particles of vegetable matter, other than burrs, in wool.

Shock: Physical or mental disturbance as a result of major physical injury or stress. Can cause failure of peripheral circulation.

Shoddy: Wool fibers recovered from used fabric. Fibers must be at least 0.5 in. (1.3 cm) in length. Must be designated reused.

Shorts: (a) A by-product of wheat milling. (b) Short pieces or locks of wool dropped out while wool is being sorted.

Shot: An injection.

Shrinkage: (a) Loss of weight in transit or in handling. (b) Loss of carcass weight in the cooler. (c) Loss of weight of grease wool in scouring and processing.

Sib: A brother or sister; a sibling.

Sire: The male parent.

Skirting: An Australian practice of removing stained or inferior parts of a fleece on the shearing floor.

Sloughing: Separation of dead and live tissue on the surface of an animal.

Sorting: Separating wool of each fleece into length, color, strength, and diameter of fiber categories at the mill.

Spermatic cord: The cord that passes from the scrotum to the body and includes the vas deferens, muscle, artery, vein, nerves, and lymphatics leading to and from the testicles.

Spermatogenesis: The formation and development of sperm in the testes.

Spinning count: A measure of the fineness of a fleece based on the length of yarn that can be spun from 1 lb (454 g) of scoured wool. It is expressed in hanks, which are 560 yards in length. Thus 64's would produce 64 hanks of yarn per pound (454 g) of scoured wool.

Spook: To scare or frighten animals. Especially those not used to being handled.

Spore: A bacterial cell with an extra thick, impervious wall that makes it resistant to sterilization.

Stag: A male castrated after secondary sexual characteristics are well developed.

Stained wool: Wool discolored by urine, manure, or other materials that will not scour out. This detracts from the value of the fleece.

Standing heat: The highest level of estrus when the female's receptivity to the male is the greatest.

Staple: (a) The length of the lock of shorn wool. (b) In the wool trade it refers to wool that averages 2.5 in. (6.3 cm) in length.

Sterile: Incapable of reproduction.

Sterilization: Process of destroying microorganisms by boiling or use of chemicals. Used mainly on surgical tools. Also to render an animal incapable of reproduction.

Stillborn: Born dead.

Stover: Stalk and leaves of harvested corn and sorghum.

Straw: The stems and leaves of grains, soybeans, etc., after threshing.

Stress: Any environmental factor causing bodily or emotional tensions that interfere with an animal's internal function.

Subcutaneous: Anything that occurs beneath the skin.

Suckle: To draw milk from a teat by sucking. To nurse.

Suint: Salts of perspiration in the raw wool fleece.

Supernumerary: Teats in excess of the two normally found on a sheep. They are usually nonfunctional.

Supplement: An ingredient added to basal feeds to improve the nutritional value.

Suppuration: Formation of pus.

Symmetry: Blending together of all parts of a well-developed animal.

Synthesis: Production of a substance by combining simpler compounds.

Synthetics: Products similar to natural ones that are produced artifically.

Tagging: Removing the dung locks from a sheep. Usually done before breeding and shearing.

Tags: Dung locks occurring in wool around and below the anus of a sheep.

Tallow: The fat of a sheep.

Tare: The weight of a container, such as a wool bag or weighing crate, that must be subtracted from the gross weight to obtain the actual weight of the product.

Tattoo: A permanent marking of an animal by insertion of indelible ink under the skin, usually in the ear or flank.

Teaser ram: A ram used to locate ewes in estrus. the ram may be vasectomized or equipped with an apron to prevent impregnating the ewes.

Tend: To care for sheep.

Tender: A weak spot or spots along the length of wool fibers caused by the growth of a small, weak fiber during sickness or poor nutrition.

Territory wool: Wool grown on the western range but not including California and Texas. The wool generally has a high shrink due to windblown soil and sand particles.

Test cross: A mating test to determine if an individual is a carrier of a recessive gene.

Test mating: A process of penning a ewe in heat with a ram to observe the libido and ability to copulate. It may also include a semen evaluation.

Testosterone: The male sex hormone produced by the testes.

Tetanus: An infectious bacterial disease that results in spasms and rigidity of muscles, especially the jaw. *Clostridium tetani* enters the body by way of wounds.

Tetany: A condition in an animal characterized by tonic muscle spasms due to a physiologic mineral imbalance. Grass tetany is caused by a magnesium deficiency.

Tether: To restrain an animal with a rope or chain so that it can only graze a limited area.

Thermogenesis: The production of heat in the body.

Thermoneutrality: An environmental state in which the animal's body is neither producing heat to overcome cold or working to dissipate excess heat. This is known as the comfort zone.

Three-way cross: A system of breeding involving the mating of males of three different breeds in rotation on crossbred females.

Thrifty: Healthy and vigorous in appearance.

Tocopherol: Fat-soluble chemical compounds with vitamin E activity.

Tonic: An appetite-stimulating agent.

Top: The continuous, untwisted strand of combed wood fibers running parallel to each other after the noils have been removed in the combing process.

Total digestible nutrients (TDN): A method of determining the energy value of feeds based on the digestibility of nutrients in the feed. TDN can also be converted to digestible energy (DE). One pound of TDN = 2000 calories DE.

Toxemia: A general, systematic toxification or poisoning caused by absorbing toxins produced by bacteria in a localized infection.

Toxic: Poisonous.

Toxoid: A toxin from a pathogenic bacteria that has been created to destroy the toxicity, but with the capability of causing the formation of antibodies when injected into the body of a healthy animal.

Trace mineral: A mineral required in micro amounts (mg/lb or less) by animals.

Tranquilizers: Drugs used to reduce physical or nervous tension of animals.

Tup: A british term for copulation or breeding a ewe.

Tupping season: The breeding season.

Twenty-eight hour law: A law regulating rail shipments, which prohibits carriers from transporting livestock more than 28 hours without feeding, watering, and resting 5 consecutive hours before resuming travel. The owner may be permitted to extend the period to 36 hours upon request.

Twist: (a) The number of turns per inch in a thread of yarn. (b) The fleshing between the hind legs.

Tying: After shearing, the fleece is spread out intact with the skin side down and the edges folded in. It is then rolled from the rear to the front and tied with paper twine.

Udder: A pendulous organ consisting of two mammary glands enclosed in a common envelope and each provided with a teat. Also called a bag.

Underfeeding: Providing inadequate energy to maintain weight or production.

Unthriftiness: Apparent lack of vigor and vitality as shown by poor condition of hair, thin body condition, and lack of growth and development.

Unwashed wool: Raw wool, unscoured wool.

Urinary bladder: The organ in the body that collects and stores urine until it is voided.

Urinary calculi: Stones formed any place in the urinary tract. Also known as kidney stones or gravel.

USDA: Abbreviation for United States Department of Agriculture.

Uterus: The part of the female reproductive tract where the fertilized ovum attaches and develops into the fetus.

Vaccination (shot): An injection of a biological agent to produce live immunity to a disease.

Vaccine: A suspension of attenuated or killed microorganism that is administered to an individual for the prevention or treatment of an infectious disease.

Vagina: The part of the female reproductive tract between the cervix and vulva where semen is deposited during copulation.

Variance: A statistical term indicating the variation within a population.

Vasectomy: Sterilization of a male by removal of a section of the spermatic cord (vas deferens). The libido is not affected.

Vectors: An organism that transmits a pathogen. Usually an insect.

Vegetable matter: Any contamination of wool from plants or plant products. Examples are seeds, burrs, chaff, and straw.

Ventral: The lower part or surface of a body near or level with the belly or abdominal area.

Vermifuge (vermicide): Any chemical or drug given to an animal to destroy internal parasites.

Viroid: A particle of nucleic acid that is much smaller than a virus but lacks the tough protein coat of the virus. It is a disease-causing particle in animals.

Virulent: Pathogenic or capable of causing disease.

Virus: Small infectious particles that can pass through a filter; they cause disease. They consist of a tough protein coat within which is enclosed either RNA or DNA, but not both. They must infect a host cell in order to reproduce.

Viscera: Internal organs of the body, including the heart, liver, and intestines.

Vitamin: Organic substances essential in minute quantities to the nutrition of animals. They function as parts of enzyme systems to regulate metabolic processes, but do not provide energy or serve as building units in the body.

Void: To excrete feces and urine.

Vomiting: To disgorge stomach contents through the mouth.

Warp: Yarns extending lengthwise in a loom.

Wasty: (a) An animal with an excessive paunch. (b) An overly finished carcass.

Wasty wool: Wool that is short, weak, and tangled in the lock, which causes waste in processing.

Weaning: Removing young animals from their dam to terminate the suckling period.

Wether: A male sheep castrated before secondary sex characteristics are well developed.

Wether wool: Any wool from a sheep after the first shearing.

Withdrawal time: Time, specified by the Food and Drug Administration, required to elapse between the application or feeding of a drug or additive and slaughter.

Woof: Filling yarn that runs the width of the loom. Usually weaker than the warp, which extends lengthways.

Wool blind: Growth of wool on the face and around the eyes so dense it obstructs the vision.

Woolens: Cloth made from woolen yarns. The fabric is not tightly woven and tends to have a soft finish. Examples are flannels and tweeds.

Woolen yarn: Yarn spun from wools that are fine but short. The wool is carded to form a weblike configuration. When the wool is spun, the yarn is not tightly twisted and the broken loose ends stick out from the thread.

Wormer: A commercial product containing an anthelmintic to control internal parasites.

Worsted: A tightly woven fabric with a hard finish. Examples are worsted and serge.

Worsted yarn: Yarn made from long-fiber wool that is combed. The fibers lie parallel before being spun into tightly twisted, smooth yarn.

Yean: To give birth.

Yearling: An animal at least 12 month of age.

Yield: (a) percent of scoured wool from grease wool.

$$\text{yield \%} = \frac{\text{wt. of scoured wool}}{\text{wt. of grease wool}} \times 100$$

(b) Percentage of live weight that becomes carcass at slaughter.

$$\text{carcass yield \%} = \frac{\text{wt. of carcass}}{\text{wt. of live lamb}} \times 100$$

Also called dressing percentage.

Yolk: Natural grease, oil, and suint in wool.

Zygote: A cell resulting from the union of male and female gametes. After further development and implantation it becomes the fetus.

Index